TELECOMMUNICATION NETWORKS
AND COMPUTER SYSTEMS

Series Editors

Mario Gerla
Aurel Lazar
Paul Kühn
Hideaki Takagi

Byeong Gi Lee   Seok Chang Kim

Scrambling Techniques for Digital Transmission

With 124 Illustrations

Springer-Verlag
London Berlin Heidelberg New York
Paris Tokyo Hong Kong
Barcelona Budapest

Byeong Gi Lee, BS, ME, PhD
Seok Chang Kim, BS, ME, PhD

Department of Electronics Engineering, Seoul National University, Seoul 151-742, Korea

Series Editors

Mario Gerla
Department of Computer Science
University of California
Los Angeles
CA 90024, USA

Paul Kühn
Institute of Communications
Switching and Data Technics
University of Stuttgart
D-70174 Stuttgart, Germany

Aurel Lazar
Department of Electrical Engineering and
Center for Telecommunications Research
Columbia University
New York, NY 10027, USA

Hideaki Takagi
IBM Japan Ltd
Tokyo Research Laboratory
5-19 Sanban-cho
Chiyoda-Ku, Tokyo 102, Japan

ISBN 3-540-19863-6 Springer-Verlag Berlin Heidelberg New York
ISBN 0-387-19863-6 Springer-Verlag New York Berlin Heidelberg

British Library Cataloguing in Publication Data
A catalogue record for this book is available from the British Library

Library of Congress Cataloging-in-Publication Data
Lee, Byeong Gi.
 Scrambling techniques for digital transmission / Byeong Gi Lee, Seok Chang Kim.
   p. cm. - - (Telecommunication networks and computer systems)
 Includes bibliographical references and index.
 ISBN 3-540-19863-6. - - ISBN 0-387-19863-6 (New York)
 1. Telecommunication systems - - Security measures. 2. Scrambling systems (Telecommunication)
 I. Kim, Seok Chang, 1964- . II. Title. III. Series.
TK5102.85.L44 1994
621.382'7 - - dc20   94-37999

Typesetting: Camera ready by authors
Printed and bound at the Athenæum Press Ltd., Gateshead, Tyne and Wear
69/3830-54321 Printed on acid-free paper

Preface

Scramblers and the constituent *shift register generators* (SRG) have been used for decades in shaping the digital transmission signals or in generating pseudo-random binary sequences for transmission applications. In recent years they have been paid more attention than ever before, and this notable change is spurred by the evolution of today's telecommunication environment. We briefly digress to take an overview of this evolving telecommunication environment.

There are two distinctive features in the trends of today's telecommunications, which are the prevalent spread of lightwave or optical transmission networks, and the integration of services and communication modes. The lightwave transmission, since its first practical application, has been applied for digital transmission as it could reduce the associated cost with the conventional copper-cable based digital transmission while drastically increasing the transmission capacity. Lightwave transmission evolved from its point-to-point application to point-to-multipoint applications, and the *synchronous optical network* (SONET) interface standard and the *synchronous digital hierarchy* (SDH) standard played a momentous role toward this evolution. In the local and metropolitan area, the lightwave transmission has been propagating to *local area network* (LAN) and *metropolitan area network* (MAN) to give birth to the *fiber-distributed data interface* (FDDI) and the *distributed-queue dual-bus* (DQDB) networks. The digital transmission capacity nowadays has reached several hundred Mbps to a few Gbps through the lightwave transmission. The optical fibers for the lightwave transmission have been replacing the backbone and trunk networks, and are spreading toward the office and the residential area in accordance with the *fiber-to-the-office* (FTTO), the *fiber-to-the-curb* (FTTC) and the *fiber-to-the-home* (FTTH) strategies.

The advent of the *integrated services digital network* (ISDN) of the 1980's introduced the concept of services integration via digital transmission, and this concept has contributed to establishing the *broadband ISDN* (BISDN) of today, which will prove to be a truly integrated network for a variety of broadband and narrowband services of the future. The BISDN employs *asynchronous transfer mode* (ATM) as the means for service integration, which again integrates the two independently developed digital communication modes −− the circuit mode and the packet mode. The integration of the two modes is embodied in the ATM cell, which corresponds to a small-size transmission frame from the circuit-mode point of view and is a fixed-size packet from the packet-mode point of view. The ATM transmission in the BISDN has been getting strong supports from lightwave transmission at 155 Mbps, 620 Mbps, and 2.4 Gbps transmission rates, and has been proliferating in the local and metropolitan area in the form of DQDB MAN.

Any form of lightwave-based high-speed digital transmission requires strong support of sophisticated scrambling techniques. Scrambling by nature is a bit-level randomizing processing applied to digital transmission signals just prior to transmission for the goal of improving the clock recovery capability and data detection reliability. The scrambling processing is an essential function in digital transmission, as it can stabilize the clock recovery in the receiving terminal, reduce jitters, and decrease inter-symbol interferences, thus effectively increasing the data detection accuracy and reliability. In the conventional low-rate *plesiochronous digital hierarchy* (PDH) systems, the transmitter signals rather relied more on sophisticated line-coding techniques such as AMI, CMI, BNZS and HDBN to achieve the above goal, getting only auxiliary supports from scrambling. However, in the lightwave transmission systems, with the rates ranging from several hundred Mbps to a few Gbps and beyond, it is not desirable to employ these complex and multi-level line-coding techniques. Fortunately, the optical cables for lightwave transmission proved to be a very reliable transmission media, and therefore a simple NRZ or RZ line-coding turned out to be satisfactory only if a well-functioning scrambling techniques are matched in front. Therefore, the role of scrambling technique has become ever more important, assuming all the responsibilities that previously belonged to line coding techniques.

The above trend also applies to the packet-mode based LAN or MAN transmission. In the past, the LAN relied only on the line-coding such as Manchester coding for its digital transmission. This used to be acceptable because the LAN remained a low-rate communication system within a small area. However, as the transmission speeds grow to hundreds of Mbps and as service integration matures to include the real-time traffics in addition to the packet data, it has to seek solutions in the lightwave transmission and regular-sized packets or frames, whose examples being the FDDI II or the DQDB. In this situation the rate increase of 25 percent that results when the Manchester coding is used is no more negligible, and the transmission reliability and efficiency can no more be neglected. Therefore, scrambling technique emerges as the dependable solution for the packet-mode LAN or MAN transmission in the high-speed lightwave environment, also.

There are three representative scrambling techniques that have been introduced to us for the past decades −− namely the *frame synchronous scrambling* (FSS), the *distributed sample scrambling* (DSS), and the *self synchronous scrambling* (SSS). They all exhibit performances satisfactory enough to fulfill the expected objectives in the lightwave transmission, but the occasions for their applications differ. The FSS shows a good performance when used for framed signals with large-size frames, the DSS is adequate for framed signals with small-size frames, and the SSS is applicable to all forms of signals, framed and unframed. The most distinctive examples of their applications can be found in the SONET and SDH transmissions for the FSS ; in the cell-based ATM transmission for the DSS ; and in the SDH-based ATM transmission for the SSS. The lightwave based packet-mode LAN or MAN transmission can also be enhanced by employing the DSS.

In place of the three scrambling techniques, the parallel scrambling techniques can be devised for applications in the high speed lightwave transmission. The parallel scrambling techniques can drastically reduce the speed of operation at the cost of increased circuit complexity. For example, the rate of scrambling operation of 2.4 Gbps for the STM-16 signal can drop to 155 Mbps, which is a sixteenth of the original rate, by employing the parallel scrambling techniques. This not only provides gain in the cost of implementation by enabling the replacement of the GaAs technology with the CMOS technology, but also enables the implementation of the scrambling operation for very high-speed signals which otherwise might be impossible with the existing technology.

This book is intended to provide a full-fledged description on the theory and applications of the three scrambling techniques — — the FSS, the DSS, and the SSS — — with an emphasis on their applications in digital transmission. The core contents of the book are taken from our research over the past ten years. As a common principle for use in discussing these scrambling techniques, the book introduces the concept of sequence space, which is a vector space representing all the recurrent sequences. The sequence space takes the central role in rigorously describing the behaviors of the SRG in the scramblers and descramblers, especially of the FSS and the DSS. It provides the means to mathematically formulate and systematically describe the behaviors of shift register generators in the serial form as well as in the parallel form. It guides us to investigate the properties of the parallel sequences for use in parallel scrambling techniques and teaches us how to implement the relevant *parallel shift register generators* (PSRG). It also unveils the secret for devising multibit-based parallel scrambling techniques for scrambling of the multibit-interleaved multiplexed signals. As such, the concept of sequence space plays the essential role in all aspects of the book, and therefore, the book may be subtitled as *The Sequence Space Theory and Its Applications to Scrambling Techniques.*

The book is organized in four parts and 21 chapters in total. Part I provides preliminary considerations on the digital transmission and the scrambling techniques. Part II introduces the concept of the sequence space and discusses various representation methods and properties of the sequence space, and the relevant SRG spaces. On this basis, it also describes how to achieve *parallel frame synchronous scrambling* (PFSS) and *multibit-parallel frame synchronous scrambling* (MPFSS), finally demonstrating their applications to the SDH/SONET signal transmission. Part III concentrates on describing the behaviors of the DSS by employing the sequence space theory developed in Part II. It investigates how the sampling and correction functions of the DSS behaves in search for the descrambler synchronization, and examines how to realize *parallel distributed sample scrambling* (PDSS) and *multibit-parallel distributed sample scrambling* (MPDSS) employing the earlier developed PSRGs. In addition it discusses the synchronization mechanisms in the errored environment, and finally demonstrates the applications of the DSS to the cell-based ATM network and the high-speed data network.

Part IV considers the SSS and the relevant signal alignment issues. It discusses how the SSS differs from the FSS and the DSS, and why only *parallel self synchronous scrambling* (PSSS), but not the *multibit-parallel self synchronous scrambling* (MPSSS), is available, in contrast to other scrambling techniques. It then considers how to realize the PSSS and how to apply it to the SDH-based ATM signal transmission. Finally it treats the signal alignment issues related to the SSS, illustrating how to achieve proper base-rate signal alignment employing the permuting operations in place of the frame formats to be overlaid on the multiplexed transmission signal.

Acknowledgements

We are thankful to the Electronics Engineering Department and the Telecommunications and Signal Processing Laboratory of the Seoul National University for providing comfortable environment and facilities to work on the book which have been crucial in completing the manuscript and the related research. We are indebted to Youngmi Joo for carefully examining the manuscript word by word and providing discussions on technical issues, especially on synchronization of the DSS. We are also grateful to Jae-Hwan Chang whose expertise in word processing helped us to finish the manuscript on time, and to Byoung Hoon Kim whose careful review helped to improve the manuscript for the second printing. Our gratitude extends to Dr. Hideaki Takagi of IBM Research in Tokyo, and Mr. Nicholas Pinfield of Springer-Verlag Publishers who recommended the publication of the book. Finally, we thank our wives and families for supporting us spiritually all through the year, sacrificing so many weekends.

Byeong Gi Lee
Seok Chang Kim

Contents

List of Notations

$\mathbf{0}$ zero vector

$\mathbf{0}_{L \times N}$ $L \times N$ zero matrix

A acquisition state

$\mathbf{A}_p$ state matrix of the parallel self synchronous scrambler

$\hat{\mathbf{A}}_p$ state matrix of the parallel self synchronous descrambler

$\mathbf{A}_{\Psi(x)}$ companion matrix, state transition matrix of MSRG

$\mathbf{A}_{\Psi(x)}^t$ transpose of $\mathbf{A}_{\Psi(x)}$, state transition matrix of SSRG

$\mathbf{A}_{\Psi_r(x)}$ companion matrix for the reciprocal polynomial $\Psi_r(x)$

B byte

$\mathbf{B}_p$ state matrix of the parallel self synchronous scrambler

$\hat{\mathbf{B}}_p$ state matrix of the parallel self synchronous descrambler

$\mathbf{B}_{\Psi(x)}$ transform matrix of the primary basis with respect to the elementary basis

$\{\, b_{j+kN} \,\}$ jth N-decimated sequence of $\{\, b_k \,\}$

$\{\, \hat{b}_{j+kN} \,\}$ jth N-decimated sequence of $\{\, \hat{b}_k \,\}$

$\{\, \tilde{b}_{j+kN} \,\}$ jth N-decimated sequence of $\{\, \tilde{b}_k \,\}$

$\{\, b_k \,\}$ input data sequence

$\{\, \hat{b}_k \,\}$ self-synchronously descrambled sequence

$\{\, \tilde{b}_k \,\}$ self-synchronously scrambled sequence

$\{\, b_k^j \,\}$ jth parallel input data sequence

$\{\, \hat{b}_k^j \,\}$ jth parallel-descrambled sequence

$\{\, \tilde{b}_k^j \,\}$ jth parallel-scrambled sequence

$\mathbf{b}_k$ signal vector

$\hat{\mathbf{b}}_k$ signal vector of the self synchronously descrambled signal, output signal vector

$\tilde{\mathbf{b}}_k$ signal vector of the self synchronously scrambled signal

C_R characteristic operator

$\mathbf{C}_p$ state matrix of the parallel self synchronous scrambler

$\hat{\mathbf{C}}_p$ state matrix of the parallel self synchronous descrambler

$\mathbf{C}_{\Psi(x)}$ transform matrix of the elementary basis with respect to the primary basis

$\mathbf{c}_i$ ith correction vector

D delay operator

$\hat{D}$ delay number-operator

D_i ith SRG sequence

$D_{M,i}$ ith MSRG sequence

$D_{M,L-1}$ terminating sequence of MSRG

$D_{S,0}$ terminating sequence of SSRG

$D_{S,i}$ ith SSRG sequence

$\mathbf{D}$ SRG sequence vector

$\mathbf{D}_p$ state matrix of the parallel self synchronous scrambler

$\hat{\mathbf{D}}_p$ state matrix of the parallel self synchronous descrambler

$d_{i,k}$ kth state value of the ith shift register in SRG

$\hat{d}_{i,k}$ kth state value of the ith shift register in the descrambler SRG

$d^m i$ number-based notation of sequence element b^i_{k-m}

$dim(\mathcal{V})$ dimension of the vector space $\mathcal{V}$

$\mathbf{d}_0$ initial state vector of SRG

$\hat{\mathbf{d}}_0$ initial state vector of the descrambler SRG

$\tilde{\mathbf{d}}_0$ synchronization initial state vector of the descrambler SRG

$\tilde{\mathbf{d}}_0^i$ ith subvector of the synchronization

$\mathbf{d}_k$ kth state vector of SRG

$\hat{\mathbf{d}}_k$ kth state vector of the descrambler SRG initial state vector $\hat{\mathbf{d}}_0$

$\tilde{\mathbf{d}}_k$ kth synchronization state vector of the descrambler SRG

$\mathcal{E}_A$ acquisition state error vector

$\mathcal{E}_S$ steady state error vector

$\mathcal{E}_V$ verification state error vector

$E^i_{\Psi(x)}$ ith elementary sequence

$\mathbf{E}_{\Psi(x)}$ elementary sequence vector

$\mathbf{e}_i$ ith basis vector

$\mathcal{F}$ field

F length of fixed-size packet

F_H length of header

F_U length of user information field

$GCD[\Psi_1(x), \Psi_2(x)]$ greatest common divisor of $\Psi_1(x)$ and $\Psi_2(x)$

$GF(q)$ Galois field

$\tilde{\mathbf{H}}$ $L \times L$ matrix formed by the state transition matrix and the generating vector of the descrambler SRG

$\mathbf{h}$ generating vector

$\hat{\mathbf{h}}$ generating vector of the descrambler

$\mathbf{h}_j$ jth generating vector of PSRG

$\mathbf{I}_{L \times L}$ $L \times L$ identity matrix

L dimension of sequence space, length of SRG

L_S observed sample number in the steady state

L_V observed sample number in the verification state

$LCM[\Psi_1(x), \Psi_2(x)]$ least common multiple of $\Psi_1(x)$ and $\Psi_2(x)$

l run length

l_i sampled signal number of the ith sample z_i

M number of multibit

$\mathbf{M}_{\Psi(x)}$ linear convolutional matrix

$\mathbf{M}^+_{\Psi(x)}$ left submatrix of linear convolutional matrix

$\mathbf{M}^-_{\Psi(x)}$ right submatrix of linear convolutional matrix

m_i transmitted signal number of the ith sample z_i

N number of parallel sequences

N_A number of tolerable errors in the acquisition state

N_S number of tolerable errors in the steady state

N_V number of tolerable errors in the verification state

O operator

$\hat{O}$ number-operator

$\tilde{\mathbf{O}}_L$ $L \times L$ upper diagonal matrix

P permuter operator

$\hat{P}$ permuter number-operator, permuter characteristic expression

$P(x)$ polynomial expression of initial state vector for the primary basis

$P_i(x)$ ith decimated polynomial

$P_{A\tilde{s}}$ probability of the false synchronization

P_{VA} probability of the false reinitialization from the verification

$P_{\tilde{V}\tilde{s}}$ probability of the false synchronization from the verification

$P^i_{\Psi(x)}$ ith primary sequence

$\mathbf{P}$ permuting matrix

$\mathbf{P}_{\Psi(x)}$ primary sequence vector

p_e bit error probability

p_i coefficient of $P(x)$

$\mathbf{p}$ initial vector for the primary basis

$Q[\Psi(x), P(x)]$ quotient polynomial of $P(x)$ divided by $\Psi(x)$

$\mathbf{Q}$ nonsingular matrix, $L \times (L+N)$ matrix $[\; 0_{L \times N}\; \mathbf{I}_{L \times L}\;]$

$\mathbf{Q}^+$ left submatrix of $\mathbf{Q}$

$\mathbf{Q}^-$ right submatrix of $\mathbf{Q}$

R total run of sequence

$R(l)$ run function of sequence

$R[\Psi(x), P(x)]$ remainder polynomial of $P(x)$ divided by $\Psi(x)$

$\mathbf{R}$ nonsingular matrix

$\tilde{\mathbf{R}}_M$ M-augmented matrix

S steady state, scrambling operator

$\hat{S}$ scrambling number-operator

$\tilde{S}$ false steady state

$S[\Psi(x), P(x)]$ polynomial expression of sequence

$\{\, s_{j+kN}\, \}$ jth N-decimated sequence of $\{\, s_k\, \}$

$\{\, s_k\, \}$ sequence, serial scrambling sequence

$\{\, \hat{s}_k\, \}$ serial descrambling sequence

$\{\, \tilde{s}_k\, \}$ estimated scrambling sequence

$\{\, s^j_k\, \}$ jth parallel scrambling sequence

$\{\,\hat{s}_k^j\,\}$ jth parallel descrambling sequence

$\{\,s_{k+m}\,\}$ m-delayed sequence of $\{\,s_k\,\}$

$\mathbf{s}$ initial vector for the elementary basis

T periodicity of sequence space, signal-detection table

$\hat{T}$ signal-detection table characteristic expression

$T(i,j)$ signal-detection value

T_{AS} average synchronization time

T_I periodicity of irreducible space

T_P signal-detection table of permuter based system

$\hat{T}_P$ signal-detection table characteristic expression of permuter based system

T_{PS} signal-detection table of permuter-scrambler based system

$\hat{T}_{PS}$ signal-detection table characteristic expression of permuter-scrambler

$T_{P_1SP_2}$ signal-detection table of permuter-scrambler-permuter based system

$\hat{T}_{P_1SP_2}$ signal-detection table characteristic expression of permuter-scrambler-permuter based system

T_p periodicity of primitive space

T_S signal-detection table of scrambler based system

$\hat{T}_S$ signal-detection table characteristic expression of scrambler based system

T_{SA} average false reinitialization time from the steady state

T_{SP} signal-detection table of scrambler-permuter based system

$\hat{T}_{SP}$ signal-detection table characteristic expression of scrambler-permuter based system

$T_{S_1PS_2}$ signal-detection table of scrambler-permuter-scrambler based system

$\hat{T}_{S_1PS_2}$ signal-detection table characteristic expression of scrambler-permuter-scrambler based system

$T_{\tilde{S}A}$ maximum average reinitialization time from the steady state

T_{VS} average synchronization time from the verification state

$T_{\tilde{V}A}$ average reinitialization time from the verification state

T_w periodicity of power space based system permuter-scrambler-permuter based system

$\mathbf{T}$ state transition matrix of SRG

$\hat{\mathbf{T}}$ state transition matrix of the descrambler SRG

$\mathbf{T}_M$ state transition matrix of MSRG

$\tilde{\mathbf{T}}_M$ M-extended state transition matrix

$\mathbf{T}_S$ state transition matrix of SSRG

$\mathbf{t}_i^t$ ith row of state transition matrix

U_i ith decimated sequence, ith parallel scrambling sequence

U nonsingular matrix

$u_{i,j}$ arbitrary binary number

$\mathcal{V}$ vector space

V verification state

$\tilde{V}$ false verification state

$V[\mathbf{A}_{\Psi(x)}, \mathbf{d}_0]$ SRG space of MSRG

$V[\mathbf{A}_{\Psi(x)}^t, \mathbf{d}_0]$ SRG space of SSRG

$V[\mathbf{T}]$ SRG maximal space, scrambling maximal space

$V[\mathbf{T}, \mathbf{d}_0]$ SRG space

$V[\mathbf{T}, \mathbf{h}]$ scrambling space

$V[\Psi(x)]$ sequence space

$V[\hat{\Psi}(x)]$ sequence subspace

$V[\Psi_I(x)]$ irreducible (sequence) space

$V[[\Psi_I(x)]^w]$ power (sequence) space

$V[\Psi_m(x)]$ minimal (sequence) space

$V[\Psi_p(x)]$ primitive (sequence) space

$V[\Psi_{\mathbf{T}}(x)]$ SRG maximal space, scrambling maximal space

$V[\Psi_{\mathbf{T},\mathbf{d}_0}(x)]$ minimal space for SRG sequences

$V[\Psi_{\mathbf{T},\mathbf{h}}(x)]$ scrambling space

$\tilde{\mathbf{V}}$ $L \times L$ matrix formed by the state transition matrix and the sampling vectors of PSRG

$\tilde{\mathbf{V}}_0$ submatrix of $\tilde{\mathbf{V}}$

$v_{i,j}$ arbitrary binary number

$\mathbf{v}_i$ ith sampling vector

$\tilde{\mathbf{v}}_i$ ith ordered sampling vector

W weight of sequence, interleaved sequence

$W[\mathbf{v}]$ weight of vector $\mathbf{v}$

$W[\Psi(x)]$ weight of the characteristic polynomial $\Psi(x)$

W_j jth multibit-parallel scrambling sequence

w power of power space

$\bar{w}$ binary-ceiling of power

$\mathcal{Y}_S$ sample-comparison vector in the steady state

$\mathcal{Y}_V$ sample-comparison vector in the verification state

$\mathcal{Z}_S$ sample vector of the error sequence $\{\ \sigma_k^S\ \}$

$\mathcal{Z}_V$ sample vector of the estimation-error sequence $\{\ \sigma_k\ \}$

z_i ith sample of the scrambling sequence

$\hat{z}_i$ ith sample of the descrambling sequence

$\tilde{z}_i$ ith ordered sample, ith received sample

$\mathbf{z}$ sample vector, transmitted sample vector

$\tilde{\mathbf{z}}$ received sample vector

α_i ith sampling time

$\hat{\alpha}_i$ ith parallel sampling time

$\tilde{\alpha}_i$ ith rearranged serial sampling time

β_i ith correction time

$\hat{\beta}_i$ ith parallel correction time

γ_i ith sample transmission time

$\hat{\gamma}_i$ ith parallel sample transmission time

$\tilde{\gamma}_i$ ith ordered sample transmission time

$\Delta_{\mathbf{T},\mathbf{d}_0}$ discrimination matrix for maximal initial state vector

$\Delta_{\mathbf{T},\mathbf{h}}$ discrimination matrix for maximal generating vector

Δ_α discrimination matrix for predictable sampling times

$\Delta_{\tilde{\alpha}}$ rearranged discrimination matrix

δ_k kth state distance vector

ε_i ith sample error

ε_i^S ith sample error in the steady state

ζ_i ith sample of the estimation-error sequence $\{\ \sigma_k\ \}$

ζ_i^S ith sample of the error sequence $\{\ \sigma_k^S\ \}$

η_i ith sample-comparison data

η_i^S ith sample-comparison data in the steady state

Θ set of integers taken from the exponents of the characteristic polynomial $\Psi(x)$

Λ correction matrix

π probability of getting into the verification state with the descrambler synchronized

σ initial error vector of the error sequence $\{ \sigma_k^S \}$

$\{ \sigma_k \}$ estimation-error sequence

$\{ \sigma_k^S \}$ error sequence in the steady state

τ period of sequence

τ_I period of irreducible sequence

τ_p period of PRBS

τ_w period of power sequence

$\phi(m)$ autocorrelation function of sequence

$\Psi(x)$ characteristic polynomial of sequence space, characteristic polynomial of self synchronous scrambler

$\Psi_I(x)$ irreducible polynomial

$\Psi_p(x)$ primitive polynomial

$\Psi_r(x)$ reciprocal polynomial of $\Psi(x)$

$\Psi_T(x)$ minimal polynomial of T

ψ_i coefficient of $\Psi(x)$

Ω two dimensional set composed of delays and signal numbers

$\sum_{j=0} V[\Psi_j(x)]$ sum space of $V[\Psi_j(x)]$'s

Part I

PRELIMINARIES

Chapter 1

Digital Transmission and Scrambling

Digital transmission is a technique that transmits information signals in a stream of digital pulses. The original signals may be in analog or digital form, and the transmission medium may be copper wire or optical fiber based. In either case, as long as the signals are conveyed in digital pulse stream over the transmission medium, it is called digital transmission.

Digital transmission has been replacing the copper cable based analog transmission since the introduction of the T1 carrier. Lightwave transmission which is prevalent in today's high-speed transmission has been mostly digital since its introduction. Data transmission for computer communications has been also digital transmission by its nature. Data transmission differs from other forms of circuit-mode digital transmission, in that it employs packet-mode with fixed or variable packet size. Therefore, digital transmission is the predominant transmission means today for the copper or optical based transmission media.

Scrambling is a bit-level(or pulse-level) processing applied to digital transmission signals just before transmission. The objective of scrambling is to improve clock recovery and data reception capability in digital transmission. While scrambling function is not limited in application to digital transmission, its role in digital transmission is quite significant. Therefore, in this introductory chapter, we will discuss some fundamentals of digital transmission in relation to the scrambling function.

1.1 Digital Transmission Systems

A digital transmission system consists of a transmitting terminal, a receiving terminal, some(or none) intermediate devices, and some transmission media, as depicted in Fig. 1.1(a). The transmitting terminal performs all functions necessary to transform the given input signals into digital data stream for digital transmission, and the receiving terminal performs the reverse functions required to reconstruct the original input signals. The intermediate device regenerates or reshapes the transmission signal which has been attenuated and distorted while traveling over the transmission medium. The medium in this case is copper or optical cable.

The transmitting terminal performs preprocessing, frameformatting, *multiplexing* (MUX), scrambling and line-coding functions on the input signals to generate the transmission signal(see Fig. 1.1(b)). The preprocessing function refers to filtering, sampling, quantizing and coding functions to be performed on analog input signals for digital conversion. If the data is already in the digital formats, this functional block becomes null. The frame-formatting function generates a prespecified frame structure and loads digital data into the payload space, while loading additional overhead data into the overhead space.[1] The digital data thus formed is called the *base-rate signal.* Multiple(or N) of base-rate signals are multiplexed to form a signal of N times the base rate, which is called the *transmission-rate signal.* Multiplexing in this case may be bit-interleaved, byte-interleaved, or any other desired type. The transmission-rate signal is then scrambled, and finally line-coded for transmission. For the scrambling function, three types of techniques are available $--$ *frame-synchronous scrambling* (FSS), *distributed sample scrambling* (DSS), and *self-synchronous scrambling* (SSS). For the line-coding function, there are a number of techniques applicable, including *alternate mark inversion* (AMI), *coded mark inversion* (CMI), *bipolar with N-zero substitution* (BNZS), *high-density bipolar with a maximum of N zeros* (HDBN), *return-to-zero* (RZ), *non-return-to-zero* (NRZ), and *two-binary one-quaternary* (2B1Q) codings. The line-coded signal is subject to media conversion processing such as *electrical-to-optical conversion* (EOC) before transmission, depending on the type of transmission

---

[1]In some conventional multiplexers such as M12 and M23 only clock-synchronizing function is done in the frame-formatting block of the figure and the frame-formatting function itself is done in conjunction with the multiplexing block.

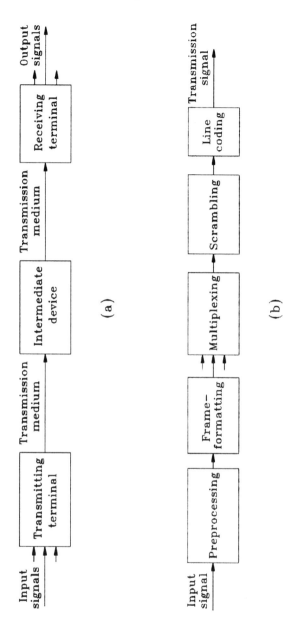

Figure 1.1. Functional blockdiagrams for digital transmission. (a) Overall digital transmission system structure(one direction), (b) signal processing functions of the transmitting terminal. (The transmission signal is subject to media conversion processing depending on the type of transmission medium.)

medium.

The receiving terminal performs the reverse of the above described functions on the transmission signal to reconstruct the original input signals, which are line-decoding, descrambling, demultiplexing(DEMUX), frame-deformatting, and postprocessing. The postprocessing is the reverse of the preprocessing function necessary for reconstructing the original analog signals. The remaining four processings are exact reverse of their transmitter counterparts. Therefore, the line-decoding and descrambling functions are performed on the received transmission-rate signal, and the demultiplexing function divides it into multiple base-rate signals.

1.2 Scrambling Functions in Digital Transmission

As indicated in Fig. 1.1(b), scrambling as well as line-coding is a function performed on the transmission-rate signal, thus yielding the effect of changing the transmission signal characteristics. In fact, the transmission-rate signal before scrambling can include some regular bit patterns which can result from periodic frame structures, and a long run of ones or zeros which can result from idle data or long-string data. If transmitted unscrambled, such regular patterns can cause pattern-dependent jitter, and the long run of ones or zeros can cause baseline wander as well as unstable clock extraction. All these collectively contribute in degrading the *inter-symbol interference* (ISI). However, if scrambling is introduced, the transmission signal characteristics are improved in such a way that the regular patterns get randomized, and the long strings get broken, so one-zero transition density gets increased. Consequently, the clock recovery in the receiving terminal gets stabilized, and jitter gets reduced, decreasing inter-symbol interference and improving data reception accuracy.

In conventional plesiochronous digital transmission systems, scrambling techniques accomplish the above functions, assisted by appropriate line-coding techniques such as AMI, CMI, BNZS and HDBN. However, in the newly emerging lightwave transmission systems with the rates ranging from several hundred *mega-bits per second* (Mbps) to a few *giga-bits per second* (Gbps), it is not desirable to employ such complex line-coding schemes. But since optical cables for lightwave transmission is a very reliable transmission medium, simple NRZ or RZ line-coding turns out to be satisfactory only if a well-functioning scrambler is matched in front. From the scrambler's point of view, this implies that it now has to fulfill all the required functions without getting assistance from line-coders. Therefore, the role of scrambling becomes ever more important in the lightwave transmission era.

It is worth noting that the above scrambling functions for digital transmission become useful because the descrambling process can undo the randomizing effect of the scrambling process. The descrambler which performs the descrambling process is nothing more than a simple replica or an input/output-reversed replica of the scrambler. In this aspect, the scrambling function may be referred as a controlled data-randomizing function.

1.3　Digital Signals and Scrambling

There are two digital hierarchies in the digital transmission −− the conventional *plesiochronous digital hierarchy* (PDH) and the recently emerged *synchronous digital hierarchy* (SDH). The PDH is subclassified into the North American system consisting of the DS-1, the DS-2, the DS-3, and the DS-4E tributary signals, and the European system consisting of the DS-1E, the DS-2E, the DS-3E, the DS-4E and the DS-5E tributary signals. The rates of these signals are as listed in Table 1.1.[2] The SDH consists of one type of *synchronous transport module* signal with level N (STM-N) for different number of N's. The *synchronous optical network* (SONET) interface signals form a variant of the SDH signals, and renders the *synchronous transport signal* with level N (STS-N) for a number of different N's. The number of interest for N are 1, 4, 16 for the STM-N, and 1, 3, 9, 12, 18, 24, 36, 48 for the STS-N, and their corresponding rates are as listed in Table 1.2.

The PDH signal in each level is obtained by multiplexing a multiple number of the PDH signals in its sublevel. This property is common to both the North American and the European systems. The number of sublevel PDH signals to multiplex differs depending on the level in the North American system, while it is fixed to four in the European system.

---

[2]There are two more signals not listed in the table, which are the DS-1C and the DS-4 signals respectively at the rates 3.152 Mbps and 274.176 Mbps. The DS-1C is obtained by multiplexing two DS-1 signals ; and the DS-4, by multiplexing six of DS-3 signals. They are not listed because one is for local use and the other is obsolete.

Table 1.1. Rates of the PDH signals

North American PDH		European PDH	
DS-1	1.544 Mbps	DS-1E	2.048 Mbps
DS-2	6.312 Mbps	DS-2E	8.448 Mbps
DS-3	44.736 Mbps	DS-3E	34.368 Mbps
DS-4E	139.264 Mbps	DS-4E	139.264 Mbps
		DS-5E	564.992 Mbps

Table 1.2. Rates of the SDH signals

SDH		SONET	
N	STM-N	N	STS-N
		1	51.840 Mbps
1	155.520 Mbps	3	155.520 Mbps
		9	466.560 Mbps
4	622.080 Mbps	12	622.080 Mbps
		18	933.120 Mbps
		24	1,244.160 Mbps
		36	1,866.240 Mbps
16	2,488.320 Mbps	48	2,488.320 Mbps

The multiplexing structure for the two PDH systems are as illustrated in Fig. 1.2. In contrast, the SDH and the SONET signals are obtained via direct synchronous multiplexing procedure from any type of the relevant PDH signals or a proper combination of them, as illustrated in Fig. 1.3. This implies, for example, that the STM-1 signal at 155.520 Mbps can be formed by directly multiplexing 84 of DS-1 signals without passing through the intermediate DS-2 or DS-3 signals.

In the process of standardizing the *broadband integrated services digital network* (BISDN), a new signal structure has been introduced under the name of *asynchronous transfer mode* (ATM) *cell*. It is a small-size frame format from the circuit-mode digital transmission's point of view, and is a fixed-size packet from the packet-mode data transmission's point of view, indicating that the conventionally different two modes are now establishing a common ground on the ATM-based BISDN. The ATM cells can be either carried over the payload space of the STM-1/STM-4 signals or self-carried in a contiguous stream, both at the rates of

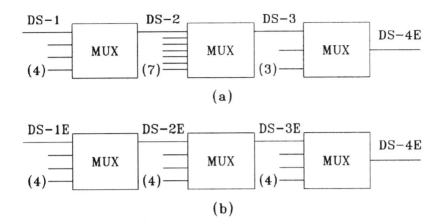

Figure 1.2. Multiplexing structure for the PDH systems. (a) North American system, (b) European system.

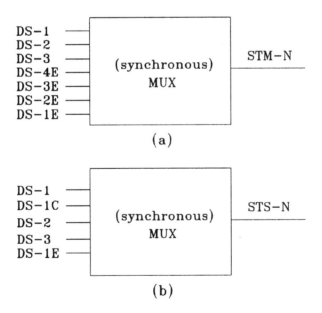

Figure 1.3. Synchronous multiplexing structure for SDH and SONET systems. (a) SDH system, (b) SONET system.

Table 1.3. Scrambling techniques for digital signals

signal			scrambling		
			type	"characteristic" polynomial	initial state
SDH	SDH	STM-N	FSS	$x^7 + x^6 + 1$	1111111
	SONET	STS-N			
ATM	SDH-based		SSS	$x^{43} + 1$	irrelevant
			FSS	$x^7 + x^6 + 1$	1111111
	cell-based		DSS	$x^{31} + x^{28} + 1$	irrelevant

155.520/622.080 Mbps. The former is called the *SDH-based ATM trans-mission*, and the latter is called the *cell-based ATM transmission*.

We now consider what kind of scrambling and/or line coding processes are applied to the above digital signals. For the PDH signals listed in Table 1.1, high data-transition line-codings such as AMI, CMI, BNZS and HDBN are employed without scrambling, while for the SDH/SONET signals in Table 1.2, frame synchronous scrambling is applied in conjunction with the NRZ line-coding. This is illustrated on the STM-N signal in Fig. 1.4(a). In the case of the SDH-based ATM transmission, two types of scrambling techniques are involved. The ATM-cell stream at 155.520(or 622.080) Mbps gets first self-synchronously scrambled in the payload space(or user information field), and is then mapped to the payload space of the STM-1(or STM-4) frame. The resulting STM-1(or STM-4) signal gets again frame-synchronously scrambled as usual STM-N signals do. This process is illustrated in Fig. 1.4(b). In the case of the cell-based ATM transmission, the ATM-cell stream is distributed-sample scrambled once, and then transmitted, as shown in Fig. 1.4(c). If we summarize the scrambling techniques for the digital signals we have discussed so far, we obtain the list in Table 1.3. The "characteristic" polynomial and the initial state in the table are two distinctive parameters characterizing a scrambler, as will become clear in the following chapters.

1.4 Packet-Mode Data Transmission and Scrambling

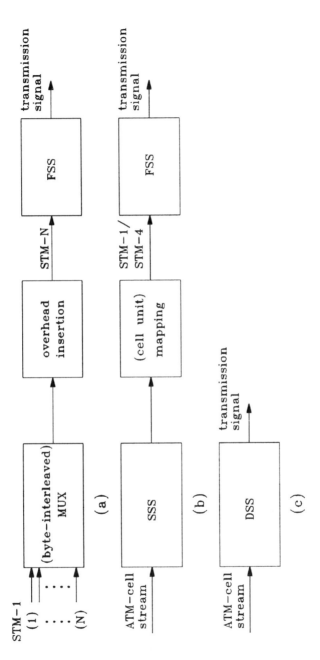

Figure 1.4. Scrambling processes applied to digital signals. (a) STM-N signal(SDH), (b) SDH-based ATM signal, (c) cell-based ATM signal.

For packet-mode data transmission, only a limited number of functional blocks in Fig. 1.1(b) have been used. In fact, packet generation, which corresponds to the frame-formatting in the figure, and line-coding used to be the main functions in packet-mode data transmission. It is because the transmission itself was of no serious concern in the *local area networks* (LAN) of few kilometers in diameter. Further, the transmission rate was low, and the target service was non-real-time(or non-isochronous) data, so the additional cost of increased transmission rate paid for the simple Manchester line-coding was no big deal. Therefore more efforts had always been exerted on fair and efficient sharing of the network than on the transmission efficiency all through its evolution stages through *carrier-sense multiple access with collision detection* (CSMA/CD), token-bus, token-ring, and *fiber-distributed data interface* (FDDI) LAN protocols.

As the service integration matures with the inclusion of real-time(or isochronous) service data, and as the network size and speed grows to the level of *metropolitan area network* (MAN), the above conventional techniques need reexamining. In the network with the speed going up to several hundred Mbps, the rate increase factor of 25 percent is no more negligible, and in networks of a hundred kilometers in diameter carrying both the real-time and non-real-time traffics, the transmission efficiency can no longer be neglected.

Fortunately, there are two rescuers in this changing environment. One is the optical cable that can reliably carry the high-speed traffic, and the other is the novel scrambling technique, *distributed sample scrambling* (DSS). As pointed out earlier, the optical fiber is so reliable a transmission medium that a simple NRZ or RZ line-coding works fine if used in conjunction with an efficient scrambling technique. DSS, as will become clear later in the book, renders the best performance among all scrambling techniques in small-size frame environments. But this is exactly the case for the high-speed integrated packet-mode network since the small fixed-size packets become the ultimate solution to guarantee the real-time requirements. This argument is readily supported by the examples of FDDI II and *distributed-queue dual-bus* (DQDB) networks, as well as in the cell-based ATM network. Therefore, the DSS technique followed by the NRZ or RZ line coding is the very structure to opt for to achieve efficient packet-mode data transmission in the high-speed lightwave transmission era.

Chapter 2
Fundamentals of Scrambling Techniques

Scrambling is a binary bit-level processing applied to the transmission rate signal in order to make the resulting binary sequence appear more random. The scrambler performing this scrambling function can be implemented simply using a few shift registers and exclusive-OR gates ; and the descramblers reconstructing the original bitstream out of the scrambled data stream has the same structure but with the reversed data flow. For a proper reconstruction of the original bitstream the shift registers in the descrambler should get synchronized to their counterparts in the scrambler. Depending on the synchronization method used, scrambling techniques are classified into three categories, namely the frame-synchronous scrambling(FSS), the distributed sample scrambling(DSS), and the self-synchronous scrambling(SSS).[1] In the FSS, the states of the scrambler and the descrambler shift registers get synchronized by being simultaneously reset to the prespecified states at the start of each frame ; in the DSS, samples taken from the scrambler shift registers are transmitted to the descrambler in a distributed manner for use

---

[1]Terminology on scrambling is flexibly used to some extent in this book. *Scrambling* in general refers to the scrambling function, and *scrambler* refers to the means that realizes the scrambling function. The reverse functions and means to achieve them are respectively referred to as *descrambling* and *descrambler*. However, the term scrambling, when used to indicate scrambling techniques, can include both the scrambling and descrambling functions. For example, the terms FSS, DSS and SSS all include the relevant scrambling and descrambling functions. Further, the terms FSS, DSS and SSS sometimes imply scramblers, while in most cases they refer scrambling techniques.

in synchronizing the descrambler shift registers ; and in the SSS, the states of the scrambler and descrambler shift registers are automatically synchronized without any additional synchronization processes.

On the other hand, examining the scrambling function in the context of digital transmission functions, we find that it is always related to the multiplexing function, or can be viewed as if so. In the existing digital transmission systems, the multiplexing function is based on bit-interleaving or byte-interleaving. In the SDH system of Fig. 1.4, for example, FSS follows the byte-interleaved multiplexing and an auxiliary overhead insertion ; in the SDH-based ATM system, SSS and FSS appear independent of multiplexing, but each of them can be viewed as if it follows a preceding imaginary bit-interleaved multiplexing which multiplexes the relevant subrate bitstreams ; and in the cell-based ATM system, the DSS can be also viewed as if it follows an artificial bit-interleaved multiplexing. As such, we can consider the scrambling function in conjunction with the preceding multiplexing function, and further noting that the scrambling function is done at the transmission rate while the multiplexing function is done on the base rate signals, we can also consider reversing the order of scrambling and multiplexing functions such that scrambling function can be equivalently carried out on the multiple parallel inputs each at the base rate. This yields the concept of *parallel scrambling*, as opposed to the original scrambling, which is called the *serial scrambling*. The parallel scrambling can be further subdivided into the *bit-parallel scrambling* and the *multibit-parallel scrambling* depending on the involved multiplexing types.

In this chapter, we will provide a fundamental overview of all available scrambling techniques. We will first consider the three scrambling techniques FSS, DSS, and SSS, one by one, comparing their distinctive characteristics. Then, we will consider the serial, bit-parallel, and multibit-parallel scrambling techniques, illustrating the structural features of each.

2.1   Frame Synchronous Scrambling

The FSS employs an autonomous system consisting of shift registers and exclusive-OR gates, which is called the *shift register generator* (SRG).[2] The SRG is organized in such a manner that it can generate a desired

---

[2]A more detailed description of SRG follows in Chapter 5.

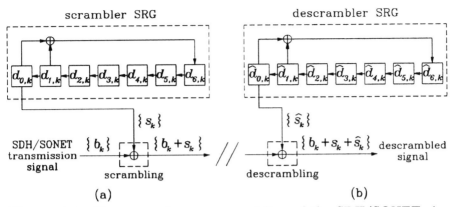

Figure 2.1. Frame synchronous scrambling of the SDH/SONET signal in the SDH/SONET transmission system. (a) Scrambling part, (b) descrambling part.

pseudo-random binary sequence (PRBS)[3] for use in scrambling. In the scrambling part, the transmission signal is scrambled by adding the PRBS to it, and in the descrambling part, the same PRBS is added to the scrambled signal for the recovery of the original signal. In order to make the PRBSs generated by the scrambler and descrambler SRGs identical, the shift register states in both SRGs should be synchronized to each other, that is, the SRGs as well as their states should be identical for the scrambler and the counterpart descrambler. To achieve this, the FSS resets the scrambler and the descrambler SRGs to some pre-specified states at the start of each frame.[4] This explains why the term "frame-synchronous" is used.

The most typical example of transmission system employing the FSS can be found in the SDH/SONET lightwave transmission system. The scrambling and the descrambling operations of this system are as depicted in Fig. 2.1.[5] In the figure, two upper dotted boxes composed of seven shift registers and one exclusive-OR gate are the scrambler and descrambler SRGs respectively generating the PRBSs { s_k } and { $\hat{s}_k$ } of period $2^7 - 1$. The transmission signal { b_k } is scrambled by adding the PRBS { s_k } generated by the scrambler SRG, and the scrambled signal { $b_k + s_k$ } is descrambled by adding the PRBS { $\hat{s}_k$ } generated

---

[3] A more detailed description of PRBS follows in Chapter 3.

[4] For a more detailed description on the operation of the FSS, refer to Chapter 3.

[5] A more detailed description on the SDH/SONET transmission systems and their FSS operations is given in Chapter 9.

by the descrambler SRG, which becomes identical to { s_k } when the descrambler SRG is synchronized to the scrambler SRG. Therefore, the descrambled signal { $b_k + s_k + \hat{s}_k$ } becomes the original transmission signal { b_k } in the synchronized state. For this synchronization, the scrambler SRG state $d_{i,k}$, $i = 0, 1, \cdots, 6$, and the descrambler SRG state $\hat{d}_{i,k}$, $i = 0, 1, \cdots, 6$, are both reset to the state "1111111" at the beginning of each SDH/SONET frame. Note that the transmission signal and the PRBSs are binary signals, and the addition is a modulo-2 based operation performed by an exclusive-OR gate.

2.2   Distributed Sample Scrambling

The DSS is basically similar to the FSS, which scrambles and descrambles the transmission signal by adding a PRBS generated by an SRG. But the DSS is different from the FSS in the method of synchronizing the state of the descrambler SRG to that of the scrambler SRG. In the DSS, the samples of the scrambler SRG state are transmitted to the descrambler in parallel with the scrambled signal, and the descrambler SRG state is corrected by them to eventually become identical to the scrambler SRG state. The samples of the scrambler SRG state are usually taken and conveyed over some available slots in the transmission frame in a distributed manner, and this is why the DSS is named so. In contrast, the SRGs in the DSS are not reset at each frame, and thus the transmission signal is scrambled by a continuous stream of PRBS. Therefore the scrambling effect of the DSS is better than that of the FSS. However, in the DSS, it should be always checked whether the SRGs stay in the synchronization state.[6]

The most typical example of transmission system employing the DSS can be found in the cell-based ATM transmission system.[7] The scrambling and the descrambling operations of this system are as depicted in Fig. 2.2. In the figure, two upper dotted boxes composed of 31 shift registers and one exclusive-OR gate are respectively the scrambler and the descrambler SRGs generating PRBSs { s_k } and { $\hat{s}_k$ } of period of $2^{31} - 1$. As in the case of the FSS, the ATM cell stream { b_k } is scrambled by adding the PRBS { s_k } generated by the scrambler SRG, and

---

[6]For a more detailed description on the operation of the DSS, refer to Chapter 10.

[7]A more detailed description on the cell-based ATM transmission and its DSS operation is given in Chapter 16.

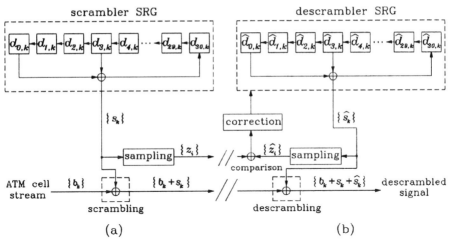

Figure 2.2. Distributed sample scrambling of the ATM cell stream in the cell-based ATM transmission system. (a) Scrambling part, (b) descrambling part.

the scrambled signal { $b_k + s_k$ } is descrambled by adding the PRBS { $\hat{s}_k$ } generated by the descrambler SRG. In order to make the descrambled signal { $b_k + s_k + \hat{s}_k$ } identical to the original transmission signal { b_k }, the descrambler SRG should be synchronized to the scrambler SRG. For this, the samples { z_i } of the scrambler SRG state are taken from the PRBS { s_k } at the prespecified sampling times, and are conveyed to the descrambler over the header error control(HEC) field in the ATM cell composed of 53 bytes. The descrambler generates its own samples { $\hat{z}_i$ } of the SRG state in the same manner and compares them to the transmitted ones. If the two sets of samples are coincident, no action takes place ; but if they are different, a correction logic is initiated to change the descrambler SRG state. This correction operation is repeated over the 31 consecutive samples, and so the SRGs in both the scrambler and the descrambler get synchronized to the same states regardless of their initial states.

2.3 Self Synchronous Scrambling

The SSS is quite different from the FSS or the DSS in the scrambling operation as well as the synchronization mechanism. While in the FSS and the DSS a PRBS signal is generated in the SRG and then added

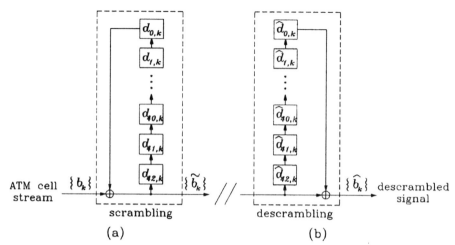

Figure 2.3. Self synchronous scrambling of the ATM cell stream in the SDH-based ATM transmission system. (a) Scrambling part, (b) descrambling part.

to the transmission signal for scrambling, in the SSS the transmission signal directly passes through a variant of SRG to get scrambled. The scrambled signal gets descrambled as it passes through the input/output-reversed replica of the scrambler. In this operation, the transmission signal itself controls the state of shift registers in the scrambler, and the scrambled signal controls the state of shift registers in the descrambler. Therefore the scrambler and descrambler are automatically synchronized without any additional synchronization mechanism, once the number of received data reaches the size of the shift register length. The term "self-synchronous" refers to this synchronization operation.[8]

The most typical example of transmission system employing the SSS can be found in the SDH-based ATM transmission system.[9] The scrambling and descrambling operations of this system are as depicted in Fig. 2.3. In the figure, the ATM cell stream $\{ b_k \}$ gets scrambled as it passes through the 43 shift registers in the transmitting part, and the scrambled signal $\{ \tilde{b}_k \}$ is descrambled as it passes through the input/output-reversed shift register block in the receiving part. The states of the shift registers in the descrambler become automatically synchronized to those

---

[8]For a more detailed description on the operation of the SSS, refer to Chapter 17.

[9]A more detailed description on the SDH-based ATM transmission and its SSS operation is given in Chapter 20.

Table 2.1. Comparison of scrambling techniques

scrambling techniques	FSS	DSS	SSS
synchro-nization	by resetting SRG at each frame start	by transmitting SRG state samples	automatically done
additional sample transmission	not necessary	necessary	not necessary
line error	no error multiplication	critical if occurs at sample data	causes error multiplication
scrambling effect	good for large-sized frames	good for small-sized frames	good in general

in the scrambler after the reception of the first 43 scrambled bits.

The comparative characteristics of the three scrambling techniques we have discussed so far are as listed in Table 2.1. Among various characteristics, the synchronization mechanism is the most distinctive one, so it is reflected in the name of each scrambling technique, The effect of line error renders an important performance criterion for scrambling techniques. In the cases of the FSS and the DSS, a bit error occurring in the scrambled data due to transmission error causes a bit error in the descrambled bitstream, while it causes multibit error, or error-multiplication, in the SSS case. Even in the DSS system, if the bit error occurs in the sample data, it can cause a serious problem.[10] Scrambling effect, or randomizing effect, is another important performance criterion for scrambling techniques. Scrambling effect is good for large-size frames in the FSS case, good for small-size frames in the DSS case, and good for either size in the SSS case.

2.4 Serial Scrambling

The serial scrambling is a term collectively referring to usual scrambling techniques in contrast to the later-introduced concept of parallel

---

[10]This problem is discussed in detail in Chapter 15.

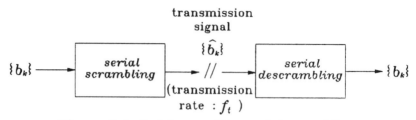

Figure 2.4. Serial scrambling and descrambling.

scrambling. The three scrambling techniques illustrated in Figs. 2.1 to 2.3 are therefore all examples of serial scrambling. If viewed in relation to the relevant multiplexing function, the serial scrambling, in its general sense, indicates the scrambling function which is applied to the multiplexed *serial* bitstream.

Fig. 2.4 depicts the serial scrambling and the descrambling functions for general cases, including the FSS, the DSS, and the SSS. The serial scrambler and descrambler in the figure could be any of those shown in Figs. 2.1 to 2.3. Note that the SRG samples or other overhead data are not indicated in the figure. The relevant multiplexing and demultiplexing functions, which are not shown in the figure, are in general either bit-interleaved or multibit-interleaved processes(see Figs. 2.5(a) and 2.6(a)). The multiplexed serial signal $\{ b_k \}$ is serially scrambled to form the transmission signal $\{ \hat{b}_k \}$ in the transmitter, and the original signal $\{ b_k \}$ is recovered through serial descrambling in the receiver. The scrambling and descrambling processings are therefore done at the transmission rate f_t in the case of serial scrambling.

2.5   Parallel Scrambling

As mentioned above, the transmission signal is, or can be viewed as, formed by multiplexing some base-rate signals based on bit- or byte-interleaving. The *base rate* refers to the rate at which the low-speed processing such as synchronization and frame-formatting are performed. The parallelly incoming base-rate signals are multiplexed to form one transmission signal, to which serial scrambling used to be applied before transmission. It is possible to apply the scrambling function to the parallel base-rate signals before multiplexing, and we call this a (*bit-*)*parallel scrambling*.

The relation between serial scrambling and parallel scrambling is as depicted in Fig. 2.5. In the case of the serial scrambling in Fig. 2.5(a), the N base-rate signals $\{\, b_k^i \,\}$, $i = 0, 1, \cdots, N - 1$, are multiplexed through bit-interleaving to form the transmission signal

$$\{\, b_k \,\} = \{\, b_0^0, b_0^1, \cdots, b_0^{N-1}, b_1^0, b_1^1, \cdots, b_1^{N-1}, \cdots \,\}, \qquad (2.1)$$

which is serially scrambled at the transmission rate f_t. The scrambled signal $\{\, \hat{b}_k \,\}$ is serially descrambled and then demultiplexed to recover the original base rate signals $\{\, b_k^i \,\}$'s. If the base rate is f_b, then the transmission rate f_t becomes N times f_b, that is, $f_t = N f_b$, so the serial scrambling rate is also $N f_b$.

In the case of the parallel scrambling in Fig. 2.5(b), the N base-rate signals $\{\, b_k^i \,\}$, $i = 0, 1, \cdots, N - 1$, are parallelly scrambled before multiplexing. The parallelly scrambled signals $\{\, \hat{b}_k^i \,\}$, $i = 0, 1, \cdots, N - 1$, are multiplexed in the form of bit-interleaving, and then transmitted. The scrambled signal $\{\, \hat{b}_k \,\}$ is demultiplexed and then parallelly descrambled to recover the original base-rate signals $\{\, b_k^i \,\}$'s. The parallel scrambling is performed at the base rate f_b, which is $1/N$ times the serial scrambling rate f_t.

For the serial and parallel scramblings in Fig. 2.5, if the parallel-scrambled transmission signal $\{\, \hat{b}_k \,\}$ in Fig. 2.5(b) is identical to the serial-scrambled transmission signal $\{\, \hat{b}_k \,\}$ in Fig. 2.5(a) for the same base-rate signals $\{\, b_k^i \,\}$'s, then the parallel scrambling is said to be *equivalent* to the serial scrambling. In this case, multiplexing of the base-rate signals followed by serial-scrambling is identical to parallel scrambling followed by multiplexing, but the scrambling rate drops from the transmission rate f_t to the base rate f_b.

2.6 Multibit-Parallel Scrambling

In some transmission systems, base-rate signals are multiplexed to form a transmission signal through byte-interleaving, instead of bit-interleaving. In this case, we call the parallel scrambling applied to the base-rate signals a *byte-parallel scrambling*. More generally, when the base-rate signals get multiplexed through multibit-interleaving, we call the corresponding parallel scrambling a *multibit-parallel scrambling*.

The relation between a serial scrambling and the relevant M-bit parallel scrambling is as depicted in Fig. 2.6. In the case of the serial scrambling in Fig. 2.6(a), the N base-rate signals $\{\, b_k^i \,\}$, $i = 0, 1, \cdots$,

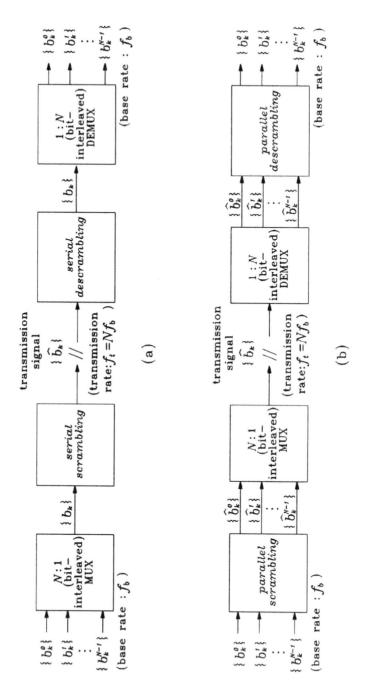

Figure 2.5. Serial and parallel scrambling. (a) Serial scrambling, (b) parallel scrambling.

$N - 1$, are multiplexed in M-bit units to form the transmission signal

$$\{ b_k \} = \{ b_0^0, \cdots, b_{M-1}^0 ; b_0^1, \cdots, b_{M-1}^1 ; b_0^{N-1} \cdots, b_{M-1}^{N-1} ;$$
$$b_M^0, \cdots, b_{2M-1}^0 ; b_M^1, \cdots, b_{2M-1}^1 ; b_M^{N-1}, \cdots, b_{2M-1}^{N-1} ; \cdots \}, \quad (2.2)$$

which is serially scrambled at the transmission rate f_t. The scrambled signal $\{ \hat{b}_k \}$ is serially descrambled and then demultiplexed to reconstruct the original base-rate signals $\{ b_k^i \}$'s. As for the bit-parallel scrambling case, the serial scrambling rate is the transmission rate f_t, which is N times the base rate f_b.

In the case of the M-bit parallel scrambling in Fig. 2.6(b), the N base-rate signals $\{ b_k^i \}$, $i = 0, 1, \cdots, N-1$, are scrambled by the M-bit parallel scrambling process before being multiplexed. The scrambled signals $\{ \hat{b}_k^i \}$, $i = 0, 1, \cdots, N-1$, are multiplexed in M-bit units, and then transmitted. The scrambled transmission signal $\{ \hat{b}_k \}$ is demultiplexed and then parallelly descrambled in M-bit units to recover the original base-rate signals $\{ b_k^i \}$'s. The M-bit parallel scrambling is performed at the base rate f_b also.

As was the case for the parallel scrambling, if the scrambled transmission signal $\{ \hat{b}_k \}$ in Fig. 2.6(b) is identical to the scrambled transmission signal $\{ \hat{b}_k \}$ in Fig. 2.5(a) for the same base-rate signals $\{ b_k^i \}$'s, then the M-bit parallel scrambling is said to be *equivalent* to the serial scrambling. After all, we find that the multibit-parallel scrambling is an M-bit extension of the (bit-)parallel scrambling, or equivalently, the latter is a special case of the former, in which $M = 1$.

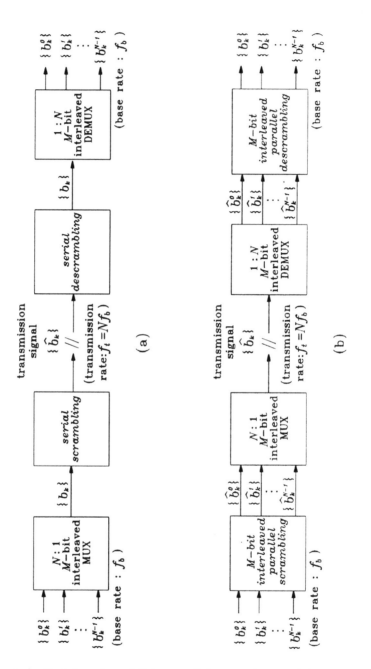

Figure 2.6. Serial and multibit-parallel scrambling. (a) Serial scrambling, (b) multibit-parallel scrambling.

Part II

FRAME SYNCHRONOUS SCRAMBLING

Chapter 3

Introduction to Frame Synchronous Scrambling

Frame synchronous scrambling (FSS) is a scrambling technique which employs SRGs generating PRBSs for use in scrambling as well as in descrambling of the data bitstream. For the synchronization of the SRGs it resets the SRG states to some prespecified states at every frame start. Due to this resetting operation, FSS is suitable for scrambling of framed signals, and the scrambling effect is better for larger-sized frames.

In this part, we concentrate our discussion on the FSS and the related background material. We first establish the sequence space and the SRG theories, and on this basis we discuss the behaviors of the serial, parallel and multibit-parallel FSSs. And in this introductory chapter, we provide fundamental descriptions on the FSS and the related elements such as scrambling sequences and the SRGs.

3.1 Operation of FSS

Fig. 3.1 is a blockdiagram of scrambling and descrambling functions in transmission systems employing the FSS. In the transmitter, transmission frame is formatted, and the input data are inserted to the frame along with overheads for *frame-alignment word* (FAW) and other *operation and maintenance* (OAM) information to form the signal { b_k }. The signal { b_k } is then scrambled through a modulo-2 addition of the scrambling sequence { s_k } which is generated by an SRG. Usually, the FAW is excluded from the scrambling so that the receiver can extract

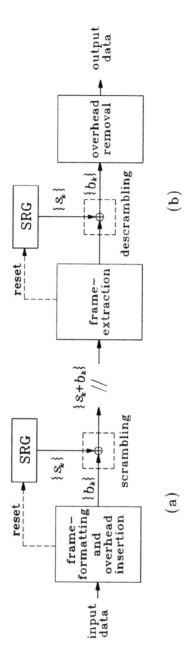

Figure 3.1. Blockdiagram for FSS based scrambling and descrambling functions. (a) Transmitter, (b) receiver.

the frame boundary by searching for the FAW. The SRG state is reset to a prespecified state at the start of each frame both in the transmitter and in the receiver, thus achieving the synchronization of SRGs. The scrambled signal $\{ b_k + s_k \}$ is descrambled in the receiver by adding the synchronized SRG sequence $\{ s_k \}$ to it. Then the original signal $\{ b_k \}$ is recovered, and finally the input data can be reconstructed in the output after overhead removal.

In general, the input data in Fig. 3.1 is formed by multiplexing some base-rate signals bit- or byte-interleaving based, and the FSS is applied to the multiplexed signal, as illustrated in Fig. 3.2(a). We call this *serial FSS*. If parallel scrambling is employed instead of serial scrambling, Fig. 3.2(a) changes into Fig. 3.2(b). The base-rate signals $\{ b_k^j \}$, $j = 0, 1, \cdots$, $N - 1$, in the figure, are then parallelly scrambled through the addition of the parallel sequences $\{ s_k^j \}$, $j = 0, 1, \cdots, N - 1$, generated by the parallel SRG. If multiplexed, the transmission signal $\{ b_k + s_k \}$ becomes identical to the serially scrambled one in Fig. 3.2(a). The descrambling illustrated in Fig. 3.2(b) is done in a similar manner. In the receiver, $\{ b_k + s_k \}$ is demultiplexed first and then parallelly descrambled to recover the original N parallel sequences. If the employed multiplexer is a bit-interleaved one, the corresponding parallel scrambling is called *parallel FSS*, and if it is a multibit-interleaved one, it is called *multibit-parallel FSS*. The serial scrambling in Fig. 3.2(a) is done at the transmission rate f_t, while the parallel scrambling in Fig. 3.2(b) is done at the base rate f_b.

3.2 Scrambling Sequences

As discussed in Chapter 1, the function of the scrambling is to randomize the transmission signal. Thus for a successful scrambling, the serial scrambling sequence $\{ s_k \}$ in Fig. 3.1 should be a sequence that makes the scrambled signal $\{ b_k + s_k \}$ sufficiently random. If the scrambling sequence $\{ s_k \}$ itself is random, the scrambled signal $\{ b_k + s_k \}$ also becomes random. Therefore, it is desirable to choose the random binary sequence as the scrambling sequence $\{ s_k \}$. But, in practice, an SRG consists of a finite number of shift registers, so the sequence generated by an SRG is always periodic, which can never be truly random. Therefore, it is desirable to investigate the properties associated with randomness, and to take the sequence possessing such properties as the scrambling

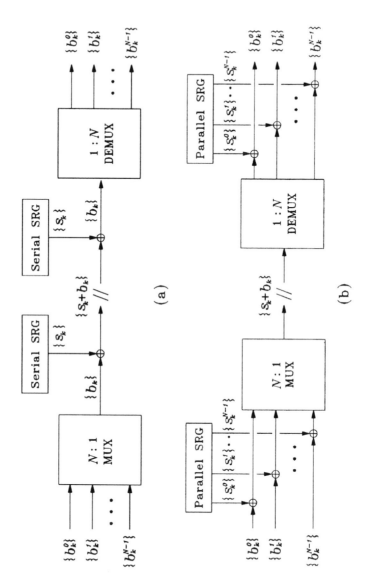

Figure 3.2. Serial and Parallel FSS. (a) Serial FSS, (b) Parallel FSS.

sequence $\{\ s_k\ \}$.

The most familiar example of a random binary sequence arises when tossing an ideal coin consecutively. The following are the properties derived from the coin-tossing experiment in association with randomness:

1. The number of occurrences of the head is approximately equal to the number of occurrences of the tail.

2. Runs of consecutive heads or of consecutive tails occur with short runs occurring more frequently than long runs. More specifically, about one-half the runs have length one, one-fourth have length two, one-eighth have length three, and so forth.

3. The resulting binary sequence indicating the sequential occurrence of heads and tails possesses a special kind of autocorrelation function, which peaks in the middle and tapers off rapidly toward the ends.

Regarding the head as 1 and the tail as 0 in this experiment, we can deduce the following properties for a *pseudo-random binary sequence* (PRBS) $\{\ s_k\ \}$ of period τ :

1. In every period, the number of 1's is nearly equal to the number of 0's.

2. In every period, half the runs have length one, one-fourth length two, one-eighth length three, and so on, as long as the number of runs so indicated exceeds one.

3. The autocorrelation function is two-valued, that is, the number of 1' in every period of the sum sequence $\{\ s_k + s_{k+m}\ \}$ is identical to all cases with $m \neq 0$, which is different from that of the sum sequence for $m = 0$.

The PRBSs possessing these three properties are the sequences for practical use in the FSS.

3.3   Shift Register Generators

The SRG is an autonomous system consisting only of shift registers and exclusive-OR gates. In the serial FSS shown in Fig. 3.2(a), the serial SRG is organized such that it can generate the serial PRBS $\{\ s_k\ \}$ for

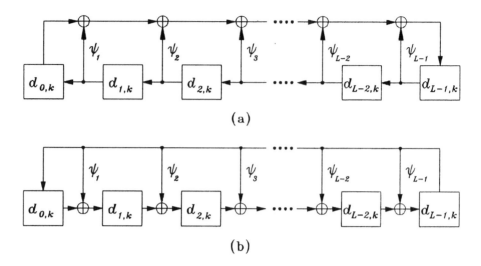

Figure 3.3. Two simplest types of SRGs. (a) Simple SRG, (b) modular SRG.

use in the desired scrambling, and in the parallel FSS shown in Fig. 3.2(b), the parallel SRG is organized to generate the equivalent parallel sequences $\{ s_k^j \}$'s.

Conventionally, two special types of SRGs -- *simple SRG* (SSRG) and *modular SRG* (MSRG) -- have been widely used for the FSS, whose circuit diagrams are as shown in Fig. 3.3. In the figure, each ψ_i for $i = 1, 2, \cdots, L - 1$, is either 0 or 1, each rectangular block denotes a shift register, and the value $d_{i,k}$ inside the block denotes the state of the shift register. In the SSRG, the state $d_{L-1,k}$ of the last shift register is controlled by the states of other shift registers. In the MSRG, in contrast, the state $d_{L-1,k}$ of the last shift register controls the states of other shift registers. In fact the SSRG and the MSRG are two simplest types of SRGs among numerous possible SRGs, as can be recognized from Fig. 3.3.

In the parallel FSS, the parallel SRGs are to generate multiple parallel sequences. A parallel SRG can be built out of an SRG by attaching a parallel sequence generating part. In this case, if the SSRG and the MSRG are employed for parallel SRGs, the resulting parallel sequence generating part can become quite complex. Therefore, it is necessary to determine parallel SRGs that require a minimal complexity in the overall circuit implementation.

3.4   Organization of the Part

The scrambling sequences and the relevant SRGs are two most important topics to understand to properly analyze and synthesize scramblers including the FSS and the DSS, and even the SSS. Therefore, in this part we will first investigate various properties of sequences and SRGs, and then consider the serial, the parallel and the multibit-parallel FSSs respectively.

The part is organized as follows : In Chapter 4, we will introduce the concept of sequence space as a means for unified approach to the description of SRGs, and in Chapter 5, we will investigate the behaviors of the sequences generated by SRGs based on this concept. In the subsequent three chapters, we will examine the scrambling sequences for FSS, and discuss how to realize SRGs for the serial, the parallel and the multibit-parallel FSSs, utilizing the basic properties of sequence spaces and SRGs. Finally, in Chapter 9, we will demonstrate how to apply these FSS techniques for low-rate efficient realizations of scramblers in the SDH/SONET transmission environment.

Chapter 4

Sequence Spaces

In this chapter, we introduce the concept of sequence space as a means to rigorously describe the behaviors of sequences and SRGs. A *sequence space* is a vector space whose elements are sequences satisfying the relation specified by a characteristic polynomial. For a sequence space, there are two bases, namely, the elementary basis and the primary basis, which form the framework of the space. In addition, the polynomial expression of sequences renders a useful tool for mathematical manipulations within the sequence spaces. Based on these new concepts, we investigate various aspects of sequence spaces, such as sequence subspaces and minimal sequence spaces.

4.1 Definition of Sequence Space

We define a *field* $\mathcal{F}$ to be the set $\{0, 1\}$ with the modulo-2 addition and multiplication operations,[1] and, based on the field $\mathcal{F}$, we define sequence space as follows :

DEFINITION 4.1 (**Sequence Space**) For a binary-coefficient polynomial $\Psi(x) = \sum_{i=0}^{L} \psi_i x^i$, $\psi_0 = \psi_L = 1$, we define the *sequence space* $V[\Psi(x)]$ to be the set

$$V[\Psi(x)] \equiv \{\{s_k, k = 0, 1, \cdots\} : \sum_{i=0}^{L} \psi_i s_{k+i} = 0 \text{ for all } k = 0, 1, \cdots\} \quad (4.1)$$

---

[1] This field is known to be *GF(2)*, which is a Galois field having two elements 0 and 1. For a more detailed description on field, refer to Section A.1 in Appendix A.

with the sequence addition $\{\ s_k\ \} + \{\ \hat{s}_k\ \} \equiv \{\ s_k + \hat{s}_k\ \}$ and the scalar multiplication $a\{\ s_k\ \} \equiv \{\ as_k\ \}$ for sequences $\{\ s_k\ \}$, $\{\ \hat{s}_k\ \}$ in $V[\Psi(x)]$ and a scalar a in $\mathcal{F}$. The polynomial $\Psi(x)$ characterizing a sequence space is called the *characteristic polynomial of the sequence space* $V[\Psi(x)]$.

Then, for a sequence space $V[\Psi(x)]$ we have the following theorem.

THEOREM 4.2 *A sequence space $V[\Psi(x)]$ is a vector space over $\mathcal{F}$.*[2]

The proof of this theorem is trivial.

We consider how to find the elements of the sequence space $V[\Psi(x)]$. According to (4.1), for a sequence $\{\ s_k\ \}$ in the sequence space $V[\Psi(x)]$ we have the recurrence relation

$$\begin{cases} s_k = \sum_{i=1}^{L} \psi_i s_{k+i}, & k = 0, 1, \cdots, \\ s_{k+L} = \sum_{i=0}^{L-1} \psi_i s_{k+i}, & k = 0, 1, \cdots, \end{cases} \tag{4.2}$$

since ψ_0 and ψ_L are all 1. This means that a sequence $\{\ s_k\ \}$ in $V[\Psi(x)]$ is completely determined once its L consecutive elements are known. So, we indicate its first L elements by an L-vector $\mathbf{s}$, that is,

$$\mathbf{s} \equiv [\ s_0\ s_1\ \cdots\ s_{L-1}\ ]^t, \tag{4.3}$$

and name it the *initial vector* of the sequence $\{\ s_k\ \}$. Then, all the sequences in $V[\Psi(x)]$ can be determined by inserting 2^L initial vectors into (4.2).

EXAMPLE 4.1 We consider the sequence space $V[x^4 + x^3 + x^2 + 1]$. Since $\Psi(x) = x^4 + x^3 + x^2 + 1$ with $L = 4$ in this case, we can obtain all the sequences contained in this sequence space by inserting the 16 initial vectors $[0\,0\,0\,0]^t$ through $[1\,1\,1\,1]^t$ into (4.2). The resulting 16 sequences are as listed in Table 4.1. From the table, we can confirm that this sequence space is a vector space over $\mathcal{F}$. ♣

4.2   Elementary Basis

We consider a fundamental basis of the sequence space $V[\Psi(x)]$, which is defined in terms of the elementary sequences.

---

[2]The definition of vector space is given in Section B.1 of Appendix B.

Table 4.1. Elements of the sequence space $V[x^4 + x^3 + x^2 + 1]$

S_0	$= \{0,0,0,0,0,0,0,0,0,0,0,\cdots\}$		
S_1	$= \{0,0,0,1,1,0,1,0,0,0,1,\cdots\}$	$= E^3_{x^4+x^3+x^2+1}$	$= P^0_{x^4+x^3+x^2+1}$
S_2	$= \{0,0,1,0,1,1,1,0,0,1,0,\cdots\}$	$= E^2_{x^4+x^3+x^2+1}$	
S_3	$= \{0,0,1,1,0,1,0,0,0,1,1,\cdots\}$	$= P^1_{x^4+x^3+x^2+1}$	
S_4	$= \{0,1,0,0,0,1,1,0,1,0,0,\cdots\}$	$= E^1_{x^4+x^3+x^2+1}$	
S_5	$= \{0,1,0,1,1,1,0,0,1,0,1,\cdots\}$		
S_6	$= \{0,1,1,0,1,0,0,0,1,1,0,\cdots\}$	$= P^2_{x^4+x^3+x^2+1}$	
S_7	$= \{0,1,1,1,0,0,1,0,1,1,1,\cdots\}$		
S_8	$= \{1,0,0,0,1,1,0,1,0,0,0,\cdots\}$	$= E^0_{x^4+x^3+x^2+1}$	
S_9	$= \{1,0,0,1,0,1,1,1,0,0,1,\cdots\}$		
S_{10}	$= \{1,0,1,0,0,0,1,1,0,1,0,\cdots\}$		
S_{11}	$= \{1,0,1,1,1,0,0,1,0,1,1,\cdots\}$		
S_{12}	$= \{1,1,0,0,1,0,1,1,1,0,0,\cdots\}$		
S_{13}	$= \{1,1,0,1,0,0,0,1,1,0,1,\cdots\}$	$= P^3_{x^4+x^3+x^2+1}$	
S_{14}	$= \{1,1,1,0,0,1,0,1,1,1,0,\cdots\}$		
S_{15}	$= \{1,1,1,1,1,1,1,1,1,1,1,\cdots\}$		

DEFINITION 4.3 (Elementary Sequences) For a sequence space $V[\Psi(x)]$ whose characteristic polynomial $\Psi(x)$ is of degree L, the sequence $\{\, s_k \,\}$ in $V[\Psi(x)]$ whose initial vector is $\mathbf{e}_i$, $i = 0, 1, \cdots, L-1$, is called the ith *elementary sequence* $E^i_{\Psi(x)}$, where $\mathbf{e}_i$ denotes the *basis vector* whose ith element is 1 and the others are all zero.

DEFINITION 4.4 (Elementary Sequence Vector) For a sequence space $V[\Psi(x)]$ whose characteristic polynomial $\Psi(x)$ is of degree L, the *elementary sequence vector* $\mathbf{E}_{\Psi(x)}$ refers to the L-vector

$$\mathbf{E}_{\Psi(x)} \equiv [\, E^0_{\Psi(x)} \; E^1_{\Psi(x)} \; \cdots \; E^{L-1}_{\Psi(x)} \,]^t. \tag{4.4}$$

Then, a sequence in the sequence space $V[\Psi(x)]$ can be expressed in terms of the elementary sequences $E^i_{\Psi(x)}$'s as the following theorem describes.

THEOREM 4.5 *The L elementary sequences $E^i_{\Psi(x)}$, $i = 0, 1, \cdots, L\text{-}1$, form a basis of the sequence space $V[\Psi(x)]$ whose characteristic polynomial $\Psi(x)$ is of degree L. Furthermore, a sequence $\{\, s_k \,\}$ in the sequence* *space $V[\Psi(x)]$ is represented by*

$$\{ s_k \} = \mathbf{s}^t \cdot \mathbf{E}_{\Psi(x)}. \tag{4.5}$$

Proof : The linear independence of the L elementary sequences $E_{\Psi(x)}^i$, $i = 0, 1, \cdots, L-1$, can be easily proved by their definition. Therefore, it suffices to show the equality in (4.5). We put $\{ \hat{s}_k \} \equiv \mathbf{s}^t \cdot \mathbf{E}_{\Psi(x)} = [s_0\, s_1\, \cdots\, s_{L-1}] \cdot \mathbf{E}_{\Psi(x)}$, and show by induction that $\hat{s}_k = s_k$, $k = 0, 1, \cdots$. By the definition of the elementary sequences, $\hat{s}_k = s_k$, $k = 0, 1, \cdots, L-1$. Now we assume that $\hat{s}_j = s_j$, $j = 0, 1, \cdots, k$; and we denote $E_{\Psi(x)}^i \equiv \{ e_k^i \}$, $i = 0, 1, \cdots, L-1$. Then, we have the relation $\hat{s}_{k+1} = \sum_{j=0}^{L-1} s_j e_{k+1}^j$, which is identical to $\sum_{j=0}^{L-1} s_j \sum_{i=0}^{L-1} \psi_i e_{k+1-L+i}^j$ due to (4.2). Interchanging the summations, we obtain $\hat{s}_{k+1} = \sum_{i=0}^{L-1} \psi_i \sum_{j=0}^{L-1} s_j e_{k+1-L+i}^j = \sum_{i=0}^{L-1} \psi_i \hat{s}_{k+1-L+i}$, which is identical to $\sum_{i=0}^{L-1} \psi_i s_{k+1-L+i}$ due to the assumption $\hat{s}_j = s_j$, $j = 0, 1, \cdots, k$. Therefore, by (4.2), $\hat{s}_{k+1} = s_{k+1}$. This completes the proof. ∎

The theorem means that the L elementary sequences $E_{\Psi(x)}^i$, $i = 0, 1, \cdots, L-1$, form a basis of the sequence space $V[\Psi(x)]$, and each sequence $\{ s_k \}$ in $V[\Psi(x)]$ is represented by (4.5). We name this the *elementary basis* for the sequence space $V[\Psi(x)]$, and we call the vector $\mathbf{s}$ the *initial vector for the elementary basis*. More specifically,

DEFINITION 4.6 (Elementary Basis) For a sequence space $V[\Psi(x)]$ whose characteristic polynomial $\Psi(x)$ is of degree L, the *elementary basis* is defined to be the set composed of the L elementary sequences $E_{\Psi(x)}^i$, $i = 0, 1, \cdots, L-1$.

EXAMPLE 4.2 We consider the sequence space $V[x^4+x^3+x^2+1]$. Then, the sequences S_8, S_4, S_2 and S_1 in Table 4.1 respectively correspond to the elementary sequences $E_{x^4+x^3+x^2+1}^0$, $E_{x^4+x^3+x^2+1}^1$, $E_{x^4+x^3+x^2+1}^2$ and $E_{x^4+x^3+x^2+1}^3$ for this sequence space. For the sequence S_5, the initial vector is $\mathbf{s}_5 = [0101]^t$, so it has the expression $S_5 = [0101] \cdot [S_8\, S_4\, S_2\, S_1]^t = S_4 + S_1$. We can confirm this using Table 4.1. ♣

From Theorem 4.5, we find that the dimension of a sequence space is as follows :

THEOREM 4.7 *The dimension of a sequence space $V[\Psi(x)]$ is identical to the degree of the characteristic polynomial $\Psi(x)$.*

We consider the expression for the delayed sequence $\{ s_{k+m} \}$ of a sequence $\{ s_k \}$ in $V[\Psi(x)]$, based on the elementary basis.

DEFINITION 4.8 (Delayed Sequence) For a sequence $\{ s_k \}$, the m-*delayed sequence* indicates the sequence $\{ s_{k+m} \}$.

THEOREM 4.9 *For a sequence $\{ s_k \}$ in the sequence space $V[\Psi(x)]$ whose initial vector is* **s**, *its m-delayed sequence $\{ s_{k+m} \}$ has the expression*

$$\{ s_{k+m} \} = \mathbf{s}^t \cdot \mathbf{A}_{\Psi(x)}^m \cdot \mathbf{E}_{\Psi(x)} \tag{4.6}$$

on the elementary basis, where $\mathbf{A}_{\Psi(x)}$ is the companion matrix for the characteristic polynomial $\Psi(x)$ defined by[3]

$$\mathbf{A}_{\Psi(x)} \equiv \begin{bmatrix} 0 & 0 & \cdots & 0 & \psi_0 \\ 1 & 0 & \cdots & 0 & \psi_1 \\ 0 & 1 & \cdots & 0 & \psi_2 \\ \vdots & \vdots & \ddots & \vdots & \vdots \\ 0 & 0 & \cdots & 1 & \psi_{L-1} \end{bmatrix}. \tag{4.7}$$

Proof : By the definition of sequence space, we can easily prove that $\{ s_{k+m} \} \in V[\Psi(x)]$ if $\{ s_k \} \in V[\Psi(x)]$. Therefore, it suffices to show that $\mathbf{s}^t \cdot \mathbf{A}_{\Psi(x)}^m = [s_m\, s_{m+1} \cdots s_{m+L-1}]$, $m = 0, 1, \cdots$, since equation (4.5) takes the form $\{ s_{k+m} \} = [s_m\, s_{m+1} \cdots s_{m+L-1}] \cdot \mathbf{E}_{\Psi(x)}$. We show this by induction. For $m = 0$, it is self-evident. We assume that $\mathbf{s}^t \cdot \mathbf{A}_{\Psi(x)}^m = [s_m\, s_{m+1} \cdots s_{m+L-1}]$ for an $m = 0, 1, \cdots$, and we show that it also holds for $m + 1$. Multiplying this by $\mathbf{A}_{\Psi(x)}$ and then applying (4.7) to this, we obtain the relation $\mathbf{s}^t \cdot \mathbf{A}_{\Psi(x)}^{m+1} = [s_{m+1} \cdots s_{m+L-1} \sum_{i=0}^{L-1} \psi_i s_{m+i}]$. Therefore, by (4.2), $\mathbf{s}^t \cdot \mathbf{A}_{\Psi(x)}^{m+1} = [s_{m+1} \cdots s_{m+L-1} s_{m+L}]$. This completes the proof. ∎

EXAMPLE 4.3 We consider the 1-delayed sequence of S_5 in the sequence space $V[x^4 + x^3 + x^2 + 1]$, whose companion matrix is

$$\mathbf{A}_{x^4+x^3+x^2+1} = \begin{bmatrix} 0 & 0 & 0 & 1 \\ 1 & 0 & 0 & 0 \\ 0 & 1 & 0 & 1 \\ 0 & 0 & 1 & 1 \end{bmatrix}. \tag{4.8}$$

---

[3]For a more detailed description on the companion matrix $\mathbf{A}_{\Psi(x)}$ and its properties, refer to Section B.3 in Appendix B.

Inserting $m = 1$ and $\mathbf{s}_5 = [0101]^t$ into (4.6), we obtain the relation $\{ s_{k+1} \} = \mathbf{s}_5^t \cdot \mathbf{A}_{x^4+x^3+x^2+1} \cdot \mathbf{E}_{x^4+x^3+x^2+1} = [1011] \cdot [S_8 S_4 S_2 S_1]^t = S_8 + S_2 + S_1$. We can confirm from Table 4.1 that the sequence $S_8 + S_2 + S_1$, which is identical to the sequence S_{11}, is the 1-delayed sequence of S_5. ♣

We consider how to express elements of sequences on the elementary basis.

THEOREM 4.10 *For a sequence $\{ s_k \}$ in the sequence space $V[\Psi(x)]$ whose initial vector is $\mathbf{s}$, its kth element s_k, $k = 0, 1, \cdots$, has the expression*

$$s_k = \mathbf{s}^t \cdot \mathbf{A}_{\Psi(x)}^k \cdot \mathbf{e}_0. \tag{4.9}$$

Proof : For an m-delayed sequence of $\{ s_k \}$, equation (4.5) takes the form $\{ s_{k+m} \} = [s_m \, s_{m+1} \, \cdots \, s_{m+L-1}] \cdot \mathbf{E}_{\Psi(x)}$, and by Theorem 4.9, $\{ s_{k+m} \} = \mathbf{s}^t \cdot \mathbf{A}_{\Psi(x)}^m \cdot \mathbf{E}_{\Psi(x)}$. Therefore, $[s_m \, s_{m+1} \, \cdots \, s_{m+L-1}] = \mathbf{s}^t \cdot \mathbf{A}_{\Psi(x)}^m$, and hence $s_m = \mathbf{s}^t \cdot \mathbf{A}_{\Psi(x)}^m \cdot \mathbf{e}_0$. ∎

EXAMPLE 4.4 We consider the sequence S_5 in the sequence space $V[x^4 + x^3 + x^2 + 1]$. We can easily confirm, by applying $\mathbf{s}_5 = [0101]^t$ and (4.8) to (4.9), that its elements are 0, 1, 0, 1, 1, 1, 0, 0, 1, 0, 1, $\cdots$. ♣

4.3 Primary Basis

We consider another useful basis for the sequence space $V[\Psi(x)]$, which is defined in terms of the so-called primary sequences.

DEFINITION 4.11 (Primary Sequences) For a sequence space $V[\Psi(x)]$ of dimension L, the ith *primary sequence* $P_{\Psi(x)}^i$, $i = 0, 1, \cdots, L-1$, is defined to be the $(i+1)$-delayed sequence of the 0th elementary sequence $E_{\Psi(x)}^0$.

DEFINITION 4.12 (Primary Sequence Vector) For a sequence space $V[\Psi(x)]$ of dimension L, the *primary sequence vector* $\mathbf{P}_{\Psi(x)}$ refers to the L-vector

$$\mathbf{P}_{\Psi(x)} \equiv [\, P_{\Psi(x)}^0 \, P_{\Psi(x)}^1 \, \cdots \, P_{\Psi(x)}^{L-1} \,]^t. \tag{4.10}$$

Then, we obtain the following two theorems.

THEOREM 4.13 *The relations between the elementary and the primary sequence vectors are*

$$\mathbf{P}_{\Psi(x)} = \mathbf{B}_{\Psi(x)} \cdot \mathbf{E}_{\Psi(x)}, \tag{4.11a}$$

$$\mathbf{E}_{\Psi(x)} = \mathbf{C}_{\Psi(x)} \cdot \mathbf{P}_{\Psi(x)}, \tag{4.11b}$$

where

$$\mathbf{B}_{\Psi(x)} \equiv \begin{bmatrix} \mathbf{e}_0^t \cdot \mathbf{A}_{\Psi(x)} \\ \mathbf{e}_0^t \cdot \mathbf{A}_{\Psi(x)}^2 \\ \vdots \\ \mathbf{e}_0^t \cdot \mathbf{A}_{\Psi(x)}^L \end{bmatrix}, \tag{4.12a}$$

$$\mathbf{C}_{\Psi(x)} \equiv \mathbf{B}_{\Psi(x)}^{-1}$$

$$= \begin{bmatrix} \psi_1 & \psi_2 & \cdots & \psi_{L-1} & \psi_L \\ \psi_2 & \psi_3 & \cdots & \psi_L & 0 \\ \vdots & \vdots & \ddots & \vdots & \vdots \\ \psi_{L-1} & \psi_L & \cdots & 0 & 0 \\ \psi_L & 0 & \cdots & 0 & 0 \end{bmatrix}. \tag{4.12b}$$

Proof : Equation (4.11a) is directly obtained by Theorem 4.9, and equation (4.11b) is obtained using the relation $\mathbf{C}_{\Psi(x)} = \mathbf{B}_{\Psi(x)}^{-1}$.[4] ∎

THEOREM 4.14 *The L primary sequences $P_{\Psi(x)}^i$, $i = 0, 1, \cdots, L\text{-}1$, form a basis of the sequence space $V[\Psi(x)]$ of dimension L. Furthermore, a sequence $\{ s_k \}$ in the sequence space $V[\Psi(x)]$ is represented by*

$$\{ s_k \} = \mathbf{p}^t \cdot \mathbf{P}_{\Psi(x)}, \tag{4.13}$$

where

$$\mathbf{p} = \mathbf{C}_{\Psi(x)} \cdot \mathbf{s}. \tag{4.14}$$

Proof : Since the transform matrix $\mathbf{B}_{\Psi(x)}$ in (4.12a) is nonsingular, the L primary sequences, $P_{\Psi(x)}^i$, $i = 0, 1, \cdots, L-1$, also form a basis of the sequence space $V[\Psi(x)]$. Equation (4.13) is a direct outcome of equations (4.5) and (4.11b). ∎

---

[4]The proof of the relation $\mathbf{C}_{\Psi(x)} = \mathbf{B}_{\Psi(x)}^{-1}$ is given in Property B.11 of Appendix B.

Theorem 4.13 describes the relations between the elementary and the primary sequences ; and Theorem 4.14 indicates that the L primary sequences $P_{\Psi(x)}^i$, $i = 0, 1, \cdots, L-1$, form a basis of the sequence space $V[\Psi(x)]$, and each sequence $\{\, s_k \,\}$ in $V[\Psi(x)]$ is represented by (4.13). We name this the *primary basis* for the sequence space $V[\Psi(x)]$, and we call the vector $\mathbf{p}$ in (4.14) the *initial vector for the primary basis*. More specifically,

DEFINITION 4.15 (Primary Basis) For a sequence space $V[\Psi(x)]$ of dimension L, the *primary basis* indicates the set composed of the L primary sequences $P_{\Psi(x)}^i$, $i = 0, 1, \cdots, L-1$.

EXAMPLE 4.5 In the case of the sequence space $V[x^4 + x^3 + x^2 + 1]$, the transform matrices $\mathbf{B}_{x^4+x^3+x^2+1}$ and $\mathbf{C}_{x^4+x^3+x^2+1}$ are respectively

$$\mathbf{B}_{x^4+x^3+x^2+1} \;=\; \begin{bmatrix} 0 & 0 & 0 & 1 \\ 0 & 0 & 1 & 1 \\ 0 & 1 & 1 & 0 \\ 1 & 1 & 0 & 1 \end{bmatrix}, \qquad (4.15a)$$

$$\mathbf{C}_{x^4+x^3+x^2+1} \;=\; \begin{bmatrix} 0 & 1 & 1 & 1 \\ 1 & 1 & 1 & 0 \\ 1 & 1 & 0 & 0 \\ 1 & 0 & 0 & 0 \end{bmatrix}. \qquad (4.15b)$$

So, the 0th primary sequence becomes $P_{x^4+x^3+x^2+1}^0 = [\,0\,0\,0\,1\,]\cdot$ $E_{x^4+x^3+x^2+1} = E_{x^4+x^3+x^2+1}^3 = S_1$. Similarly, we can identify that $P_{x^4+x^3+x^2+1}^1 = S_3$, $P_{x^4+x^3+x^2+1}^2 = S_6$, and $P_{x^4+x^3+x^2+1}^3 = S_{13}$, as indicated in Table 4.1, thus getting $\mathbf{P}_{x^4+x^3+x^2+1} = [\,S_1\,S_3\,S_6\,S_{13}\,]^t$. For the sequence S_5, $\mathbf{s}_5 = [\,0\,1\,0\,1\,]^t$, and so by (4.14) $\mathbf{p}_5 = \mathbf{C}_{x^4+x^3+x^2+1}\cdot \mathbf{s}_5 = [\,0\,1\,1\,0\,]^t$. Therefore by (4.13) $S_5 = \mathbf{p}_5^t\cdot \mathbf{P}_{x^4+x^3+x^2+1} = S_3 + S_6$, which we can confirm using Table 4.1. ♣

For the expression of the m-delayed sequence $\{\, s_{k+m} \,\}$ of a sequence $\{\, s_k \,\}$ in $V[\Psi(x)]$ based on the primary basis, we have the following theorem.

THEOREM 4.16 *For a sequence $\{\, s_k \,\}$ in the sequence space $V[\Psi(x)]$ whose initial vector for the primary basis is $\mathbf{p}$, its m-delayed sequence $\{\, s_{k+m} \,\}$ has the expression*

$$\{\, s_{k+m} \,\} \;=\; \mathbf{p}^t\cdot(\mathbf{A}_{\Psi(x)}^t)^m\cdot \mathbf{P}_{\Psi(x)} \qquad (4.16)$$

on the primary basis.

Proof : By (4.14) and (4.12b), $\mathbf{s}^t = \mathbf{p}^t \cdot \mathbf{B}_{\Psi(x)}$. Inserting this along with (4.11b) into (4.6), we obtain $\{ s_{k+m} \} = \mathbf{p}^t \cdot \mathbf{B}_{\Psi(x)} \cdot \mathbf{A}_{\Psi(x)}^m \cdot \mathbf{C}_{\Psi(x)} \cdot \mathbf{P}_{\Psi(x)}$. Therefore, by the relation $\mathbf{B}_{\Psi(x)} \cdot \mathbf{A}_{\Psi(x)}^m \cdot \mathbf{C}_{\Psi(x)} = (\mathbf{A}_{\Psi(x)}^t)^m$, we have (4.16).[5] ∎

EXAMPLE 4.6 In the case of the sequence S_5 in the sequence space $V[x^4 + x^3 + x^2 + 1]$, the 1-delayed sequence of S_5 has the expression $\mathbf{p}_5^t \cdot \mathbf{A}_{x^4+x^3+x^2+1}^t \cdot \mathbf{P}_{x^4+x^3+x^2+1} = [\,0\,0\,1\,1\,] \cdot [\,S_1\,S_3\,S_6\,S_{13}\,]^t = S_6 + S_{13} = S_{11}$, which we have already confirmed in Example 4.3. ♣

For the expression of elements of sequences on the primary basis, we have the following theorem :

THEOREM 4.17 *For a sequence $\{ s_k \}$ in the sequence space $V[\Psi(x)]$ whose initial vector for the primary basis is* $\mathbf{p}$, *its kth element* s_k, $k = 0, 1, \cdots$, *has the expression*

$$s_k = \mathbf{p}^t \cdot (\mathbf{A}_{\Psi(x)}^t)^k \cdot \mathbf{e}_{L-1}. \tag{4.17}$$

Proof : Inserting $\mathbf{s}^t = \mathbf{p}^t \cdot \mathbf{B}_{\Psi(x)}$ into (4.9), we obtain $s_k = \mathbf{p}^t \cdot \mathbf{B}_{\Psi(x)} \cdot \mathbf{A}_{\Psi(x)}^k \cdot \mathbf{e}_0$, which becomes $\mathbf{p}^t \cdot \mathbf{B}_{\Psi(x)} \cdot \mathbf{A}_{\Psi(x)}^k \cdot \mathbf{C}_{\Psi(x)} \cdot \mathbf{e}_{L-1}$ since we have $\mathbf{e}_0 = \mathbf{C}_{\Psi(x)} \cdot \mathbf{e}_{L-1}$ by (4.12b). Therefore, by Property B.14 in Appendix B we have (4.17). ∎

EXAMPLE 4.7 Applying $\mathbf{p}_5 = [\,0\,1\,1\,0\,]^t$, $\mathbf{e}_3 = [\,0\,0\,0\,1\,]^t$ and (4.8) to (4.17), we can easily check that the elements of the sequence S_5 are 0, 1, 0, 1, 1, 1, 0, 0, 1, 0, 1, $\cdots$. ♣

4.4 Polynomial Expression of Sequences

The elementary and primary bases we have discussed in the previous two sections render vector-based expressions for sequences. As opposed to this linear algebraic approach, we can also take an abstract algebraic approach, which yields polynomial-based expressions for sequences.

---

[5] For the companion matrix $\mathbf{A}_{\Psi(x)}$, we have the relation $\mathbf{B}_{\Psi(x)} \cdot \mathbf{A}_{\Psi(x)}^i \cdot \mathbf{C}_{\Psi(x)} = (\mathbf{A}_{\Psi(x)}^t)^i$ for $i = 0, 1, \cdots$. Refer to Property B.14 in Appendix B.

DEFINITION 4.18 (Polynomial Expression of Sequence) The expression $S[\Psi(x), P(x)]$ is called the *polynomial expression* for the sequence $\{ s_k \} = \mathbf{p}^t \cdot \mathbf{P}_{\Psi(x)}$, if $P(x)$ is the *polynomial expression of the initial vector* $\mathbf{p}$, that is, $P(x) = \sum_{i=0}^{L-1} p_i x^i$ and $\mathbf{p}^t = [\, p_0 \; p_1 \; \cdots \; p_{L-1} \,]$.

The first element s_0 of a sequence $S[\Psi(x), P(x)]$ is obtained by the following lemma.

LEMMA 4.19 *If* $\{ s_k \} = S[\Psi(x), P(x)]$, *then* $s_0 = Q[\Psi(x), xP(x)]$, *where* $Q[\Psi(x), xP(x)]$ *denotes the quotient polynomial of* $xP(x)$ *divided by* $\Psi(x)$.

Proof : This lemma is obtained by inserting $k = 0$ into (4.17). ■

For expressions of the m-delayed sequence and the elements of $\{ s_k \}$, we have the following two theorems.

THEOREM 4.20 *If a sequence* $\{ s_k \}$ *in the sequence space* $V[\Psi(x)]$ *has the polynomial expression* $S[\Psi(x), P(x)]$, *then its m-delayed sequence* $\{ s_{k+m} \}$ *has the polynomial expression*

$$\{ s_{k+m} \} = S[\Psi(x), R[\Psi(x), x^m P(x)]], \tag{4.18}$$

where $R[\Psi(x), x^m P(x)]$ *is the remainder polynomial of* $x^m P(x)$ *divided by* $\Psi(x)$.

Proof : We show this by induction. For $m = 0$, equality (4.18) is self-evident. We assume that $\{ s_{k+m} \} = S[\Psi(x), R[\Psi(x), x^m P(x)]]$ for an $m = 0, 1, \cdots$, and we show that it also holds for $m + 1$. We put $R[\Psi(x), x^m P(x)] = \sum_{i=0}^{L-1} \hat{p}_i x^i$, and $\hat{\mathbf{p}} = [\, \hat{p}_0 \; \hat{p}_1 \; \cdots \; \hat{p}_{L-1} \,]^t$. Then, the sequence $\{ s_{k+m} \}$ has the expression $\hat{\mathbf{p}}^t \cdot \mathbf{P}_{\Psi(x)}$ on the primary basis. By Theorem 4.16, $\{ s_{k+m+1} \}$, which is the 1-delayed sequence of $\{ s_{k+m} \}$, has the expression $\hat{\mathbf{p}}^t \cdot \mathbf{A}^t_{\Psi(x)} \cdot \mathbf{P}_{\Psi(x)}$, and it is identical to $[\, \psi_0 \hat{p}_{L-1} \; \hat{p}_0 + \psi_1 \hat{p}_{L-1} \; \cdots \; \hat{p}_{L-2} + \psi_{L-1} \hat{p}_{L-1} \,] \cdot \mathbf{P}_{\Psi(x)}$ by (4.7). Therefore, $\{ s_{k+m+1} \}$ has the polynomial expression $S[\Psi(x), \psi_0 \hat{p}_{L-1} + (\hat{p}_0 + \psi_1 \hat{p}_{L-1})x + \cdots + (\hat{p}_{L-2} + \psi_{L-1} \hat{p}_{L-1})x^{L-1}]$. However, since $R[\Psi(x), x^{m+1} P(x)] = R[\Psi(x), xR[\Psi(x), x^m P(x)]] = R[\sum_{i=0}^{L} \psi_i x^i, x \sum_{i=0}^{L-1} \hat{p}_i x^i] = \psi_0 \hat{p}_{L-1} + (\hat{p}_0 + \psi_1 \hat{p}_{L-1})x + \cdots + (\hat{p}_{L-2} + \psi_{L-1} \hat{p}_{L-1})x^{L-1}$, we have the relation $\{ s_{k+m+1} \} = S[\Psi(x), R[\Psi(x), x^{m+1} P(x)]]$. This completes the proof. ■

Theorem 4.21 *If a sequence* $\{\, s_k \,\}$ *in the sequence space* $V[\Psi(x)]$ *has the polynomial expression* $S[\Psi(x), P(x)]$, *then for* $j = 0, 1, \cdots$,

$$Q[\Psi(x), x^{j+1}P(x)] = \sum_{i=0}^{j} s_{j-i} x^i. \tag{4.19}$$

Proof : We prove the theorem by induction. Lemma 4.19 shows (4.19) for $j = 0$. We assume that (4.19) holds for a $j = 0, 1, \cdots$, and we prove that it is also valid for $j + 1$. Since (4.19) implies the equality $x^{j+1}P(x) = (\sum_{i=0}^{j} s_{j-i} x^i)\Psi(x) + R[\Psi(x), x^{j+1}P(x)]$, we obtain, by multiplying x to this, the relation $Q[\Psi(x), x^{j+2}P(x)] = x \sum_{i=0}^{j} s_{j-i} x^i + Q[\Psi(x), xR[\Psi(x), x^{j+1}P(x)]]$. However, by Theorem 4.20, $\{\, s_{k+j+1} \,\} = S[\Psi(x), R[\Psi(x), x^{j+1}P(x)]]$, and so applying Lemma 4.19 to this, we have the relation $Q[\Psi(x), xR[\Psi(x), x^{j+1}P(x)]] = s_{j+1}$. Therefore, $Q[\Psi(x), x^{j+2}P(x)] = x \sum_{i=0}^{j} s_{j-i} x^i + s_{j+1} = \sum_{i=0}^{j+1} s_{j+1-i} x^i$. Thus we have (4.19) for $j + 1$, and this completes the proof. ∎

EXAMPLE 4.8 We consider the sequence S_5 in Table 4.1, whose polynomial expression is $S[x^4 + x^3 + x^2 + 1, x^2 + x]$ since $\mathbf{p}_5 = [0\,1\,1\,0]^t$. Then, the 4-delayed sequence of S_5 has the polynomial expression $S[x^4 + x^3 + x^2 + 1, R[x^4 + x^3 + x^2 + 1, x^4(x^2 + x)]] = S[x^4 + x^3 + x^2 + 1, x^3 + 1]$, which is equivalent to the primary basis expression $[1\,0\,0\,1] \cdot \mathbf{P}_{x^4+x^3+x^2+1} = P^0_{x^4+x^3+x^2+1} + P^3_{x^4+x^3+x^2+1} = S_1 + S_{13}$. We can confirm from Table 4.1 that this is equivalent to S_{12}, which is the 4-delayed sequence of S_5. On the other hand, for $j = 3$, equation (4.19) provides that $\sum_{i=0}^{3} s_{3-i} x^i = Q[x^4 + x^3 + x^2 + 1, x^4(x^2 + x)] = 0x^3 + 1x^2 + 0x + 1$. This implies that the 0th, the 1st, the 2nd and the 3rd elements of S_5 are respectively 0, 1, 0 and 1. ♣

We consider the period of a sequence in relation to its polynomial expression.

Definition 4.22 (Period of Sequence) For a sequence $\{\, s_k \,\}$, the *period* τ is the smallest integer such that $s_k = s_{k+\tau}$ for all $k = 0, 1, \cdots$.

Theorem 4.23 *A sequence* $\{\, s_k \,\}$ *of period* τ *is an element of the sequence space* $V[x^\tau + 1]$, *and its polynomial expression is* $S[x^\tau + 1, \sum_{i=0}^{\tau-1} s_{\tau-1-i} x^i]$.

Proof : If the period of a sequence $\{\, s_k \,\}$ is τ, we have the relation $s_j + s_{j+\tau} = 0$, $j = 0, 1, \cdots$. Therefore by (4.1), $\{\, s_k \,\} \in V[x^\tau + 1]$. On the other hand, by (4.5), the sequence $\{\, s_k \,\}$ has the expression $[\, s_0\, s_1 \cdots s_{\tau-1}\,] \cdot \mathbf{E}_{x^\tau+1}$ on the elementary basis, and its expression on the primary basis becomes $[\, s_{\tau-1}\, s_{\tau-2} \cdots s_0\,] \cdot \mathbf{P}_{x^\tau+1}$ due to (4.11b) and (4.12b).[Note that ψ_i is 1 for $i = 0$ and $L(=\tau)$, and the others are 0 in this case.] Since $P(x)$ is the polynomial expression of the initial vector $\mathbf{p} = [\, s_{\tau-1}\, s_{\tau-2} \cdots s_0\,]^t$, the polynomial expression of $\{\, s_k \,\}$ is $S[x^\tau + 1, \sum_{i=0}^{\tau-1} s_{\tau-1-i}x^i]$. ∎

EXAMPLE *4.9* We consider the sequence S_5 in Table 4.1. Then, its period τ is 7. Therefore, by the theorem S_5 can be represented by the polynomial expression $S[x^7 + 1, x^5 + x^3 + x^2 + x]$. ♣

4.5   Sequence Subspaces

We consider the subspaces of a sequence space and the related properties.

DEFINITION 4.24 (Sequence Subspace)   For a sequence space $V[\Psi(x)]$, a *sequence subspace* $V[\hat{\Psi}(x)]$ is defined to be a sequence space which is a subspace of the sequence space $V[\Psi(x)]$.

Then we have the following theorem.

THEOREM 4.25 *A sequence space* $V[\hat{\Psi}(x)]$ *is a sequence subspace of a sequence space* $V[\Psi(x)]$ *if and only if* $\hat{\Psi}(x)$ *divides* $\Psi(x)$.

Proof : We first show the "if" part of the theorem. Let $\Psi(x) = \sum_{i=0}^{L} \psi_i x^i$ and $\{\, s_k \,\} \in V[\hat{\Psi}(x)]$. Then, by (4.9), we have the relation $\sum_{i=0}^{L} \psi_i s_{k+i} = \mathbf{s}^t \cdot (\psi_0 + \psi_1 \mathbf{A}_{\hat{\Psi}(x)} + \cdots + \psi_L \mathbf{A}^L_{\hat{\Psi}(x)}) \cdot \mathbf{A}^k_{\hat{\Psi}(x)} \cdot \mathbf{e}_0 = \mathbf{s}^t \cdot \Psi(\mathbf{A}_{\hat{\Psi}(x)}) \cdot \mathbf{A}^k_{\hat{\Psi}(x)} \cdot \mathbf{e}_0$, $k = 0, 1, \cdots$. Therefore, if $\hat{\Psi}(x)$ divides $\Psi(x)$, $\Psi(\mathbf{A}_{\hat{\Psi}(x)}) = 0,$[6] and hence $\sum_{i=0}^{L} \psi_i s_{k+i} = 0$, $k = 0, 1, \cdots$. This means $\{\, s_k \,\} \in V[\Psi(x)]$ by (4.1).

Now we prove the "only if" part. If $V[\hat{\Psi}(x)]$ of dimension $\hat{L}$ is a sequence subspace of $V[\Psi(x)]$ of dimension $L \geq \hat{L}$, each ith elementary sequence $E^i_{\hat{\Psi}(x)}$, $i = 0, 1, \cdots, \hat{L} - 1$, is an element of the sequence space

---

[6]For the companion matrix $\mathbf{A}_{\hat{\Psi}(x)}$, if $\hat{\Psi}(x)$ divides $\Psi(x)$, then $\Psi(\mathbf{A}_{\hat{\Psi}(x)}) = 0$. Refer to Property B.10 in Appendix B.

$V[\Psi(x)]$. We denote $E^i_{\hat{\Psi}(x)} \equiv \{ \hat{e}^i_k \}$, $i = 0, 1, \cdots, \hat{L} - 1$. Then, by (4.1), we have the relation $\sum^L_{l=0} \psi_l \hat{e}^i_{k+l} = 0$, $k = 0, 1, \cdots$, and so by (4.9) $\mathbf{e}^t_i \cdot \Psi(\mathbf{A}_{\hat{\Psi}(x)}) \cdot \mathbf{A}^k_{\hat{\Psi}(x)} \cdot \mathbf{e}_0 = 0$, $k = 0, 1, \cdots$, which can be rewritten, by the relation $\mathbf{A}^k_{\hat{\Psi}(x)} \cdot \mathbf{e}_0 = \mathbf{e}_k$,[7] in the form of $\mathbf{e}^t_i \cdot \Psi(\mathbf{A}_{\hat{\Psi}(x)}) \cdot \mathbf{e}_k = 0$ for $k = 0, 1, \cdots, \hat{L} - 1$. Since this relation is satisfied for arbitrary choice of $i = 0, 1, \cdots, \hat{L} - 1$, it implies that $\Psi(\mathbf{A}_{\hat{\Psi}(x)}) = 0$. Therefore, $\hat{\Psi}(x)$ divides $\Psi(x)$.[8] This completes the proof. ∎

EXAMPLE 4.10 We consider the sequence space $V[x^3 + x + 1]$, whose elements are shown in Table 4.2. Then the characteristic polynomial $x^3 + x + 1$ divides the characteristic polynomial $x^4 + x^3 + x^2 + 1$ of the sequence space $V[x^4 + x^3 + x^2 + 1]$. Therefore, by the theorem, $V[x^3 + x + 1]$ is a sequence subspace of $V[x^4 + x^3 + x^2 + 1]$. We can confirm this by comparing Tables 4.1 and 4.2. ♣

The following theorem describes the relation between the polynomial expressions of a sequence in a sequence space and in its sequence subspace.

THEOREM 4.26 *For a nonzero polynomial $a(x)$,[9]*

$$S[\Psi(x), P(x)] = S[a(x)\Psi(x), a(x)P(x)]. \tag{4.20}$$

Proof : We denote by $\{ s_k \}$ and $\{ \hat{s}_k \}$, respectively, the sequence $S[\Psi(x), P(x)]$ and $S[a(x)\Psi(x), a(x)P(x)]$. Then, it suffices to show that $s_k = \hat{s}_k$, $k = 0, 1, \cdots$. By (4.19), we have the relations $Q[\Psi(x), x^{k+1}P(x)] = \sum^k_{i=0} s_{k-i} x^i$ and $Q[a(x)\Psi(x), x^{k+1}a(x)P(x)] = \sum^k_{i=0} \hat{s}_{k-i} x^i$. But the two quotients are equal, that is, $Q[\Psi(x), x^{k+1}P(x)] = Q[a(x)\Psi(x), x^{k+1} a(x)P(x)]$. Therefore, we have $s_k = \hat{s}_k$. ∎

This theorem means that a sequence $S[\Psi(x), P(x)]$ in $V[\Psi(x)]$ is identical to the sequence $S[a(x)\Psi(x), a(x)P(x)]$ in $V[a(x)\Psi(x)]$. In other

---

[7]For the companion matrix $\mathbf{A}_{\Psi(x)}$ for a degree-L characteristic polynomial $\Psi(x)$, $\mathbf{A}^i_{\Psi(x)} \cdot \mathbf{e}_0 = \mathbf{e}_i$, $i = 0, 1, \cdots, L - 1$. Refer to Property B.8 in Appendix B.

[8]For the companion matrix $\mathbf{A}_{\hat{\Psi}(x)}$, if $\Psi(\mathbf{A}_{\hat{\Psi}(x)}) = 0$, then $\hat{\Psi}(x)$ divides $\Psi(x)$. Refer to Property B.10 in Appendix B.

[9]In the theorem, it is implicitly assumed that $a(x)$ is a nonzero polynomial with $a(0) = 1$. Otherwise, the constant term of the characteristic polynomial $a(x)\Psi(x)$ can not be 1, thus violating the definition of the sequence space.

Table 4.2. Elements of the sequence space $V[x^3 + x + 1]$

$\hat{S}_0$	$= \{0,0,0,0,0,0,0,0,0,0,0,\cdots\}$
$\hat{S}_1$	$= \{0,0,1,0,1,1,1,0,0,1,0,\cdots\} = E^2_{x^3+x+1} = P^0_{x^3+x+1}$
$\hat{S}_2$	$= \{0,1,0,1,1,1,0,0,1,0,1,\cdots\} = E^1_{x^3+x+1} = P^1_{x^3+x+1}$
$\hat{S}_3$	$= \{0,1,1,1,0,0,1,0,1,1,1,\cdots\}$
$\hat{S}_4$	$= \{1,0,0,1,0,1,1,1,0,0,1,\cdots\} = E^0_{x^3+x+1}$
$\hat{S}_5$	$= \{1,0,1,1,1,0,0,1,0,1,1,\cdots\} = P^2_{x^3+x+1}$
$\hat{S}_6$	$= \{1,1,0,0,1,0,1,1,1,0,0,\cdots\}$
$\hat{S}_7$	$= \{1,1,1,0,0,1,0,1,1,1,0,\cdots\}$

words, if two constituent polynomials in a polynomial expression of a sequence share $a(x)$ in common, then the sequence is an element in the sequence space $V[a(x)\Psi(x)]$ as well as in its subspace $V[\Psi(x)]$.

EXAMPLE 4.11 We consider the sequence S_5 in Table 4.1, whose polynomial expression is $S_5 = S[x^4+x^3+x^2+1, x^2+x]$. By (4.20), the expression can be simplified to $S[(x+1)(x^3+x+1), (x+1)x] = S[x^3+x+1, x]$. This implies that $S_5 = [010] \cdot \mathbf{P}_{x^3+x+1} = P^1_{x^3+x+1} = \hat{S}_2$. We can confirm this using Table 4.2. ♣

We consider the subset of sequence spaces obtained by taking the intersection of them. The following theorem describes this.

THEOREM 4.27 *The intersection of two sequence spaces $V[\Psi_1(x)]$ and $V[\Psi_2(x)]$ is a sequence space whose characteristic polynomial is the greatest common divisor of the two characteristic polynomials $\Psi_1(x)$ and $\Psi_2(x)$, that is,*

$$V[\Psi_1(x)] \cap V[\Psi_2(x)] = V[GCD[\Psi_1(x), \Psi_2(x)]]. \qquad (4.21)$$

Proof : To prove the theorem, we let $\Psi(x) = GCD[\Psi_1(x), \Psi_2(x)]$. We first show that $V[\Psi(x)] \subset V[\Psi_1(x)] \cap V[\Psi_2(x)]$. Since $\Psi(x)$ divides both $\Psi_1(x)$ and $\Psi_2(x)$, by Theorem 4.25 $V[\Psi(x)] \subset V[\Psi_1(x)]$ and $V[\Psi(x)] \subset V[\Psi_2(x)]$. Therefore, $V[\Psi(x)] \subset V[\Psi_1(x)] \cap V[\Psi_2(x)]$.

Now we prove that $V[\Psi_1(x)] \cap V[\Psi_2(x)] \subset V[\Psi(x)]$. If a sequence $\{s_k\}$ belongs to the set $V[\Psi_1(x)] \cap V[\Psi_2(x)]$, we can represent the sequence $\{s_k\}$ by polynomial expressions $S[\Psi_1(x), P_1(x)]$ and $S[\Psi_2(x), P_2(x)]$ for appropriate polynomials $P_1(x)$ and $P_2(x)$. Let $\Psi_1(x) = a_1(x)\Psi(x)$ and $\Psi_2(x) = a_2(x)\Psi(x)$, where $a_1(x)$ and $a_2(x)$ are relatively prime. Then, by Theorem 4.26, $\{\, s_k \,\} = S[a_1(x)\Psi(x), P_1(x)] = S[a_1(x)a_2(x)\Psi(x), a_2(x)P_1(x)]$, and also $\{\, s_k \,\} = S[a_2(x)\Psi(x), P_2(x)] = S[a_1(x)a_2(x)\Psi(x), a_1(x)P_2(x)]$. Therefore, $a_2(x)P_1(x) = a_1(x)P_2(x)$. However, since $a_1(x)$ is relatively prime to $a_2(x)$, $a_1(x)$ divides $P_1(x)$. Therefore, by Theorem 4.26, $\{\, s_k \,\} = S[\Psi_1(x), P_1(x)] = S[\Psi(x), P_1(x)/a_1(x)]$. This means that $\{\, s_k \,\}$ is an element of $V[\Psi(x)]$, and so $V[\Psi_1(x)] \cap V[\Psi_2(x)] \subset V[\Psi(x)]$. This completes the proof. ∎

EXAMPLE 4.12 In the case of the two sequence spaces $V[x^4 + x^3 + x^2 + 1]$ and $V[x^3 + x + 1]$, the greatest common divisor of the two corresponding characteristic polynomials is $x^3 + x + 1$. Therefore, by the theorem, the intersection of the two sequence spaces is $V[x^3 + x + 1]$. We can confirm this using Tables 4.1 and 4.2. ♣

The concept of sequence space also helps to prove the relation between the periodicity of a sequence space and the period of its constituent sequences.

DEFINITION 4.28 (Periodicity of Sequence Space) For a sequence space $V[\Psi(x)]$, we define the *periodicity* T to be the smallest integer such that the characteristic polynomial $\Psi(x)$ divides $x^T + 1$.

THEOREM 4.29 *The period τ of a sequence $\{\, s_k \,\}$ in the sequence space $V[\Psi(x)]$ is a submultiple of the periodicity T of the sequence space $V[\Psi(x)]$.*

    Proof : Let $\{\, s_k \,\} \in V[\Psi(x)]$. Then, since $V[\Psi(x)]$ is a subspace of $V[x^T + 1]$ due to Theorem 4.25, $\{\, s_k \,\} \in V[x^T + 1]$. Therefore, by (4.1) $s_j + s_{j+T} = 0$ for all $j = 0, 1, \cdots$, and hence the period τ of $\{\, s_k \,\}$ divides T. ∎

EXAMPLE 4.13 We consider the sequence space $V[x^3 + x + 1]$. The periodicity T of this sequence space is 7, and the period τ of each sequence $\{\, s_k \,\}$ in $V[x^3 + x + 1]$ is 1 or 7, as can be observed from Table 4.2. Therefore we find that τ divides T. ♣

4.6   Minimal Sequence Spaces

Finally, we consider the smallest-dimensional sequence space for a single sequence or a group of sequences.

DEFINITION 4.30 (Common Sequence Space) For N sequences { s_k^i }, $i = 0, 1, \cdots, N - 1$, a *common sequence space* (in short, *common space*) refers to a sequence space containing the N sequences.

DEFINITION 4.31 (Minimal Sequence Space) For N sequences { s_k^i }, $i = 0, 1, \cdots, N-1$, a *minimal sequence space* (in short, *minimal space*) refers to the smallest-dimensional common sequence space for the N sequences.

Then, the minimal space has the properties described by the following two theorems.

THEOREM 4.32 *The minimal space for any N sequences { s_k^i }, $i = 0, 1, \cdots, N-1$, is unique.*

Proof : We show this by contradiction. We suppose that two different sequence spaces $V[\Psi_1(x)]$ and $V[\Psi_2(x)]$ are minimal spaces for the N sequences { s_k^i }, $i = 0, 1, \cdots, N-1$. Then, { s_k^i } $\in V[\Psi_1(x)] \cap V[\Psi_2(x)]$, and so by Theorem 4.27 { s_k^i } $\in V[\Psi(x)]$, $i = 0, 1, \cdots, N-1$, for $\Psi(x) = GCD[\Psi_1(x), \Psi_2(x)]$. Therefore, the sequence space $V[\Psi(x)]$ is a common space for the N sequences. However, since $\Psi_1(x)$ and $\Psi_2(x)$ have the same degree but are different, degree of $\Psi(x)$ is less than those of both $\Psi_1(x)$ and $\Psi_2(x)$. Therefore, by Theorem 4.7 the dimension of $V[\Psi(x)]$ is smaller than those of both $V[\Psi_1(x)]$ and $V[\Psi_2(x)]$. But this is a contradiction to the assumption that $V[\Psi_1(x)]$ and $V[\Psi_2(x)]$ are minimal spaces.   ■

THEOREM 4.33 *Let a sequence space $V[\Psi_m(x)]$ be the minimal space for N sequences { s_k^i }, $i = 0, 1, \cdots, N-1$. Then, a sequence space $V[\Psi(x)]$ is a common space for the N sequences if and only if $\Psi_m(x)$ divides $\Psi(x)$.*

Proof : We first show the "if" part of the theorem. If $\Psi_m(x)$ divides $\Psi(x)$, then by Theorem 4.25, $V[\Psi_m(x)]$ is a subspace of $V[\Psi(x)]$. Therefore, $V[\Psi(x)]$ is a common space for the N sequences { s_k^i }, $i = 0, 1, \cdots, N - 1$.

Now we prove the "only if" part. If a sequence space $V[\Psi(x)]$ is a common space for the N sequences $\{ s_k^i \}$, $i = 0, 1, \cdots, N-1$, then $\{ s_k^i \} \in V[\Psi_m(x)] \cap V[\Psi(x)]$, $i = 0, 1, \cdots, N-1$, and so by Theorem 4.27, $\{ s_k^i \} \in V[\Psi_c(x)]$, $i = 0, 1, \cdots, N-1$, for $\Psi_c(x) = GCD[\Psi_m(x), \Psi(x)]$. That is, the sequence space $V[\Psi_c(x)]$ is also a common space. However, since $\Psi_c(x)$ divides $\Psi_m(x)$, $V[\Psi_c(x)]$ is a subspace of the minimal space $V[\Psi_m(x)]$ due to Theorem 4.25. But the minimal space is unique by Theorem 4.32. Therefore $V[\Psi_c(x)] = V[\Psi_m(x)]$, or equivalently, $\Psi_c(x) = \Psi_m(x)$, which means that $\Psi_m(x)$ divides $\Psi(x)$. ∎

According to Theorem 4.32, a minimal space is unique ; and according to Theorem 4.33, the characteristic polynomial of a common space is a multiple of the characteristic polynomial of its minimal space. Therefore, combining the two theorems we can determine all common spaces for a known minimal space.

The following two theorems describe how to determine the minimal space for a single sequence.

THEOREM 4.34 *For a sequence $\{ s_k \}$ whose polynomial expression is $S[\Psi(x), P(x)]$, let $d(x)$ be the greatest common divisor of $\Psi(x)$ and $P(x)$. Then, the minimal space for the sequence $\{ s_k \}$ is the sequence space $V[\Psi(x)/d(x)]$, and the sequence $\{ s_k \}$ has the polynomial expression*

$$\{ s_k \} = S[\Psi(x)/d(x), P(x)/d(x)] \tag{4.22}$$

for the minimal space.

Proof : Let a sequence space $V[\Psi_m(x)]$ be the minimal space for $\{ s_k \}$; and let $\Psi_c(x) = \Psi(x)/d(x)$ and $P_c(x) = P(x)/d(x)$. Then, it suffices to show that $\Psi_m(x) = \Psi_c(x)$. By Theorem 4.26, $\{ s_k \} = S[\Psi_c(x), P_c(x)]$, and so the sequence space $V[\Psi_c(x)]$ is a common space for the sequence $\{ s_k \}$. Therefore, by Theorem 4.33 $\Psi_m(x)$ divides $\Psi_c(x)$. To complete the proof, we now show that $\Psi_c(x)$ divides $\Psi_m(x)$. If $\{ s_k \}$ has the polynomial expression $S[\Psi_m(x), P_m(x)]$ for an appropriate polynomial $P_m(x)$, then by Theorem 4.26, $\{ s_k \} = S[\Psi_c(x), P_m(x)\Psi_c(x)/\Psi_m(x)]$. Therefore, $P_m(x)\Psi_c(x)/\Psi_m(x) = P_c(x)$, that is, $P_m(x)\Psi_c(x) = P_c(x)\Psi_m(x)$. However, since $\Psi_c(x)$ and $P_c(x)$ are relatively prime to each other, $\Psi_c(x)$ divides $\Psi_m(x)$. This completes the proof. ∎

THEOREM 4.35 *Given a sequence $\{ s_k \}$ of period τ, let $d(x)$ be the greatest common divisor of $x^\tau + 1$ and $\sum_{i=0}^{\tau-1} s_{\tau-1-i} x^i$. Then, the minimal* *space for the sequence* $\{ s_k \}$ *is the sequence space* $V[(x^\tau + 1)/d(x)]$, *and the sequence* $\{ s_k \}$ *has the polynomial expression*

$$\{ s_k \} = S[(x^\tau + 1)/d(x), (\sum_{i=0}^{\tau-1} s_{\tau-1-i} x^i)/d(x)] \qquad (4.23)$$

for the minimal space.

Proof : This theorem is a direct outcome of Theorems 4.23 and 4.34.   ∎

The following theorem describes how to obtain the minimal space for multiple sequences.

THEOREM 4.36 *Let* $V[\Psi_i(x)]$, $i = 0, 1, \cdots, N$-1, *be the minimal space for the sequence* $\{ s_k^i \}$; *and let* $V[\Psi_m(x)]$ *be the minimal space for the* N *sequences* $\{ s_k^i \}$, $i = 0, 1, \cdots, N$-1. *Then,* $\Psi_m(x)$ *is the least common multiple of the* N *polynomials* $\Psi_i(x)$, $i = 0, 1, \cdots, N$-1, *that is,*

$$V[\Psi_m(x)] = V[LCM[\Psi_0(x), \Psi_1(x), \cdots, \Psi_{N-1}(x)]]. \qquad (4.24)$$

Proof : Let $\Psi_c(x) = LCM[\Psi_0(x), \Psi_1(x), \cdots, \Psi_{N-1}(x)]$. Then, it suffices to show that $\Psi_c(x) = \Psi_m(x)$. Since $\Psi_i(x)$, $i = 0, 1, \cdots, N-1$, divides $\Psi_c(x)$, by Theorem 4.33 $V[\Psi_c(x)]$ is a common space for each sequence $\{ s_k^i \}$. Therefore, $V[\Psi_c(x)]$ is a common space for the N sequences $\{ s_k^i \}$, $i = 0, 1, \cdots, N - 1$, and so by Theorem 4.33 $\Psi_m(x)$ divides $\Psi_c(x)$. To complete the proof, we show that $\Psi_c(x)$ divides $\Psi_m(x)$. Since $V[\Psi_m(x)]$ is a common space for each sequence $\{ s_k^i \}$, $i = 0, 1, \cdots, N - 1$, by Theorem 4.33 $\Psi_i(x)$ divides $\Psi_m(x)$. Therefore, the least common multiple $\Psi_c(x)$ of $\Psi_i(x)$'s also divides $\Psi_m(x)$. This completes the proof.   ∎

EXAMPLE 4.14 We determine the minimal space of the sequence S_5 in Table 4.1, whose polynomial expression is $S[x^4 + x^3 + x^2 + 1, x^2 + x]$. The greatest common divisor $d(x)$ of $x^4 + x^3 + x^2 + 1$ and $x^2 + x$ is $x + 1$. Therefore, by Theorem 4.34, the minimal space for the sequence S_5 is $V[x^3 + x + 1]$, and S_5 has the polynomial expression $S[x^3 + x + 1, x]$ for the minimal space.

For another example, we determine the minimal space of the sequence S_4 in Table 4.1, whose period is 7. The greatest common divisor

$d(x)$ of $x^7 + 1$ and $x^5 + x + 1$ is $x^3 + x^2 + 1$ in this case. Therefore, by Theorem 4.35, the minimal space for the sequence S_4 is $V[x^4 + x^3 + x^2 + 1]$, and S_4 has the polynomial expression $S[x^4 + x^3 + x^2 + 1, x^2 + x + 1]$ for the minimal space.

Now we determine the minimal space of the two sequences S_4 and S_5. The minimal spaces of the sequence S_4 and S_5 are respectively $V[x^4 + x^3 + x^2 + 1]$ and $V[x^3 + x + 1]$ as determined above, and the least common multiple of the two corresponding characteristic polynomials is $x^4 + x^3 + x^2 + 1$. Therefore, by Theorem 4.36, the minimal space of the two sequences S_4 and S_5 is $V[x^4 + x^3 + x^2 + 1]$. ♣

The following theorem describes the relation between the period of a sequence and the periodicity of its minimal space.

THEOREM 4.37 *The period τ of a sequence $\{ s_k \}$ is identical to the periodicity T of its minimal space $V[\Psi_m(x)]$.*

Proof : By Theorem 4.29, $\tau \leq T$. Therefore, it suffices to show that $T \leq \tau$. By Theorem 4.23, $\{ s_k \} \in V[x^\tau + 1]$, that is, $V[x^\tau + 1]$ is a common space for the sequence $\{ s_k \}$. Therefore by Theorem 4.33 $\Psi_m(x)$ divides $x^\tau + 1$, which means $T \leq \tau$ by definition of periodicity. This completes the proof. ∎

EXAMPLE 4.15 We consider the sequence S_5 in Table 4.1. Then, the minimal space for this sequence is $V[x^3 + x + 1]$, and its periodicity T is 7. Therefore, by the theorem, the period of the sequence S_5 is 7, which we can confirm using Table 4.1. ♣

Chapter 5
Shift Register Generators

Shift register generator (SRG) is an autonomous system consisting of shift registers and exclusive-OR gates. In this chapter, we investigate the behaviors of SRGs in view of the sequence space theory developed in the previous chapter. For a given SRG, we first consider how to determine the sequence space generated by SRG sequences with a fixed initial state vector (SRG space), and then consider how to find the largest-dimensional sequence space that can be obtained by varying the initial state vectors (SRG maximal space). Conversely, for a given sequence space we consider how to find the minimum-sized SRGs that can generate the sequence space (basic SRG). Finally, we examine the two typical SRGs, namely the simple SRG and the modular SRG, in view of the concept of basic SRGs.

5.1   Shift Register Generator

We first define shift register generator(SRG) and its related terms such as the length, the state vector and the state transition matrix.

DEFINITION 5.1 (SRG) A *shift register generator* (SRG) refers to an autonomous and time-invariant system consisting of exclusive-OR gates and shift registers whose current state is determined by a modulo-2 linear sum of the previous state according to the connection of the exclusive-OR gates.

DEFINITION 5.2 (Length) For an SRG, the *length L* is defined to be the number of shift registers in the SRG.

DEFINITION 5.3 (State Vector) For an SRG of length L, the kth *state vector* $\mathbf{d}_k$ is defined to be an L-vector representing the state of shift registers in the SRG at time k, that is,

$$\mathbf{d}_k \equiv [\, d_{0,k} \;\; d_{1,k} \;\; \cdots \;\; d_{L-1,k} \,]^t, \tag{5.1}$$

where the state $d_{i,k}$, $i = 0, 1, \cdots, L - 1$, denotes the value of the ith shift register in the SRG at time k.

DEFINITION 5.4 (Initial State Vector) The 0th state vector $\mathbf{d}_0$ is called the *initial state vector.*

DEFINITION 5.5 (State Transition Matrix) For an SRG of length L, the *state transition matrix* $\mathbf{T}$ is defined to be an $L \times L$ matrix representing the relation between the state vectors $\mathbf{d}_k$ and $\mathbf{d}_{k-1}$, or more specifically,

$$\mathbf{d}_k = \mathbf{T} \cdot \mathbf{d}_{k-1}. \tag{5.2}$$

An SRG is represented by a state transition matrix $\mathbf{T}$ and an initial state vector $\mathbf{d}_0$. The state transition matrix $\mathbf{T}$ governs the structure of the SRG, and the initial state vector $\mathbf{d}_0$ controls the initial point of the generated sequences. To be more rigorous, the configuration of an SRG is *uniquely determined* by its state transition matrix $\mathbf{T}$, and the sequences generated by the SRG is *uniquely determined* when the initial state vector $\mathbf{d}_0$ is additionally furnished.

Among the SRGs formed according to Definition 5.1, there are included some SRGs which are *redundant* in the sense that the states of one or more shift registers are dependent on the states of the other shift registers. In terms of state transition matrices, this implies that some row vectors of the corresponding state transition matrix are represented by linear sums of other row vectors. That is, the state transition matrices for the redundant SRGs are *singular*. Therefore, we exclude such redundant SRGs from our discussion, so *we will assume that state transition matrices are all nonsingular in this book unless explicitly negated.*

EXAMPLE 5.1 Fig. 5.1 shows an SRG of length 5. The squared boxes denote the shift registers, and the values $d_{i,k}$, $i = 0, 1, 2, 3, 4$, inside the boxes indicate the state of the five shift registers. The state transition

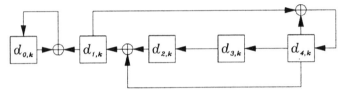

Figure 5.1. An example of shift register generator.

matrix $\mathbf{T}$ of this SRG is

$$\mathbf{T} = \begin{bmatrix} 1 & 1 & 0 & 0 & 0 \\ 0 & 0 & 1 & 0 & 1 \\ 0 & 0 & 0 & 1 & 0 \\ 0 & 0 & 0 & 0 & 1 \\ 0 & 1 & 0 & 0 & 1 \end{bmatrix}. \quad \clubsuit \qquad (5.3)$$

5.2 Minimal Spaces for SRG Sequences

We consider the sequences generated by an SRG and examine how to associate it with sequence spaces.

DEFINITION 5.6 (SRG Sequences) For an SRG of length L, the ith *SRG sequence* D_i, $i = 0, 1, \cdots, L - 1$, refers to the sequence generated by the ith shift register, that is,

$$D_i \equiv \{ d_{i,k} \}. \qquad (5.4)$$

DEFINITION 5.7 (SRG Sequence Vector) For an SRG of length L, the *SRG sequence vector* $\mathbf{D}$ designates the L-vector

$$\mathbf{D} \equiv [\, D_0 \ \ D_1 \ \ \cdots \ \ D_{L-1} \,]^t. \qquad (5.5)$$

Then, due to the relations in (5.2) and (5.4), the SRG sequence vector $\mathbf{D}$ is uniquely determined once its state transition matrix $\mathbf{T}$ and initial state vector $\mathbf{d}_0$ are determined.

EXAMPLE 5.2 We consider the SRG in Fig. 5.1 with the state transition matrix $\mathbf{T}$ in (5.3). For this SRG, if the initial state vector is $\mathbf{d}_0 = [\, 1\, 0\, 0\, 1\, 1 \,]^t$, by (5.2) we obtain the state vectors $\mathbf{d}_1 = [\, 1\, 1\, 1\, 1\, 1 \,]^t$, $\mathbf{d}_2 = [\, 0\, 0\, 1\, 1\, 0 \,]^t$, $\cdots$. Therefore, we obtain the SRG sequences D_i's listed in

Table 5.1. The SRG sequences generated by the SRG in Fig. 5.1 for the initial state vector $\mathbf{d}_0 = [\,1\,0\,0\,1\,1\,]^t$

$D_0 = \{1,1,0,0,1,0,1,1,1,0,0,\cdots\}$
$D_1 = \{0,1,0,1,1,1,0,0,1,0,1,\cdots\}$
$D_2 = \{0,1,1,1,0,0,1,0,1,1,1,\cdots\}$
$D_3 = \{1,1,1,0,0,1,0,1,1,1,0,\cdots\}$
$D_4 = \{1,1,0,0,1,0,1,1,1,0,0,\cdots\}$

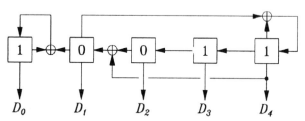

Figure 5.2. An example of SRG, whose initial state vector is $\mathbf{d}_0 = [\,1\,0\,0\,1\,1\,]^t$, and the generated SRG sequences are D_0 through D_4 in Table 5.1.

Table 5.1. Note that the ith sequence D_i for $i = 0, 1, 2, 3, 4$, is obtained by taking output from the ith shift register, as illustrated in Fig. 5.2. The value 1 or 0 inside each shift register denotes the initial state of the shift register. ♣

The following two theorems describe how to obtain the common and minimal spaces for the SRG sequences.

THEOREM 5.8 *A sequence space $V[\Psi(x)]$ is a common space for the L SRG sequences D_i, $i = 0, 1, \cdots, L\text{-}1$, generated by the SRG of length L with the state transition matrix $\mathbf{T}$ and the initial state vector $\mathbf{d}_0$, if and only if the characteristic polynomial $\Psi(x)$ meets the condition*[1]

$$\Psi(\mathbf{T}) \cdot \mathbf{d}_0 = 0. \tag{5.6}$$

Proof : We first prove the "if" part of the theorem. Let $\Psi(x) = \sum_{i=0}^{\hat{L}} \psi_i x^i$ be a polynomial such that $\Psi(\mathbf{T}) \cdot \mathbf{d}_0 = 0$. Then, by (5.1)

---

[1] Note that the state transition matrix $\mathbf{T}$ is assumed nonsingular in the book, unless stated otherwise.

and (5.2), for each $l = 0, 1, \cdots, L - 1$, $d_{l,k} = \mathbf{e}_l^t \cdot \mathbf{T}^k \cdot \mathbf{d}_0$, and hence $\sum_{i=0}^{\hat{L}} \psi_i d_{l,k+i} = \sum_{i=0}^{\hat{L}} \psi_i \mathbf{e}_l^t \cdot \mathbf{T}^{k+i} \cdot \mathbf{d}_0 = \mathbf{e}_l^t \cdot \mathbf{T}^k \cdot \Psi(\mathbf{T}) \cdot \mathbf{d}_0$, which is 0 due to the assumption $\Psi(\mathbf{T}) \cdot \mathbf{d}_0 = 0$. Therefore, by (4.1) and (5.4), $D_l \in V[\Psi(x)]$, $l = 0, 1, \cdots, L - 1$, that is, $V[\Psi(x)]$ is a common space for the SRG sequences D_l's.

Now we prove the "only if" part. Let $V[\Psi(x)]$ of dimension $\hat{L}$ be a common space for the SRG sequences D_l, $l = 0, 1, \cdots, L - 1$. Then, inserting $L = \hat{L}$, $k = 0$ and $s_{k+i} = d_{l,k+i}$ into (4.1), we obtain $\sum_{i=0}^{\hat{L}} \psi_i d_{l,i} = 0$, and applying the relation $d_{l,i} = \mathbf{e}_l^t \cdot \mathbf{T}^i \cdot \mathbf{d}_0$ to this, we have the relation $\mathbf{e}_l^t \cdot \Psi(\mathbf{T}) \cdot \mathbf{d}_0 = 0$, $l = 0, 1, \cdots, L - 1$. Therefore, $\Psi(\mathbf{T}) \cdot \mathbf{d}_0 = 0$. ■

THEOREM 5.9 *For an SRG of length L with the state transition matrix* **T** *and the initial state vector* $\mathbf{d}_0$, *let* $\Psi_{\mathbf{T},\mathbf{d}_0}(x)$ *denote the lowest-degree polynomial that meets (5.6). Then, the sequence space* $V[\Psi_{\mathbf{T},\mathbf{d}_0}(x)]$ *is the minimal space for the SRG sequences* D_i, $i = 0, 1, \cdots, L\text{-}1$.

Proof : This theorem is directly obtained by Theorem 5.8 and the definition of the minimal space. ■

By Theorem 4.32, the minimal space is unique. Therefore, the lowest-degree polynomial $\Psi_{\mathbf{T},\mathbf{d}_0}(x)$ that meets (5.6) is also unique.

EXAMPLE 5.3 In the case of the SRG in Fig. 5.2 with the state transition matrix **T** in (5.3) and the initial state vector $\mathbf{d}_0 = [1\,0\,0\,1\,1]^t$, the lowest-degree polynomial $\Psi_{\mathbf{T},\mathbf{d}_0}(x)$ that makes $\Psi_{\mathbf{T},\mathbf{d}_0}(\mathbf{T}) \cdot \mathbf{d}_0 = 0$ is $x^3 + x + 1$. Therefore, by Theorem 5.9, the minimal space for the SRG sequences D_i, $i = 0, 1, 2, 3, 4$, is the sequence space $V[x^3 + x + 1]$. In fact, applying Theorems 4.35 and 4.36 to the SRG sequences D_i's in Table 5.1, we can confirm that the minimal space for the SRG sequences D_i's is $V[x^3 + x + 1]$. ♣

Now we consider the expression of the SRG sequence vector for the minimal space.

THEOREM 5.10 *Let the sequence space* $V[\Psi_{\mathbf{T},\mathbf{d}_0}(x)]$ *of dimension* $\hat{L}$ *be the minimal space for the SRG sequences* D_i, $i = 0, 1, \cdots, L\text{-}1$, *generated by the SRG of length L with the state transition matrix* **T** *and the initial state vector* $\mathbf{d}_0$. *Then, the SRG sequence vector* **D** *has the expression*

$$\mathbf{D} = [\, \mathbf{d}_0 \quad \mathbf{T} \cdot \mathbf{d}_0 \quad \cdots \quad \mathbf{T}^{\hat{L}-1} \cdot \mathbf{d}_0 \,] \cdot \mathbf{E}_{\Psi_{\mathbf{T},\mathbf{d}_0}(x)} \tag{5.7}$$

on the elementary basis of the minimal space.

Proof : Let $\mathbf{s}_i$, $i = 0, 1, \cdots, L - 1$, be the initial vector for the ith SRG sequence D_i. Then, by (5.1) and (5.2), $\mathbf{s}_i = [\,\mathbf{e}_i^t \cdot \mathbf{d}_0 \quad \mathbf{e}_i^t \cdot \mathbf{T} \cdot \mathbf{d}_0 \quad \cdots \quad \mathbf{e}_i^t \cdot \mathbf{T}^{\hat{L}-1} \cdot \mathbf{d}_0\,]^t$. Therefore, by (4.5) $D_i = [\,\mathbf{e}_i^t \cdot \mathbf{d}_0 \quad \mathbf{e}_i^t \cdot \mathbf{T} \cdot \mathbf{d}_0 \quad \cdots \quad \mathbf{e}_i^t \cdot \mathbf{T}^{\hat{L}-1} \cdot \mathbf{d}_0\,] \cdot \mathbf{E}_{\Psi_{\mathbf{T},\mathbf{d}_0}}(x)$. This corresponds to the ith component of (5.7). ∎

EXAMPLE 5.4 In the case of the SRG in Fig. 5.2 with the state transition matrix $\mathbf{T}$ in (5.3) and the initial state vector $\mathbf{d}_0 = [10011]^t$, the minimal space for the SRG sequences D_i, $i = 0, 1, 2, 3, 4$, is $V[x^3 + x + 1]$. Therefore, by Theorem 5.10, the SRG sequence vector has the expression

$$\mathbf{D} = \begin{bmatrix} 1 & 1 & 0 \\ 0 & 1 & 0 \\ 0 & 1 & 1 \\ 1 & 1 & 1 \\ 1 & 1 & 0 \end{bmatrix} \cdot \mathbf{E}_{x^3+x+1} \tag{5.8}$$

on the elementary basis of the minimal space. From Tables 4.2 and 5.1, we can confirm this. ♣

5.3   SRG Spaces

In the previous section we have considered common and minimal sequence spaces for the SRG sequences D_i's. In order to complement this we consider in this section a more direct approach which generates the sequence space by taking linear combinations of D_i's.

DEFINITION 5.11 (SRG Space) For the SRG sequences D_i, $i = 0, 1, \cdots, L - 1$, generated by an SRG of length L with the state transition matrix $\mathbf{T}$ and the initial state vector $\mathbf{d}_0$, the *SRG space* $V[\mathbf{T}, \mathbf{d}_0]$ is defined to be the vector space formed by D_i's. More specifically,

$$V[\mathbf{T}, \mathbf{d}_0] \equiv \{ \sum_{i=0}^{L-1} a_i D_i : a_i \in \mathcal{F}, \ i = 0, 1, \cdots, L - 1 \}. \tag{5.9}$$

According to this definition, the SRG space $V[\mathbf{T}, \mathbf{d}_0]$ is the minimal-dimensional sequence space containing the SRG sequences D_i's. Therefore, an arbitrary sequence in the SRG space $V[\mathbf{T}, \mathbf{d}_0]$ can be represented by a linear sum of the SRG sequences D_i's. However a sequence which is not contained in the SRG space $V[\mathbf{T}, \mathbf{d}_0]$ can not be represented by a linear sum of D_i's.

THEOREM 5.12 *For an SRG of length L with the state transition matrix $\mathbf{T}$ and the initial state vector $\mathbf{d}_0$, the SRG space $V[\mathbf{T}, \mathbf{d}_0]$ is identical to the minimal space $V[\Psi_{\mathbf{T}, \mathbf{d}_0}(x)]$ for the SRG sequences D_i, $i = 0, 1, \cdots, L\text{-}1$.*

Proof : Since the SRG space is the minimal-dimensional vector space containing the SRG sequences D_i, $i = 0, 1, \cdots, L - 1$, $V[\mathbf{T}, \mathbf{d}_0]$ is a subspace of the minimal space $V[\Psi_{\mathbf{T}, \mathbf{d}_0}(x)]$. Therefore, it suffices to show that the dimension of $V[\mathbf{T}, \mathbf{d}_0]$ is identical to the dimension $\hat{L}$ of $V[\Psi_{\mathbf{T}, \mathbf{d}_0}(x)]$. By definition of $\Psi_{\mathbf{T}, \mathbf{d}_0}(x)$, the $\hat{L}$ vectors $\mathbf{T}^i \cdot \mathbf{d}_0$, $i = 0, 1, \cdots, \hat{L} - 1$, are linearly independent, and so the rank of the $L \times \hat{L}$ matrix $[\, \mathbf{d}_0 \quad \mathbf{T} \cdot \mathbf{d}_0 \quad \cdots \quad \mathbf{T}^{\hat{L}-1} \cdot \mathbf{d}_0 \,]$ is $\hat{L}$. Hence by (5.7) $\hat{L}$ among the L sequences $\mathbf{D}_i$, $i = 0, 1, \cdots, L - 1$, are linearly independent. Therefore, the dimension of $V[\mathbf{T}, \mathbf{d}_0]$ is $\hat{L}$, and the proof is complete. ∎

The theorem means that the minimal-dimensional vector space $V[\mathbf{T}, \mathbf{d}_0]$ containing the SRG sequences D_i's is the minimal-dimensional sequence space $V[\Psi_{\mathbf{T}, \mathbf{d}_0}(x)]$ containing them.

EXAMPLE 5.5 In the case of the SRG in Fig. 5.2 with the state transition matrix $\mathbf{T}$ in (5.3) and the initial state vector $\mathbf{d}_0 = [10011]^t$, the minimal space for the SRG sequences $D_0, D_1, \cdots, D_4$, is $V[x^3 + x + 1]$. Therefore, by the theorem, any sequence in the minimal space $V[x^3 + x + 1]$ can be represented by a linear sum of the SRG sequences D_i's, which we can confirm using Tables 4.2 and 5.1. ♣

5.4 SRG Maximal Spaces

In this section, we consider the largest-dimensional SRG space which can be generated by an SRG.

DEFINITION 5.13 (SRG Maximal Space) For an SRG with the state transition matrix $\mathbf{T}$, an *SRG maximal space* $V[\mathbf{T}]$ refers to the largest-dimensional SRG space of all SRG spaces $V[\mathbf{T}, \mathbf{d}_0]$'s obtained by varying the initial state vectors $\mathbf{d}_0$'s.

THEOREM 5.14 *The SRG maximal space $V[\mathbf{T}]$ is unique, and is identical to the sequence space $V[\Psi_{\mathbf{T}}(x)]$ for the minimal polynomial $\Psi_{\mathbf{T}}(x)$ of $\mathbf{T}$.*[2]

---

[2] The minimal polynomial $\Psi_{\mathbf{T}}(x)$ of a matrix $\mathbf{T}$ is the lowest-degree polynomial that makes $\Psi_{\mathbf{T}}(\mathbf{T}) = 0$. Refer to Section B.2 in Appendix B.

Proof : We first prove that $V[\Psi_{\mathbf{T}}(x)]$ is an SRG maximal space. Since $\Psi_{\mathbf{T}}(x)$ is the minimal polynomial of $\mathbf{T}$, there exists an initial state vector $\mathbf{d}_0$ such that the lowest-degree polynomial $\Psi_{\mathbf{T},\mathbf{d}_0}(x)$ satisfying $\Psi_{\mathbf{T},\mathbf{d}_0}(\mathbf{T}) \cdot \mathbf{d}_0 = 0$ is $\Psi_{\mathbf{T}}(x)$.[3] Therefore, by Theorems 5.9 and 5.12, the SRG space $V[\mathbf{T}, \mathbf{d}_0]$ for this initial state vector $\mathbf{d}_0$ is $V[\Psi_{\mathbf{T}}(x)]$. To complete the proof, it suffices to prove that the SRG space $V[\mathbf{T}, \mathbf{d}_0]$ for an arbitrary initial state vector $\mathbf{d}_0$ is a subspace of $V[\Psi_{\mathbf{T}}(x)]$. By Theorems 5.8 and 5.12, $V[\Psi_{\mathbf{T}}(x)]$ and $V[\mathbf{T}, \mathbf{d}_0]$ are respectively common and minimal spaces for the SRG sequences D_i's. Therefore, by Theorems 4.25 and 4.33, $V[\mathbf{T}, \mathbf{d}_0]$ is a subspace of $V[\Psi_{\mathbf{T}}(x)]$.

Now we prove the uniqueness of the SRG maximal space $V[\mathbf{T}]$. Let $V[\Psi(x)]$ be an SRG maximal space, and let $\mathbf{d}_0$ be an initial state vector such that $V[\mathbf{T}, \mathbf{d}_0] = V[\Psi(x)]$. Then, by Theorems 5.8 and 5.12, $V[\Psi_{\mathbf{T}}(x)]$ and $V[\Psi(x)]$ are respectively common and minimal spaces for the SRG sequences D_i's. Therefore, by Theorem 4.33, $\Psi(x)$ divides $\Psi_{\mathbf{T}}(x)$. However since the degrees of $\Psi(x)$ and $\Psi_{\mathbf{T}}(x)$ are the same, $\Psi(x)$ is identical to $\Psi_{\mathbf{T}}(x)$.   ∎

EXAMPLE 5.6 In the case of the SRG in Fig. 5.1 with the state transition matrix $\mathbf{T}$ in (5.3), the minimal polynomial of $\mathbf{T}$ is $\Psi_{\mathbf{T}}(x) = x^4 + x^3 + x^2 + 1$. Therefore, by the theorem, the SRG maximal space is $V[x^4 + x^3 + x^2 + 1]$. In fact, by Theorems 5.9 and 5.12, we can obtain the SRG spaces $V[\mathbf{T}, \mathbf{d}_0]$'s by varying initial state vectors $\mathbf{d}_0$'s, as shown in Table 5.2, and from the table we can confirm that the SRG maximal space is $V[\mathbf{T}] = V[x^4 + x^3 + x^2 + 1]$. ♣

Now we consider the initial state vector for an SRG which makes the SRG space identical to the SRG maximal space.

DEFINITION 5.15 (Maximal Initial State Vector) For an SRG with the state transition matrix $\mathbf{T}$, a *maximal initial state vector* $\mathbf{d}_0$ refers to an initial state vector that makes the SRG space $V[\mathbf{T}, \mathbf{d}_0]$ identical to the SRG maximal space $V[\mathbf{T}]$.

In general, there can exist multiple maximal initial state vectors. For example, in the case of the SRG in Fig. 5.1, there are 21 maximal initial state vectors, as can be confirmed from Table 5.2.

---

[3] Refer to Property B.6 in Appendix B.

Table 5.2. The SRG spaces of the SRG in Fig. 5.1 for various nitial state vectors

initial state vector $\mathbf{d}_0$	SRG space $V[\mathbf{T}, \mathbf{d}_0]$
$[0\,0\,0\,0\,0]^t$	$V[1]$
$[0\,0\,0\,0\,1]^t$	$V[x^4 + x^3 + x^2 + 1]$
$[0\,0\,0\,1\,0]^t$	$V[x^4 + x^3 + x^2 + 1]$
$[0\,0\,0\,1\,1]^t$	$V[x^4 + x^3 + x^2 + 1]$
$[0\,0\,1\,0\,0]^t$	$V[x^4 + x^3 + x^2 + 1]$
$[0\,0\,1\,0\,1]^t$	$V[x^4 + x^3 + x^2 + 1]$
$[0\,0\,1\,1\,0]^t$	$V[x^3 + x + 1]$
$[0\,0\,1\,1\,1]^t$	$V[x + 1]$
$[0\,1\,0\,0\,0]^t$	$V[x^4 + x^3 + x^2 + 1]$
$[0\,1\,0\,0\,1]^t$	$V[x^4 + x^3 + x^2 + 1]$
$[0\,1\,0\,1\,0]^t$	$V[x^3 + x + 1]$
$[0\,1\,0\,1\,1]^t$	$V[x^4 + x^3 + x^2 + 1]$
$[0\,1\,1\,0\,0]^t$	$V[x^3 + x + 1]$
$[0\,1\,1\,0\,1]^t$	$V[x^4 + x^3 + x^2 + 1]$
$[0\,1\,1\,1\,0]^t$	$V[x^4 + x^3 + x^2 + 1]$
$[0\,1\,1\,1\,1]^t$	$V[x^4 + x^3 + x^2 + 1]$
$[1\,0\,0\,0\,0]^t$	$V[x + 1]$
$[1\,0\,0\,0\,1]^t$	$V[x^4 + x^3 + x^2 + 1]$
$[1\,0\,0\,1\,0]^t$	$V[x^4 + x^3 + x^2 + 1]$
$[1\,0\,0\,1\,1]^t$	$V[x^3 + x + 1]$
$[1\,0\,1\,0\,0]^t$	$V[x^4 + x^3 + x^2 + 1]$
$[1\,0\,1\,0\,1]^t$	$V[x^3 + x + 1]$
$[1\,0\,1\,1\,0]^t$	$V[x^4 + x^3 + x^2 + 1]$
$[1\,0\,1\,1\,1]^t$	$V[x + 1]$
$[1\,1\,0\,0\,0]^t$	$V[x^4 + x^3 + x^2 + 1]$
$[1\,1\,0\,0\,1]^t$	$V[x^3 + x + 1]$
$[1\,1\,0\,1\,0]^t$	$V[x^4 + x^3 + x^2 + 1]$
$[1\,1\,0\,1\,1]^t$	$V[x^4 + x^3 + x^2 + 1]$
$[1\,1\,1\,0\,0]^t$	$V[x^4 + x^3 + x^2 + 1]$
$[1\,1\,1\,0\,1]^t$	$V[x^4 + x^3 + x^2 + 1]$
$[1\,1\,1\,1\,0]^t$	$V[x^4 + x^3 + x^2 + 1]$
$[1\,1\,1\,1\,1]^t$	$V[x^3 + x + 1]$

THEOREM 5.16 *For an SRG with the state transition matrix* $\mathbf{T}$, *let* $\hat{L}$ *be the dimension of the SRG maximal space* $V[\mathbf{T}]$. *Then, an initial state vector* $\mathbf{d}_0$ *is a maximal initial state vector, if and only if the discrimination matrix* $\Delta_{\mathbf{T},\mathbf{d}_0}$ *defined by*

$$\Delta_{\mathbf{T},\mathbf{d}_0} \equiv [\, \mathbf{d}_0 \ \ \mathbf{T} \cdot \mathbf{d}_0 \ \cdots \ \mathbf{T}^{\hat{L}-1} \cdot \mathbf{d}_0 \,] \qquad (5.10)$$

is of rank $\hat{L}$.

Proof : We first prove the "if" part of the theorem. Let $\mathbf{d}_0$ be an initial state vector such that the rank of $\Delta_{\mathbf{T},\mathbf{d}_0}$ is $\hat{L}$. Then, the $\hat{L}$ vectors $\mathbf{T}^i \cdot \mathbf{d}_0$, $i = 0, 1, \cdots, \hat{L} - 1$, are linearly independent. Therefore, the degree of the lowest-degree polynomial $\Psi_{\mathbf{T},\mathbf{d}_0}(x)$ that meets (5.6) is $\hat{L}$ or higher, and so by Theorem 5.12, $V[\mathbf{T}, \mathbf{d}_0]$ is of dimension $\hat{L}$ or higher. However, since the dimension of the SRG maximal space $V[\mathbf{T}]$ is $\hat{L}$ and the SRG maximal space $V[\mathbf{T}]$ is unique, $V[\mathbf{T}, \mathbf{d}_0]$ should be the SRG maximal space $V[\mathbf{T}]$. That is, $\mathbf{d}_0$ is a maximal initial state vector.

Now we prove the "only if" part by contradiction. Suppose that there exists a maximal initial state vector $\mathbf{d}_0$ such that the rank of $\Delta_{\mathbf{T},\mathbf{d}_0}$ is less than $\hat{L}$. Then, the $\hat{L}$ vectors $\mathbf{T}^i \cdot \mathbf{d}_0$, $i = 0, 1, \cdots, \hat{L} - 1$, are linearly dependent. Therefore, the degree of the lowest-degree polynomial $\Psi_{\mathbf{T},\mathbf{d}_0}(x)$ that meets (5.6) is less than $\hat{L}$, and so by Theorems 4.7 and 5.12, the dimension of the SRG space $V[\mathbf{T}, \mathbf{d}_0]$ is less than $\hat{L}$. However, this is a contradiction to the assumption that $\mathbf{d}_0$ is a maximal initial state vector. ∎

EXAMPLE 5.7 In the case of the SRG in Fig. 5.1 with the state transition matrix $\mathbf{T}$ in (5.3), the SRG maximal space is $V[x^4 + x^3 + x^2 + 1]$, whose dimension is 4. If we choose $\mathbf{d}_0 = [\,0\,0\,0\,0\,1\,]^t$, by (5.10) we obtain the discrimination matrix

$$\Delta_{\mathbf{T},\mathbf{d}_0} = \begin{bmatrix} 0 & 0 & 1 & 0 \\ 0 & 1 & 1 & 1 \\ 0 & 0 & 1 & 1 \\ 0 & 1 & 1 & 0 \\ 1 & 1 & 0 & 1 \end{bmatrix}, \qquad (5.11)$$

whose rank is 4. Therefore, by Theorem 5.16, this $\mathbf{d}_0$ is a maximal initial state vector, which we can confirm using Table 5.2. ♣

The following theorem describes the relation between the SRG maximal space and the SRG space.

THEOREM 5.17 *For an SRG with the state transition matrix* $\mathbf{T}$, *the SRG space* $V[\mathbf{T}, \mathbf{d}_0]$ *for an initial state vector* $\mathbf{d}_0$ *is a subspace of the SRG maximal space* $V[\mathbf{T}]$.

We omit the proof of this theorem as it is already incorporated in the proof of Theorem 5.14.

The theorem implies that the SRG space $V[\mathbf{T}, \mathbf{d}_0]$ for an arbitrary initial state vector $\mathbf{d}_0$ is a subspace of the SRG maximal space $V[\mathbf{T}]$. Therefore, combining this with Theorem 5.14, we can state that for an SRG with the state transition matrix $\mathbf{T}$, the maximal SRG space becomes $V[\Psi_{\mathbf{T}}(x)]$ for the minimal polynomial $\Psi_{\mathbf{T}}(x)$ of $\mathbf{T}$, and hence the SRG space $V[\mathbf{T}, \mathbf{d}_0]$ is a subspace of the SRG maximal space $V[\Psi_{\mathbf{T}}(x)]$. It further implies that an arbitrary sequence in the SRG maximal space $V[\Psi_{\mathbf{T}}(x)]$ can be represented by a linear sum of the SRG sequences D_i's for a maximal initial state vector $\mathbf{d}_0$, but a sequence not contained in the SRG maximal space can not be represented by a linear sum of the SRG sequences for any initial state vector.

EXAMPLE 5.8 In the case of the SRG in Fig. 5.1, we can confirm from Table 5.2 that any SRG space $V[\mathbf{T}, \mathbf{d}_0]$ for an initial state vector $\mathbf{d}_0$ is a subspace of the SRG maximal space $V[\mathbf{T}] = V[x^4 + x^3 + x^2 + 1]$. ♣

5.5   Basic SRGs

In the previous sections, we considered sequence spaces for the SRG sequences generated by SRGs. In this section we consider the reverse direction. That is, we consider SRGs which can generate the SRG sequences that form a given sequence space.

DEFINITION 5.18 (Basic SRG) For a sequence space $V[\Psi(x)]$, a *basic SRG* (BSRG) refers to an SRG of the smallest length whose SRG maximal space is identical to the sequence space $V[\Psi(x)]$.

That is, a BSRG for a sequence space $V[\Psi(x)]$ is a smallest-length SRG which can generate the sequence space $V[\Psi(x)]$. Then, the length, the state transition matrix, and the maximal initial state vector of a BSRG are determined by the following three theorems.

THEOREM 5.19 *The length of a BSRG for a sequence space is the same as the dimension of the sequence space.*

Proof : Let $\hat{L}$ and $\mathbf{T}$ be respectively the length and state transition matrix of a BSRG for a sequence space $V[\Psi(x)]$ of dimension L. Then, by Theorem 5.14 and the definition of BSRG, the minimal polynomial $\Psi_{\mathbf{T}}(x)$ of $\mathbf{T}$ is $\Psi(x)$, that is, the degree of $\Psi_{\mathbf{T}}(x)$ is L. Therefore, $L \leq \hat{L}$, since the degree L of the minimal polynomial $\Psi_{\mathbf{T}}(x)$ is not greater than $\hat{L}$. To complete the proof, we now prove that $L \geq \hat{L}$. We consider the SRG of length L whose state transition matrix is the companion matrix $\mathbf{A}_{\Psi(x)}$ in (4.7). Then, since the minimal polynomial of $\mathbf{A}_{\Psi(x)}$ is $\Psi(x)$, this SRG maximal space is $V[\Psi(x)]$ due to Theorem 5.14. Therefore, $L \geq \hat{L}$, since a BSRG is a smallest-length SRG whose SRG maximal space is $V[\Psi(x)]$. This completes the proof. ∎

THEOREM 5.20 *A matrix $\mathbf{T}$ is the state transition matrix of a BSRG for the sequence space $V[\Psi(x)]$, if and only if it is similar to the companion matrix $\mathbf{A}_{\Psi(x)}$.*[4]

Proof : To prove the theorem, let the dimension of the sequence space $V[\Psi(x)]$ be L. We first prove the "if" part of the theorem. If the state transition matrix $\mathbf{T}$ of an SRG is similar to the companion matrix $\mathbf{A}_{\Psi(x)}$, then its minimal polynomial $\Psi_{\mathbf{T}}(x)$ of $\mathbf{T}$ is identical to that of $\mathbf{A}_{\Psi(x)}$, that is, $\Psi_{\mathbf{T}}(x) = \Psi(x)$. Thus, by Theorem 5.14, this SRG maximal space $V[\mathbf{T}]$ is $V[\Psi(x)]$. Further, the length of this SRG is L, which is identical to the dimension of the sequence space $V[\Psi(x)]$. Therefore, by Theorem 5.19, this SRG is a BSRG for the sequence space $V[\Psi(x)]$.

Now we prove the "only if" part. If $\mathbf{T}$ is an $L \times L$ state transition matrix of a BSRG for the sequence space $V[\Psi(x)]$, then the minimal polynomial $\Psi_{\mathbf{T}}(x)$ is $\Psi(x)$ due to Theorem 5.14. But $\mathbf{A}_{\Psi(x)}$ is also an $L \times L$ matrix whose minimal polynomial is $\Psi(x)$. Therefore, $\mathbf{T}$ is similar to $\mathbf{A}_{\Psi(x)}$ (refer to Reference [9]). ∎

THEOREM 5.21 *An initial state vector $\mathbf{d}_0$ is a maximal initial state vector of the BSRG for a sequence space $V[\Psi(x)]$, if and only if the*

---

[4]A matrix $\mathbf{M}$ is said *similar* to a matrix $\hat{\mathbf{M}}$ if there exists a nonsingular matrix $\mathbf{Q}$ such that $\mathbf{M} = \mathbf{Q} \cdot \hat{\mathbf{M}} \cdot \mathbf{Q}^{-1}$. Refer to Section B.4 in Appendix B.

corresponding discrimination matrix $\Delta_{\mathbf{T},\mathbf{d}_0}$ *in (5.10) is nonsingular for the state transition matrix* $\mathbf{T}$ *taken similar to* $\mathbf{A}_{\Psi(x)}$ *in (4.7).*[5]

Proof : The theorem is directly obtained by Theorem 5.16. ∎

Theorem 5.19 implies that the length of an SRG should be L if it is to become a BSRG for an L-dimensional sequence space. Theorems 5.20 and 5.21 provide two important guidelines on choosing the state transition matrix and the initial state vector for a BSRG : First, the state transition matrix $\mathbf{T}$ should be chosen to be similar to $\mathbf{A}_{\Psi(x)}$, and secondly, the initial state vector $\mathbf{d}_0$ should be chosen such that $\Delta_{\mathbf{T},\mathbf{d}_0}$ in (5.10) is nonsingular.

EXAMPLE 5.9 We design a BSRG for the sequence space $V[x^4 + x^3 + x^2 + 1]$. We choose the length $L = 4$, which is identical to the dimension of $V[x^4 + x^3 + x^2 + 1]$, and the state transition matrix

$$\mathbf{T} = \begin{bmatrix} 0 & 1 & 0 & 0 \\ 0 & 0 & 1 & 0 \\ 1 & 0 & 1 & 1 \\ 1 & 1 & 1 & 0 \end{bmatrix}, \tag{5.12}$$

which is similar to the companion matrix $\mathbf{A}_{x^4+x^3+x^2+1}$ in (4.8). Then, the resulting BSRG is as shown in Fig. 5.3. For this BSRG, if we choose the initial state vector $\mathbf{d}_0 = [\,0\,1\,0\,0\,]^t$, then by (5.10) we obtain the discrimination matrix

$$\Delta_{\mathbf{T},\mathbf{d}_0} = \begin{bmatrix} 0 & 1 & 0 & 0 \\ 1 & 0 & 0 & 0 \\ 0 & 0 & 0 & 1 \\ 0 & 1 & 1 & 0 \end{bmatrix}, \tag{5.13}$$

which is nonsingular. Therefore, due to Theorem 5.21, this $\mathbf{d}_0$ becomes a maximal initial state vector for the BSRG in Fig. 5.3. ♣

THEOREM 5.22 *For a state transition matrix* $\mathbf{T}$ *and a maximal initial state vector* $\mathbf{d}_0$ *of a BSRG for the sequence space* $V[\Psi(x)]$, *the following*

---

[5]Note that in the case of BSRG, $\hat{L}$ in (5.10), which is the dimension of the BSRG maximal space, is identical to the length of the BSRG due to Theorem 5.19. Therefore, the discrimination matrix $\Delta_{\mathbf{T},\mathbf{d}_0}$ becomes a square matrix.

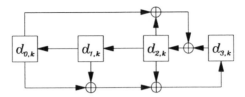

Figure 5.3. An example of BSRG for the sequence space $V[x^4 + x^3 + x^2 + 1]$.

relations hold for the relevant nonsingular discrimination matrix $\Delta_{\mathbf{T},\mathbf{d}_0}$

$$\mathbf{T} \;=\; \Delta_{\mathbf{T},\mathbf{d}_0} \cdot \mathbf{A}_{\Psi(x)} \; \Delta_{\mathbf{T},\mathbf{d}_0}^{-1}, \tag{5.14a}$$

$$\mathbf{d}_0 \;=\; \Delta_{\mathbf{T},\mathbf{d}_0} \cdot \mathbf{e}_0, \tag{5.14b}$$

$$\mathbf{D} \;=\; \Delta_{\mathbf{T},\mathbf{d}_0} \cdot \mathbf{E}_{\Psi(x)}. \tag{5.14c}$$

Proof : To prove (5.14a), we let $\Psi(x) = \sum_{i=0}^{L} \psi_i x^i$ and consider the matrix $\Delta_{\mathbf{T},\mathbf{d}_0} \cdot \mathbf{A}_{\Psi(x)}$. Then, by (5.10) and (4.7), $\Delta_{\mathbf{T},\mathbf{d}_0} \cdot \mathbf{A}_{\Psi(x)} = [\,\mathbf{T} \cdot \mathbf{d}_0 \;\; \mathbf{T}^2 \cdot \mathbf{d}_0 \;\; \cdots \;\; \mathbf{T}^{L-1} \cdot \mathbf{d}_0 \;\; \sum_{i=0}^{L-1} \psi_i \mathbf{T}^i \cdot \mathbf{d}_0\,]$ (note that $\hat{L}$ in (5.10) becomes L in this case). Since $\mathbf{T}$ is similar to $\mathbf{A}_{\Psi(x)}$ due to Theorem 5.20, $\sum_{i=0}^{L} \psi_i \mathbf{T}^i = 0$, that is, $\sum_{i=0}^{L-1} \psi_i \mathbf{T}^i = \mathbf{T}^L$. Thus, $\Delta_{\mathbf{T},\mathbf{d}_0} \cdot \mathbf{A}_{\Psi(x)} = [\,\mathbf{T} \cdot \mathbf{d}_0 \;\; \mathbf{T}^2 \cdot \mathbf{d}_0 \;\; \cdots \;\; \mathbf{T}^{L-1} \cdot \mathbf{d}_0 \;\; \mathbf{T}^L \cdot \mathbf{d}_0\,] = \mathbf{T} \cdot \Delta_{\mathbf{T},\mathbf{d}_0}$. Therefore, we have the equation (5.14a).

Equation (5.14b) is a direct outcome of (5.10), and equation (5.14c) is obtained by inserting (5.10) and $\Psi_{\mathbf{T},\mathbf{d}_0}(x) = \Psi(x)$ into (5.7). ∎

The theorem describes how the state transition matrix $\mathbf{T}$, the maximal initial state vector $\mathbf{d}_0$ and the SRG sequence vector $\mathbf{D}$ of a BSRG are related with the discrimination matrix $\Delta_{\mathbf{T},\mathbf{d}_0}$. While it is useful to know the relations in (5.14), they are not directly applicable in designing SRGs that can generate sequences in a prespecified sequence space due to the recursive relations tying $\mathbf{T}$, $\mathbf{d}_0$ and $\Delta_{\mathbf{T},\mathbf{d}_0}$. Therefore, in the following we consider a more general set of expressions that enables us to determine $\mathbf{T}$, $\mathbf{d}_0$ and $\mathbf{D}$ directly for the given sequence space $V[\Psi(x)]$.

THEOREM 5.23 *For a nonsingular matrix* $\mathbf{Q}$, *an SRG with the state transition matrix*

$$\mathbf{T} \;=\; \mathbf{Q} \cdot \mathbf{A}_{\Psi(x)} \cdot \mathbf{Q}^{-1} \tag{5.15a}$$

is a BSRG for the sequence space $V[\Psi(x)]$ *; the initial state vector of
the form*

$$\mathbf{d}_0 \;=\; \mathbf{Q} \cdot \mathbf{e}_0 \qquad\qquad (5.15b)$$

*is a maximal initial state vector for this BSRG ; and the BSRG sequence
vector* $\mathbf{D}$ *for this maximal initial state vector becomes*

$$\mathbf{D} \;=\; \mathbf{Q} \cdot \mathbf{E}_{\Psi(x)}. \qquad\qquad (5.15c)$$

Proof : The first part of the theorem is directly obtained by Theo-
rem 5.20. To prove the second part, we insert $\mathbf{T} = \mathbf{Q} \cdot \mathbf{A}_{\Psi(x)} \cdot \mathbf{Q}^{-1}$ and
$\mathbf{d}_0 = \mathbf{Q} \cdot \mathbf{e}_0$ into (5.10). Then, we obtain $\Delta_{\mathbf{T},\mathbf{d}_0} = [\, \mathbf{Q} \cdot \mathbf{e}_0 \;\; \mathbf{Q} \cdot \mathbf{A}_{\Psi(x)} \cdot \mathbf{e}_0 \;\; \cdots$
$\mathbf{Q} \cdot \mathbf{A}_{\Psi(x)}^{L-1} \cdot \mathbf{e}_0 \,]$, which is identical to $[\, \mathbf{Q} \cdot \mathbf{e}_0 \;\;\; \mathbf{Q} \cdot \mathbf{e}_1 \;\;\; \cdots \;\;\; \mathbf{Q} \cdot \mathbf{e}_{L-1} \,]$
due to the relation $\mathbf{A}_{\Psi(x)}^{i} \cdot \mathbf{e}_0 = \mathbf{e}_i$, $i = 0, 1, \cdots, L - 1$. That is, $\Delta_{\mathbf{T},\mathbf{d}_0}$
$= \mathbf{Q}$, which is nonsingular by assumption. Thus by Theorem 5.21, $\mathbf{d}_0$
$= \mathbf{Q} \cdot \mathbf{e}_0$ is a maximal initial state vector. Finally, by Theorem 5.10 and
(5.10), $\mathbf{D} = \mathbf{Q} \cdot \mathbf{E}_{\Psi(x)}$. ∎

The theorem explains how to determine BSRGs for a sequence space
$V[\Psi(x)]$. That is, the SRG with the state transition matrix $\mathbf{T} = \mathbf{Q} \cdot$
$\mathbf{A}_{\Psi(x)} \cdot \mathbf{Q}^{-1}$ for an arbitrary nonsingular matrix $\mathbf{Q}$ is a BSRG for the
sequence space $V[\Psi(x)]$, and the initial state vector $\mathbf{d}_0 = \mathbf{Q} \cdot \mathbf{e}_0$ is a
maximal initial state vector for this BSRG. It should be noted that $\mathbf{Q}$
itself now becomes the discrimination matrix $\Delta_{\mathbf{T},\mathbf{d}_0}$ for those $\mathbf{T}$ and $\mathbf{d}_0$,
and therefore no separate singularity test is necessary on it. In fact this
is reflected on equation (5.15c).

Noting that $\mathbf{Q}$ is nonsingular in the BSRG sequence vector expression in (5.15c), we find that the BSRG sequences D_i's form a basis of
the sequence space $V[\Psi(x)]$. Therefore, applying Theorem 5.23, we can
determine the state transition matrix $\mathbf{T}$ and the initial state vector $\mathbf{d}_0$
of the BSRG which generates any desired basis of the sequence space
$V[\Psi(x)]$. More specifically, in order to generate a particular set of sequences B_i, $i = 0, 1, \cdots, L - 1$, which form a basis of $V[\Psi(x)]$, we
may form the nonsingular matrix $\mathbf{Q}$ such that its ith row is identical to
the initial state vector of the ith basis sequence B_i. By inserting this
matrix $\mathbf{Q}$ into (5.15a) and (5.15b) respectively, we can obtain the state
transition matrix $\mathbf{T}$ and the initial state vector $\mathbf{d}_0$ of the BSRG whose
ith BSRG sequence D_i, $i = 0, 1, \cdots, L - 1$, becomes the basis sequence
B_i.

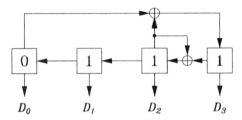

Figure 5.4. Another example of BSRG for the sequence space $V[x^4 + x^3 + x^2 + 1]$.

EXAMPLE 5.10 We determine the BSRG for the sequence space $V[x^4 + x^3 + x^2 + 1]$ that generates the four sequences $B_0 = [0\,1\,1\,0] \cdot \mathbf{E}_{x^4+x^3+x^2+1}$, $B_1 = [1\,1\,0\,1] \cdot \mathbf{E}_{x^4+x^3+x^2+1}$, $B_2 = [1\,0\,1\,0] \cdot \mathbf{E}_{x^4+x^3+x^2+1}$, and $B_3 = [1\,1\,1\,0] \cdot \mathbf{E}_{x^4+x^3+x^2+1}$, which form a basis of the sequence space $V[x^4 + x^3 + x^2 + 1]$. We first make the nonsingular matrix $\mathbf{Q}$ whose rows are identical to the initial state vectors of the basis sequences B_i's, that is,

$$\mathbf{Q} = \begin{bmatrix} 0 & 1 & 1 & 0 \\ 1 & 1 & 0 & 1 \\ 1 & 0 & 1 & 0 \\ 1 & 1 & 1 & 0 \end{bmatrix}. \tag{5.16}$$

Then, inserting this into (5.15a) and (5.15b) respectively, we obtain the following state transition matrix and initial state vector :

$$\mathbf{T} = \begin{bmatrix} 0 & 1 & 0 & 0 \\ 0 & 0 & 1 & 0 \\ 0 & 0 & 1 & 1 \\ 1 & 0 & 1 & 0 \end{bmatrix}, \tag{5.17a}$$

$$\mathbf{d}_0 = [\,0\,1\,1\,1\,]^t. \tag{5.17b}$$

The resulting circuit diagram for this BSRG is as shown in Fig. 5.4. In the figure, the numbers inside the shift registers indicate the initial state vector $\mathbf{d}_0$, and the BSRG sequences D_i, $i = 0, 1, 2, 3$, are identical to the basis sequences B_i's respectively. ♣

5.6   Simple and Modular SRGs

We now analyze the behavior of the two popular SRGs –– the simple SRG(SSRG) and the modular SRG(MSRG).

DEFINITION 5.24 (Simple SRG) *Simple SRG* (SSRG) is the SRG whose state transition matrix is

$$\mathbf{T}_S = \mathbf{A}^t_{\Psi(x)} \tag{5.18}$$

for a sequence space $V[\Psi(x)]$.[6]

DEFINITION 5.25 (Modular SRG) *Modular SRG* (MSRG) is the SRG whose state transition matrix is

$$\mathbf{T}_M = \mathbf{A}_{\Psi(x)} \tag{5.19}$$

for a sequence space $V[\Psi(x)]$.[7]

The configurations of the SSRG and MSRG for the sequence space $V[\Psi(x)] = V[\sum_{i=0}^{L} \psi_i x^i]$ are as shown in Fig. 5.5.[8] If viewed from the concept of BSRG, these two SRGs turn out to be two special types of BSRGs as the following theorem states :

THEOREM 5.26 *The SSRG and MSRG for the sequence space $V[\Psi(x)]$ are BSRGs for the sequence space $V[\Psi(x)]$.*

Proof : The state transition matrices $\mathbf{T}_S$ in (5.18) is similar to the companion matrix $\mathbf{A}_{\Psi(x)}$ since $\mathbf{A}^t_{\Psi(x)} = \mathbf{B}_{\Psi(x)} \cdot \mathbf{A}_{\Psi(x)} \cdot \mathbf{C}_{\Psi(x)}$ for the transform matrices $\mathbf{B}_{\Psi(x)}$ in (4.12a) and $\mathbf{C}_{\Psi(x)} = \mathbf{B}^{-1}_{\Psi(x)}$ in (4.12b). The state transition matrix $\mathbf{T}_M$ in (5.19) is obviously similar to the companion matrix $\mathbf{A}_{\Psi(x)}$. Therefore, we have the theorem. ∎

The theorem means that the SSRG and the MSRG are two special types of BSRGs whose state transition matrices are respectively the transposed companion matrix $\mathbf{A}^t_{\Psi(x)}$ and the companion matrix

---

[6]There are many different ways of defining the SSRG. Conventionally, the SSRG has been characterized by the "characteristic" polynomial $C(x)$ (or the "reset" polynomial), while in this book it is defined on the sequence space $V[\Psi(x)]$. The polynomials $\Psi(x)$ and $C(x)$ are related by the reciprocal relation $\Psi(x) = x^L C(x^{-1})$.

[7]There are various ways of defining the MSRG. Conventionally, the MSRG has been characterized by the "generating" polynomial $G(x)$. The relation between the polynomials $\Psi(x)$ and $G(x)$ is $\Psi(x) = G(x)$.

[8]Note that the corresponding "characteristic" and "generating" polynomials of the two configurations in the figure are respectively $C(x) = x^L \Psi(x^{-1}) = \sum_{i=0}^{L} \psi_{L-i} x^i$ and $G(x) = \Psi(x) = \sum_{i=0}^{L} \psi_i x^i$.

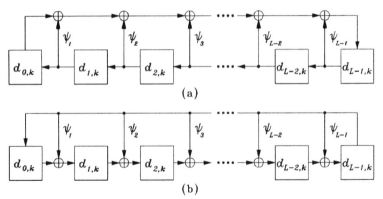

Figure 5.5. Configurations of (a) the SSRG and (b) the MSRG for the sequence space $V[\sum_{i=0}^{L} \psi_i x^i]$.

$\mathbf{A}_{\Psi(x)}$. As the companion matrix $\mathbf{A}_{\Psi(x)}$ in (4.7) is uniquely determined for a characteristic polynomial $\Psi(x)$, so are the SSRG and the MSRG uniquely determined for a sequence space $V[\Psi(x)]$. In fact, the SSRG and the MSRG are the two *simplest types of BSRGs* for the sequence space $V[\Psi(x)]$ as can be observed from Fig. 5.5. Hence the SSRG and MSRG sequences can be represented on the elementary and primary bases, as the following two theorems describe :

THEOREM 5.27 *For the SSRG of the sequence space $V[\Psi(x)]$ of dimension L, the ith SSRG sequence $D_{S,i}$, $i = 0, 1, \cdots, L-1$, for an initial state vector $\mathbf{d}_0$, has the expression*

$$D_{S,i} = \mathbf{d}_0^t \cdot \mathbf{A}_{\Psi(x)}^i \cdot \mathbf{E}_{\Psi(x)} \tag{5.20}$$

on the elementary basis of the sequence space $V[\Psi(x)]$.

Proof : Since $\Psi(\mathbf{A}_{\Psi(x)}^t) = 0$, we have $\Psi(\mathbf{A}_{\Psi(x)}^t) \cdot \mathbf{d}_0 = 0$ for an arbitrary initial state vector $\mathbf{d}_0$. Therefore, by Theorem 5.8, the sequence space $V[\Psi(x)]$ is a common space for the SSRG sequences $D_{S,i}$, $i = 0$, $1, \cdots, L-1$, that is, $D_{S,i} \in V[\Psi(x)]$, $i = 0, 1, \cdots, L-1$. Let $\mathbf{s}_i$, $i = 0, 1, \cdots, L-1$, be the initial vector of the ith SSRG sequence $D_{S,i}$. Then, by (5.1), (5.2) and (5.4) along with $\mathbf{T} = \mathbf{A}_{\Psi(x)}^t$, we obtain $\mathbf{s}_i^t = \mathbf{e}_i^t \cdot [\, \mathbf{d}_0 \quad \mathbf{A}_{\Psi(x)}^t \cdot \mathbf{d}_0 \quad \cdots \quad (\mathbf{A}_{\Psi(x)}^t)^{L-1} \cdot \mathbf{d}_0 \,]$. But $[\, \mathbf{d}_0 \quad \mathbf{A}_{\Psi(x)}^t \cdot \mathbf{d}_0 \quad \cdots$ $(\mathbf{A}_{\Psi(x)}^t)^{L-1} \cdot \mathbf{d}_0 \,] = [\, \mathbf{d}_0 \quad \mathbf{A}_{\Psi(x)}^t \cdot \mathbf{d}_0 \quad \cdots \quad (\mathbf{A}_{\Psi(x)}^t)^{L-1} \cdot \mathbf{d}_0 \,]^{t}$.[9] Therefore,

---

[9]For an arbitrary L-vector $\mathbf{v}$, $[\, \mathbf{v} \quad \mathbf{A}_{\Psi(x)}^t \cdot \mathbf{v} \quad \cdots \quad (\mathbf{A}_{\Psi(x)}^t)^{L-1} \cdot \mathbf{v} \,] = [\, \mathbf{v} \quad \mathbf{A}_{\Psi(x)}^t \cdot$ $\mathbf{v} \quad \cdots \quad (\mathbf{A}_{\Psi(x)}^t)^{L-1} \cdot \mathbf{v} \,]^{t}$. Refer to Property B.15 in Appendix B.

we get $\mathbf{s}_i^t = \mathbf{d}_0^t \cdot \mathbf{A}_{\Psi(x)}^i$, and the proof is complete due to (4.5). ∎

THEOREM 5.28 *For the MSRG of the sequence space $V[\Psi(x)]$ of dimension L, the ith MSRG sequence $D_{M,i}$, $i = 0, 1, \cdots, L\text{-}1$, for an initial state vector $\mathbf{d}_0$, has the expression*

$$D_{M,i} = \mathbf{d}_0^t \cdot \left(\sum_{j=i+1}^{L} \psi_j (\mathbf{A}_{\Psi(x)}^t)^{j-(i+1)} \right) \cdot \mathbf{P}_{\Psi(x)} \qquad (5.21)$$

on the primary basis of the sequence space $V[\Psi(x)]$.

Proof : We prove the theorem in a similar manner. Since $\Psi(\mathbf{A}_{\Psi(x)}) = 0$, we have $\Psi(\mathbf{A}_{\Psi(x)}) \cdot \mathbf{d}_0 = 0$ for an arbitrary initial state vector $\mathbf{d}_0$. Therefore, by Theorem 5.8, the sequence space $V[\Psi(x)]$ is a common space for the MSRG sequences $D_{M,i}$, $i = 0, 1, \cdots, L-1$, that is, $D_{M,i} \in V[\Psi(x)]$, $i = 0, 1, \cdots, L-1$. Let $\mathbf{s}_i$ and $\mathbf{p}_i$, $i = 0, 1, \cdots, L-1$, be the initial vectors of the ith MSRG sequence $D_{M,i}$ respectively on the elementary and primary bases. Then, by (5.1), (5.2) and (5.4) along with $\mathbf{T} = \mathbf{A}_{\Psi(x)}$, we obtain $\mathbf{s}_i^t = \mathbf{e}_i^t \cdot [\mathbf{d}_0 \ \mathbf{A}_{\Psi(x)} \cdot \mathbf{d}_0 \ \cdots \ \mathbf{A}_{\Psi(x)}^{L-1} \cdot \mathbf{d}_0]$, and by (4.14) we get $\mathbf{p}_i^t = \mathbf{e}_i^t \cdot [\mathbf{d}_0 \ \mathbf{A}_{\Psi(x)} \cdot \mathbf{d}_0 \ \cdots \ \mathbf{A}_{\Psi(x)}^{L-1} \cdot \mathbf{d}_0] \cdot \mathbf{C}_{\Psi(x)}$. But $[\mathbf{d}_0 \ \mathbf{A}_{\Psi(x)} \cdot \mathbf{d}_0 \ \cdots \ \mathbf{A}_{\Psi(x)}^{L-1} \cdot \mathbf{d}_0] \cdot \mathbf{C}_{\Psi(x)} = \mathbf{C}_{\Psi(x)} \cdot [\mathbf{d}_0 \ \mathbf{A}_{\Psi(x)} \cdot \mathbf{d}_0 \ \cdots \ \mathbf{A}_{\Psi(x)}^{L-1} \cdot \mathbf{d}_0]^t$,[10] and hence $\mathbf{p}_i^t = \mathbf{e}_i^t \cdot \mathbf{C}_{\Psi(x)} \cdot [\mathbf{d}_0 \ \mathbf{A}_{\Psi(x)} \cdot \mathbf{d}_0 \ \cdots \ \mathbf{A}_{\Psi(x)}^{L-1} \cdot \mathbf{d}_0]^t$. Therefore $\mathbf{p}_i^t = \mathbf{d}_0^t \cdot \sum_{j=i+1}^{L} \psi_j (\mathbf{A}_{\Psi(x)}^t)^{j-(i+1)}$ by (4.12b), and the proof is complete. ∎

Inserting $i = 0$ and $i = L-1$ respectively to (5.20) and (5.21), we obtain the relations

$$D_{S,0} = \mathbf{d}_0^t \cdot \mathbf{E}_{\Psi(x)}, \qquad (5.22a)$$
$$D_{M,L-1} = \mathbf{d}_0^t \cdot \mathbf{P}_{\Psi(x)}. \qquad (5.22b)$$

These are called the *terminating sequences* $D_{S,0}$ and $D_{M,L-1}$ respectively of the SSRG and the MSRG. For an arbitrary sequence $\{ s_k \}$ in the sequence space $V[\Psi(x)]$ we can make the terminating sequences $D_{S,0}$ and $D_{M,L-1}$ identical to the sequence $\{ s_k \}$. That is, the terminating sequence $D_{S,0}$ of the SSRG becomes the sequence $\{ s_k \}$ for the initial

---

[10] For an arbitrary L-vector $\mathbf{v}$, $[\mathbf{v} \ \mathbf{A}_{\Psi(x)} \cdot \mathbf{v} \ \cdots \ \mathbf{A}_{\Psi(x)}^{L-1} \cdot \mathbf{v}] \cdot \mathbf{C}_{\Psi(x)} = \mathbf{C}_{\Psi(x)} \cdot [\mathbf{v} \ \mathbf{A}_{\Psi(x)} \cdot \mathbf{v} \ \cdots \ \mathbf{A}_{\Psi(x)}^{L-1} \cdot \mathbf{v}]^t$. Refer to Property B.15 in Appendix B.

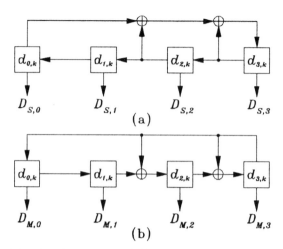

Figure 5.6. (a) The SSRG and (b) the MSRG of the sequence space
$V[x^4 + x^3 + x^2 + 1]$.

state vector $\mathbf{d}_0$ chosen identical to the initial vector $\mathbf{s}$ for the elementary
basis. Similarly, the terminating sequence $D_{M,L-1}$ of the MSRG be-
comes the sequence $\{ s_k \}$ for the initial state vector $\mathbf{d}_0$ chosen identical
to the initial vector $\mathbf{p}$ for the primary basis.

EXAMPLE 5.11 For the SSRG and the MSRG of the sequence space
$V[x^4 + x^3 + x^2 + 1]$, whose configurations are shown in Fig. 5.6, we make
the terminating sequences $D_{S,0}$ and $D_{M,3}$ identical to the sequence S_5
in Table 4.1. Then, the initial vectors of the sequence S_5 for the elementary and primary bases are respectively $\mathbf{s} = [\,0\,1\,0\,1\,]^t$ and $\mathbf{p} =
[\,0\,1\,1\,0\,]^t$. Therefore, for the SSRG in Fig. 5.6(a) the terminating se-
quence $D_{S,0}$ becomes the sequence S_5 if we choose the initial state vector
$\mathbf{d}_0 = [\,0\,1\,0\,1\,]^t$, and similarly for the MSRG in Fig. 5.6(b) the terminating sequence $D_{M,3}$ becomes the sequence S_5 if we choose the initial
state vector $\mathbf{d}_0 = [\,0\,1\,1\,0\,]^t$. ♣

In order to consider the maximalness of initial state vectors for the
SSRG and the MSRG as well as their relevant minimal spaces, we consider the relations between the SRG spaces and the terminating sequences.

THEOREM 5.29 *For the SSRG of the sequence space $V[\Psi(x)]$, the SRG
space $V[\mathbf{A}_{\Psi(x)}^t, \mathbf{d}_0]$ for an initial state vector $\mathbf{d}_0$ is identical to the minimal space for the terminating sequence $D_{S,0}$ in (5.22a).*

Proof : For an initial state vector $\mathbf{d}_0$, let $V[\Psi_m(x)]$ be the minimal space for the terminating sequence $D_{S,0}$. By Theorem 5.12, the SRG space $V[\mathbf{A}^t_{\Psi(x)}, \mathbf{d}_0]$ is the minimal space for the L SSRG sequences $D_{S,i}$, $i = 0, 1, \cdots, L-1$, and hence $V[\mathbf{A}^t_{\Psi(x)}, \mathbf{d}_0]$ is a common space for the terminating sequence $D_{S,0}$. By Theorems 4.25 and 4.33, $V[\Psi_m(x)]$ is therefore a subspace of $V[\mathbf{A}^t_{\Psi(x)}, \mathbf{d}_0]$. To complete the proof, we now prove that $V[\mathbf{A}^t_{\Psi(x)}, \mathbf{d}_0]$ is a subspace of $V[\Psi_m(x)]$. Since $D_{S,i}$, $i = 0, 1, \cdots, L-1$, is the i-delayed sequence of $D_{S,0}$ due to (5.20) and Theorem 4.9, $V[\Psi_m(x)]$ is a common space for the L SSRG sequences $D_{S,i}$, $i = 0, 1, \cdots, L-1$. Therefore, by Theorems 4.25 and 4.33, $V[\mathbf{A}^t_{\Psi(x)}, \mathbf{d}_0]$ is a subspace of $V[\Psi_m(x)]$, and the proof is complete. ∎

THEOREM 5.30 *For the MSRG of the sequence space $V[\Psi(x)]$, the SRG space $V[\mathbf{A}_{\Psi(x)}, \mathbf{d}_0]$ for an initial state vector $\mathbf{d}_0$ is identical to the minimal space for the terminating sequence $D_{M,L-1}$ in (5.22b).*

Proof : This theorem can be proved in a manner similar to the proof of Theorem 5.29. ∎

The theorems indicate that the SSRG and MSRG spaces respectively correspond to the minimal spaces for their terminating sequences. Therefore, in regard to the maximalness of initial state vectors for the SSRG and the MSRG, we have the following two theorems.

THEOREM 5.31 *For the SSRG of the sequence space $V[\Psi(x)]$ of dimension L, an initial state vector $\mathbf{d}_0$ is a maximal initial state vector, if and only if the minimal space for the terminating sequence $D_{S,0}$ in (5.22a) is identical to the sequence space $V[\Psi(x)]$.*

THEOREM 5.32 *For the MSRG of the sequence space $V[\Psi(x)]$ of dimension L, an initial state vector $\mathbf{d}_0$ is a maximal initial state vector, if and only if the minimal space for the terminating sequence $D_{M,L-1}$ in (5.22b) is identical to the sequence space $V[\Psi(x)]$.*

Theorems 5.27, 5.28, 5.31 and 5.32 can be combined to provide the following statements : For the SSRG of the sequence space $V[\Psi(x)]$, if we choose an initial state vector $\mathbf{d}_0$ such that the minimal space for the sequence $\mathbf{d}_0^t \cdot \mathbf{E}_{\Psi(x)}$ is $V[\Psi(x)]$, then the initial state vector $\mathbf{d}_0$ is a maximal initial state vector, and the terminating sequence $D_{S,0}$ becomes the sequence $\mathbf{d}_0^t \cdot \mathbf{E}_{\Psi(x)}$. Likewise, for the MSRG of the sequence space

$V[\Psi(x)]$, if we choose an initial state vector $\mathbf{d}_0$ such that the minimal space for the sequence $\mathbf{d}_0^t \cdot \mathbf{P}_{\Psi(x)}$ is $V[\Psi(x)]$, then the initial state vector $\mathbf{d}_0$ is a maximal initial state vector, and the terminating sequence $D_{M,L-1}$ becomes the sequence $\mathbf{d}_0^t \cdot \mathbf{P}_{\Psi(x)}$. Therefore the choice of initial state vectors is very influential in sequence generation.

EXAMPLE 5.12 In the case of the sequence space $V[x^4 + x^3 + x^2 + 1]$, the minimal space for the sequence S_4 in Table 4.1 is $V[x^4 + x^3 + x^2 + 1]$, and its initial vectors for the elementary and primary bases are respectively $\mathbf{s} = [\,0\,1\,0\,0\,]^t$ and $\mathbf{p} = [\,1\,1\,1\,0\,]^t$ (see Example 4.14). Therefore, the initial state vector $\mathbf{d}_0 = [\,0\,1\,0\,0\,]^t$ is a maximal initial state vector for the SSRG of the sequence space $V[x^4 + x^3 + x^2 + 1]$ in Fig. 5.6(a), and the initial state vector $\mathbf{d}_0 = [\,1\,1\,1\,0\,]^t$ is a maximal initial state vector for the MSRG of the sequence space $V[x^4 + x^3 + x^2 + 1]$ in Fig. 5.6(b). ♣

The following two theorems show that the SSRG and MSRG can respectively generate the primary and elementary bases for some appropriate initial state vectors.

THEOREM 5.33 *For the SSRG of the sequence space $V[\Psi(x)]$ of dimension L, if the initial state vector is $\mathbf{d}_0 = \mathbf{e}_{L-1}$, then $D_{S,i} = P^i_{\Psi(x)}$, $i = 0, 1, \cdots, L\text{-}1$.*

Proof : Inserting $\mathbf{d}_0 = \mathbf{e}_{L-1}$ into (5.20), we have $D_{S,i} = \mathbf{e}_{L-1}^t \cdot \mathbf{A}^i_{\Psi(x)} \cdot \mathbf{E}_{\Psi(x)}$, and since $\mathbf{e}_{L-1}^t = \mathbf{e}_0^t \cdot \mathbf{A}_{\Psi(x)}$ by (4.7), we obtain $D_{S,i} = \mathbf{e}_0^t \cdot \mathbf{A}^{i+1}_{\Psi(x)} \cdot \mathbf{E}_{\Psi(x)}$. Employing (4.12a), we can rewrite it as $D_{S,i} = \mathbf{e}_i^t \cdot \mathbf{B}_{\Psi(x)} \cdot \mathbf{E}_{\Psi(x)}$. Therefore, by (4.11a) $D_{S,i} = \mathbf{e}_i^t \cdot \mathbf{P}_{\Psi(x)} = P^i_{\Psi(x)}$. ∎

THEOREM 5.34 *For the MSRG of the sequence space $V[\Psi(x)]$ of dimension L, if the initial state vector is $\mathbf{d}_0 = \mathbf{e}_0$, then $D_{M,i} = E^i_{\Psi(x)}$, $i = 0, 1, \cdots, L\text{-}1$.*

Proof : Inserting $\mathbf{d}_0 = \mathbf{e}_0$ into (5.21), we have $D_{M,i} = \mathbf{e}_0^t \cdot (\sum_{j=i+1}^{L} \psi_j (\mathbf{A}^t_{\Psi(x)})^{j-(i+1)}) \cdot \mathbf{P}_{\Psi(x)}$, and since $\mathbf{e}_0^t \cdot (\mathbf{A}^t_{\Psi(x)})^j = \mathbf{e}_j^t$, $j = 0, 1, \cdots, L-1$,[11] we have $D_{M,i} = (\sum_{j=i+1}^{L} \psi_j \mathbf{e}_{j-(i+1)}^t) \cdot \mathbf{P}_{\Psi(x)}$. Employing (4.12b), we can rewrite it as $D_{M,i} = \mathbf{e}_i^t \cdot \mathbf{C}_{\Psi(x)} \cdot \mathbf{P}_{\Psi(x)}$. Therefore, by (4.11b) $D_{M,i} = \mathbf{e}_i^t \cdot \mathbf{E}_{\Psi(x)} = E^i_{\Psi(x)}$. ∎

---

[11] For the companion matrix $\mathbf{A}_{\Psi(x)}$ for a degree-L characteristic polynomial $\Psi(x)$, $\mathbf{e}_0^t \cdot (\mathbf{A}^t_{\Psi(x)})^j = \mathbf{e}_j^t$, $j = 0, 1, \cdots, L-1$. Refer to Property B.8 in Appendix B.

Theorems 5.33 and 5.34 indicate that the SSRG sequences for the initial state vector $\mathbf{d}_0 = \mathbf{e}_{L-1}$ are identical to the primary sequences, and that the MSRG sequences for an initial state vector $\mathbf{d}_0 = \mathbf{e}_0$ are identical to the elementary sequences.

EXAMPLE 5.13 For the SSRG of the sequence space $V[x^4 + x^3 + x^2 + 1]$ shown in Fig. 5.6(a), if we choose the initial state vector $\mathbf{d}_0 = [0\,0\,0\,1]^t$, then $D_{S,0} = P^0_{x^4+x^3+x^2+1}$, $D_{S,1} = P^1_{x^4+x^3+x^2+1}$, $D_{S,2} = P^2_{x^4+x^3+x^2+1}$, $D_{S,3} = P^3_{x^4+x^3+x^2+1}$; and for the MSRG of the sequence space $V[x^4 + x^3 + x^2 + 1]$ shown in Fig. 5.6(b), if we choose the initial state vector $\mathbf{d}_0 = [1\,0\,0\,0]^t$, then $D_{M,0} = E^0_{x^4+x^3+x^2+1}$, $D_{M,1} = E^1_{x^4+x^3+x^2+1}$, $D_{M,2} = E^2_{x^4+x^3+x^2+1}$, $D_{M,3} = E^3_{x^4+x^3+x^2+1}$. ♣

Chapter 6

Serial Frame Synchronous Scrambling

In the *frame synchronous scrambling* (FSS), the input data bitstream is scrambled by adding a *pseudo-random binary sequence* (PRBS) to this, and the scrambled data bitstream is descrambled by adding the same PRBS to this. In this chapter, we examine the behaviors of PRBSs in view of the sequence space theory developed in Chapter 4, and consider how to realize SRGs that generate PRBSs based on the SRG theory developed in Chapter 5. We will first define PRBS systematically using the mathematical definitions of weight, run and autocorrelation of sequences. Then, we will consider a special class of sequence spaces, namely the primitive sequence spaces, whose characteristic polynomials are primitive, showing that the sequences in the primitive sequence spaces are PRBSs. Finally, we will consider how to realize SRGs that generate PRBSs in the primitive sequence spaces.

6.1 Pseudo-Random Binary Sequences

In order to systematically define pseudo-random binary sequence described in Chapter 3, we first define the weight, run and autocorrelation of sequences.

DEFINITION 6.1 (Weight) For a sequence $\{ s_k \}$ of period τ, the *weight* W refers to the number of 1's in one period.

DEFINITION 6.2 (Run) For a sequence $\{ s_k \}$ of period τ, a consecutive 1's or 0's is called a *run*, and the number of 1's or 0's in a run is called the *run length*. The *run function* $R(l)$, $l = 0, 1, \cdots$, is defined to be the number of the runs of run length l in one period, and the *total run* R indicates the total number of the runs in one period, that is, $R = \sum_l R(l)$.

DEFINITION 6.3 (Autocorrelation) For a sequence $\{ s_k \}$ of period τ, the *autocorrelation function* $\phi(m)$, $m = 0, 1, \cdots, \tau - 1$, is defined to be the weight of the sequence $\{ s_k + s_{k+m} \}$ obtained by summing the sequence $\{ s_k \}$ and its m-delayed sequence $\{ s_{k+m} \}$.[1]

EXAMPLE 6.1 We consider the sequence S_6 in Table 4.1, whose period is $\tau = 7$. Since $s_1 = s_2 = s_4 = 1$ for the sequence, we have the weight $W = 3$. In one period, there are four runs 11, 0, 1, and 000,[2] and so we have the run functions $R(1) = 2$, $R(2) = 1$, $R(3) = 1$ and the total run $R = 4$. For this sequence, $\{ s_k + s_k \} = \{ 0, 0, 0, 0, 0, 0, 0, \cdots \}$, $\{ s_k + s_{k+1} \} = \{ 1, 0, 1, 1, 1, 0, 0, \cdots \}$, $\{ s_k + s_{k+2} \} = \{ 1, 1, 0, 0, 1, 0, 1, \cdots \}$, $\{ s_k + s_{k+3} \} = \{ 0, 0, 1, 0, 1, 1, 1, \cdots \}$, $\{ s_k + s_{k+4} \} = \{ 1, 1, 1, 0, 0, 1, 0, \cdots \}$, $\{ s_k + s_{k+5} \} = \{ 0, 1, 1, 1, 0, 0, 1, \cdots \}$, $\{ s_k + s_{k+6} \} = \{ 0, 1, 0, 1, 1, 1, 0, \cdots \}$. Therefore, we obtain the autocorrelation functions $\phi(0) = 0$, $\phi(1) = \phi(2) = \phi(3) = \phi(4) = \phi(5) = \phi(6) = 4$. ♣

Using the above definitions, we can reorganize the definition of PRBS in Chapter 3 in mathematical expressions as follows :

DEFINITION 6.4 (PRBS) A periodic sequence $\{ s_k \}$ of period τ_p is a *pseudo-random binary sequence* (PRBS) if it meets the following three randomness characteristics :

R1. The weight is nearly a half of the period, i.e.,

$$|2W - \tau_p| \leq 1. \tag{6.1}$$

R2. Half the runs are of length one, one-fourth are of length two, one-eighth are of length three, etc., i.e.,

$$R(l) = R/2^l, \quad l = 1, 2, \cdots. \tag{6.2}$$

---

[1]Note that this differs from the conventional definition of autocorrelation in which the two sequences are multiplied, not summed, term by term. Nontheless, the two-value property holds with $\phi(0) = 0$, instead of $\phi(0) = W$.

[2]Note that $s_0 = 0$ is not a run since $s_5 = s_6 = 0$ and $s_7 (= s_0) = 0$. That is, $s_5 s_6 s_7 = 000$ becomes a run.

R3. The autocorrelation function $\phi(m)$ is two-valued, i.e.,

$$\phi(m) = \begin{cases} 0, & m = 0, \\ K, & m = 1, 2, \cdots, \tau_p - 1. \end{cases} \tag{6.3}$$

EXAMPLE 6.2 We observed in Example 6.1 that the sequence S_6 in Table 4.1 satisfies the above three randomness characteristics. Therefore, it is a PRBS. ♣

6.2   Primitive Sequence Spaces

We consider a special class of sequence spaces whose characteristic polynomials are primitive, and investigate their properties in relation to the randomness characteristics of PRBSs.

DEFINITION 6.5 (Primitive Polynomial) An irreducible polynomial $\Psi_p(x)$ of degree L is called a *primitive polynomial*, if $\Psi_p(x)$ divides $x^i + 1$ for $i = 2^L - 1$ but not for $i = 0, 1, \cdots, 2^L - 2$.[3]

DEFINITION 6.6 (Primitive Sequence Space) The *primitive sequence space* (in short, *primitive space*) $V[\Psi_p(x)]$ refers to the sequence space whose characteristic polynomial $\Psi_p(x)$ is a primitive polynomial.

EXAMPLE 6.3 The irreducible polynomial $x^3 + x + 1$ divides $x^i + 1$ for $i = 7$ but not for $i = 0, 1, \cdots, 6$, so it is primitive. Therefore, the sequence space $V[x^3 + x + 1]$ in Table 4.2 is a primitive space. ♣

We first consider the periodicity of a primitive space. The following theorem describes this.

THEOREM 6.7 *The periodicity T_p of a primitive space $V[\Psi_p(x)]$ of dimension L is*

$$T_p = 2^L - 1. \tag{6.4}$$

Proof : This theorem is directly obtained by the definitions of primitive polynomial and periodicity.   ∎

---

[3]The term *primitive* implies the fact that the primitive polynomial has a *primitive root* of the polynomial $x^{2^L-1} + 1$ over Galois field $GF(2^L)$. Refer to Section A.4 in Appendix A.

EXAMPLE 6.4 The primitive space $V[x^3 + x + 1]$ in Table 4.2 has the periodicity $T_p = 2^3 - 1 = 7$. ♣

For a sequence in a primitive space, its minimal space and period are respectively determined by the following two theorems :

THEOREM 6.8 *If a non-zero sequence { s_k } is an element of a primitive space $V[\Psi_p(x)]$, then its minimal space $V[\Psi_m(x)]$ is identical to the primitive space, or equivalently,*

$$V[\Psi_m(x)] = V[\Psi_p(x)]. \qquad (6.5)$$

Proof : Let a non-zero sequence { s_k } in the primitive space $V[\Psi_p(x)]$ has the polynomial expression $S[\Psi_p(x), P(x)]$ for a non-zero polynomial $P(x)$. Then, since the primitive polynomial $\Psi_p(x)$ is irreducible, $P(x)$ is relatively prime to $\Psi_p(x)$. Therefore, by Theorem 4.34 the minimal space for { s_k } is the primitive space $V[\Psi_p(x)]$. ■

THEOREM 6.9 *If a non-zero sequence { s_k } is an element of a primitive space $V[\Psi_p(x)]$ of dimension L, then its period τ_p is*

$$\tau_p = 2^L - 1. \qquad (6.6)$$

Proof : This theorem is a direct outcome of Theorem 4.37. ■

EXAMPLE 6.5 We consider a non-zero sequence { s_k } in the primitive space $V[x^3+x+1]$. Then, by Theorem 6.8 its minimal space is $V[\Psi_m(x)]$ $= V[x^3 + x + 1]$. We can confirm this by applying Theorem 4.35 to the sequences $\hat{S}_1$ through $\hat{S}_7$ in Table 4.2. ♣

EXAMPLE 6.6 Every non-zero sequence in the primitive space $V[x^3 + x + 1]$, has the period $\tau_p = 2^3 - 1 = 7$ due to Theorem 6.9. In fact, we observe from Table 4.2 that the sequences $\hat{S}_1$ through $\hat{S}_7$ in $V[x^3+x+1]$ have period 7. ♣

The following theorem shows that the delayed sequences of a sequence in a primitive space can form a primitive space.

THEOREM 6.10 *If a non-zero sequence { s_k } is an element of a primitive space $V[\Psi_p(x)]$ of dimension L, its delayed sequences { s_{k+m} }, m = 0, 1, $\cdots$, $2^L - 2$, and a zero sequence also form the primitive space $V[\Psi_p(x)]$.*

Proof : Let S be the set composed of a zero sequence and the delayed sequences $\{ s_{k+m} \}$, $m = 0, 1, \cdots, 2^L - 2$. Then, it suffices to show that $S = V[\Psi_p(x)]$. Since $\{ s_k \}$ is assumed to be in $V[\Psi_p(x)]$, $\{ s_{k+m} \}$ $\in V[\Psi_p(x)]$, $m = 0, 1, \cdots, 2^L - 2$ by Theorem 4.9. Therefore, $S \subset V[\Psi_p(x)]$. To complete the proof, it suffices to show that the number of elements in S is 2^L, which is the number of elements in $V[\Psi_p(x)]$. Since the period of $\{ s_k \}$ is $2^L - 1$ due to Theorem 6.9, the delayed sequences $\{ s_{k+m} \}$, $m = 0, 1, \cdots, 2^L - 2$, are all different. Therefore, the total number of elements in S becomes 2^L including the zero sequence. This completes the proof. ♣

EXAMPLE *6.7* We consider the sequence $\hat{S}_2$ in the primitive space $V[x^3 + x + 1]$ of Table 4.2. Then, the 0-, 1-, 2-, 3-, 4-, 5-, and 6-delayed sequences are respectively the sequences $\hat{S}_2$, $\hat{S}_5$, $\hat{S}_3$, $\hat{S}_7$, $\hat{S}_6$, $\hat{S}_4$, and $\hat{S}_1$ in Table 4.2. Therefore, these seven sequences and the zero sequence $\hat{S}_0$ form the primitive space $V[x^3 + x + 1]$. ♣

Now we consider the weight, run and autocorrelation for sequences in the primitive space. The following three theorems describe these.

THEOREM 6.11 *If a non-zero sequence $\{ s_k \}$ is an element of a primitive space $V[\Psi_p(x)]$ of dimension L, then its weight W is*

$$W = 2^{L-1}. \tag{6.7}$$

Proof : By Theorem 6.10, the weight W of a sequence $\{ s_k \}$ in the primitive space $V[\Psi_p(x)]$ of dimension L is identical to the number of the sequences in $V[\Psi_p(x)]$ with the 0th elements of 1. Therefore, $W = 2^{L-1}$, since half of the 2^L sequences in $V[\Psi_p(x)]$ have 1 as the 0th element. ■

THEOREM 6.12 *If a non-zero sequence $\{ s_k \}$ is an element of a primitive space $V[\Psi_p(x)]$ of dimension L, then its run function $R(l)$ and total run R are*

$$R = 2^{L-1}, \tag{6.8}$$

$$R(l) = \begin{cases} 0, & l = L+1, L+2, \cdots, \\ R/2^{L-1}, & l = L, \\ R/2^l, & l = L-1, L-2, \cdots, 1. \end{cases} \tag{6.9}$$

Proof : For a non-zero sequence $\{\, s_k\,\}$ in the primitive space $V[\Psi_p(x)]$ of dimension L, whose period is $2^L - 1$ due to Theorem 6.9, we consider the following $2^L - 1$ groups obtained by sliding the L-element window by one element each time : $s_0 s_1 \cdots s_{L-1}, s_1 s_2 \cdots s_L, \cdots,$ $s_{2^L-2} s_{2^L-1} \cdots s_{2^L+L-3}$. Then, by Theorem 6.10 these $2^L - 1$ groups correspond by one-to-one to the initial vectors of all the non-zero sequences in $V[\Psi_p(x)]$, that is, $00\cdots001, 00\cdots010, 00\cdots011, \cdots, 11\cdots111$.

The group of L consecutive ones should be preceded by the group of a zero and $L - 1$ consecutive ones and followed by the group of $L - 1$ consecutive ones and a zero, since otherwise there would appear another group of L ones, which contradicts the one-to-one property. Therefore, there exist no runs of $L + 1$ or more ones, and only one run of L ones. This also implies that there exists no run of $L-1$ ones, since it would be always either preceded or followed by the group of L ones. The number of runs of l ones for $l = 1, 2, \cdots, L - 2$, is the same as the number of the groups which consist of a zero, l consecutive ones, and a zero, followed by $L - l - 2$ mixture of ones and zeros. Therefore, by the one-to-one property there exist 2^{L-l-2} runs of l ones.

Now we consider the runs of zeros. It is evident that there is no run of L or more zeros, since the group of L consecutive zeros do not appear. The group of $L - 1$ consecutive zeros followed by a one should be preceded by the group consisting of a one and $L-1$ consecutive zeros, since otherwise the all-zero group would appear. Therefore, there is one run of $L - 1$ zeros. The number of runs of l zeros for $l = 1, 2, \cdots, L - 2$, is 2^{L-l-2} as for the ones case.

Combining the two cases, we obtain $R(l) = 0, l = L + 1, L + 2, \cdots,$ $R(L) = 1, R(L - 1) = 1$, and $R(l) = 2^{L-l-1}, l = 1, 2, \cdots, L - 2$, and therefore $R = \sum_l R(l) = 2^{L-1}$. This proves the theorem. ■

THEOREM 6.13 *If a non-zero sequence $\{\, s_k\,\}$ is an element of a primitive space $V[\Psi_p(x)]$ of dimension L, then its autocorrelation function $\phi(m)$ is*

$$\phi(m) = \begin{cases} 0, & m = 0, \\ 2^{L-1}, & m = 1, 2, \cdots, 2^L - 2. \end{cases} \tag{6.10}$$

Proof : If a non-zero sequence $\{\, s_k\,\}$ is an element of the primitive space $V[\Psi_p(x)]$ of dimension L, its m-delayed sequence $\{\, s_{k+m}\,\}$ is also an element of $V[\Psi_p(x)]$ due to Theorem 4.9, and hence the summed sequence $\{\, s_k + s_{k+m}\,\}$ is also an element of $V[\Psi_p(x)]$. When $m = 0$, the summed sequence obviously becomes a zero sequence, so $\phi(0) = 0$. For $m = 1, 2, \cdots, 2^L - 2$, the sequence $\{ s_k + s_{k+m} \}$ is a non-zero sequence, so, by Theorem 6.11, its weight is 2^{L-1}, that is, $\phi(m) = 2^{L-1}$.

∎

EXAMPLE 6.8 We consider a sequence $\{ s_k \}$ in the primitive space $V[x^3 + x + 1]$. Then, by Theorem 6.11 its weight is $W = 2^{3-1} = 4$; by Theorem 6.12 its total run and run functions are $R = 2^{3-1} = 4$ and $R(3) = 4/2^2 = 1$, $R(2) = 4/2^2 = 1$, $R(1) = 4/2^1 = 2$; and by Theorem 6.13 its autocorrelation function is $\phi(0) = 0$ and $\phi(m) = 2^{3-1} = 4$ for $m = 1, 2, \cdots, 6$. We can confirm these using the sequences $\hat{S}_1$ through $\hat{S}_7$ in Table 4.2. ♣

Using Theorems 6.9 and 6.11, we find that a non-zero sequence $\{ s_k \}$ in a primitive space satisfies the randomness characteristic R1 since $2W - \tau_p = 2^L - (2^L - 1) = 1$, and from Theorems 6.12 and 6.13 we observe that the sequence $\{ s_k \}$ also meets the randomness characteristics R2 and R3. Therefore, we can put them together in one statement as follows :

THEOREM 6.14 *If a non-zero sequence $\{ s_k \}$ is an element of a primitive space $V[\Psi_p(x)]$ of dimension L, it is a PRBS of period $2^L - 1$.*

The theorem implies that *all* the non-zero sequences in a primitive space are PRBSs.

EXAMPLE 6.9 The sequences $\hat{S}_1$ through $\hat{S}_7$ in Table 4.2 are non-zero sequences in the primitive space $V[x^3 + x + 1]$. Therefore, they are PRBSs. ♣

EXAMPLE 6.10 We consider the sequence space $V[x^4 + x + 1]$, whose elements are listed in Table 6.1. Since the characteristic polynomial $\Psi_p(x) = x^4 + x + 1$ is primitive, by Theorem 6.14, the non-zero sequences $\tilde{S}_1$ through $\tilde{S}_{15}$ in the table are all PRBSs of period $\tau_p = 15$. ♣

6.3 PRBS Generators

We now consider how to realize SRGs that generate PRBSs in a primitive space. We call such an SRG a *PRBS generator*. We will examine how to determine the length, the state transition matrix, and the initial state vector of PRBS generators for a given primitive space.

Table 6.1. Elements of the primitive space $V[x^4 + x + 1]$. All the non-zero sequences $\tilde{S}_1$ through $\tilde{S}_{15}$ in $V[x^4 + x + 1]$ are PRBSs of period 15

$\tilde{S}_0$	$= [0000] \cdot \mathbf{E}_{x^4+x+1}$	$= \{0,0,0,0,0,0,0,0,0,0,0,0,0,0,0,\cdots\}$
$\tilde{S}_1$	$= [0001] \cdot \mathbf{E}_{x^4+x+1}$	$= \{0,0,0,1,0,0,1,1,0,1,0,1,1,1,1,\cdots\}$
$\tilde{S}_2$	$= [0010] \cdot \mathbf{E}_{x^4+x+1}$	$= \{0,0,1,0,0,1,1,0,1,0,1,1,1,1,0,\cdots\}$
$\tilde{S}_3$	$= [0011] \cdot \mathbf{E}_{x^4+x+1}$	$= \{0,0,1,1,0,1,0,1,1,1,1,0,0,0,1,\cdots\}$
$\tilde{S}_4$	$= [0100] \cdot \mathbf{E}_{x^4+x+1}$	$= \{0,1,0,0,1,1,0,1,0,1,1,1,1,0,0,\cdots\}$
$\tilde{S}_5$	$= [0101] \cdot \mathbf{E}_{x^4+x+1}$	$= \{0,1,0,1,1,1,1,0,0,0,1,0,0,1,1,\cdots\}$
$\tilde{S}_6$	$= [0110] \cdot \mathbf{E}_{x^4+x+1}$	$= \{0,1,1,0,1,0,1,1,1,1,0,0,0,1,0,\cdots\}$
$\tilde{S}_7$	$= [0111] \cdot \mathbf{E}_{x^4+x+1}$	$= \{0,1,1,1,1,0,0,0,1,0,0,1,1,0,1,\cdots\}$
$\tilde{S}_8$	$= [1000] \cdot \mathbf{E}_{x^4+x+1}$	$= \{1,0,0,0,1,0,0,1,1,0,1,0,1,1,1,\cdots\}$
$\tilde{S}_9$	$= [1001] \cdot \mathbf{E}_{x^4+x+1}$	$= \{1,0,0,1,1,0,1,0,1,1,1,1,0,0,0,\cdots\}$
$\tilde{S}_{10}$	$= [1010] \cdot \mathbf{E}_{x^4+x+1}$	$= \{1,0,1,0,1,1,1,1,0,0,0,1,0,0,1,\cdots\}$
$\tilde{S}_{11}$	$= [1011] \cdot \mathbf{E}_{x^4+x+1}$	$= \{1,0,1,1,1,1,0,0,0,1,0,0,1,1,0,\cdots\}$
$\tilde{S}_{12}$	$= [1100] \cdot \mathbf{E}_{x^4+x+1}$	$= \{1,1,0,0,0,1,0,0,1,1,0,1,0,1,1,\cdots\}$
$\tilde{S}_{13}$	$= [1101] \cdot \mathbf{E}_{x^4+x+1}$	$= \{1,1,0,1,0,1,1,1,1,0,0,0,1,0,0,\cdots\}$
$\tilde{S}_{14}$	$= [1110] \cdot \mathbf{E}_{x^4+x+1}$	$= \{1,1,1,0,0,0,1,0,0,1,1,0,1,0,1,\cdots\}$
$\tilde{S}_{15}$	$= [1111] \cdot \mathbf{E}_{x^4+x+1}$	$= \{1,1,1,1,0,0,0,1,0,0,1,1,0,1,0,\cdots\}$

THEOREM 6.15 *A PRBS generator for the primitive space* $V[\Psi_p(x)]$ *of dimension* L *has length* L *or more.*

Proof : We show this by contradiction. Suppose that an SRG with the state transition matrix $\mathbf{T}$ and the length $\hat{L} < L$ does generate a PRBS in the primitive space $V[\Psi_p(x)]$. Then, the primitive space $V[\Psi_p(x)]$ is a subspace of the SRG space $V[\mathbf{T}]$, which is identical to the sequence space $V[\Psi_{\mathbf{T}}(x)]$ due to Theorem 5.14. Therefore, by Theorem 4.25 $\Psi_p(x)$ divides the minimal polynomial $\Psi_{\mathbf{T}}(x)$ of $\mathbf{T}$. However, this is a contradiction since the degree of $\Psi_p(x)$ is L and the degree of $\Psi_{\mathbf{T}}(x)$ is $\hat{L}$ or less. ∎

For an efficient realization of PRBS generators we concentrate on the PRBS generators of length L.[4] We call such a length-L PRBS generator

---

[4]For realizations of PRBS generators of length more than L, refer to Example 7.25 in Section 7.8.

a *PRBS minimal generator*. More formally,

DEFINITION 6.16 (PRBS Minimal Generator) A *PRBS minimal generator* refers to an SRG of length L that generates a PRBS in the primitive space $V[\Psi_p(x)]$ of dimension L.

The state transition matrix of a PRBS minimal generator is determined by the following theorem.

THEOREM 6.17 *The state transition matrix* $\mathbf{T}$ *of a PRBS minimal generator for the primitive space* $V[\Psi_p(x)]$ *is similar to the companion matrix* $\mathbf{A}_{\Psi_p(x)}$ *for the characteristic polynomial* $\Psi_p(x)$.

Proof : Let $\mathbf{T}$ be the state transition matrix of a PRBS minimal generator generating a PRBS in the primitive space $V[\Psi_p(x)]$ of dimension L. Then, $V[\Psi_p(x)]$ is a subspace of the SRG maximal space $V[\mathbf{T}]$. Since $\mathbf{T}$ is an $L \times L$ matrix, the dimension of $V[\mathbf{T}]$, which is identical to the sequence space $V[\Psi_{\mathbf{T}}]$ by Theorem 5.14, is L or less. Therefore, $V[\mathbf{T}] = V[\Psi_p(x)]$, and hence the PRBS minimal generator is a BSRG for the primitive space $V[\Psi_p(x)]$. Consequently, $\mathbf{T}$ is similar to $\mathbf{A}_{\Psi_p(x)}$ due to Theorem 5.20. ∎

In order to generate a PRBS in the primitive space $V[\Psi_p(x)]$, we are now supposed to take a PRBS minimal generator whose state transition matrix $\mathbf{T}$ is similar to the companion matrix $\mathbf{A}_{\Psi_p(x)}$. We then consider how to choose initial state vectors of such a PRBS minimal generator to generate a PRBS. In case the state transition matrix $\mathbf{T}$ of a PRBS minimal generator is similar to $\mathbf{A}_{\Psi_p(x)}$, its SRG maximal space $V[\mathbf{T}]$ becomes the primitive space $V[\Psi_p(x)]$ due to Theorem 5.14. Therefore, in order to generate a PRBS we should choose the initial state vector $\mathbf{d}_0$ of a PRBS minimal generator to be a maximal one. In the case of the PRBS minimal generator, the maximal initial state vectors turn out arbitrarily selectable as the following theorem describes.

THEOREM 6.18 *An arbitrary non-zero initial state vector* $\mathbf{d}_0$ *is a maximal initial state vector for a PRBS minimal generator.*

Proof : Let $\mathbf{d}_0$ be an arbitrary non-zero initial state vector of the PRBS minimal generator with the state transition matrix $\mathbf{T}$ chosen similar to $\mathbf{A}_{\Psi_p(x)}$. Then, by Theorem 5.17 the SRG space $V[\mathbf{T}, \mathbf{d}_0]$ is a subspace of the SRG maximal space $V[\mathbf{T}]$, which is the primitive space $V[\Psi_p(x)]$ due to Theorems 6.17 and 5.14. Since $\Psi_p(x)$ is a primitive polynomial, by Theorem 4.25 the SRG space $V[\mathbf{T}, \mathbf{d}_0]$ must be $V[1]$ or $V[\Psi_p(x)]$. But $V[\mathbf{T}, \mathbf{d}_0]$ can not be $V[1]$ since the SRG sequence D_i's for a non-zero initial state vector $\mathbf{d}_0$ are all non-zero sequences. Therefore, $V[\mathbf{T}, \mathbf{d}_0] = V[\Psi_p(x)] = V[\mathbf{T}]$, and hence the arbitrarily chosen non-zero initial state vector $\mathbf{d}_0$ is maximal.   ∎

Now that the state transition matrix $\mathbf{T}$ of a PRBS minimal generator is similar to the companion matrix $\mathbf{A}_{\Psi_p(x)}$ by Theorem 6.17, and that the corresponding maximal initial state vector $\mathbf{d}_0$ is any non-zero vector, we can state the following theorem.

THEOREM 6.19 *For a PRBS minimal generator with the state transition matrix* $\mathbf{T}$ *similar to* $\mathbf{A}_{\Psi_p(x)}$ *and a non-zero initial state vector* $\mathbf{d}_0$, *each SRG sequence* D_i, $i = 0, 1, \cdots, L\text{-}1$, *becomes a PRBS in the primitive space* $V[\Psi_p(x)]$.

Proof : Each SRG sequence D_i, $i = 0, 1, \cdots, L - 1$, is an element of the SRG maximal space $V[\mathbf{T}]$, which becomes the primitive space $V[\Psi_p(x)]$ due to Theorem 5.14 ; and if the initial state vector $\mathbf{d}_0$ is not a zero vector, D_i is not a zero sequence. Therefore, by Theorem 6.14, it is a PRBS in the primitive space $V[\Psi_p(x)]$.   ∎

The theorem means that *all* the SRG sequences generated by a PRBS minimal generator with a non-zero initial state vector are PRBSs. Combining this theorem with Theorems 6.17 and 6.18, we can establish the following guideline : *In order to generate a PRBS in the primitive space* $V[\Psi_p(x)]$ *of dimension L, the PRBS generator should be chosen such that its length is L, and its state transition matrix is similar to the companion matrix* $\mathbf{A}_{\Psi_p(x)}$, *but its initial state vector may be arbitrarily chosen to be any non-zero vector.*

EXAMPLE 6.11 We consider how to realize a PRBS minimal generator for the primitive space $V[x^4 + x + 1]$. In accordance with Theorem 6.17, we take the state transition matrix such that

$$
\mathbf{T} = \begin{bmatrix} 1 & 0 & 0 & 0 \\ 0 & 1 & 0 & 0 \\ 0 & 0 & 1 & 1 \\ 0 & 0 & 0 & 1 \end{bmatrix} \cdot \mathbf{A}_{x^4+x+1} \cdot \begin{bmatrix} 1 & 0 & 0 & 0 \\ 0 & 1 & 0 & 0 \\ 0 & 0 & 1 & 1 \\ 0 & 0 & 0 & 1 \end{bmatrix}^{-1}
$$

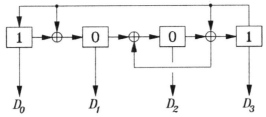

Figure 6.1. An example of a PRBS minimal generator for the primitive space $V[x^4 + x + 1]$. Its SRG sequences D_0, D_1, D_2, D_3 respectively become the PRBSs $\tilde{S}_{14}$, $\tilde{S}_1$, $\tilde{S}_4$, $\tilde{S}_{12}$ in Table 6.1.

$$= \begin{bmatrix} 0 & 0 & 0 & 1 \\ 1 & 0 & 0 & 1 \\ 0 & 1 & 1 & 1 \\ 0 & 0 & 1 & 1 \end{bmatrix}. \qquad (6.11)$$

Then, by Theorem 6.18 any choice of a non-zero initial state vector will bring about the PRBSs in Table 6.1. If we choose the initial state vector

$$\mathbf{d}_0 = [1\,0\,0\,1]^t, \qquad (6.12)$$

we obtain the PRBS minimal generator shown Fig. 6.1. We can confirm that the SRG sequences D_0, D_1, D_2, D_3 respectively become the PRBSs $\tilde{S}_{14}$, $\tilde{S}_1$, $\tilde{S}_4$, $\tilde{S}_{12}$ in Table 6.1. ♣

In the following, we consider how to realize PRBS minimal generators to generate the desired specific PRBSs.

DEFINITION 6.20 (Generating Vector) For a sequence $\{\,s_k\,\}$ obtained by a linear sum of the SRG sequences D_i, $i = 0, 1, \cdots, L - 1$, generated by an SRG of length L, the *generating vector* $\mathbf{h}$ refers to an L-vector representing the relation between the sequence $\{\,s_k\,\}$ and the SRG sequence vector $\mathbf{D}$, or more specifically,

$$\{\,s_k\,\} = \mathbf{h}^t \cdot \mathbf{D}. \qquad (6.13)$$

THEOREM 6.21 *For a PRBS* $\{\,s_k\,\} = \mathbf{s}^t \cdot \mathbf{E}_{\Psi_p(x)}$ *in the primitive space* $V[\Psi_p(x)]$, *let* $\mathbf{T}$ *and* $\mathbf{d}_0$ *be respectively the state transition matrix similar to* $\mathbf{A}_{\Psi_p(x)}$ *and the non-zero initial state vector of a PRBS minimal generator. Then, the generating vector* $\mathbf{h}$ *generating the PRBS* $\{\,s_k\,\}$ *is*

$$\mathbf{h} = (\Delta_{\mathbf{T},\mathbf{d}_0}^t)^{-1} \cdot \mathbf{s}. \qquad (6.14)$$

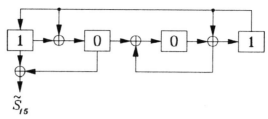

Figure 6.2. An example of a PRBS minimal generator generating the PRBS $\tilde{S}_{15}$ in Table 6.1.

Proof : Let $\mathbf{h}$ be the generating vector generating the sequence $\{ s_k \} = \mathbf{s}^t \cdot \mathbf{E}_{\Psi_p(x)}$. Then, by (6.13) $\{ s_k \} = \mathbf{h}^t \cdot \mathbf{D}$, and by inserting (5.14c) for $\Psi(x) = \Psi_p(x)$ into this, we obtain $\{ s_k \} = \mathbf{h}^t \cdot \Delta_{\mathbf{T},\mathbf{d}_0} \cdot \mathbf{E}_{\Psi_p(x)}$. Therefore, $\mathbf{s}^t = \mathbf{h}^t \cdot \Delta_{\mathbf{T},\mathbf{d}_0}$, and hence $\mathbf{h} = (\Delta^t_{\mathbf{T},\mathbf{d}_0})^{-1} \cdot \mathbf{s}$, since $\Delta_{\mathbf{T},\mathbf{d}_0}$ is nonsingular for the maximal initial state vector $\mathbf{d}_0$.   ■

The theorem means that in order to generate a prespecified PRBS $\{ s_k \} = \mathbf{s}^t \cdot \mathbf{E}_{\Psi_p(x)}$, the generating vector $\mathbf{h}$ should be chosen as in (6.14) for the PRBS minimal generator.

EXAMPLE 6.12 We consider how to generate the PRBS $\tilde{S}_{15} = [\,1\,1\,1\,1\,] \cdot \mathbf{E}_{x^4+x+1}$ in Table 6.1 using the PRBS minimal generator in Fig. 6.1. For the PRBS minimal generator, we get, by (6.11) and (6.12), the discrimination matrix

$$\Delta_{\mathbf{T},\mathbf{d}_0} = \begin{bmatrix} 1 & 1 & 1 & 0 \\ 0 & 0 & 0 & 1 \\ 0 & 1 & 0 & 0 \\ 1 & 1 & 0 & 0 \end{bmatrix}. \tag{6.15}$$

Therefore, inserting this and $\mathbf{s} = [\,1\,1\,1\,1\,]^t$ into (6.14), we obtain the generating vector $\mathbf{h} = [\,1\,1\,0\,0\,]^t$, whose circuit diagram is as shown in Fig. 6.2. ♣

As described in the previous chapter, the SSRG and the MSRG are two simplest types of BSRGs whose state transition matrices are respectively $\mathbf{T}_S = \mathbf{A}^t_{\Psi_p(x)}$ and $\mathbf{T}_M = \mathbf{A}_{\Psi_p(x)}$, and their terminating sequences are respectively represented by $D_{S,0} = \mathbf{d}_0^t \cdot \mathbf{E}_{\Psi_p(x)}$ and $D_{M,L-1} = \mathbf{d}_0^t \cdot \mathbf{P}_{\Psi_p(x)}$ for the initial state vector $\mathbf{d}_0$. Therefore, PRBS minimal generators can be realized based on the SSRG and the MSRG as the following two theorems describe.

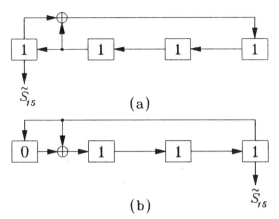

Figure 6.3. Examples of PRBS minimal generators generating the PRBS $\tilde{S}_{15} = [1\,1\,1\,1]^t \cdot \mathbf{E}_{x^4+x+1} = [0\,1\,1\,1]^t \cdot \mathbf{P}_{x^4+x+1}$ in Table 6.1. (a) SSRG-based realization, (b) MSRG-based realization.

THEOREM 6.22 *A PRBS with the expression $\{\,s_k\,\} = \mathbf{s}^t \cdot \mathbf{E}_{\Psi_p(x)}$ on the elementary basis can be generated by the SSRG with the initial state vector $\mathbf{d}_0 = \mathbf{s}$ and the generating vector $\mathbf{h} = \mathbf{e}_0$.*

THEOREM 6.23 *A PRBS with the expression $\{\,s_k\,\} = \mathbf{p}^t \cdot \mathbf{P}_{\Psi_p(x)}$ on the primary basis can be generated by the MSRG with the initial state vector $\mathbf{d}_0 = \mathbf{p}$ and the generating vector $\mathbf{h} = \mathbf{e}_{L-1}$.*

EXAMPLE 6.13 The PRBS $\tilde{S}_{15}$ in Table 6.1 is represented by $[1\,1\,1\,1] \cdot \mathbf{E}_{x^4+x+1}$ on the elementary basis and $[0\,1\,1\,1] \cdot \mathbf{P}_{x^4+x+1}$ on the primary basis respectively. Therefore, by Theorem 6.22 this PRBS is generated by the SSRG of the primitive space $V[x^4 + x + 1]$ with the initial state vector $\mathbf{d}_0 = [1\,1\,1\,1]^t$ and the generating vector $\mathbf{h} = \mathbf{e}_0$; and by Theorem 6.23 it can be also generated by the MSRG of the primitive space $V[x^4 + x + 1]$ with the initial state vector $\mathbf{d}_0 = [0\,1\,1\,1]^t$ and the generating vector $\mathbf{h} = \mathbf{e}_3$. The resulting PRBS minimal generators are as depicted in Fig. 6.3. ♣

Chapter 7

Parallel Frame Synchronous Scrambling

In the *parallel frame synchronous scrambling* (PFSS), the parallel input data bitstreams are scrambled before multiplexing, and the scrambled data bitstream is descrambled after demultiplexing. In this chapter, we discuss the behaviors of parallel scrambling sequences for use in the parallel scrambling in view of the sequence space theory developed in Chapter 4, and consider how to realize SRGs generating parallel scrambling sequences using the SRG theory developed in Chapter 5. We will first show that the parallel scrambling sequences are the decimated sequences of the serial scrambling sequence. Then, we will discuss how to decompose the serial sequence into a linear sum of the so-called *irreducible sequence* and *power sequence*, and consider how to determine the decimated sequences of the irreducible and power sequences. We will also examine how to obtain the decimated sequences of the original serial sequences decomposed into the irreducible and power sequences. Finally, we will discuss how to realize *parallel SRGs* to generate parallel sequences, and consider how to achieve their minimal realizations.

7.1 Parallel Scrambling Sequences

If we redraw the general blockdiagram of serial and parallel scramblings in Fig. 2.2 for the frame synchronous scrambling case, we get the block-diagram in Fig. 7.1. In the serial scrambling, N parallel input data bitstreams $\{ b_k^i \}$, $i = 0, 1, \cdots, N - 1$, are multiplexed, and then the

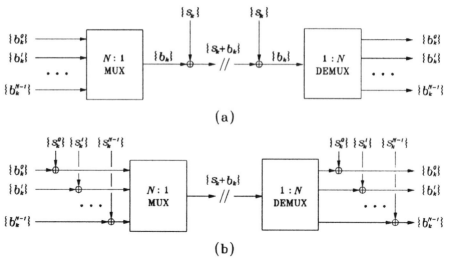

Figure 7.1. Serial and parallel frame synchronous scrambling. (a) Serial scrambling, (b) parallel scrambling.

multiplexed bitstream $\{ b_k \}$ is scrambled by adding a PRBS $\{ s_k \}$. In the receiving part, the scrambled bitstream $\{ b_k + s_k \}$ is descrambled by adding the same PRBS, and then the descrambled data $\{ b_k \}$ are demultiplexed for the recovery of the parallel input data bitstreams $\{ b_k^i \}$, $i = 0, 1, \cdots, N - 1$. But in the parallel scrambling, N parallel input data bitstreams $\{ b_k^i \}$, $i = 0, 1, \cdots, N - 1$, are scrambled before multiplexing by adding the parallel scrambling sequences $\{ s_k^i \}$, $i = 0, 1, \cdots, N - 1$, respectively to the parallel input bitstreams $\{ b_k^i \}$, $i = 0, 1, \cdots, N - 1$, and the scrambled bitstream $\{ b_k + s_k \}$ is descrambled after demultiplexing by adding the same parallel scrambling sequences to the demultiplexed bitstreams.

In the parallel scrambling, the scrambled bitstream $\{ b_k + s_k \}$ in Fig. 7.1(b) should be identical to the serial-scrambled bitstream $\{ b_k + s_k \}$ in Fig. 7.1(a) for the same parallel input data bitstreams $\{ b_k^i \}$, $i = 0, 1, \cdots, N - 1$. Therefore, the multiplexed signal of the parallel scrambling sequences $\{ s_k^i \}$, $i = 0, 1, \cdots, N - 1$, should be identical to the serial scrambling sequence $\{ s_k \}$, and so we have the relation

$$\{ s_k^i \} = \{ s_{i+kN} \}, \quad i = 0, 1, \cdots, N - 1. \qquad (7.1)$$

To mathematically describe the relation between the serial and parallel scrambling sequences in (7.1), we define the decimated sequences

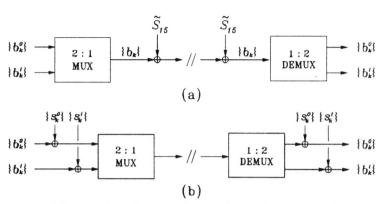

Figure 7.2. The serial and parallel scrambling for a 2:1 multiplexer and the PRBS $\tilde{S}_{15}$ in Table 6.1. (a) Serial scrambling, (b) parallel scrambling.

as follows :

DEFINITION 7.1 (Decimated Sequences) For a sequence $\{\, s_k \,\}$, the ith n-*decimated sequence* U_i, $i = 0, 1, \cdots, n - 1$, refers to the sequence $\{\, s_{i+kn} \,\}$, or equivalently,

$$U_i \equiv \{\, s_i, s_{i+n}, s_{i+2n}, \cdots \}. \tag{7.2}$$

Then, the parallel scrambling sequences $\{\, s_k^i \,\}$, $i = 0, 1, \cdots, N - 1$, in Fig. 7.1 respectively correspond to the N-decimated sequences U_i, $i = 0, 1, \cdots, N - 1$, of the serial sequence $\{\, s_k \,\}$.[1]

EXAMPLE 7.1 We consider the serial scrambling in Fig. 7.2(a) which employs a 2:1 multiplexing and the PRBS $\tilde{S}_{15}$ in Table 6.1. In this case, the 2-decimated sequences U_0 and U_1 are respectively

$$\begin{cases} U_0 = \{\, 1, 1, 0, 0, 0, 1, 0, 0, 1, 1, 0, 1, 0, 1, 1, \cdots \}, \\ U_1 = \{\, 1, 1, 0, 1, 0, 1, 1, 1, 1, 0, 0, 0, 1, 0, 0, \cdots \}. \end{cases} \tag{7.3}$$

Therefore, for the parallel scrambling in Fig. 7.2(b), if the parallel sequences $\{\, s_k^0 \,\}$ and $\{\, s_k^1 \,\}$ are taken to be U_0 and U_1 respectively, then the resulting scrambled and descrambled sequences become identical to those in Fig. 7.2(a). ♣

---

[1]Note that the index n in Definition 7.1 is for general purpose. That is, $n = N$ for the bit-parallel scrambling in Fig. 7.1, but $n = MN$ for the M-bit parallel scrambling to be discussed in Chapter 8.

7.2   Decimated Sequences

We first consider the periods of the decimated sequences of a sequence
$\{ s_k \}$ whose decimation factor n is relatively prime to the period of the
sequence $\{ s_k \}$. The following theorem describes this.

THEOREM 7.2 *Let U_i, $i = 0, 1, \cdots, n\text{-}1$, be the ith n-decimated sequence of a sequence $\{ s_k \}$, and let the decimation factor n be relatively
prime to the period τ of the sequence $\{ s_k \}$. Then, the period of each
n-decimated sequence U_i, $i = 0, 1, \cdots, n\text{-}1$, is also τ ; and the ith decimated sequence U_i is the (im)-delayed sequence of the 0th decimated
sequence U_0, where m is the smallest integer that satisfies*

$$mn = 1 \text{ modulo } \tau. \tag{7.4}$$

Proof : To show the first part of the theorem, we denote by $\hat{\tau}$ the
period of the ith decimated sequence $U_i = \{ s_{i+kn} \}$. Then, since the
period of the sequence $\{ s_k \}$ is τ, we have the relation $s_{i+kn} = s_{i+kn+\tau n}$
$= s_{i+(k+\tau)n}$. Therefore, the period $\hat{\tau}$ of the sequence $\{ s_{i+kn} \}$ divides τ.
But since the period of the sequence $\{ s_{i+kn} \}$ is $\hat{\tau}$, we have the relation
$s_{i+kn} = s_{i+(k+\hat{\tau})n} = s_{i+kn+\hat{\tau}n}$. Futhermore, since τ is relatively prime to
n, we get the relation $s_k = s_{k+\hat{\tau}n}$, $k = 0, 1, 2, \cdots$. Therefore the period
τ divides $\hat{\tau}n$, and hence τ divides $\hat{\tau}$. Therefore, $\hat{\tau} = \tau$.

To prove the second part, it suffices to show that the sequence U_{i+1}
is the m-delayed sequence of the sequence U_i, $i = 0, 1, \cdots, n - 2$. We
denote by V_i the m-delayed sequence of the sequence U_i. Then, $V_i = \{$
$s_{i+(k+m)n} \} = \{ s_{i+kn+mn} \}$. But since the period of the sequence $\{ s_k$
$\}$ is τ, we obtain, by (7.4), $V_i = \{ s_{i+1+kn} \}$, or equivalently, $V_i = U_{i+1}$.
This completes the proof. ∎

The theorem means that if the decimation factor n is relatively prime
to the period τ of a sequence $\{ s_k \}$, then all the decimated sequences
U_i's have the same period τ, and further each decimated sequence is a
delayed sequence of the 0th decimated sequence U_0. In the case that the
decimation factor n is not relatively prime to the period τ the periods
of the decimated sequences become shorter, as the following theorem
describes.

THEOREM 7.3 *Let U_i, $i = 0, 1, \cdots, n\text{-}1$, be the ith n-decimated sequence of a sequence $\{ s_k \}$, and let the decimation factor n be not relatively prime to the period τ of the sequence $\{ s_k \}$. Then, the period of
each n-decimated sequence U_i, $i = 0, 1, \cdots, n\text{-}1$, divides $LCM[\tau, n]/n$.*

Proof : Since the period of the sequence $\{s_k\}$ is τ, we have the relation $s_{i+kn} = s_{i+kn+LCM[\tau,n]} = s_{i+(k+LCM[\tau,n]/n)n}$. Therefore, the period of the sequence $U_i = \{s_{i+kn}\}$ divides $LCM[\tau, n]/n$. ■

EXAMPLE 7.2 We consider the sequence $\tilde{S}_{15}$ in Table 6.1, whose period is $\tau = 15$. If we choose the decimation factor $n = 2$, it is relatively prime to the period $\tau = 15$. Therefore, by Theorem 7.2 the 2-decimated sequences U_0 and U_1 also have period 15, and U_1 is the 8-delayed sequence of U_0 since the smallest integer m meeting (7.4) is 8. We can confirm these using the element expressions of U_i's in (7.3). ♣

EXAMPLE 7.3 In the case of the sequence $\tilde{S}_{15}$ in Table 6.1, if we choose the decimation factor $n = 3$, the decimation factor $n = 3$ is not relatively prime to the period $\tau = 15$ of the sequence $\tilde{S}_{15}$. Therefore, by Theorem 7.3 all the periods of the 3-decimated sequences U_0, U_1 and U_2 divides 5 since $LCM[15,3]/3 = 5$ in this case. In fact, the 3-decimated sequences are respectively

$$\begin{cases} U_0 = \{1, 1, 0, 0, 0, 1, 1, 0, 0, 0, 1, 1, 0, 0, 0, \cdots\}, \\ U_1 = \{1, 0, 1, 1, 1, 1, 0, 1, 1, 1, 1, 0, 1, 1, 1, \cdots\}, \\ U_2 = \{1, 0, 0, 1, 0, 1, 0, 0, 1, 0, 1, 0, 0, 1, 0, \cdots\}, \end{cases} \qquad (7.5)$$

and the period is 5 for each of them. ♣

We consider how to determine the decimated sequences for a given sequence. In order to handle this issue in terms of polynomial expressions, we first define the decimated polynomials and consider how to determine them.

DEFINITION 7.4 (Decimated Polynomials) For a polynomial $P(x)$, the *i*th *n-decimated polynomials* $P_i(x)$, $i = 0, 1, \cdots, n-1$, are defined to be the polynomials meeting the relation

$$P(x) = \sum_{i=0}^{n-1} x^{n-1-i} P_i(x^n). \qquad (7.6)$$

THEOREM 7.5 *Let* $P_i(x)$ *be the ith n-decimated polynomial of a polynomial* $P(x) = \sum_{j=0}^{nL-1} p_j x^j$.[2] *Then,*

$$P_i(x) = \sum_{j=0}^{L-1} p_{(n-1-i)+jn} x^j. \qquad (7.7)$$

---

[2]In general, the coefficients p_j could be zero for some j including $j = nL - 1$.

Proof : Let $P_i(x) = \sum_{l=0}^{L-1} \hat{p}_l^i x^l$. Then, inserting this into (7.6) along with $P(x) = \sum_{j=0}^{nL-1} p_j x^j$, we get the relation

$$\sum_{j=0}^{nL-1} p_j x^j = \sum_{i=0}^{n-1} \sum_{l=0}^{L-1} \hat{p}_l^i x^{(n-1-i)+ln}. \tag{7.8}$$

Arranging the left-hand side of this by putting $j = (n-1-i) + ln$, we obtain

$$\sum_{j=0}^{nL-1} p_j x^j = \sum_{i=0}^{n-1} \sum_{l=0}^{L-1} p_{(n-1-i)+ln} x^{(n-1-i)+ln}. \tag{7.9}$$

Therefore, comparing the coefficients of (7.8) and (7.9), we obtain the relation $p_{(n-1-i)+ln} = \hat{p}_l^i$, $i = 0, 1, \cdots, n-1, l = 0, 1, \cdots, L-1$. This proves the theorem. ∎

The theorem indicates that the decimated polynomials for a given polynomial have the coefficients which are taken from the given polynomial in a regular pattern. We observe from (7.6) and (7.7) that the degree of polynomial drops from $nL-1$ to $L-1$ after the n-decimation processing.

EXAMPLE 7.4 We consider two polynomials $P(x) = x^7 + x^6 + x^5 + x^4 + x$ and $\hat{P}(x) = x^{11} + x^{10} + x^9 + x^7 + x^6 + x^3 + x$. Then, the 2-decimated polynomials $P_0(x)$ and $P_1(x)$ of $P(x)$ are respectively

$$\begin{cases} P_0(x) = & x^3 + x^2 + 1, \\ P_1(x) = & x^3 + x^2, \end{cases} \tag{7.10}$$

and the 3-decimated polynomials $\hat{P}_0(x)$, $\hat{P}_1(x)$ and $\hat{P}_2(x)$ of $\hat{P}(x)$ are respectively

$$\begin{cases} \hat{P}_0(x) = & x^3, \\ \hat{P}_1(x) = & x^3 + x^2 + 1, \\ \hat{P}_2(x) = & x^3 + x^2 + x. \end{cases} \tag{7.11}$$ ♣

The polynomial expressions of the decimated sequences are related to the polynomial expression of the original sequence in the following manner :

THEOREM 7.6 *Let* $S[\Psi(x^n), P(x)]$ *be the polynomial expression of a given sequence* $\{ s_k \}$. *Then, the polynomial expression of the ith n-decimated sequence* U_i, $i = 0, 1, \cdots, n\text{-}1$, *is*

$$U_i = S[\Psi(x), P_i(x)], \tag{7.12}$$

where $P_i(x)$ is the ith n-decimated polynomial of $P(x)$.

Proof : We denote by $\{\ t_k^i\ \}$, $i = 0, 1, \cdots, n - 1$, the sequence $S[\Psi(x), P_i(x)]$ respectively. Then, it suffices to show $s_{i+kn} = t_k^i$ for $i = 0, 1, \cdots, n - 1$ and $k = 0, 1, \cdots$.

From Theorem 4.21, we have the relations

$$Q[\Psi(x^n), x^{kn}P(x)] = \sum_{l=0}^{kn-1} s_{kn-1-l}x^l, \qquad (7.13)$$

and

$$Q[\Psi(x), x^k P_i(x)] = \sum_{l=0}^{k-1} t_{k-1-l}^i x^l, \qquad (7.14)$$

for $i = 0, 1, \cdots, n-1$. Inserting (7.6) into (7.13) and then recalling that the degree of the remainder $R[\Psi(x^n), x^{kn}P_i(x^n)]$, $i = 0, 1, \cdots, n-1$, is not more than n times $(L$ - $1)$, where L is the degree of $\Psi(x)$, we obtain the relation

$$\sum_{i=0}^{n-1} x^{n-1-i}Q[\Psi(x^n), x^{kn}P_i(x^n)] = \sum_{l=0}^{kn-1} s_{kn-1-l}x^l\ ; \qquad (7.15)$$

and substituting x^n for x in (7.14), we have

$$Q[\Psi(x^n), x^{kn}P_i(x^n)] = \sum_{l=0}^{k-1} t_{k-1-l}^i x^{ln}. \qquad (7.16)$$

Inserting (7.16) into (7.15), we obtain

$$\sum_{i=0}^{n-1}\sum_{l=0}^{k-1} t_{k-1-l}^i x^{(n-1-i)+ln} = \sum_{l=0}^{kn-1} s_{kn-1-l}x^l. \qquad (7.17)$$

Finally, putting $l = (n - 1 - i) + jn$ in the right-hand side of (7.17), we obtain

$$\sum_{i=0}^{n-1}\sum_{l=0}^{k-1} t_{k-1-l}^i x^{(n-1-i)+ln} = \sum_{i=0}^{n-1}\sum_{j=0}^{k-1} s_{i+(k-1-j)n} x^{(n-1-i)+jn}, \qquad (7.18)$$

and therefore we have $t_{k-1-l}^i = s_{i+(k-1-l)n}$, or $t_j^i = s_{i+jn}$. ∎

EXAMPLE 7.5 We consider the 2-decimated sequences U_0 and U_1 of the sequence $\tilde{S}_{15}$ in Table 6.1. By (4.14), the sequence $\tilde{S}_{15}$ has the polynomial expression $\tilde{S}_{15} = S[x^4 + x + 1, x^3 + x^2 + x]$, and applying (4.20) to this along with $a(x) = x^4 + x + 1$, we have the polynomial expression $\tilde{S}_{15} = S[x^8 + x^2 + 1, x^7 + x^6 + x^5 + x^4 + x]$. Therefore, by the theorem and (7.10), the 2-decimated sequences take the polynomial expressions

$$\left\{ \begin{array}{rl} U_0 = & S[x^4 + x + 1, x^3 + x^2 + 1], \\ U_1 = & S[x^4 + x + 1, x^3 + x^2], \end{array} \right. \tag{7.19}$$

which respectively become the sequences $\tilde{S}_{12}$ and $\tilde{S}_{13}$ in Table 6.1. From (7.3) we can confirm that $U_0 = \tilde{S}_{12}$ and $U_1 = \tilde{S}_{13}$. ♣

EXAMPLE 7.6 In the case of the 3-decimated sequences U_0, U_1 and U_2 of the sequence $\tilde{S}_{15}$ in Table 6.1, we apply (4.20) to the polynomial expression $\tilde{S}_{15} = S[x^4 + x + 1, x^3 + x^2 + x]$ along with $a(x) = x^8 + x^4 + x^2 + x + 1$, to obtain the expression $\tilde{S}_{15} = S[x^{12} + x^9 + x^6 + x^3 + 1, x^{11} + x^{10} + x^9 + x^7 + x^6 + x^3 + x]$. Therefore, by the theorem and (7.11), we obtain the following polynomial expressions :

$$\left\{ \begin{array}{rl} U_0 = & S[x^4 + x^3 + x^2 + x + 1, x^3], \\ U_1 = & S[x^4 + x^3 + x^2 + x + 1, x^3 + x^2 + 1], \\ U_2 = & S[x^4 + x^3 + x^2 + x + 1, x^3 + x^2 + x]. \end{array} \right. \tag{7.20}$$

We can confirm them by applying Theorem 4.21 to the decimated sequences U_i's of period 5 in (7.5). ♣

7.3   Decomposition of Sequences

In order to apply Theorem 7.6 in identifying the n-decimated sequences of a sequence $\{ s_k \}$, we first have to find the polynomial expression $S[\Psi(x^n), P(x)]$ of the sequence $\{ s_k \}$ such that the characteristic polynomial $\Psi(x^n)$ is a polynomial of x^n. But it is not easy to find such a polynomial expression $S[\Psi(x^n), P(x)]$ for an arbitrary sequence $\{ s_k \}$. To get around this problem, we rather consider decomposing the given sequence into a linear sum of sequences whose decimated sequences can be more easily determined. For this, we first consider the vector space which is obtained by linear sums of sequence spaces.

DEFINITION 7.7 (Sum Space, Sum Sequence)   For J sequence spaces $V[\Psi_j(x)]$, $j = 0, 1, \cdots, J - 1$, we define by the *sum space*

$\sum_{j=0}^{J-1} V[\Psi_j(x)]$ the vector space for the *sum sequence* which is formed by linear sums of the sequences in the J sequence spaces, or more specifically,

$$\sum_{j=0}^{J-1} V[\Psi_j(x)] \equiv \{\sum_{j=0}^{J-1} \alpha_j S_j : \alpha_j \in \mathcal{F}, S_j \in V[\Psi_j(x)], j = 0, 1, \cdots, J-1\}.$$

$$(7.21)$$

Then, the sum space can be determined by the following theorem.

THEOREM 7.8 *The sum space $\sum_{j=0}^{J-1} V[\Psi_j(x)]$ of the J sequence spaces $V[\Psi_j(x)]$, $j = 0, 1, \cdots, J$-1, is the sequence space whose characteristic polynomial is the least common multiple of the J characteristic polynomials $\Psi_j(x)$, $j = 0, 1, \cdots, J$-1, that is,*

$$\sum_{j=0}^{J-1} V[\Psi_j(x)] = V[LCM[\Psi_0(x), \Psi_1(x), \cdots, \Psi_{J-1}(x)]]. \qquad (7.22)$$

Proof : To prove the theorem, we first show that for two arbitrary sequence spaces $V[\Psi(x)]$ and $V[\hat{\Psi}(x)]$, $V[\Psi(x)] + V[\hat{\Psi}(x)] = V[LCM[\Psi(x), \hat{\Psi}(x)]]$. Since both the characteristic polynomials $\Psi(x)$ and $\hat{\Psi}(x)$ divide $LCM[\Psi(x), \hat{\Psi}(x)]$, by Theorem 4.25 both the sequence spaces $V[\Psi(x)]$ and $V[\hat{\Psi}(x)]$ are subspaces of $V[LCM[\Psi(x), \hat{\Psi}(x)]]$. Therefore, their sum space $V[\Psi(x)] + V[\hat{\Psi}(x)]$ is a subset of $V[LCM[\Psi(x), \hat{\Psi}(x)]]$. However, by Theorem 4.27, the intersection of the two sequence spaces $V[\Psi(x)]$ and $V[\hat{\Psi}(x)]$ becomes $V[GCD[\Psi(x), \hat{\Psi}(x)]]$, so the dimension of $V[\Psi(x)] + V[\hat{\Psi}(x)]$ is $deg\{\Psi(x)\} + deg\{\hat{\Psi}(x)\} - deg\{GCD[\Psi(x), \hat{\Psi}(x)]\}$,[3] which is identical to the dimension of $V[LCM[\Psi(x), \hat{\Psi}(x)]]$ due to Theorem 4.7 and the relation $deg\{LCM[\Psi(x), \hat{\Psi}(x)]\} = deg\{\Psi(x)\} + deg\{\hat{\Psi}(x)\} - deg\{GCD[\Psi(x), \hat{\Psi}(x)]\}$. Therefore, $V[\Psi(x)] + V[\hat{\Psi}(x)] = V[LCM[\Psi(x), \hat{\Psi}(x)]]$.

Applying the relation $V[\Psi(x)] + V[\hat{\Psi}(x)] = V[LCM[\Psi(x), \hat{\Psi}(x)]]$ to $\sum_{j=0}^{J-1} V[\Psi_j(x)]$ repeatedly, we obtain (7.22). ∎

According to the theorem the sum space of sequence spaces is the sequence space whose characteristic polynomial is the least common multiple of the characteristic polynomials of the sequence spaces. Therefore,

---

[3]Note that the dimension of the vector space obtained by linear sums of two vector spaces $\mathcal{V}$ and $\hat{\mathcal{V}}$ is identical to the sum of the dimensions of $\mathcal{V}$ and $\hat{\mathcal{V}}$ less the dimension of the vector space obtained by their intersection. Refer to (B.1) in Appendix B.

using the theorem, we can decompose a sequence space into sequence subspaces whose characteristic polynomials are irreducible polynomials or powers of irreducible polynomials, as the following theorem describes.

THEOREM 7.9 *For a sequence space* $V[\Psi(x)]$, *let* $\Psi(x) = [\Psi_0(x)]^{w_0}$ $[\Psi_1(x)]^{w_1} \cdots [\Psi_{J-1}(x)]^{w_{J-1}}$ *for integers* w_0, w_1, $\cdots$, w_{J-1}, *and irreducible polynomials* $\Psi_j(x)$, $j = 0, 1, \cdots, J$-1, *which are different from one another. Then, the sequence space* $V[\Psi(x)]$ *is represented by the sum space*

$$V[\Psi(x)] = \sum_{j=0}^{J-1} V[[\Psi_j(x)]^{w_j}]. \qquad (7.23)$$

Proof : This theorem is evident in view of Theorem 7.8, since $LCM[[\Psi_0(x)]^{w_0}, [\Psi_1(x)]^{w_1}, \cdots, [\Psi_{J-1}(x)]^{w_{J-1}}] = \Psi(x).$ ∎

EXAMPLE 7.7 We consider the sequence space $V[x^8 + x^7 + x^4 + x + 1]$. Then, the characteristic polynomial is factorized into the form $x^8 + x^7 + x^4 + x + 1 = (x^2 + x + 1)^2(x^4 + x^3 + x^2 + x + 1)$. Therefore, by the theorem this sequence space can be decomposed into $V[(x^2 + x + 1)^2] + V[x^4 + x^3 + x^2 + x + 1]$. ♣

The sum space in (7.23), in fact, turns out to be the minimal space for the sequence obtained by a linear sum of sequences, as the following theorem describes.

THEOREM 7.10 *Let* $\{ s_k \}$ *be a sum sequence of the* J *sequences* $\{ s_k^j \}$, $j = 0, 1, \cdots, J$-1, *whose minimal space is respectively* $V[[\Psi_j(x)]^{w_j}]$ *for irreducible polynomials* $\Psi_j(x)$, $j = 0, 1, \cdots, J$-1, *which are different from one another. Then, the minimal space* $V[\Psi_m(x)]$ *for the sequence* $\{ s_k \}$ *is*

$$V[\Psi_m(x)] = V[\prod_{j=0}^{J-1}[\Psi_j(x)]^{w_j}]. \qquad (7.24)$$

Proof : Let each sequence $\{ s_k^j \}$, $j = 0, 1, \cdots, J-1$, have the polynomial expression $S[[\Psi_j(x)]^{w_j}, P_j(x)]$, where $P_j(x)$ is relatively prime to $\Psi_j(x)$; and let $\Psi(x) = [\Psi_0(x)]^{w_0}[\Psi_1(x)]^{w_1} \cdots [\Psi_{J-1}(x)]^{w_{J-1}}$. Then, by (4.20), $\{ s_k^j \} = S[\Psi(x), P_j(x)\Psi(x)/[\Psi_j(x)]^{w_j}], j = 0, 1, \cdots, J-1$, and so $\{ s_k \} = S[\Psi(x), \sum_{j=0}^{J-1} P_j(x)\Psi(x)/[\Psi_j(x)]^{w_j}]$. Noting that $\Psi_j(x)$'s are irreducible polynomials different from each other, we can easily show that $\sum_{j=0}^{J-1} P_j(x)\Psi(x)/[\Psi_j(x)]^{w_j}$ is relatively prime to $\Psi(x)$. Therefore, by Theorem 4.34, the minimal space for the sequence $\{ s_k \}$ is $V[\Psi(x)]$, or $V[[\Psi_0(x)]^{w_0}[\Psi_1(x)]^{w_1}\cdots[\Psi_{J-1}(x)]^{w_{J-1}}]$. ∎

The theorem means that if the minimal space for each sequence $\{ s_k^j \}$, $j = 0, 1, \cdots, J-1$, is $V[[\Psi_j(x)]^{w_j}]$, then the minimal space for the sum sequence $\{ s_k \} = \sum_{j=0}^{J-1}\{ s_k^j \}$ is the sequence space $V[\prod_{j=0}^{J-1}[\Psi_j(x)]^{w_j}]$.

EXAMPLE 7.8 We consider the two sequences S_{15} in Table 4.1 and $\tilde{S}_{15}$ in Table 6.1. The minimal spaces for S_{15} and $\tilde{S}_{15}$ are respectively $V[x + 1]$ and $V[x^4 + x + 1]$. Therefore, by the theorem the minimal space for the sum sequence $S_{15} + \tilde{S}_{15}$ is the sequence space $V[(x+1)(x^4 + x + 1)]$. ♣

We now consider how to decompose a sequence into a linear sum of sequences whose characteristic polynomials are irreducible polynomials or powers of irreducible polynomials.

THEOREM 7.11 *Let $V[\Psi_m(x)]$ be the minimal space for a sequence $\{ s_k \}$, and let $\Psi_m(x) = [\Psi_0(x)]^{w_0}[\Psi_1(x)]^{w_1}\cdots[\Psi_{J-1}(x)]^{w_{J-1}}$, where $\Psi_j(x)$, $j = 0, 1, \cdots, J\text{-}1$, are irreducible polynomials different from one another. Then, the sequence $\{ s_k \}$ is uniquely represented by the linear sum of J sequences $\{ s_k^j \}$, $j = 0, 1, \cdots, J\text{-}1$, whose minimal spaces are respectively the sequence spaces $V[[\Psi_j(x)]^{w_j}]$.*

Proof : We show the theorem in three steps : First, we show that the sequence $\{ s_k \}$ can be represented by a linear sum of J sequences $\{ s_k^j \} \in V[[\Psi_j(x)]^{w_j}]$, $j = 0, 1, \cdots, J-1$, and then show that the minimal spaces of such sequences $\{ s_k^j \}$, $j = 0, 1, \cdots, J-1$, are respectively the sequence spaces $V[[\Psi_j(x)]^{w_j}]$, $j = 0, 1, \cdots, J-1$. Finally, we prove that the sequences $\{ s_k^j \}$ in $V[[\Psi_j(x)]^{w_j}]$, $j = 0, 1, \cdots, J-1$, such that $\{ s_k \} = \sum_{j=0}^{J-1}\{ s_k^j \}$, are unique.

The first step is directly obtained by the definition of the sum space and Theorem 7.9. So we prove the second step. Let $\{ s_k^j \} \in V[[\Psi_j(x)]^{w_j}]$, $j = 0, 1, \cdots, J-1$, be the sequences such that $\{ s_k \} = \sum_{j=0}^{J-1}\{ s_k^j \}$; and let $V[\Psi_{m,j}(x)]$, $j = 0, 1, \cdots, J-1$, be respectively the minimal space for the sequence $\{ s_k^j \}$. Then, by Theorem 4.33, $\Psi_{m,j}(x) = [\Psi_j(x)]^{m_j}$, $m_j \le w_j$, since $\Psi_j(x)$ is an irreducible polynomial. But by Theorem 7.10, the minimal space for the sequence $\{ s_k \} = \sum_{j=0}^{J-1}\{ s_k^j \}$ is $V[[\Psi_0(x)]^{m_0}[\Psi_1(x)]^{m_1}\cdots[\Psi_{J-1}(x)]^{m_{J-1}}]$. Therefore, $m_j = w_j$, $j = 0, 1, \cdots, J-1$. This proves the second step.

Finally, we prove that the sequences $\{ s_k^j \}$ in $V[[\Psi_j(x)]^{w_j}]$, $j = 0, 1, \cdots, J-1$, such that $\{ s_k \} = \sum_{j=0}^{J-1}\{ s_k^j \}$, are unique. Suppose that $\{ s_k^j \}$ and $\{ \hat{s}_k^j \}$ are two such sequences, and consider the minimal space $V[\Psi_{m,j}(x)]$ for the sum sequence $\{ s_k^j \} + \{ \hat{s}_k^j \}$. Then, $\{ s_k^j \} + \{ \hat{s}_k^j \} \in V[[\Psi_j(x)]^{w_j}]$, and hence by Theorem 4.33, $\Psi_{m,j}(x) = [\Psi_j(x)]^{m_j}$, $m_j \leq w_j$, since $\Psi_j(x)$ is an irreducible polynomial. Thus the minimal space for the sequence $\sum_{j=0}^{J-1}(\{ s_k^j \} + \{ \hat{s}_k^j \})$ is $V[[\Psi_0(x)]^{m_0}[\Psi_1(x)]^{m_1} \cdots [\Psi_{J-1}(x)]^{m_{J-1}}]$ due to Theorem 7.10. But $\sum_{j=0}^{J-1}(\{ s_k^j \} + \{ \hat{s}_k^j \}) = \{ s_k \} + \{ s_k \} = \{ 0 \}$, so the corresponding minimal space is $V[1]$. Therefore, $m_j = 0$, $j = 0, 1, \cdots, m_{J-1}$, which means that $\{ s_k^j \} = \{ \hat{s}_k^j \}$. This completes the proof. ∎

EXAMPLE 7.9 In the case of the sequence $\{ s_k \}$ whose minimal space is $V[x^8+x^7+x^4+x+1]$, by the theorem and Example 7.7, the sequence can be represented as a linear sum of two sequences whose minimal spaces are $V[(x^2 + x + 1)^2]$ and $V[x^4 + x^3 + x^2 + x + 1]$. ♣

7.4 Decimation of Irreducible Sequences

We now consider how to determine the decimated sequences of a sequence whose characteristic polynomial is irreducible. For this, we first define the irreducible sequence space and the irreducible sequence as follows :

DEFINITION 7.12 (Irreducible Sequence Space) An *irreducible sequence space* (in short, *irreducible space*) $V[\Psi_I(x)]$ refers to a sequence space whose characteristic polynomial $\Psi_I(x)$ is *irreducible*.

DEFINITION 7.13 (Irreducible Sequence) A non-zero sequence in an irreducible space is called an *irreducible sequence*.

Then, the periodicity of an irreducible space, the minimal space of an irreducible sequence, and the period of an irreducible sequence are determined by the following three theorems.

THEOREM 7.14 *The periodicity T_I of an irreducible space $V[\Psi_I(x)]$ of dimension L divides $2^L - 1$.*

Proof : To prove the theorem, we consider the period τ of the sequence $\{\, s_k \,\} = S[\Psi_I(x), 1]$ in the irreducible space $V[\Psi_I(x)]$. Then, by Theorem 4.34 the minimal space for the sequence $\{\, s_k \,\}$ becomes $V[\Psi_I(x)]$, and hence the period τ of $\{\, s_k \,\}$ is identical to the periodicity T_I of the minimal space $V[\Psi_I(x)]$ due to Theorem 4.37. On the other hand, since $\Psi_I(x)$ is an irreducible polynomial of degree L, $\Psi_I(x)$ divides $x^{2^L-1} + 1$,[4] so by Theorem 4.25 the irreducible space $V[\Psi_I(x)]$ is a subspace of $V[x^{2^L-1} + 1]$. Therefore, the sequence $\{\, s_k \,\}$ is an element of $V[x^{2^L-1} + 1]$, and hence by Theorem 4.29 its period τ divides the periodicity $2^L - 1$ of $V[x^{2^L-1} + 1]$. This proves the theorem. ∎

THEOREM 7.15 *The minimal space* $V[\Psi_m(x)]$ *for an irreducible sequence* $\{\, s_k \,\}$ *is identical to the irreducible space* $V[\Psi_I(x)]$, *i.e.,*

$$V[\Psi_m(x)] = V[\Psi_I(x)]. \tag{7.25}$$

Proof : Let an irreducible sequence $\{\, s_k \,\}$ in the irreducible space $V[\Psi_I(x)]$ have the polynomial expression $S[\Psi_I(x), P(x)]$ for a non-zero polynomial $P(x)$. Then, since $\Psi_I(x)$ is irreducible, $P(x)$ is relatively prime to $\Psi_I(x)$. Therefore, by Theorem 4.34 the irreducible space $V[\Psi_I(x)]$ is the minimal space for $\{\, s_k \,\}$. ∎

THEOREM 7.16 *The period* τ_I *of an irreducible sequence* $\{\, s_k \,\}$ *is identical to the periodicity* T_I *of the irreducible space* $V[\Psi_I(x)]$, *i.e.,*

$$\tau_I = T_I. \tag{7.26}$$

Proof : This theorem is a direct outcome of Theorems 4.37 and 7.15. ∎

By the previous three theorems, we know that the period τ_I of an irreducible sequence is identical to the periodicity T_I of the corresponding irreducible space $V[\Psi_I(x)]$, which is a submultiple of $2^L - 1$, and that the irreducible space $V[\Psi_I(x)]$ itself is the minimal space for the irreducible sequence.

EXAMPLE 7.10 The sequence space $V[x^4+x^3+x^2+x+1]$, whose elements are listed in Table 7.1, is an irreducible space of dimension $L = 4$, since

---

[4]An irreducible polynomial of degree L divides $x^{2^L-1} + 1$. Refer to Property A.4 in Appendix A.

Table 7.1. Elements of the irreducible space $V[x^4 + x^3 + x^2 + x + 1]$

S_0	=	$[0000] \cdot \mathbf{E}_{x^4+x^3+x^2+x+1}$	=	$\{0,0,0,0,0,0,0,0,0,\cdots\}$
S_1	=	$[0001] \cdot \mathbf{E}_{x^4+x^3+x^2+x+1}$	=	$\{0,0,0,1,1,0,0,0,1,\cdots\}$
S_2	=	$[0010] \cdot \mathbf{E}_{x^4+x^3+x^2+x+1}$	=	$\{0,0,1,0,1,0,0,1,0,\cdots\}$
S_3	=	$[0011] \cdot \mathbf{E}_{x^4+x^3+x^2+x+1}$	=	$\{0,0,1,1,0,0,0,1,1,\cdots\}$
S_4	=	$[0100] \cdot \mathbf{E}_{x^4+x^3+x^2+x+1}$	=	$\{0,1,0,0,1,0,1,0,0,\cdots\}$
S_5	=	$[0101] \cdot \mathbf{E}_{x^4+x^3+x^2+x+1}$	=	$\{0,1,0,1,0,0,1,0,1,\cdots\}$
S_6	=	$[0110] \cdot \mathbf{E}_{x^4+x^3+x^2+x+1}$	=	$\{0,1,1,0,0,0,1,1,0,\cdots\}$
S_7	=	$[0111] \cdot \mathbf{E}_{x^4+x^3+x^2+x+1}$	=	$\{0,1,1,1,1,0,1,1,1,\cdots\}$
S_8	=	$[1000] \cdot \mathbf{E}_{x^4+x^3+x^2+x+1}$	=	$\{1,0,0,0,1,1,0,0,0,\cdots\}$
S_9	=	$[1001] \cdot \mathbf{E}_{x^4+x^3+x^2+x+1}$	=	$\{1,0,0,1,0,1,0,0,1,\cdots\}$
S_{10}	=	$[1010] \cdot \mathbf{E}_{x^4+x^3+x^2+x+1}$	=	$\{1,0,1,0,0,1,0,1,0,\cdots\}$
S_{11}	=	$[1011] \cdot \mathbf{E}_{x^4+x^3+x^2+x+1}$	=	$\{1,0,1,1,1,1,0,1,1,\cdots\}$
S_{12}	=	$[1100] \cdot \mathbf{E}_{x^4+x^3+x^2+x+1}$	=	$\{1,1,0,0,0,1,1,0,0,\cdots\}$
S_{13}	=	$[1101] \cdot \mathbf{E}_{x^4+x^3+x^2+x+1}$	=	$\{1,1,0,1,1,1,1,0,1,\cdots\}$
S_{14}	=	$[1110] \cdot \mathbf{E}_{x^4+x^3+x^2+x+1}$	=	$\{1,1,1,0,1,1,1,1,0,\cdots\}$
S_{15}	=	$[1111] \cdot \mathbf{E}_{x^4+x^3+x^2+x+1}$	=	$\{1,1,1,1,0,1,1,1,1,\cdots\}$

the characteristic polynomial $\Psi_I(x) = x^4 + x^3 + x^2 + x + 1$ is irreducible. Therefore, by Theorem 7.14 its periodicity T_I is a submultiple of $2^L - 1 = 15$. In fact, its periodicity is $T_I = 5$. ♣

EXAMPLE 7.11 For an irreducible sequence $\{\, s_k \,\}$ in the irreducible space $V[x^4 + x^3 + x^2 + x + 1]$, its minimal space is $V[\Psi_m(x)] = V[x^4 + x^3 + x^2 + x + 1]$ by Theorem 7.15, and its period is $\tau_I = T_I = 5$ by Theorem 7.16. We can confirm these using Theorem 4.35 and the irreducible sequences S_1 through S_{15} in Table 7.1. ♣

Now we consider how to determine the decimated sequences of an irreducible sequence. A decimated sequence is determined if its minimal space is known along with the involved initial vectors. But the initial vectors are easily determined by (7.2). Therefore, the main interest for determining a decimated sequence is in identifying its minimal space.[5]

---

[5]In fact, the minimal space for the decimated sequences plays an important role in the realization of the SRGs generating the parallel sequences or the decimated sequences, as will become clear in Sections 7.8 through 7.10.

THEOREM 7.17 *Let U_i, $i = 0, 1, \cdots, n\text{-}1$, be the ith n-decimated sequence of an irreducible sequence $\{ s_k \}$ in the irreducible space $V[\Psi_I(x)]$ of dimension L. Then, the minimal space $V[\Psi_m(x)]$ for U_i, $i = 0, 1, \cdots, n\text{-}1$, is an irreducible space whose characteristic polynomial $\Psi_m(x)$ divides $x^{2^L-1} + 1$ and its extension $\Psi_m(x^n)$ is divided by $\Psi_I(x)$.*

Proof : We first show the existence of the irreducible polynomial $\Psi_m(x)$ such that $\Psi_m(x)$ divides $x^{2^L-1} + 1$ and its extension $\Psi_m(x^n)$ is divided by $\Psi_I(x)$. Let $f_j(x)$, $j = 0, 1, \cdots, J-1$, be the irreducible factors of $x^{2^L-1} + 1$. Then, since $\Psi_I(x)$ is an irreducible polynomial of degree L, $\Psi_I(x)$ divides $x^{2^L-1} + 1$, and hence $\Psi_I(x)$ also divides $x^{n(2^L-1)} + 1 = f_0(x^n)f_1(x^n)\cdots f_{J-1}(x^n)$. Therefore, there exists a factor $f_i(x)$ of $x^{2^L-1} + 1$ such that its extension $f_i(x^n)$ is divided by $\Psi_I(x)$.

Now we prove that the minimal space for the decimated sequences U_i, $i = 0, 1, \cdots, n-1$, is the irreducible space $V[\Psi_m(x)]$. Let $\Psi_m(x^n) = a(x)\Psi_I(x)$ and $\{ s_k \} = S[\Psi_I(x), P(x)]$ for a non-zero polynomial $P(x)$. Then, by Theorem 4.26 $\{ s_k \} = S[\Psi_m(x^n), a(x)P(x)]$, and by Theorem 7.6 $U_i = S[\Psi_m(x), P_i(x)]$, where $P_i(x)$ is the ith n-decimated polynomial of $a(x)P(x)$. Therefore, $U_i \in V[\Psi_m(x)]$, $i = 0, 1, \cdots, n-1$, and hence by Theorem 7.15 the minimal space for the decimated sequences U_i's is the irreducible space $V[\Psi_m(x)]$. ∎

The theorem means that if an irreducible sequence $\{ s_k \}$ is an element of the irreducible space $V[\Psi_I(x)]$, then its n-decimated sequences U_i's have the minimal space $V[\Psi_m(x)]$ whose characteristic polynomial $\Psi_m(x)$ is a factor of $x^{2^L-1} + 1$ such that its extension $\Psi_m(x^n)$ is a multiple of the irreducible characteristic polynomial $\Psi_I(x)$.

EXAMPLE 7.12 We consider the 5-decimated sequences U_i, $i = 0, 1, 2, 3, 4$, of the sequence S_1 in Table 7.1. The sequence S_1 is an irreducible sequence in the irreducible space $V[x^4 + x^3 + x^2 + x + 1]$ of dimension $L = 4$, and the polynomial $x^{2^L-1} + 1 = x^{15} + 1$ has the five factors $x + 1$, $x^2 + x + 1$, $x^4 + x + 1$, $x^4 + x^3 + 1$ and $x^4 + x^3 + x^2 + x + 1$. For the factor $\Psi_m(x) = x + 1$ the characteristic polynomial $\Psi_I(x) = x^4 + x^3 + x^2 + x + 1$ divides $\Psi_m(x^n) = x^5 + 1$. Therefore, by the theorem the minimal space for the 5-decimated sequences U_i's is $V[\Psi_m(x)] = V[x + 1]$. By (7.2) and the element expression of S_1 in Table 7.1, the initial vectors of U_i's are $s_0 = s_1 = s_2 = [\,0\,]^t$ and $s_3 = s_4 = [\,1\,]^t$, and hence the decimated sequences become $U_0 = U_1 = U_2 = [\,0\,] \cdot \mathbf{E}_{x+1} = \{\, 0, 0, 0, \cdots \,\}$ and $U_3 = U_4 = [\,1\,] \cdot \mathbf{E}_{x+1} = \{\, 1, 1, 1, \cdots \,\}$. ♣

In case n is a large number, it is not easy to apply Theorem 7.17 to find the factor $\Psi_m(x)$ of $x^{2^L-1}+1$ such that its extension $\Psi_m(x^n)$ is a multiple of $\Psi_I(x)$. To circumvent this difficulty, we introduce the following alternative, which has an equally general coverage.

THEOREM 7.18 *Let U_i, $i = 0, 1, \cdots, n\text{-}1$, be the ith n-decimated sequence of an irreducible sequence $\{ s_k \}$ in the irreducible space $V[\Psi_I(x)]$ of dimension L, and let the decimation factor be of the form $n = 2^j \hat{n}$ for an integer j. Then, the minimal space for the decimated sequences U_i, $i = 0, 1, \cdots, n\text{-}1$, is the irreducible space $V[\Psi_m(x)]$ whose characteristic polynomial $\Psi_m(x)$ divides $x^{2^L-1}+1$ and its extension $\Psi_m(x^{\hat{n}})$ is divided by $\Psi_I(x)$.*

Proof : Let $\Psi_m(x)$ be an irreducible polynomial such that $\Psi_m(x)$ divides $x^{2^L-1}+1$ and $\Psi_I(x)$ divides $\Psi_m(x^{\hat{n}})$. Then, since $\Psi_m(x^n) = [\Psi_m(x^{\hat{n}})]^{2^j}$,[6] $\Psi_I(x)$ also divides $\Psi_m(x^n)$. Therefore, by Theorem 7.17 the irreducible space $V[\Psi_m(x)]$ is the minimal space for the n-decimated sequences U_i's.   ∎

The theorem implies that for an irreducible sequence $\{ s_k \}$ the minimal space for its $(2^j\hat{n})$-decimated sequences is identical to that for its $\hat{n}$-decimated sequences. That is, the term 2^j in the decimation factor n does not affect the minimal space for the decimated sequences.

EXAMPLE 7.13 We consider the 10-decimated sequences U_i, $i = 0, 1, \cdots, 9$, of the sequence S_1 in Table 7.1. The sequence S_1 is an irreducible sequence in the irreducible space $V[x^4 + x^3 + x^2 + x + 1]$, and the decimation factor n is 2×5. Therefore, the minimal space for the 10-decimated sequences U_i's is identical to the minimal space for the 5-decimated sequences, which is $V[x + 1]$ according to Example 7.12. ♣

In the following, we consider the special case when the decimation factor is relatively prime to the period of the original irreducible sequence.

THEOREM 7.19 *Let U_i, $i = 0, 1, \cdots, n\text{-}1$, be the ith n-decimated sequence of an irreducible sequence $\{ s_k \}$ in the irreducible space $V[\Psi_I(x)]$*

---

[6] For a binary polynomial $a(x)$, $a(x^{2^j}) = [a(x)]^{2^j}$, $j = 0, 1, \cdots$. Refer to Property A.7 in Appendix A.

of dimension L, and let the decimation factor n be relatively prime to the period τ of the sequence $\{ s_k \}$. Then, the minimal space for each decimated sequence U_i, $i = 0$, 1, $\cdots$, $n\text{-}1$, is the irreducible space $V[\Psi_m(x)]$ of dimension L whose characteristic polynomial $\Psi_m(x)$ divides $x^{2^L-1} + 1$ and its extension $\Psi_m(x^n)$ is divided by $\Psi_I(x)$.

Proof : Let $V[\Psi_i(x)]$, $i = 0$, 1, $\cdots$, $n - 1$, be the minimal space for the ith n-decimated sequence U_i of an irreducible sequence $\{ s_k \}$ in the irreducible space $V[\Psi_I(x)]$ of dimension L, and let $\Psi_m(x)$ be the irreducible polynomial such that $\Psi_m(x)$ divides $x^{2^L-1} + 1$ and $\Psi_I(x)$ divides $\Psi_m(x^n)$. Then, it suffices to show that $\Psi_i(x) = \Psi_m(x)$, $i = 0$, 1, $\cdots$, $n - 1$, and that the degree of $\Psi_m(x)$ is L.

By Theorem 7.2, U_i, $i = 0, 1, \cdots, n-1$, is the delayed sequence of U_0, and hence its minimal space $V[\Psi_i(x)]$ is identical to the minimal space $V[\Psi_0(x)]$ for U_0, that is, $\Psi_0(x) = \Psi_1(x) = \cdots = \Psi_{n-1}(x)$. Therefore, by Theorem 4.36 the minimal space for the decimated sequences U_i's is the sequence space $V[\Psi_0(x)]$. On the hand hand, by Theorem 7.17, the minimal space for the decimated sequences U_i's is the irreducible space $V[\Psi_m(x)]$. Therefore, by Theorem 4.32, $\Psi_0(x) = \Psi_m(x)$, and hence $\Psi_i(x) = \Psi_m(x)$, $i = 0, 1, \cdots, n - 1$.

To complete the proof, we prove that the degree of $\Psi_m(x)$ is equal to L. Since $\Psi_m(x)$ is a factor of the polynomial $x^{2^L-1} + 1$, its degree $\hat{L}$ is L or less,[7] that is, $\hat{L} \leq L$. To prove $\hat{L} \geq L$, we consider the m-decimated sequences $\hat{U}_i$, $i = 0, 1, \cdots, m - 1$, of the sequence U_0 in the irreducible space $V[\Psi_m(x)]$ of dimension $\hat{L}$ for the smallest integer m meeting (7.4), and their minimal space $V[\hat{\Psi}_m(x)]$. Then, the 0th decimated sequence $\hat{U}_0$ becomes the original irreducible sequence $\{ s_k \}$, and by Theorem 7.17 the minimal space $V[\hat{\Psi}_m(x)]$ for the decimated sequences $\hat{U}_i$'s has degree $\hat{L}$ or less since $\hat{\Psi}_m(x)$ is an irreducible polynomial dividing $x^{2^L-1} + 1$. Therefore, by Theorem 4.33 the characteristic polynomial $\Psi_I(x)$ of the minimal space for $\hat{U}_0$ divides the characteristic polynomial $\hat{\Psi}_m(x)$ of a common space for $\hat{U}_0$, and hence $L \leq \hat{L}$. This completes the proof. ∎

The theorem means that if the decimation factor n is relatively prime to the period of an irreducible sequence $\{ s_k \}$ in the irreducible space $V[\Psi_I(x)]$ of dimension L, each n-decimated sequence U_i has the minimal

---

[7]Each factor of the polynomial $x^{2^L-1} + 1$ has degree L or less. Refer to Property A.5 in Appendix A.

space $V[\Psi_m(x)]$ whose characteristic polynomial $\Psi_m(x)$ is a degree-L factor of $x^{2^L-1}+1$ such that its extension $\Psi_m(x^n)$ is divided by $\Psi_I(x)$.

EXAMPLE 7.14 We consider the 3-decimated sequences U_0, U_1 and U_2 of the sequence S_1 in Table 7.1. The sequence S_1 is an irreducible sequence in the irreducible space $V[x^4+x^3+x^2+x+1]$ of dimension $L=4$ and period $\tau_I = 5$, and the decimation factor $n = 3$ is relatively prime to the period. The polynomial $x^{2^L-1}+1 = x^{15}+1$ has three degree-4 factors x^4+x+1, x^4+x^3+1 and $x^4+x^3+x^2+x+1$, and for the factor $\Psi_m(x) = x^4+x^3+x^2+x+1$ the characteristic polynomial $\Psi_I(x) = x^4+x^3+x^2+x+1$ divides $\Psi_m(x^n) = x^{12}+x^9+x^6+x^3+1$. Therefore, by the theorem the minimal space for each sequence U_i, $i = 0, 1, 2$, is $V[\Psi_m(x)] = x^4+x^3+x^2+x+1$. By (7.2), the initial vectors of U_i's are $s_0 = [0\,1\,0\,1]^t$, $s_1 = [0\,1\,0\,0]^t$ and $s_2 = [0\,0\,1\,0]^t$, and therefore the decimated sequences become respectively

$$\begin{cases} U_0 = [0\,1\,0\,1] \cdot \mathbf{E}_{x^4+x^3+x^2+x+1} = S_5, \\ U_1 = [0\,1\,0\,0] \cdot \mathbf{E}_{x^4+x^3+x^2+x+1} = S_4, \\ U_2 = [0\,0\,1\,0] \cdot \mathbf{E}_{x^4+x^3+x^2+x+1} = S_2, \end{cases} \tag{7.27}$$

for the sequences S_5, S_4 and S_2 in Table 7.1. ♣

In case the decimation factor is a power of 2, it always falls within the special category of Theorem 7.19, as described in the following theorem.

THEOREM 7.20 *Let U_i, $i = 0, 1, \cdots, n\text{-}1$, be the ith n-decimated sequence of an irreducible sequence $\{ s_k \}$ in the irreducible space $V[\Psi_I(x)]$ of dimension L, and let the decimation factor n be a power of 2, that is, $n = 2^j$ for an integer j. Then, the minimal space for each decimated sequence U_i, $i = 0, 1, \cdots, n\text{-}1$, is the irreducible space $V[\Psi_I(x)]$.*

Proof : By Theorem 7.14, the periodicity T_I of the irreducible space $V[\Psi_I(x)]$ divides the odd number $2^L - 1$, and by Theorem 7.16, the period τ_I of the irreducible sequence $\{ s_k \}$ is identical to the periodicity T_I. Therefore, the decimation factor $n = 2^j$ is relatively prime to the period $\tau_I = T_I$, which is odd. Further, since $\Psi_I(x^{2^j}) = [\Psi_I(x)]^{2^j}$, $\Psi_I(x)$ divides $\Psi_I(x^n)$. Therefore, by Theorem 7.19, we have the theorem. ∎

The theorem indicates that the case of power-of-two decimation factor is a special case in which the minimal space for each n-decimated sequence U_i of an irreducible sequence $\{ s_k \}$ always becomes identical to the irreducible space containing the irreducible sequence $\{ s_k \}$ itself.

EXAMPLE 7.15 In the case of the 2-decimated sequences U_0 and U_1 of the sequence S_1 in Table 7.1, the sequence S_1 is an irreducible sequence in the irreducible space $V[x^4 + x^3 + x^2 + x + 1]$, and the decimation factor $n = 2$ is a power of 2. Therefore, by the theorem the minimal space for the decimated sequences U_0 and U_1 is the irreducible space $V[x^4 + x^3 + x^2 + x + 1]$. By (7.2), the initial vectors of U_0 and U_1 are $s_0 = [\,0\,0\,1\,0\,]^t$ and $s_1 = [\,0\,1\,0\,0\,]^t$, and hence the decimated sequences become respectively

$$\begin{cases} U_0 = [\,0\,0\,1\,0\,] \cdot \mathbf{E}_{x^4 + x^3 + x^2 + x + 1} = S_2, \\ U_1 = [\,0\,1\,0\,0\,] \cdot \mathbf{E}_{x^4 + x^3 + x^2 + x + 1} = S_4, \end{cases} \qquad (7.28)$$

for the sequences S_2 and S_4 in Table 7.1. ♣

7.5 Decimation of Power Sequences

To complement the discussions on irreducible sequences in the previous section, we consider how to determine the decimated sequences of a sequence whose characteristic polynomial is a power of an irreducible polynomial.

DEFINITION 7.21 (Power Sequence Space) A *power sequence space* (in short, *power space*) is defined to be a sequence space $V[[\Psi_I(x)]^w]$ whose characteristic polynomial $[\Psi_I(x)]^w$ is a power of an irreducible polynomial $\Psi_I(x)$, and the superscript w in the characteristic polynomial $[\Psi_I(x)]^w$ is called the *power* of the power space $V[[\Psi_I(x)]^w]$.

DEFINITION 7.22 (Power Sequence) A non-zero sequence in a power space is called a *power sequence*.

For simple notations, we define the binary-ceiling of a decimal number as follows :

DEFINITION 7.23 (Binary-ceiling) The *binary-ceiling* $\bar{w}$ of a decimal number w is the smallest power-of-two number larger than or equal to w.

For example, the binary-ceiling numbers for some decimal numbers are as listed in in Table 7.2.

Then the periodicity of a power space can be described as follows :

Table 7.2. Relation between a number w and its binary-ceiling $\bar{w}$

w	1	2	3 - 4	5 - 8	9 - 16	17 - 32	33 - 64	$\cdots$
$\bar{w}$	1	2	4	8	16	32	64	$\cdots$

THEOREM 7.24 *Let T_I be the periodicity of an irreducible space $V[\Psi_I(x)]$. Then, the periodicity T_w of the power space $V[[\Psi_I(x)]^w]$ is*

$$T_w = \bar{w}T_I \tag{7.29}$$

for the binary-ceiling number $\bar{w}$ of the power w.

Proof : To prove the theorem, we first prove that T_I divides T_w and T_w divides $\bar{w}T_I$. To do this, we consider the irreducible sequence $\{ s_k \} = S[\Psi_I(x), 1]$ and the power sequence $\{ \hat{s}_k \} = S[[\Psi_I(x)]^w, 1]$. Then, by Theorem 4.34, their minimal spaces are respectively the irreducible space $V[\Psi_I(x)]$ and the power space $V[[\Psi_I(x)]^w]$, and hence by Theorem 4.37, their periods τ_I and τ_w are identical to the periodicities T_I and T_w of the sequence spaces $V[\Psi_I(x)]$ and $V[[\Psi_I(x)]^w]$ respectively. On the other hand, since $\Psi_I(x)$ divides $x^{T_I} + 1$ and $w \leq \bar{w}$, $\Psi_I(x)$ divides $[\Psi_I(x)]^w$ and also $[\Psi_I(x)]^w$ divides $(x^{T_I} + 1)^{\bar{w}} = x^{\bar{w}T_I} + 1$.[8] Hence, by Theorem 4.25 the irreducible sequence $\{ s_k \}$ and the power sequence $\{ \hat{s}_k \}$ are elements of the sequence spaces $V[[\Psi_I(x)]^w]$ and $V[x^{\bar{w}T_I} + 1]$ respectively, and by Theorem 4.29, their periods τ_I and τ_w respectively divide the periodicities T_w and $\bar{w}T_I$ of $V[[\Psi_I(x)]^w]$ and $V[x^{\bar{w}T_I} + 1]$. Therefore, T_I and T_w respectively divide T_w and $\bar{w}T_I$.

To complete the proof, we put $\bar{w} = 2^l$ and $T_w = 2^{\hat{l}}T_I$ for an $\hat{l} \leq l$, noting that T_I divides T_w and T_w divides $\bar{w}T_I = 2^l T_I$. Then, by the definition of periodicity, $[\Psi_I(x)]^w$ divides $x^{2^{\hat{l}}T_I} + 1$, that is, $[\Psi_I(x)]^w$ divides $(x^{T_I} + 1)^{2^{\hat{l}}}$. If $\hat{l} < l$, then $w > 2^{\hat{l}}$, and hence $[\Psi_I(x)]^2 = \Psi_I(x^2)$ divides $(x^{T_I} + 1)$. But it is impossible since T_I is an odd number due to Theorem 7.14. Therefore, $\hat{l} = l$. This proves the theorem. ∎

The theorem means that if an irreducible characteristic polynomial is raised in power by w, then the corresponding power space is raised

---

[8]Note that for a binary polynomial $a(x)$, $a(x^{2^j}) = [a(x)]^{2^j}$, $j = 0, 1, \cdots$. Refer to Property A.7 in Appendix A.

Table 7.3. Elements of the power space $V[(x^2 + x + 1)^2]$

elements	minimal spaces
$\hat{S}_0 \;= [0000] \cdot \mathbf{E}_{(x^2+x+1)^2} = \{0,0,0,0,0,0,0,0,0,0,\cdots\}$	$V[1]$
$\hat{S}_1 \;= [0001] \cdot \mathbf{E}_{(x^2+x+1)^2} = \{0,0,0,1,0,1,0,0,0,1,\cdots\}$	$V[(x^2+x+1)^2]$
$\hat{S}_2 \;= [0010] \cdot \mathbf{E}_{(x^2+x+1)^2} = \{0,0,1,0,1,0,0,0,1,0,\cdots\}$	$V[(x^2+x+1)^2]$
$\hat{S}_3 \;= [0011] \cdot \mathbf{E}_{(x^2+x+1)^2} = \{0,0,1,1,1,1,0,0,1,1,\cdots\}$	$V[(x^2+x+1)^2]$
$\hat{S}_4 \;= [0100] \cdot \mathbf{E}_{(x^2+x+1)^2} = \{0,1,0,0,0,1,0,1,0,0,\cdots\}$	$V[(x^2+x+1)^2]$
$\hat{S}_5 \;= [0101] \cdot \mathbf{E}_{(x^2+x+1)^2} = \{0,1,0,1,0,0,0,1,0,1,\cdots\}$	$V[(x^2+x+1)^2]$
$\hat{S}_6 \;= [0110] \cdot \mathbf{E}_{(x^2+x+1)^2} = \{0,1,1,0,1,1,0,1,1,0,\cdots\}$	$V[x^2+x+1]$
$\hat{S}_7 \;= [0111] \cdot \mathbf{E}_{(x^2+x+1)^2} = \{0,1,1,1,1,0,0,1,1,1,\cdots\}$	$V[(x^2+x+1)^2]$
$\hat{S}_8 \;= [1000] \cdot \mathbf{E}_{(x^2+x+1)^2} = \{1,0,0,0,1,0,1,0,0,0,\cdots\}$	$V[(x^2+x+1)^2]$
$\hat{S}_9 \;= [1001] \cdot \mathbf{E}_{(x^2+x+1)^2} = \{1,0,0,1,1,1,1,0,0,1,\cdots\}$	$V[(x^2+x+1)^2]$
$\hat{S}_{10}= [1010] \cdot \mathbf{E}_{(x^2+x+1)^2} = \{1,0,1,0,0,0,1,0,1,0,\cdots\}$	$V[(x^2+x+1)^2]$
$\hat{S}_{11}= [1011] \cdot \mathbf{E}_{(x^2+x+1)^2} = \{1,0,1,1,0,1,1,0,1,1,\cdots\}$	$V[x^2+x+1]$
$\hat{S}_{12}= [1100] \cdot \mathbf{E}_{(x^2+x+1)^2} = \{1,1,0,0,1,1,1,1,0,0,\cdots\}$	$V[(x^2+x+1)^2]$
$\hat{S}_{13}= [1101] \cdot \mathbf{E}_{(x^2+x+1)^2} = \{1,1,0,1,1,0,1,1,0,1,\cdots\}$	$V[x^2+x+1]$
$\hat{S}_{14}= [1110] \cdot \mathbf{E}_{(x^2+x+1)^2} = \{1,1,1,0,0,1,1,1,1,0,\cdots\}$	$V[(x^2+x+1)^2]$
$\hat{S}_{15}= [1111] \cdot \mathbf{E}_{(x^2+x+1)^2} = \{1,1,1,1,0,0,1,1,1,1,\cdots\}$	$V[(x^2+x+1)^2]$

in periodicity by $\bar{w}$. In view of Theorem 4.37, the period of the corresponding power sequence also increases accordingly. That is,

THEOREM 7.25 *For a power sequence $\{\,s_k\,\}$ whose minimal space is the power space $V[[\Psi_I(x)]^w]$, the period τ_w is*

$$\tau_w \;=\; \bar{w}T_I. \tag{7.30}$$

EXAMPLE 7.16 We consider the power space $V[(x^2 + x + 1)^2]$, whose elements are as listed in Table 7.3. In this case $w = 2$, $\bar{w} = 2$, and the periodicity of the irreducible space $V[x^2 + x + 1]$ is $T_I = 3$. Therefore, by Theorem 7.24 the periodicity of the power space $V[(x^2 + x + 1)^2]$ is $T_w = 6$, and by Theorem 7.25 a power sequence $\{\,s_k\,\}$ whose minimal space is the power space $V[(x^2 + x + 1)^2]$ has period $\tau_w = 6$. We can confirm these using Table 7.3, in which all sequences other than $\hat{S}_0$, $\hat{S}_6$, $\hat{S}_{11}$ and $\hat{S}_{13}$ are the corresponding power sequences. ♣

Now we consider how to determine the decimated sequences of a power sequence. The following theorem describes this for an arbitrarily sized decimation factor.

THEOREM 7.26 *For a power sequence* $\{ s_k \}$, *let its decimation factor* n *be* $2^j \hat{n}$ *for an integer* j *and an odd number* $\hat{n}$, *and let its minimal space be the power space* $V[[\Psi_I(x)]^w]$ *for the irreducible polynomial* $\Psi_I(x)$ *of degree* L. *Then, the minimal space for the n-decimated sequences* U_i, $i = 0, 1, \cdots, n\text{-}1$, *is the power space* $V[[\Psi_m(x)]^{w_m}]$ *whose irreducible polynomial* $\Psi_m(x)$ *divides* $x^{2^L - 1} + 1$ *and its extension* $\Psi_m(x^{\hat{n}})$ *is divided by* $\Psi_I(x)$, *and the power* w_m *is the integer in the interval*

$$2^{-j} w \leq w_m < 2^{-j} w + 1. \tag{7.31}$$

Proof : Let the polynomial expression of the power sequence $\{ s_k \}$ be $S[[\Psi_I(x)]^w, P(x)]$ for a polynomial $P(x)$ relatively prime to the irreducible polynomial $\Psi_I(x)$. Then, since $\Psi_I(x)$ divides $\Psi_m(x^{\hat{n}})$ and $w \leq 2^j w_m$ by assumption, $[\Psi_I(x)]^w$ divides $[\Psi_m(x^{\hat{n}})]^{2^j w_m} = [\Psi_m(x^{2^j \hat{n}})]^{w_m} = [\Psi_m(x^n)]^{w_m}$. Hence by Theorem 4.26 $\{ s_k \} = S[[\Psi_m(x^n)]^{w_m}, P(x) [\Psi_m(x^n)]^{w_m}/[\Psi_I(x)]^w]$. Therefore, by Theorem 7.6

$$U_i = S[[\Psi_m(x)]^{w_m}, P_i(x)], \quad i = 0, 1, \cdots, n - 1, \tag{7.32}$$

where $P_i(x)$ is the ith n-decimated polynomial of $P(x)[\Psi_m(x^n)]^{w_m}/ [\Psi_I(x)]^w$. This means that the power space $V[[\Psi_m(x)]^{w_m}]$ is a common space for the decimated sequences U_i's. Let the minimal space for the decimated sequences U_i's be the power space $V[[\Psi_m(x)]^{\hat{w}}]$. Then $\hat{w} \leq w_m$, since $\Psi_m(x)$ is an irreducible polynomial and the characteristic polynomial of the minimal space divides that of a common space due to Theorem 4.33. Further, let each decimated sequence U_i, $i = 0, 1, \cdots$, $n-1$, have the polynomial expression $S[[\Psi_m(x)]^{\hat{w}}, \hat{P}_i(x)]$ on the minimal space $V[[\Psi_m(x)]^{\hat{w}}]$. Then, by Theorem 4.26

$$U_i = S[[\Psi_m(x)]^{w_m}, \hat{P}_i(x)[\Psi_m(x)]^{w_m - \hat{w}}], \quad i = 0, 1, \cdots, n - 1. \tag{7.33}$$

Therefore, comparing this with (7.32), we have the relation $P_i(x) = \hat{P}_i(x)[\Psi_m(x)]^{w_m - \hat{w}}$, $i = 0, 1, \cdots, n-1$. Since $P_i(x)$'s are the n-decimated polynomials of $P(x)[\Psi_m(x^n)]^{w_m}/[\Psi_I(x)]^w$, by (7.6) we obtain the relation $P(x)[\Psi_m(x^n)]^{w_m}/[\Psi_I(x)]^w = \sum_{i=0}^{n-1} x^{n-1-i} \hat{P}_i(x^n)[\Psi_m(x^n)]^{w_m - \hat{w}}$, that is, $P(x) = [\Psi_I(x)]^w \sum_{i=0}^{n-1} x^{n-1-i} \hat{P}_i(x^n)/[\Psi_m(x^n)]^{\hat{w}}$. Therefore, $[\Psi_I(x)]^w$ divides $[\Psi_m(x^n)]^{\hat{w}} = [\Psi_m(x^{\hat{n}})]^{2^j \hat{w}}$, since $P(x)$ is relatively prime to $\Psi_I(x)$. If $\hat{w} < w_m$, then $w > 2^j \hat{w}$, so $[\Psi_I(x)]^2 = \Psi_I(x^2)$ divides $\Psi_m(x^{\hat{n}})$. In this case $\Psi_I(x^2)$ divides $x^{\hat{n}(2^L-1)} + 1$, since $\Psi_m(x^{\hat{n}})$ divides $x^{\hat{n}(2^L-1)} + 1$. But this is a contradiction because both $\hat{n}$ and $2^L - 1$ are odd numbers. Therefore, $\hat{w} = w_m$. This proves the theorem. ∎

The theorem means that for a power sequence $\{ s_k \}$ whose minimal space is the power space $V[[\Psi_I(x)]^w]$ for the irreducible polynomial $\Psi_I(x)$ of degree L, the minimal space of its $(2^j \hat{n})$-decimated sequences U_i's is the power space $V[[\Psi_m(x)]^{w_m}]$, where $\Psi_m(x)$ is the factor of $x^{2^L-1} + 1$ such that its extension $\Psi_m(x^{\hat{n}})$ is a multiple of $\Psi_I(x)$, and the power w_m is determined by (7.31). This result is generally applicable to an arbitrary decimation factor since $2^j \hat{n}$ for $j = 0, 1, \cdots$ and $\hat{n} = 1, 2, \cdots$, can represent any positive integer number. Theorem 7.18 is a special case of this theorem in which $w = 1$ and hence $w_m = 1$. Note that the irreducible polynomial $\Psi_m(x)$ embedded in the minimal space $V[[\Psi_m(x)]^{w_m}]$ for the n-decimated sequences depends only on the original irreducible polynomial $\Psi_I(x)$ and the decimation factor n, and the power w_m depends only on the original power w and the decimation factor n.

EXAMPLE 7.17 We consider the 6-decimated sequences U_i, $i = 0, 1, \cdots, 5$, of the sequence $\hat{S}_1$ in Table 7.3. In this case, the minimal space for the sequence $\hat{S}_1$ is the power space $V[[\Psi_I(x)]^w]$ with the irreducible polynomial $\Psi_I(x) = x^2 + x + 1$ and the power $w = 2$, and the decimation factor n is 2×3, that is, $j = 1$ and $\hat{n} = 3$. For the factor $\Psi_m(x) = x + 1$ of $x^{2^L-1} + 1 = x^3 + 1$, $\Psi_I(x) = x^2 + x + 1$ divides $\Psi_m(x^{\hat{n}}) = x^3 + 1$; and w_m meeting (7.31) is 1. Therefore, by the theorem, the minimal space for the 6-decimated sequences U_i's is the power space $V[x + 1]$. By (7.2), the initial vectors of U_i's are $s_0 = s_1 = s_2 = s_4 = [\,0\,]^t$ and $s_3 = s_5 = [\,1\,]^t$, and hence the decimated sequences become $U_0 = U_1 = U_2 = U_4 = [\,0\,] \cdot E_{x+1} = \{ 0, 0, 0, 0, 0, 0, \cdots \}$ and $U_3 = U_5 = [\,1\,] \cdot E_{x+1} = \{ 1, 1, 1, 1, 1, 1, \cdots \}$. ♣

EXAMPLE 7.18 In the case of the 10-decimated sequences U_i, $i = 0, 1, \cdots, 9$, of a power sequence $\{ s_k \}$ whose minimal space is the power space $V[(x^4 + x^3 + x^2 + x + 1)^3]$, we have $j = 1$, $\hat{n} = 5$, $w = 3$ and $w_m = 2$. Therefore, by the theorem and Example 7.13, the minimal space for the 10-decimated sequences U_i's becomes the power space $V[(x + 1)^2]$. ♣

In the following, we consider two special cases in which the decimation factor is an odd number or a power of 2.

THEOREM 7.27 *For a power sequence* { s_k }, *let its decimation factor n be an odd number, and let its minimal space be the power space $V[[\Psi_I(x)]^w]$ for the irreducible polynomial $\Psi_I(x)$ of degree L. Then, the minimal space for the n-decimated sequences U_i, $i = 0, 1, \cdots, n\text{-}1$, is the power space $V[[\Psi_m(x)]^w]$ whose irreducible polynomial $\Psi_m(x)$ divides $x^{2^L-1} + 1$ and its extension $\Psi_m(x^n)$ is divided by $\Psi_I(x)$.*

Proof : If n is an odd number, j and $\hat{n}$ in Theorem 7.26 becomes 0 and n respectively. Therefore, by (7.31) $w_m = w$, and hence we have the theorem.   ∎

THEOREM 7.28 *For a power sequence* { s_k }, *let its decimation factor n be a power of 2, and let its minimal space be the power space $V[[\Psi_I(x)]^w]$. Then, the minimal space for the n-decimated sequences U_i, $i = 0, 1, \cdots, n\text{-}1$, is the power space $V[[\Psi_I(x)]^{w_m}]$ whose power w_m is the integer satisfying*

$$w/n \leq w_m < w/n + 1. \tag{7.34}$$

Proof : In this case, $n = 2^j$, $j = 0, 1, \cdots$, and $\hat{n} = 1$. Hence, $\Psi_I(x)$ divides $\Psi_I(x^{\hat{n}})$. Also, for $\hat{n} = 1$, (7.31) turns into (7.34). Therefore, by Theorem 7.26, we have the theorem.   ∎

According to the two theorems, we find that in the minimal space $V[[\Psi_m(x)]^{w_m}]$ for the n-decimated sequences U_i's, the original power w is preserved in case the decimation factor n is an odd number ; while the original irreducible polynomial $\Psi_I(x)$ is preserved in case the decimation factor n is a power of 2.

EXAMPLE 7.19 In the case of the 3-decimated sequences U_0, U_1 and U_2 of the sequence $\hat{S}_1$ in Table 7.3, the minimal space for the sequence $\hat{S}_1$ is the power space $V[[\Psi_I(x)]^w]$ with $\Psi_I(x) = x^2 + x + 1$ and $w = 2$, and the decimation factor $n = 3$ is an odd number. For the factor $\Psi_m(x) = x + 1$ of $x^{2^L-1} + 1 = x^3 + 1$, $\Psi_I(x)$ divides $\Psi_m(x^n) = x^3 + 1$. Therefore, by Theorem 7.27, the minimal space for the 3-decimated sequences U_i's is the power space $V[(x + 1)^2]$. By (7.2), the initial vectors of U_i's are $s_0 = s_2 = [\,0\ 1\,]^t$ and $s_1 = [\,0\ 0\,]^t$, and hence the decimated sequences become $U_0 = U_2 = [\,0\ 1\,] \cdot E_{(x+1)^2} = \{\,0, 1, 0, 1, 0, 1, \cdots\,\}$ and $U_1 = [\,0\ 0\,] \cdot E_{(x+1)^2} = \{\,0, 0, 0, 0, 0, 0, \cdots\,\}$. ♣

EXAMPLE 7.20 In the case of the 2-decimated sequences U_0 and U_1 of the sequence $\hat{S}_1$ in Table 7.3, the decimation factor $n = 2$ is a power of 2, and the w_m meeting (7.34) is 1. Therefore, by Theorem 7.28, the minimal space for the 2-decimated sequences U_i's is the power space $V[[\Psi_I(x)]^{w_m}] = V[x^2 + x + 1]$. By (7.2), the initial vectors of U_0 and U_1 are $s_0 = [0\,0\,0]^t$ and $s_1 = [0\,1\,1]^t$, and hence the decimated sequences become $U_0 = [0\,0\,0] \cdot \mathbf{E}_{x^2+x+1} = \hat{S}_0$ and $U_1 = [0\,1\,1] \cdot \mathbf{E}_{x^2+x+1} = \hat{S}_6$ in Table 7.3. ♣

EXAMPLE 7.21 We consider the n-decimated sequences of a power sequence $\{\,s_k\,\}$ whose minimal space is the power space $V[(x^4 + x^3 + x^2 + x + 1)^3]$. If the decimation factor n is 5, by Theorem 7.27 and Example 7.12 the minimal space for the 5-decimated sequences of $\{\,s_k\,\}$ is the power space $V[(x + 1)^3]$. On the other hand, if the decimation factor n is 2, w_m meeting (7.34) is 2, and hence by Theorem 7.28 the minimal space for the 2-decimated sequences of $\{\,s_k\,\}$ is the power space $V[(x^4 + x^3 + x^2 + x + 1)^2]$. ♣

7.6 Decimation of PRBS

In this section, we consider the decimated sequences of a PRBS in a primitive space, which is a special class of irreducible sequences. Note that a PRBS is the sequence which is practically used in the serial scrambling.

We first consider the case when the decimation factor is relatively prime to the period of a PRBS, and then consider the case when it is a power of 2.

THEOREM 7.29 *Let U_i, $i = 0, 1, \cdots, n\text{-}1$, be the ith n-decimated sequence of a PRBS $\{\,s_k\,\}$ in a primitive space $V[\Psi_p(x)]$ of dimension L, and let the decimation factor n be relatively prime to the period $\tau_p = 2^L - 1$ of the PRBS $\{\,s_k\,\}$. Then, each decimated sequence U_i, $i = 0, 1, \cdots, n\text{-}1$, is a PRBS in the primitive space $V[\Psi_m(x)]$ of dimension L whose characteristic polynomial $\Psi_m(x)$ divides $x^{2^L-1} + 1$ and its extension $\Psi_m(x^n)$ is divided by $\Psi_p(x)$.*

Proof : Let $\Psi_m(x)$ be the irreducible polynomial such that $\Psi_m(x)$ divides $x^{2^L-1} + 1$ and $\Psi_p(x)$ divides $\Psi_m(x^n)$. Then, by Theorem 7.19, the minimal space for each n-decimated sequence U_i, $i = 0, 1, \cdots, n - 1$, is the irreducible space $V[\Psi_m(x)]$, and the degree of $\Psi_m(x)$ is L. Therefore, to prove the theorem, it suffices to show that $\Psi_m(x)$ is a primitive polynomial. For this, we consider the 0th n-decimated sequence U_0. By Theorem 7.2, the period of U_0 is $2^L - 1$, and by Theorem 7.15 its minimal space is the irreducible space $V[\Psi_m(x)]$. Therefore, by Theorem 4.37 the periodicity of the irreducible space $V[\Psi_m(x)]$ is $2^L - 1$, that is, the smallest integer T for which $\Psi_m(x)$ divides $x^T + 1$ is $2^L - 1$. Therefore, $\Psi_m(x)$ is a primitive polynomial since $\Psi_m(x)$ is a degree-L polynomial.[9] ∎

The theorem means that if the decimation factor n is relatively prime to the period of a PRBS in a primitive space, each n-decimated sequence is a PRBS in a primitive space of the same dimension.

EXAMPLE 7.22 We consider the 7-decimated sequences U_i, $i = 0, 1, \cdots$, 6, of the sequence $\tilde{S}_{15}$ in Table 6.1. Then, the sequence $\tilde{S}_{15}$ is an element of the primitive space $V[x^4 + x + 1]$ of dimension $L = 4$, and hence its period is $\tau_p = 2^L - 1 = 15$, which is relatively prime to the decimation factor $n = 7$. For the degree-4 primitive factor $\Psi_m(x) = x^4 + x^3 + 1$ of the polynomial $x^{2^L-1} + 1 = x^{15} + 1$, the primitive characteristic polynomial $\Psi_p(x) = x^4 + x + 1$ divides $\Psi_m(x^n) = x^{28} + x^{21} + 1$. Therefore, by the theorem each 7-decimated sequence U_i, $i = 0, 1, \cdots, 6$, is a PRBS in the primitive space $V[\Psi_m(x)] = V[x^4 + x^3 + 1]$ of dimension 4. By (7.2), the initial vectors of U_i's are $s_0 = [1\,1\,0\,0]^t$, $s_1 = [1\,0\,1\,1]^t$, $s_2 = [1\,0\,1\,0]^t$, $s_3 = [1\,1\,1\,0]^t$, $s_4 = [0\,1\,1\,1]^t$, $s_5 = [0\,0\,0\,1]^t$ and $s_6 = [0\,1\,0\,0]^t$, and hence the decimated sequences become respectively

$$\begin{cases} U_0 = [1\,1\,0\,0] \cdot E_{x^4+x^3+1}, \\ U_1 = [1\,0\,1\,1] \cdot E_{x^4+x^3+1}, \\ U_2 = [1\,0\,1\,0] \cdot E_{x^4+x^3+1}, \\ U_3 = [1\,1\,1\,0] \cdot E_{x^4+x^3+1}, \\ U_4 = [0\,1\,1\,1] \cdot E_{x^4+x^3+1}, \\ U_5 = [0\,0\,0\,1] \cdot E_{x^4+x^3+1}, \\ U_6 = [0\,1\,0\,0] \cdot E_{x^4+x^3+1}. \end{cases} \qquad (7.35)$$
♣

THEOREM 7.30 *Let U_i, $i = 0, 1, \cdots, n-1$, be the ith n-decimated sequence of a PRBS $\{ s_k \}$ in a primitive space $V[\Psi_p(x)]$ of dimension L,*

---

[9]For a polynomial $a(x)$ of degree L, if the smallest integer T for which $a(x)$ divides $x^T + 1$ is $2^L - 1$, then $a(x)$ is a primitive polynomial. Refer to Property A.6 in Appendix A.

and let the decimation factor n be a power of 2. Then, each decimated sequence U_i, $i = 0, 1, \cdots, n\text{-}1$, is a PRBS in the same primitive space $V[\Psi_p(x)]$.

Proof : This theorem is directly obtained from Theorem 7.20. ∎

According to this theorem, the original primitive space $V[\Psi_p(x)]$ for a PRBS is preserved for the n-decimated sequences in case the decimation factor n is a power of 2.

EXAMPLE 7.23 In the case of the 4-decimated sequences U_0, U_1, U_2 and U_3 of the sequence $\tilde{S}_{15}$ in Table 6.1, the sequence $\tilde{S}_{15}$ is a PRBS in the primitive space $V[x^4 + x + 1]$, and the decimation factor $n = 4$ is a power of 2. Therefore, by the theorem each 4-decimated sequence U_i, $i = 0$, 1, 2, 3, is a PRBS in the original primitive space $V[x^4 + x + 1]$. By (7.2), the initial vectors of U_i's are $s_0 = [\,1\,0\,0\,0\,]^t$, $s_1 = [\,1\,0\,0\,1\,]^t$, $s_2 = [\,1\,0\,1\,0\,]^t$ and $s_3 = [\,1\,1\,1\,1\,]^t$, and hence the decimated sequences become respectively

$$\begin{cases} U_0 = [\,1\,0\,0\,0\,] \cdot \mathbf{E}_{x^4+x+1} = \tilde{S}_8, \\ U_1 = [\,1\,0\,0\,1\,] \cdot \mathbf{E}_{x^4+x+1} = \tilde{S}_9, \\ U_2 = [\,1\,0\,1\,0\,] \cdot \mathbf{E}_{x^4+x+1} = \tilde{S}_{10}, \\ U_3 = [\,1\,1\,1\,1\,] \cdot \mathbf{E}_{x^4+x+1} = \tilde{S}_{15}, \end{cases} \qquad (7.36)$$

for the sequences $\tilde{S}_8$, $\tilde{S}_9$, $\tilde{S}_{10}$ and $\tilde{S}_{15}$ in Table 6.1. ♣

7.7   Decimation of Sum Sequences

In Section 7.3, we have considered how to decompose a sequence into irreducible and power sequences, and in Sections 7.4 through 7.6, we have considered how to determine the decimated sequences of the irreducible and power sequences as well as the primitive sequences. Therefore, to complete the discussion on the decimation of sequences we consider in this section how to determine the decimated sequences of the sum sequence for the irreducible and the power sequences.

We first consider the common space and the minimal space of decimated sequences.

THEOREM 7.31 Let U_i, $i = 0, 1, \cdots, n\text{-}1$, be the ith n-decimated sequence of a sequence $\{\, s_k \,\}$ whose minimal space is the sequence space $V[\Psi_m(x)]$. Then, the sequence space $V[\hat{\Psi}(x)]$ is a common space for the

decimated sequences U_i, $i = 0, 1, \cdots, n-1$, if and only if $\Psi_m(x)$ divides
$\hat{\Psi}(x^n)$. Furthermore, the minimal space for the decimated sequences U_i,
$i = 0, 1, \cdots, n-1$, is the sequence space $V[\hat{\Psi}_m(x)]$ whose characteris-
tic polynomial $\hat{\Psi}_m(x)$ is the lowest-degree polynomial such that $\Psi_m(x)$
divides $\hat{\Psi}_m(x^n)$.

Proof : For this theorem, it suffices to show the first part for the
common space, since the second part for the minimal space is directly
obtained by the definition of the minimal space. To prove the "if" part,
we let $\{ s_k \} = S[\Psi_m(x), P(x)]$. If $\Psi_m(x)$ divides $\hat{\Psi}(x^n)$, by Theorem
4.26 we have $\{ s_k \} = S[\hat{\Psi}(x^n), P(x)\hat{\Psi}(x^n)/\Psi_m(x)]$, and hence by Theorem 7.6 we obtain $U_i = S[\hat{\Psi}(x), P_i(x)]$, $i = 0, 1, \cdots, n-1$, where $P_i(x)$
is the ith n-decimated polynomial of $P(x)\hat{\Psi}(x^n)/\Psi_m(x)$. Therefore, U_i
$\in V[\hat{\Psi}(x)]$, $i = 0, 1, \cdots, n-1$. This proves the "if" part.

Now to prove the "only if" part, we let $U_i = S[\hat{\Psi}(x), \hat{P}_i(x)]$, $i = 0$,
$1, \cdots, n-1$, for a polynomial $\hat{P}_i(x)$, and consider the sequence $\{ \hat{s}_k \}$
$= S[\hat{\Psi}(x^n), \sum_{i=0}^{n-1} x^{n-1-i}\hat{P}_i(x^n)]$ in the sequence space $V[\hat{\Psi}(x^n)]$. Then,
by Theorem 7.6, the ith n-decimated sequence of $\{ \hat{s}_k \}$ becomes the
sequence U_i, $i = 0, 1, \cdots, n-1$. Therefore, $\{ \hat{s}_k \} = \{ s_k \}$, and so
$V[\hat{\Psi}(x^n)]$ is a common space for $\{ s_k \}$. Hence by Theorem 4.33 $\Psi_m(x)$
divides $\hat{\Psi}(x^n)$. This completes the proof.   ∎

The theorem means that if the minimal space of a sequence $\{ s_k \}$ is $V[\Psi_m(x)]$, then the minimal space for its n-decimated sequences
is the sequence space $V[\hat{\Psi}_m(x)]$ whose characteristic polynomial is the
smallest-degree polynomial such that its extension $\hat{\Psi}_m(x^n)$ is a multiple
of $\Psi_m(x)$.

Now we consider how to determine the minimal space for the decimated sequences of a sum sequence from the minimal spaces for the decimated sequences of the decomposed irreducible and power sequences.
The following theorem describes this.

THEOREM 7.32 *Let a sequence $\{ s_k \}$ be a linear sum of the J irre-*
ducible or power sequences $\{ s_k^j \}$, $j = 0, 1, \cdots, J$-1, whose minimal
spaces are respectively $V[[\Psi_j(x)]^{w_j}]$ for irreducible polynomials $\Psi_j(x)$, j
$= 0, 1, \cdots, J$-1, which are different from one another. Then, the min-
imal space $V[\hat{\Psi}_m(x)]$ for the n-decimated sequences U_i, $i = 0, 1, \cdots,$
n-1, of the sum sequence $\{ s_k \}$ is the sum space of the minimal spaces
$V[[\hat{\Psi}_j(x)]^{\hat{w}_j}]$, $j = 0, 1, \cdots, J$-1, for the n-decimated sequences $\hat{U}_{j,i}$, $i =$
$0, 1, \cdots, n$-1, of the decomposed irreducible or power sequences $\{ s_k^j \}$.

Proof : Let $V[\Psi_m(x)]$ be the minimal space for the sum sequence $\{\ s_k\ \}$, and let $V[\hat{\Psi}(x)]$ be the sum space of the minimal spaces $V[[\hat{\Psi}_j(x)]^{\hat{w}_j}]$, $j = 0, 1, \cdots, J-1$. Then, by Theorem 7.10

$$\Psi_m(x) = [\Psi_0(x)]^{w_0}[\Psi_1(x)]^{w_1}\cdots[\Psi_{J-1}(x)]^{w_{J-1}}, \qquad (7.37)$$

and by Theorem 7.8

$$\hat{\Psi}(x) = LCM[[\hat{\Psi}_0(x)]^{\hat{w}_0}, [\hat{\Psi}_1(x)]^{\hat{w}_1}, \cdots, [\hat{\Psi}_{J-1}(x)]^{\hat{w}_{J-1}}]. \qquad (7.38)$$

But by Theorem 7.31 $[\Psi_j(x)]^{w_j}$ divides $[\hat{\Psi}_j(x^n)]^{\hat{w}_j}$, $j = 0, 1, \cdots, J-1$, since $V[[\Psi_j(x)]^{w_j}]$ and $V[[\hat{\Psi}_j(x)]^{\hat{w}_j}]$ are respectively the minimal spaces for the sequence $\{\ s_k^j\ \}$ and its n-decimated sequences $\hat{U}_{j,i}$, $i = 0, 1, \cdots$, $n-1$. Thus by (7.37) and (7.38) $\Psi_m(x)$ divides $\hat{\Psi}(x^n)$. Therefore, by Theorem 7.31, $\hat{\Psi}(x)$ is a common space for the n-decimated sequences U_i's, and hence by Theorem 4.33 $\hat{\Psi}_m(x)$ divides $\hat{\Psi}(x)$.

Now to complete the proof, we prove that $\hat{\Psi}(x)$ divides $\hat{\Psi}_m(x)$. By Theorem 7.31, $\Psi_m(x)$ divides $\hat{\Psi}_m(x^n)$ since $V[\Psi_m(x)]$ and $V[\hat{\Psi}_m(x)]$ are respectively the minimal spaces for the sequence $\{\ s_k\ \}$ and its n-decimated sequences U_i, $i = 0, 1, \cdots, n-1$. Thus by (7.37) $[\Psi_j(x)]^{w_j}$, $j = 0, 1, \cdots, J-1$, divides $\hat{\Psi}_m(x^n)$. Therefore, by Theorem 7.31 $\hat{\Psi}_m(x)$ is a common space for the n-decimated sequences $\hat{U}_{j,i}$'s, and by Theorem 4.33 $[\hat{\Psi}_j(x)]^{\hat{w}_j}$, $j = 0, 1, \cdots, J-1$, divides $\hat{\Psi}_m(x)$. Hence by (7.38) $\hat{\Psi}(x)$ divides $\hat{\Psi}_m(x)$. ∎

The theorem means that if a sequence $\{\ s_k\ \}$ is decomposed into irreducible or power sequences $\{\ s_k^j\ \}$'s, the minimal space for the decimated sequences of the sequence $\{\ s_k\ \}$ is the sum space of the minimal spaces for the decimated sequences of the decomposed sequences $\{\ s_k^j\ \}$'s. According to the discussions in Section 7.3 an arbitrary sequence can be decomposed into irreducible and power sequences, and, further, the minimal spaces for the decimated sequences of the irreducible and power sequences can be determined according to the theorems provided in Sections 7.4 and 7.5. Therefore, it is now possible to determine the minimal spaces for the decimated sequences of any arbitrary sequences.

EXAMPLE 7.24 We consider the 3-decimated sequences U_0, U_1 and U_2 of a sequence $\{\ s_k\ \}$ whose minimal space is the sequence space $V[x^8 + x^7 + x^4 + x + 1]$. Then, the sequence $\{\ s_k\ \}$ is uniquely represented by a linear sum of the two sequences whose minimal spaces are respectively

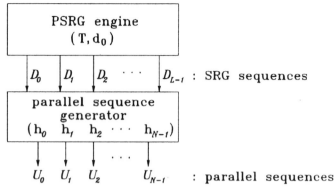

Figure 7.3. Blockdiagram of parallel shift register generator.

$V[(x^2+x+1)^2]$ and $V[x^4+x^3+x^2+x+1]$ (refer to Example 7.9), and the minimal spaces for the 3-decimated sequences of the two sequences are respectively $V[(x+1)^2]$ (refer to Example 7.19) and $V[x^4+x^3+x^2+x+1]$ (refer to Example 7.14). Therefore by the theorem the minimal space for the 3-decimated sequences U_0, U_1 and U_2 is the sum space $V[(x+1)^2]$ $+ V[x^4 + x^3 + x^2 + x + 1]$, or $V[(x + 1)^2(x^4 + x^3 + x^2 + x + 1)]$. ♣

7.8 Parallel Shift Register Generators

Based on the theorems we have discussed so far, we now consider how to realize an SRG generating parallel sequences, which are the decimated sequences of a serial sequence. We call such an SRG a *parallel shift register generator* (PSRG).

A PSRG is composed of two parts —— the *PSRG engine part* and the *parallel sequence generator part*, as shown in Fig. 7.3. The PSRG engine part generates the SRG sequences D_i, $i = 0, 1, \cdots, L-1$; and the parallel sequence generator part forms the N parallel sequences U_j, $j = 0, 1, \cdots$, $N - 1$, by linearly summing the SRG sequences D_i's. Since the SRG sequences D_i's are uniquely determined by the state transition matrix $\mathbf{T}$ and the initial state vector $\mathbf{d}_0$, the PSRG engine can be uniquely described by the state transition matrix $\mathbf{T}$ and the initial state vector $\mathbf{d}_0$; and since the summations of the SRG sequences D_i's are uniquely determined by the generating vector $\mathbf{h}$ defined in (6.13), the parallel sequence generator can be uniquely described by the generating vectors $\mathbf{h}_j$, $j = 0, 1, \cdots, N - 1$.

Therefore, in order to realize a PSRG generating the desired parallel sequences, we need to determine the state transition matrix $\mathbf{T}$ and the initial state vector $\mathbf{d}_0$ of the PSRG engine part as well as the generating vectors $\mathbf{h}_j$'s of the parallel sequence generator part. The PSRG engine part is determined by the following theorem.

THEOREM 7.33 *Let* $\mathbf{T}$ *and* $\mathbf{d}_0$ *be respectively the state transition matrix and the initial state vector of a PSRG generating the* N *parallel sequences* U_j, $j = 0, 1, \cdots, N\text{-}1$, *whose minimal space is the sequence space* $V[\Psi_m(x)]$. *Then, both the SRG maximal space* $V[\mathbf{T}]$ *and the SRG space* $V[\mathbf{T}, \mathbf{d}_0]$ *have the minimal space* $V[\Psi_m(x)]$ *as a subspace.*

Proof : If a PSRG engine with the state transition matrix $\mathbf{T}$ and the initial state vector $\mathbf{d}_0$ can generate the N parallel sequences U_j's using some generating vectors $\mathbf{h}_j$'s, then the SRG space $V[\mathbf{T}, \mathbf{d}_0]$ is a common space for the parallel sequences U_j's. Therefore, by Theorems 4.33 and 4.25 the minimal space $V[\Psi_m(x)]$ for the sequences U_i's is a subspace of the common space $V[\mathbf{T}, \mathbf{d}_0]$. Further, by Theorem 5.17 the SRG space $V[\mathbf{T}, \mathbf{d}_0]$ is a subspace of the SRG maximal space $V[\mathbf{T}]$. Hence the minimal space $V[\Psi_m(x)]$ is also a subspace of the SRG maximal space $V[\mathbf{T}]$. ∎

The theorem means that if the minimal space for the parallel sequences is $V[\Psi_m(x)]$, the state transition matrix $\mathbf{T}$ and the initial state vector $\mathbf{d}_0$ of the PSRG engine part should be chosen such that the minimal space $V[\Psi_m(x)]$ is a subspace of both the SRG maximal space $V[\mathbf{T}]$ and the SRG space $V[\mathbf{T}, \mathbf{d}_0]$. Therefore, by Theorem 5.14 the state transition matrix $\mathbf{T}$ should be chosen such that its minimal polynomial $\Psi_{\mathbf{T}}(x)$ is a multiple of the characteristic polynomial $\Psi_m(x)$, and by Theorems 5.12 and 5.9 the initial state vector $\mathbf{d}_0$ should be chosen such that the lowest-degree polynomial $\Psi_{\mathbf{T}, \mathbf{d}_0}(x)$ meeting (5.6) is also a multiple of $\Psi_m(x)$. Note that the PSRG engine part depends only on the minimal space for the parallel sequences, independently of their initial vectors.

Now we consider how to determine the generating vectors of the parallel sequence generator part which is to operate in conjunction with the PSRG engine part meeting Theorem 7.33. The following theorem describes this.

THEOREM 7.34 *Let* $\mathbf{h}_j$, $j = 0, 1, \cdots, N\text{-}1$, *be the generating vectors of a PSRG generating the* N *parallel sequences* U_j, $j = 0, 1, \cdots, N\text{-}1$, *whose initial vectors are respectively* s_j, $j = 0, 1, \cdots, N\text{-}1$. *Then,*

$$[\, d_0 \quad T \cdot d_0 \quad \cdots \quad T^{\hat{L}-1} \cdot d_0\,]^t \cdot h_j \;=\; s_j \qquad (7.39)$$

for the state transition matrix T *and the initial state vector* d_0 *meeting Theorem 7.33, where* $\hat{L}$ *is the dimension of the SRG space* $V[T, d_0]$.

Proof : Let $V[\Psi_{T,d_0}(x)]$ be the minimal space for the SRG sequences D_i's. Then, by (5.7) and (6.13) we have the relation $U_j = h_j^t \cdot [\, d_0 \quad T \cdot d_0 \quad \cdots \quad T^{\hat{L}-1} \cdot d_0\,] \cdot E_{\Psi_{T,d_0}}(x)$, $j = 0, 1, \cdots, N-1$. But by Theorem 5.12 $V[T, d_0] = V[\Psi_{T,d_0}(x)]$, and hence $U_j = s_j^t \cdot E_{\Psi_{T,d_0}}(x)$. Therefore, we have the relation (7.39).

To complete the proof, we prove the existence of h_j meeting (7.39). Since the dimension of the SRG space $V[T, d_0]$ is $\hat{L}$, by Theorems 5.12 and 5.9 the rank of the matrix $[\, d_0 \quad T \cdot d_0 \quad \cdots \quad T^{\hat{L}-1} \cdot d_0\,]^t$ is $\hat{L}$. Therefore, there exists a generating vector h_j meeting (7.39), since s_j is an $\hat{L}$-vector. ∎

As a consequence of the above two theorems, we can realize PSRGs in the following manner : We first find a state transition matrix T whose minimal polynomial $\Psi_T(x)$ is a multiple of the characteristic polynomial $\Psi_m(x)$ of the minimal space for the parallel sequences, and then choose an initial state vector d_0 such that the lowest-degree polynomial $\Psi_{T,d_0}(x)$ meeting (5.6) is also a multiple of $\Psi_m(x)$. Finally, we select generating vectors h_j, $j = 0, 1, \cdots, N-1$, meeting (7.39).

EXAMPLE 7.25 We realize a PSRG generating the 4 parallel sequences U_i, $i = 0, 1, 2, 3$, in (7.36), which are the 4-decimated sequences of the PRBS $\tilde{S}_{15}$ in Table 6.1. Note that their minimal space is $V[\Psi_m(x)] = V[x^4 + x + 1]$, and their initial vectors are respectively

$$\begin{cases} s_0 = [\,1\,0\,0\,0\,]^t, \\ s_1 = [\,1\,0\,0\,1\,]^t, \\ s_2 = [\,1\,0\,1\,0\,]^t, \\ s_3 = [\,1\,1\,1\,1\,]^t. \end{cases} \qquad (7.40)$$

We first choose the state transition matrix

$$T = \begin{bmatrix} 0 & 0 & 0 & 0 & 1 \\ 1 & 0 & 0 & 0 & 0 \\ 0 & 1 & 0 & 0 & 1 \\ 0 & 0 & 1 & 1 & 0 \\ 0 & 0 & 0 & 1 & 0 \end{bmatrix}. \qquad (7.41)$$

Then, its minimal polynomial is $\Psi_T(x) = x^5 + x^4 + x^2 + 1 = (x + 1)(x^4 + x + 1)$, which is a multiple of $\Psi_m(x) = x^4 + x + 1$. Therefore, the minimal space $V[\Psi_m(x)]$ is a subspace of the SRG maximal space $V[T]$, thus meeting Theorem 7.33. For this SRG, we choose the initial state vector

$$\mathbf{d}_0 = [\,1\,1\,0\,0\,0\,]^t. \tag{7.42}$$

Then, the lowest-degree polynomial $\Psi_{T,\mathbf{d}_0}(x)$ meeting (5.6) is $x^4 + x + 1$, which is also a multiple of $\Psi_m(x)$. Therefore, the minimal space $V[\Psi_m(x)]$ is a subspace of the SRG space $V[T, \mathbf{d}_0]$, which meets Theorem 7.33. Finally, to determine the generating vectors $\mathbf{h}_j$'s, we insert (7.41) and (7.42) into (7.39). Then, we have the relation

$$\begin{bmatrix} 1 & 1 & 0 & 0 & 0 \\ 0 & 1 & 1 & 0 & 0 \\ 0 & 0 & 1 & 1 & 0 \\ 0 & 0 & 0 & 0 & 1 \end{bmatrix} \cdot \mathbf{h}_j = \mathbf{s}_j.$$

Inserting the initial vectors $\mathbf{s}_j$'s in (7.40) into this, we obtain the generating vectors

$$\begin{cases} \mathbf{h}_0 = [\,1\,0\,0\,0\,0\,]^t, \\ \mathbf{h}_1 = [\,1\,0\,0\,0\,1\,]^t, \\ \mathbf{h}_2 = [\,1\,0\,0\,1\,0\,]^t, \\ \mathbf{h}_3 = [\,1\,0\,1\,0\,1\,]^t. \end{cases} \tag{7.43}$$

Using $\mathbf{T}$, $\mathbf{d}_0$ and $\mathbf{h}_j$'s in (7.41) through (7.43), we can build the PSRG depicted in Fig. 7.4. Note that this PSRG generates the parallel scrambling sequences which can be combined with a 4:1 multiplexer, to generate the serial scrambling sequence $\tilde{S}_{15}$ in Table 6.1. ♣

7.9   Minimal Realizations of PSRG

Now, we consider the minimum-length realizations of PSRG which generates the desired parallel sequences. We call such a PSRG a *minimal PSRG*. For a set of desired parallel sequences there does not exist any PSRG whose length is less than the minimal PSRG.

We first consider the length of minimal PSRGs.

THEOREM 7.35 *The length of a minimal PSRG generating the desired parallel sequences is the same as the dimension of the minimal space for the parallel sequences.*

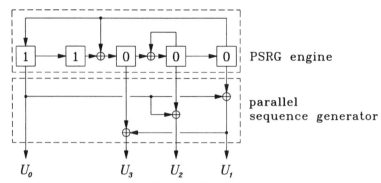

Figure 7.4. An example of the PSRG generating the 4 parallel sequences in (7.36), which are the 4-decimated sequences of the PRBS $\tilde{S}_{15}$ in Table 6.1.

Proof : Let L be the dimension of the minimal space $V[\Psi_m(x)]$ for the parallel sequences ; and let $\hat{L}$ be the length of a minimal PSRG generating the parallel sequences. Then, the state transition matrix $\mathbf{T}$ of the minimal PSRG is an $\hat{L} \times \hat{L}$ matrix, and so the degree of its minimal polynomial $\Psi_{\mathbf{T}}(x)$ is $\hat{L}$ or less.[10] Thus, by Theorem 5.14 the dimension of the SRG maximal space $V[\mathbf{T}]$ is $\hat{L}$ or less. On the other hand, by Theorem 7.33 the minimal space $V[\Psi_m(x)]$ for the parallel sequences is a subspace of the SRG maximal space $V[\mathbf{T}]$. Therefore, the dimension L of the minimal space $V[\Psi_m(x)]$ is $\hat{L}$ or less, that is, $L \leq \hat{L}$.

To complete the proof, we now prove that $\hat{L} \leq L$. For this, we consider a BSRG for the minimal space $V[\Psi_m(x)]$. Then, by Theorem 5.19 its length is L, and it can generate any sequence in $V[\Psi_m(x)]$ with a maximal initial state vector. That is, this BSRG of length L can be a PSRG generating the parallel sequences with appropriate generating vectors. Therefore, $\hat{L} \leq L$, since a minimal PSRG is a minimum-length PSRG. This completes the proof.   ■

According to the theorem, the length of the minimal PSRG is L if the minimal space for the desired parallel sequences is of dimension L. So any PSRG of length less than L can not generate the parallel sequences in an L-dimensional minimal space.

The following two theorems describe how to determine the PSRG engine and the parallel sequence generator of a minimal PSRG.

---

[10] If a matrix $\mathbf{M}$ is an $L \times L$ matrix, the degree of its minimal polynomial $\Psi_{\mathbf{M}}(x)$ can not be larger than L. Refer to Property B.4 in Appendix B.

THEOREM 7.36 *Let* $\mathbf{T}$ *and* $\mathbf{d}_0$ *be respectively the state transition matrix and the initial state vector of a minimal PSRG generating the N parallel sequences* U_j, $j = 0, 1, \cdots, N\text{-}1$, *whose minimal space is* $V[\Psi_m(x)]$. *Then,* $\mathbf{T}$ *is a state transition matrix of a BSRG for the minimal space* $V[\Psi_m(x)]$, *and* $\mathbf{d}_0$ *is a maximal initial state vector for this BSRG.*

Proof : Let L be the dimension of the minimal space $V[\Psi_m(x)]$. Then, by Theorems 7.35, $\mathbf{T}$ is an $L \times L$ matrix, and hence its minimal polynomial $\Psi_{\mathbf{T}}(x)$ has degree L or less. Therefore, by Theorem 5.14 the dimension of the SRG maximal space $V[\mathbf{T}]$ is also L or less. But by Theorem 7.33 the minimal space $V[\Psi_m(x)]$ of dimension L is a subspace of $V[\mathbf{T}]$. Therefore, we have the relation $V[\mathbf{T}] = V[\Psi_m(x)]$, and hence by Theorem 5.19 this SRG with the state transition matrix $\mathbf{T}$ is a BSRG for the sequence space $V[\Psi_m(x)]$.

For an initial state vector $\mathbf{d}_0$, $V[\Psi_m(x)]$ is a subspace of the SRG space $V[\mathbf{T}, \mathbf{d}_0]$ due to Theorem 7.33, and the SRG space $V[\mathbf{T}, \mathbf{d}_0]$ is a subspace of the SRG maximal space $V[\mathbf{T}]$ due to Theorem 5.17. But since $V[\mathbf{T}] = V[\Psi_m(x)]$, we have the relation $V[\mathbf{T}, \mathbf{d}_0] = V[\mathbf{T}]$. This means that $\mathbf{d}_0$ is a maximal initial state vector. ∎

THEOREM 7.37 *Let* $\mathbf{h}_j$, $j = 0, 1, \cdots, N\text{-}1$, *be the generating vectors of a minimal PSRG generating the N parallel sequences* U_j, $j = 0, 1, \cdots, N\text{-}1$, *whose initial vectors are* $\mathbf{s}_j$, $j = 0, 1, \cdots, N\text{-}1$, *respectively. Then,*

$$\mathbf{h}_j = (\Delta_{\mathbf{T},\mathbf{d}_0}^t)^{-1} \cdot \mathbf{s}_j \qquad (7.44)$$

for the state transition matrix $\mathbf{T}$ *and the initial state vector* $\mathbf{d}_0$ *meeting Theorem 7.36, where* $\Delta_{\mathbf{T},\mathbf{d}_0}$ *is the discrimination matrix defined in (5.10).*

Proof : Let $V[\Psi_m(x)]$ of dimension L be the minimal space for the parallel sequences. Then, by Theorem 7.36 $V[\mathbf{T}] = V[\Psi_m(x)]$ and $V[\mathbf{T}, \mathbf{d}_0] = V[\mathbf{T}]$. Therefore, setting the $\hat{L}$ in (7.39) to L, we get the relation $\Delta_{\mathbf{T},\mathbf{d}_0}^t \cdot \mathbf{h}_j = \mathbf{s}_j$. But the discrimination matrix $\Delta_{\mathbf{T},\mathbf{d}_0}$ is nonsingular for the maximal initial state vector $\mathbf{d}_0$ of a BSRG due to Theorem 5.21. Therefore, we have (7.44). ∎

Theorem 7.36 states that the state transition matrix $\mathbf{T}$ of a minimal PSRG engine which generates any desired parallel sequences is one of the BSRGs for the minimal space $V[\Psi_m(x)]$ of the parallel sequences, and the initial state vector d_0 is a maximal initial state vector for this BSRG ; and Theorem 7.37 states that the generating vectors h_j's of a minimal PSRG sequence generator is determined by (7.44) for the minimal PSRG engine meeting Theorem 7.36. Therefore, combining these two with Theorems 5.20 and 5.21 we can realize a minimal PSRG in the following manner : *We first find a state transition matrix* T *which is similar to the companion matrix* $A_{\Psi_m(x)}$, *and then choose an initial state vector* d_0 *making the discrimination matrix* Δ_{T,d_0} *in (5.10) nonsingular. Finally, we determine the generating vectors* h_j *'s according to (7.44).*

In practice, it is not simple to find a state transition matrix T similar to $A_{\Psi_m(x)}$ and its maximal initial state vector d_0 for a PSRG engine. However, the similarity of T and the maximalness of d_0 can be easily achieved by employing the SSRG or the MSRG as a PSRG engine. That is, in the case of the SSRG engine the state transition matrix of the SSRG for the minimal space $V[\Psi_m(x)]$ is $T_S = A_{\Psi_m(x)}^t$, which is similar to $A_{\Psi_m(x)}$, and its maximal initial state vector d_0 can be obtained by Theorem 5.31 ; and in the case of the MSRG engine the state transition matrix of the MSRG for the minimal space $V[\Psi_m(x)]$ is $T_M = A_{\Psi_m(x)}$, which is also similar to $A_{\Psi_m(x)}$, and its maximal initial state vector d_0 can be obtained by Theorem 5.32.

EXAMPLE 7.26 We consider realizing a minimal PSRG generating the 4 parallel sequences U_i, $i = 0, 1, 2, 3$, in (7.36) based on the SSRG. Since the minimal space for the parallel sequences U_i's is $V[\Psi_m(x)] = V[x^4+x+1]$, we take the SSRG for the minimal space $V[x^4+x+1]$. That is, we take the state transition matrix $T_S = A_{x^4+x+1}^t$. For this SSRG we take the maximal initial state vector $d_0 = e_0$, noting that any non-zero initial state vector becomes a maximal one in this primitive space case due to Theorems 5.31 and 6.8. Then, the discrimination matrix Δ_{T_S,d_0} becomes

$$\Delta_{T_S,d_0} = \begin{bmatrix} 1 & 0 & 0 & 0 \\ 0 & 0 & 0 & 1 \\ 0 & 0 & 1 & 0 \\ 0 & 1 & 0 & 0 \end{bmatrix}.$$

Inserting this into (7.44) along with the initial vectors s_j's in (7.40), we obtain the generating vectors $h_0 = e_0$, $h_1 = e_0 + e_1$, $h_2 = e_0 + e_2$, and $h_3 = e_0 + e_1 + e_2 + e_3$. The resulting circuit diagram is as shown in Fig. 7.5. ♣

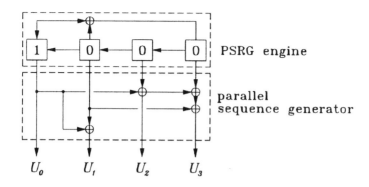

Figure 7.5. An SSRG-based minimal PSRG which generates the 4 parallel sequences in (7.36).

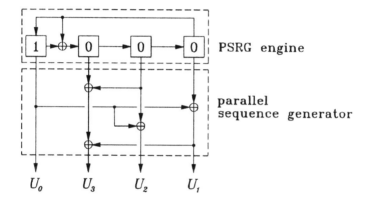

Figure 7.6. An MSRG-based minimal PSRG which generates the 4 parallel sequences in (7.36).

EXAMPLE 7.27 In dualism to the previous example, we consider the MSRG-based realization of the minimal PSRG generating the 4 parallel sequences U_i, $i = 0, 1, 2, 3$, in (7.36). We take the MSRG for the minimal space $V[x^4 + x + 1]$, that is, the state transition matrix $\mathbf{T}_M = \mathbf{A}_{x^4+x+1}$, and for this MSRG we take the maximal initial state vector $\mathbf{d}_0 = \mathbf{e}_0$ in compliance with Theorem 5.32. Then, the discrimination matrix becomes $\Delta_{\mathbf{T}_M, \mathbf{d}_0} = \mathbf{I}$. Therefore, by (7.44) we obtain the generating vectors $\mathbf{h}_i = \mathbf{s}_i$, $i = 0, 1, 2, 3$, and hence we obtain the PSRG diagram shown in Fig. 7.6. ♣

7.10   Simple Realizations of Minimal PSRG

According to Theorems 7.36 and 7.37, there can exist a number of minimal PSRGs for any desired parallel sequences. Therefore, in this section we consider how to find simple-structured PSRGs among them.

As far as the PSRG engine is concerned, the SSRG and the MSRG are the simplest realizations since they are themselves the simplest type of SRGs. However, if we employ the SSRG or the MSRG for the PSRG engine, the sequence generator could become quite complex in generating the desired parallel sequences. The complexity in this case is represented by the number of exclusive-OR gates. Therefore, it could be a better choice, from the viewpoint of overall complexity, to take a PSRG other than the SSRG or MSRG if it renders the attached parallel sequence generator part rather simple. In this regard, we consider efficient realizations of PSRGs. Noting that a minimal PSRG engine is a BSRG, we can combine Theorems 7.36 and 7.37 into the following theorem.

THEOREM 7.38 *Let the N parallel sequences U_j, $j = 0, 1, \cdots, N\text{-}1$, have the expression $\mathbf{s}_j \cdot \mathbf{E}_{\Psi_m(x)}$ for initial vectors $\mathbf{s}_j$'s on their minimal space $V[\Psi_m(x)]$. Then, for a nonsingular matrix $\mathbf{Q}$, the PSRG with the state transition matrix $\mathbf{T}$, the initial state vector $\mathbf{d}_0$, and the generating vectors $\mathbf{h}_j$, $j = 0, 1, \cdots, N\text{-}1$, such that*

$$\left\{ \begin{array}{ll} \mathbf{T} & = \mathbf{Q} \cdot \mathbf{A}_{\Psi_m(x)} \cdot \mathbf{Q}^{-1}, \\ \mathbf{d}_0 & = \mathbf{Q} \cdot \mathbf{e}_0, \\ \mathbf{h}_j & = (\mathbf{Q}^t)^{-1} \cdot \mathbf{s}_j, \end{array} \right. \tag{7.45}$$

is a minimal PSRG generating the parallel sequences U_j's.

Proof : By Theorem 7.36, the minimal PSRG engine with a state transition matrix $\mathbf{T}$ and an initial state vector $\mathbf{d}_0$ is a BSRG for the minimal space $V[\Psi_m(x)]$. Therefore, by Theorem 5.23, $\mathbf{T}$ and $\mathbf{d}_0$ have the expressions $\mathbf{T} = \mathbf{Q} \cdot \mathbf{A}_{\Psi_m(x)} \cdot \mathbf{Q}^{-1}$ and $\mathbf{d}_0 = \mathbf{Q} \cdot \mathbf{e}_0$ for a nonsingular matrix $\mathbf{Q}$. Further, in this case the discrimination matrix $\Delta_{\mathbf{T},\mathbf{d}_0}$ becomes $\mathbf{Q}$ (refer to the proof of Theorem 5.23). Therefore inserting $\Delta_{\mathbf{T},\mathbf{d}_0} = \mathbf{Q}$ into (7.44), we obtain $\mathbf{h}_j = (\mathbf{Q}^t)^{-1} \cdot \mathbf{s}_j$. This proves the theorem. ∎

According to the theorem, an arbitrary nonsingular matrix $\mathbf{Q}$ can be materialized to build a minimal PSRG generating the parallel sequences U_i's whose minimal space is $V[\Psi_m(x)]$ and initial vectors are $\mathbf{s}_j$'s. If the matrix is combined with the relevant companion matrix $\mathbf{A}_{\Psi_m(x)}$ and the initial vectors $\mathbf{s}_j$'s, it renders the state transition matrix $\mathbf{T}$, the initial state vector $\mathbf{d}_0$ and the generating vector $\mathbf{h}_j$'s as specified in (7.45). Therefore, in this case it is of our main interest to determine a nonsingular matrix $\mathbf{Q}$ which can make the state transition matrix $\mathbf{T}$ and the generating vector $\mathbf{h}_j$ simple. The following two theorems describe how to achieve this.

THEOREM 7.39 *Let* $\mathbf{q}_i^t$, $i = 0, 1, \cdots, L\text{-}1$, *be the ith row of the nonsingular matrix* $\mathbf{Q}$ *in (7.45). If* $\mathbf{q}_{i+1}^t = \mathbf{q}_i^t \cdot \mathbf{A}_{\Psi_m(x)}$ *for an* $i = 0, 1, \cdots, L\text{-}2$, *then the ith row,* $\mathbf{t}_i^t$, *of the state transition matrix* $\mathbf{T}$ *in (7.45) becomes*

$$\mathbf{t}_i^t = \mathbf{e}_{i+1}^t. \tag{7.46}$$

Proof : By (7.45), $\mathbf{t}_i^t = \mathbf{e}_i^t \cdot \mathbf{T} = \mathbf{q}_i^t \cdot \mathbf{A}_{\Psi_m(x)} \cdot \mathbf{Q}^{-1}$, and by assumption $\mathbf{q}_{i+1}^t = \mathbf{q}_i^t \cdot \mathbf{A}_{\Psi_m(x)}$, so $\mathbf{t}_i^t = \mathbf{q}_{i+1}^t \cdot \mathbf{Q}^{-1}$. Therefore, $\mathbf{t}_i^t = \mathbf{e}_{i+1}^t$ since $\mathbf{Q} \cdot \mathbf{Q}^{-1} = \mathbf{I}$. ∎

THEOREM 7.40 *Let* $\mathbf{q}_i^t$, $i = 0, 1, \cdots, L\text{-}1$, *be the ith row of the nonsingular matrix* $\mathbf{Q}$ *in (7.45). If* $\mathbf{q}_i^t$ *is identical to the jth initial vector* $\mathbf{s}_j^t$ *for a* $j = 0, 1, \cdots, N\text{-}1$, *then the generating vector* $\mathbf{h}_j$ *in (7.45) becomes*

$$\mathbf{h}_j = \mathbf{e}_i. \tag{7.47}$$

Proof : If $\mathbf{q}_i^t = \mathbf{s}_j^t$, then $\mathbf{e}_i^t \cdot \mathbf{Q} = \mathbf{s}_j^t$, or $\mathbf{Q}^t \cdot \mathbf{e}_i = \mathbf{s}_j$. Therefore, by (7.45) $\mathbf{h}_j = (\mathbf{Q}^t)^{-1} \cdot \mathbf{s}_j = \mathbf{e}_i$. ∎

Theorem 7.39 states that if we choose the nonsingular matrix $\mathbf{Q}$ such that its ith and $(i+1)$th rows have the relation $\mathbf{q}_{i+1}^t = \mathbf{q}_i^t \cdot \mathbf{A}_{\Psi_m(x)}$, the ith row $\mathbf{t}_i^t$ of the state transition matrix $\mathbf{T}$ becomes $\mathbf{e}_{i+1}^t$. If viewed from the SRG structure's point, this corresponds to the implementation that the ith shift register is directly connected to the output of the $(i+1)$th shift register, as depicted in Fig. 7.7. On the other hand, Theorem 7.40 means that if we choose the nonsingular matrix $\mathbf{Q}$ such that its ith row is identical to the transpose of the jth initial vector $\mathbf{s}_j$, the jth generating vector $\mathbf{h}_j$ becomes $\mathbf{e}_i$. In terms of SRG structure, this corresponds to the implementation that the jth parallel sequence U_j is directly taken out of the ith shift register, as indicated in Fig. 7.8.

In the following, we demonstrate a number of simple PSRG realizations that can be achieved by applying Theorems 7.39 and 7.40.

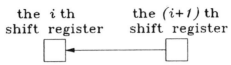

Figure 7.7. The SRG arrangement implementing the relation $\mathbf{t}_i^t = \mathbf{e}_{i+1}^t$.

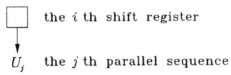

Figure 7.8. The SRG arrangement implementing the relation $\mathbf{h}_j = \mathbf{e}_i$.

EXAMPLE 7.28 We consider a simple minimal PSRG generating the 4 parallel sequences U_i, $i = 0, 1, 2, 3$, in (7.36). Noting that the initial vectors $\mathbf{s}_j$'s $j = 0, 1, 2, 3$, in this case are linearly independent (see (7.40)), we choose the nonsingular matrix $\mathbf{Q}$ such that $\mathbf{q}_i^t = \mathbf{s}_i^t$, $i = 0, 1, 2, 3$, i.e.,

$$\mathbf{Q} = \begin{bmatrix} 1 & 0 & 0 & 0 \\ 1 & 0 & 0 & 1 \\ 1 & 0 & 1 & 0 \\ 1 & 1 & 1 & 1 \end{bmatrix}. \tag{7.48}$$

Then, by Theorem 7.40, the generating vectors are $\mathbf{h}_j = \mathbf{e}_j$, $j = 0, 1, 2, 3$. Inserting (7.48) into (7.45), we obtain the state transition matrix

$$\mathbf{T} = \begin{bmatrix} 1 & 1 & 0 & 0 \\ 0 & 1 & 1 & 0 \\ 0 & 0 & 1 & 1 \\ 1 & 1 & 0 & 1 \end{bmatrix},$$

and the initial state vector $\mathbf{d}_0 = \mathbf{e}_0 + \mathbf{e}_1 + \mathbf{e}_2 + \mathbf{e}_3$. Therefore, we obtain the minimal PSRG shown in Fig. 7.9. This PSRG has a rather complicated SRG engine part but a very simple parallel sequence generator part. Simplification in the parallel sequence generator part therefore is the gain obtained at the sacrifice of the complexity in the SRG engine part. The overall complexity, in terms of the number of exclusive-OR gates, drops in this implementation, when compared to the SSRG-based minimal PSRG in Fig. 7.5 or the MSRG-based minimal PSRG in Fig. 7.6. ♣

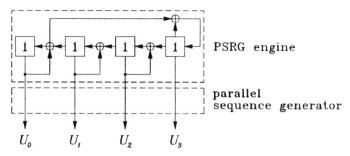

Figure 7.9. A simplified minimal PSRG generating the 4 parallel sequences in (7.36) obtained by taking $q_i^t = s_i^t$, $i = 0, 1, 2$ and 3.

EXAMPLE 7.29 For the 4 parallel sequences U_i, $i = 0, 1, 2, 3$, in (7.36), we now choose the nonsingular matrix $\mathbf{Q}$ such that its ith rows q_i^t's are $q_0^t = s_0^t$, $q_1^t = s_0^t \cdot \mathbf{A}_{x^4+x+1}$, $q_2^t = s_2^t$ and $q_3^t = s_2^t \cdot \mathbf{A}_{x^4+x+1}$, i.e.,

$$\mathbf{Q} = \begin{bmatrix} 1 & 0 & 0 & 0 \\ 0 & 0 & 0 & 1 \\ 1 & 0 & 1 & 0 \\ 0 & 1 & 0 & 1 \end{bmatrix}. \tag{7.49}$$

Then, by Theorem 7.39, the 0th and the 2nd rows of the state transition matrix $\mathbf{T}$ are respectively $t_0^t = e_1^t$ and $t_2^t = e_3^t$, and by Theorem 7.40, the 0th and the 2nd generating vectors are respectively $\mathbf{h}_0 = \mathbf{e}_0$ and $\mathbf{h}_2 = \mathbf{e}_2$. Inserting (7.49) into (7.45), we obtain the 1st and the 3rd rows of the state transition matrix $t_1^t = e_0^t + e_2^t$ and $t_3^t = e_1^t + e_2^t$; the initial state vector $\mathbf{d}_0 = \mathbf{e}_0 + \mathbf{e}_2$; and the 1st and the 3rd generating vectors $\mathbf{h}_1 = \mathbf{e}_0 + \mathbf{e}_1$ and $\mathbf{h}_3 = \mathbf{e}_2 + \mathbf{e}_3$. Therefore, we obtain the minimal PSRG shown in Fig. 7.10. This implementation also requires less number of exclusive-OR gates than the minimal PSRGs in Figs. 7.5 or 7.6. We observe that the complexity is evenly shared by the SRG engine part and the parallel sequence generator part in this example. ♣

EXAMPLE 7.30 We consider another type of minimal PSRG realization for generating the parallel sequences U_0, U_1 and U_2 in (7.5), which are the 3-decimated sequences of the PRBS $\tilde{S}_{15}$ in Table 6.1. The minimal space for the parallel sequences U_i's is $V[\Psi_m(x)] = V[x^4+x^3+x^2+x+1]$ (refer to Example 7.6), and their initial vectors are $s_0 = [1\,1\,0\,0\,]^t$, $s_1 = [1\,0\,1\,1\,]^t$ and $s_2 = [1\,0\,0\,1\,]^t$. If we choose the nonsingular matrix $\mathbf{Q}$

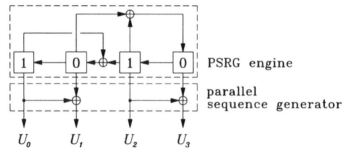

Figure 7.10. Another simplified minimal PSRG generating the 4 parallel sequences in (7.36) obtained by taking $\mathbf{q}_0^t = \mathbf{s}_0^t$, $\mathbf{q}_1^t = \mathbf{s}_0^t \cdot \mathbf{A}_{x^4+x+1}$, $\mathbf{q}_2^t = \mathbf{s}_2^t$ and $\mathbf{q}_3^t = \mathbf{s}_2^t \cdot \mathbf{A}_{x^4+x+1}$.

such that $\mathbf{q}_i^t = \mathbf{s}_0^t \cdot \mathbf{A}_{x^4+x^3+x^2+x+1}^i$, $i = 0, 1, 2, 3$, i.e.,

$$\mathbf{Q} = \begin{bmatrix} 1 & 1 & 0 & 0 \\ 1 & 0 & 0 & 0 \\ 0 & 0 & 0 & 1 \\ 0 & 0 & 1 & 1 \end{bmatrix}, \tag{7.50}$$

then by Theorem 7.39, we obtain $\mathbf{t}_i^t = \mathbf{e}_{i+1}^t$, $i = 0, 1, 2$, and by Theorem 7.40, the 0th generating vector is $\mathbf{h}_0 = \mathbf{e}_0$. Inserting (7.50) into (7.45), we get $\mathbf{t}_3^t = \mathbf{e}_0^t + \mathbf{e}_1^t + \mathbf{e}_2^t + \mathbf{e}_3^t$; $\mathbf{d}_0 = \mathbf{e}_0 + \mathbf{e}_1$; and $\mathbf{h}_1 = \mathbf{e}_1 + \mathbf{e}_3$, $\mathbf{h}_2 = \mathbf{e}_1 + \mathbf{e}_2$. Therefore, we obtain the minimal PSRG shown in Fig. 7.11 whose SRG engine is identical to the SSRG for the minimal space $V[x^4 + x^3 + x^2 + x + 1]$, that is, $\mathbf{T}_S = \mathbf{A}_{x^4+x^3+x^2+x+1}^t$.[11]

To obtain a simple PSRG realization, we choose the nonsingular matrix $\mathbf{Q}$ whose rows are $\mathbf{q}_0^t = \mathbf{s}_0^t$, $\mathbf{q}_1^t = \mathbf{s}_0^t \cdot \mathbf{A}_{x^4+x^3+x^2+x+1}$, $\mathbf{q}_2^t = \mathbf{s}_2^t$ and $\mathbf{q}_3^t = \mathbf{s}_2^t \cdot \mathbf{A}_{x^4+x^3+x^2+x+1}$, i.e.,

$$\mathbf{Q} = \begin{bmatrix} 1 & 1 & 0 & 0 \\ 1 & 0 & 0 & 0 \\ 1 & 0 & 0 & 1 \\ 0 & 0 & 1 & 0 \end{bmatrix}. \tag{7.51}$$

Then by Theorem 7.39, we have $\mathbf{t}_0^t = \mathbf{e}_1^t$ and $\mathbf{t}_2^t = \mathbf{e}_3^t$, and by Theorem 7.40, $\mathbf{h}_0 = \mathbf{e}_0$ and $\mathbf{h}_2 = \mathbf{e}_2$. Inserting (7.51) into (7.45), we get $\mathbf{t}_1^t = \mathbf{e}_1^t +$

---

[11]In fact, if we choose the nonsingular matrix $\mathbf{Q}$ whose ith row is $\mathbf{q}_i^t = \mathbf{v}^t \cdot \mathbf{A}_{\Psi_m(x)}^i$ for a vector $\mathbf{v}$, then we always obtain the SSRG for the minimal space $V[\Psi_m(x)]$ by Theorem 7.39. Note that in the configuration of SSRG the ith shift register is directly connected to the $(i+1)$th shift register.

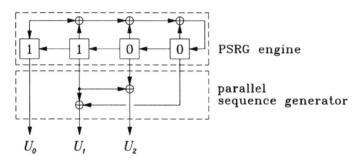

Figure 7.11. A minimal PSRG generating the 3 parallel sequences in (7.5), which are the 3-decimated sequences of the PRBS $\tilde{S}_{15}$ in Table 6.1.

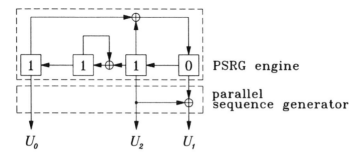

Figure 7.12. A simplified minimal PSRG generating the 3 parallel sequences in (7.5) obtained by taking $\mathbf{q}_0^t = \mathbf{s}_0^t$, $\mathbf{q}_1^t = \mathbf{s}_0^t \cdot \mathbf{A}_{x^4+x^3+x^2+x+1}$, $\mathbf{q}_2^t = \mathbf{s}_2^t$ and $\mathbf{q}_3^t = \mathbf{s}_2^t \cdot \mathbf{A}_{x^4+x^3+x^2+x+1}$.

$\mathbf{e}_2^t, \mathbf{t}_3^t = \mathbf{e}_0^t + \mathbf{e}_2^t$; $\mathbf{d}_0 = \mathbf{e}_0 + \mathbf{e}_1 + \mathbf{e}_2$; and $\mathbf{h}_1 = \mathbf{e}_2 + \mathbf{e}_3$. Therefore, we obtain the minimal PSRG shown in Fig. 7.12. The number of exclusive-OR gates drops to three in this realization, which used to be five in the SSRG-based realization in Fig. 7.11. Note that the minimal PSRGs in Fig. 7.11 or Fig. 7.12 can be used for the parallel scrambling with a 3:1 multiplexer, in case the corresponding original serial scrambling sequence is the PRBS $\tilde{S}_{15}$ in Table 6.1. ♣

EXAMPLE 7.31 Lastly, we consider a simple minimal PSRG realization generating the parallel sequences U_0 and U_1 in (7.3), which are the 2-decimated sequences of the PRBS $\tilde{S}_{15}$ in Table 6.1. The minimal space for the parallel sequences U_i's is $V[\Psi_m(x)] = V[x^4 + x + 1]$ (refer to Example 7.5), and their initial vectors are $\mathbf{s}_0 = [\,1\,1\,0\,0\,]^t$ and $\mathbf{s}_1 =$

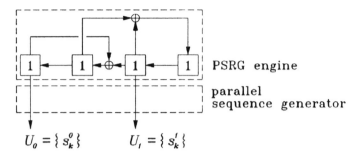

$$U_o = \{ s_k^o \} \qquad\qquad U_i = \{ s_k' \}$$

Figure 7.13. A minimal PSRG generating the 2 parallel sequences in (7.3). This PSRG can be used for generating the parallel sequences $\{ s_k^0 \}$ and $\{ s_k^1 \}$ for use in Fig. 7.2(b).

$[1\,1\,0\,1]^t$. We choose the nonsingular matrix $\mathbf{Q}$ whose rows are $\mathbf{q}_0^t = \mathbf{s}_0^t$, $\mathbf{q}_1^t = \mathbf{s}_0^t \cdot \mathbf{A}_{x^4+x+1}$, $\mathbf{q}_2^t = \mathbf{s}_1^t$ and $\mathbf{q}_3^t = \mathbf{s}_1^t \cdot \mathbf{A}_{x^4+x+1}$, i.e.,

$$\mathbf{Q} = \begin{bmatrix} 1 & 1 & 0 & 0 \\ 1 & 0 & 0 & 0 \\ 1 & 1 & 0 & 1 \\ 1 & 0 & 1 & 0 \end{bmatrix}. \qquad (7.52)$$

Then we have $\mathbf{t}_0^t = \mathbf{e}_1^t$ and $\mathbf{t}_2^t = \mathbf{e}_3^t$; and $\mathbf{h}_0 = \mathbf{e}_0$ and $\mathbf{h}_1 = \mathbf{e}_2$. Inserting (7.52) into (7.45), we get $\mathbf{t}_1^t = \mathbf{e}_0^t + \mathbf{e}_2^t$, $\mathbf{t}_3^t = \mathbf{e}_1^t + \mathbf{e}_2^t$; $\mathbf{d}_0 = \mathbf{e}_0 + \mathbf{e}_1 + \mathbf{e}_2 + \mathbf{e}_3$. Therefore, we obtain the minimal PSRG shown in Fig. 7.13. It has no sequence generator part and its PSRG engine is identical to that in Fig. 7.10. This PSRG can be directly used for generating the parallel sequences $\{ s_k^0 \}$ and $\{ s_k^1 \}$ for use in Fig. 7.2(b). ♣

Chapter 8
Multibit-Parallel Frame Synchronous Scrambling

Multibit-parallel frame synchronous scrambling (MPFSS) is a generalization of the *parallel frame synchronous scrambling* (PFSS) in which multiplexing is done multibit based. In this sense the PFSS we have considered so far is a special case of the MPFSS since in the PFSS multiplexing is done single bit based. Multibit-interleaved multiplexing is a scheme widely used these days in lightwave-based digital transmissions such as SDH/SONET (see Chapter 9), and therefore the parallel scrambling in this environment −− that is, the MPFSS −− becomes very important. As in the case of the PFSS, the parallel input data bitstreams are scrambled before multiplexing, and the scrambled data bitstream is descrambled after demultiplexing in the MPFSS. However, since the multiplexer and demultiplexer employed in the MPFSS operate multibit based, the parallel scrambling sequences for use in the MPFSS is quite different from those employed in the PFSS.

In this chapter, we will discuss the behaviors of parallel scrambling sequences for use in the MPFSS, and then examine how to realize PSRGs generating the parallel scrambling sequences. We will show that the multibit-parallel scrambling sequences can be obtained by decimating the serial scrambling sequences and then interleaving the decimated sequences. For this, we will consider how to identify the sequences obtained by interleaving multiple sequences, and then consider how to determine the minimal space for the parallel sequences for use in the MPFSS. Finally, we will discuss how to realize PSRGs generating the

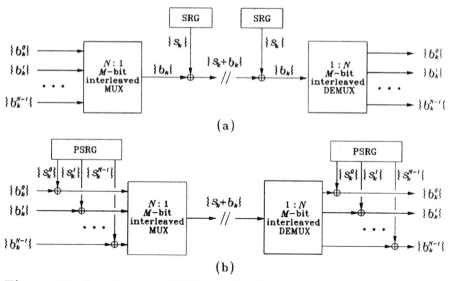

Figure 8.1. Serial and multibit-parallel frame synchronous scrambling. (a) Serial scrambling, (b) multibit-parallel scrambling.

parallel sequences.

8.1   Multibit-Parallel Sequences

We first consider the relation between the serial scrambling sequence and the multibit-parallel scrambling sequences.

The serial FSS and the multibit-parallel FSS can be contrasted as depicted in Fig. 8.1. In the MPFSS, N parallel input data bit streams $\{ b_k^i \}$, $i = 0, 1, \cdots, N - 1$, are scrambled before M-bit interleaved multiplexing by adding the parallel sequences $\{ s_k^i \}$, $i = 0, 1, \cdots, N - 1$, respectively to them ; and the scrambled bitstream, which is identical to the serially-scrambled bitstream $\{ b_k + s_k \}$, is descrambled by the same parallel sequences $\{ s_k^i \}$, $i = 0, 1, \cdots, N - 1$, after M-bit interleaved demultiplexing to reconstruct the demultiplexed bitstreams respectively.

Noting that the M-bit interleaved bitstream of the parallel-scrambled bitstream $\{ b_k + s_k \}$ in Fig. 8.1(b) is identical to the serial-scrambled bitstream $\{ b_k + s_k \}$ in Fig. 8.1(a) for the same parallel input data bitstreams $\{ b_k^i \}$, $i = 0, 1, \cdots, N - 1$, we find that the M-bit interleaved signal of the parallel scrambling sequences $\{ s_k^i \}$, $i = 0, 1, \cdots, N - 1$, should be identical to the serial scrambling sequence $\{ s_k \}$, i.e.,

$$
\begin{cases}
\{s_k^0\} = \{s_0, \cdots, s_{M-1}; s_{MN}, \cdots, s_{M(N+1)-1}; \cdots\}, \\
\{s_k^1\} = \{s_M, \cdots, s_{2M-1}; s_{M(N+1)}, \cdots, s_{M(N+2)-1}; \cdots\}, \\
\qquad\qquad \vdots \\
\{s_k^{N-1}\} = \{s_{M(N-1)}, \cdots, s_{MN-1}; s_{M(2N-1)}, \cdots, s_{2MN-1}; \cdots\}.
\end{cases}
\tag{8.1}
$$

To mathematically describe the relation between the serial and multibit-parallel scrambling sequences in (8.1), we define the interleaved sequence as follows :

DEFINITION 8.1 (Interleaved Sequence) For n sequences $\{ t_k^i \}$, $i = 0, 1, \cdots, n-1$, we define the *interleaved sequence* W to be the sequence

$$
W \equiv \{ t_0^0, t_0^1, \cdots, t_0^{n-1} ; t_1^0, t_1^1, \cdots, t_1^{n-1} ; \cdots\cdots\}.
\tag{8.2}
$$

Then, the parallel sequences $\{ s_k^i \}$'s in (8.1) become the interleaved sequences of the the MN-decimated sequences of the serial sequence $\{ s_k \}$. To be more specific, we first MN-decimate the serial sequence $\{ s_k \}$ to obtain the MN-decimated sequences

$$
\begin{cases}
U_0 = \{ s_0, s_{MN}, s_{2MN}, \cdots \}, \\
\qquad \vdots \\
U_{M-1} = \{ s_{M-1}, s_{MN+M-1}, s_{2MN+M-1}, \cdots \}, \\
U_M = \{ s_M, s_{MN+M}, s_{2MN+M}, \cdots \}, \\
\qquad \vdots \\
U_{2M-1} = \{ s_{2M-1}, s_{MN+2M-1}, s_{MN+2M-1}, \cdots \}, \\
\qquad \vdots \\
U_{M(N-1)} = \{ s_{M(N-1)}, s_{MN+M(N-1)}, s_{2MN+M(N-1)}, \cdots \}, \\
\qquad \vdots \\
U_{MN-1} = \{ s_{MN-1}, s_{2MN-1}, s_{3MN-1}, \cdots \},
\end{cases}
\tag{8.3}
$$

and interleave each adjacent M decimated sequences U_{iM} through $U_{(i+1)M-1}$, $i = 0, 1, \cdots, N - 1$. Then, the resulting sequences are the parallel sequences $\{ s_k^i \}$, $i = 0, 1, \cdots, N - 1$, in (8.1). Note that interleaving is the reverse process of decimation. That is, for the interleaved sequence W in (8.2) the sequence $\{ t_k^i \}$, $i = 0, 1, \cdots, n - 1$, is the ith n-decimated sequence of W.

In the PFSS, the parallel sequences $\{ s_k^i \}$, $i = 0, 1, \cdots, N-1$, in (7.1) depend only on the parameter N indicating the number of parallel input data sequences ; but in the MPFSS, the parallel sequences $\{ s_k^i \}$, $i = 0, 1, \cdots, N - 1$, in (8.1) depend on the parameter N and the parameter M indicating the number of bits taken for multibit interleaving. In this regards, we call the parallel sequences $\{ s_k^i \}$, $i = 0, 1, \cdots, N - 1$, in (8.1) the (M, N) *parallel sequences* for the serial sequence $\{ s_k \}$. Note that the parallel sequences $\{ s_k^i \}$, $i = 0, 1, \cdots, N - 1$, in (7.1) used in the PFSS therefore correspond the $(1, N)$ parallel sequences for the serial sequence $\{ s_k \}$.

EXAMPLE 8.1 We consider the serial scrambling shown in Fig. 8.2(a), which employs a 2:1 2-bit interleaved multiplexer and the PRBS $\tilde{S}_{15}$ in Table 6.1. In this case, $M = 2$ and $N = 2$, so we need the (2,2) parallel sequences W_0 and W_1 for the corresponding 2-bit parallel scrambling. In order to generate them, we first 4-decimate the PRBS $\tilde{S}_{15}$. Then, the decimated sequences become respectively $U_0 = \tilde{S}_8$, $U_1 = \tilde{S}_9$, $U_2 = \tilde{S}_{10}$ and $U_3 = \tilde{S}_{15}$ in Table 6.1 (refer to Example 7.23). Interleaving the sequences U_0 and U_1, and the sequences U_2 and U_3, we obtain the (2,2) parallel sequences

$$\left\{ \begin{array}{l} W_0 = \{\, 1, 1, 0, 0, 0, 0, 0, 1, 1, 1, 0, 0, 0, 1, 1, \\ \qquad\quad 0, 1, 1, 0, 1, 1, 1, 0, 1, 1, 0, 1, 0, 1, 0, \cdots \}, \\ W_1 = \{\, 1, 1, 0, 1, 1, 1, 0, 1, 1, 0, 1, 0, 1, 0, 1, \\ \qquad\quad 1, 0, 0, 0, 0, 0, 1, 1, 1, 0, 0, 0, 1, 1, 0, \cdots \}. \end{array} \right. \qquad (8.4)$$

Therefore, the parallel sequences $\{ s_k^0 \}$ and $\{ s_k^1 \}$ for the 2-bit parallel scrambling in Fig. 8.2(b) should be W_0 and W_1 in (8.4) respectively. If we interleave the sequences W_0 and W_1 in (8.4) 2-bit based, we obtain the serial sequence $\tilde{S}_{15}$ in Table 6.1. ♣

8.2 Interleaved Sequences

In the previous section, we have seen that the parallel sequences can be obtained by applying decimation and interleaving processes on the serial sequences. Since we have fully investigated the properties of the decimated sequences in the previous chapter, we concentrate on the properties of the interleaved sequences in this section.

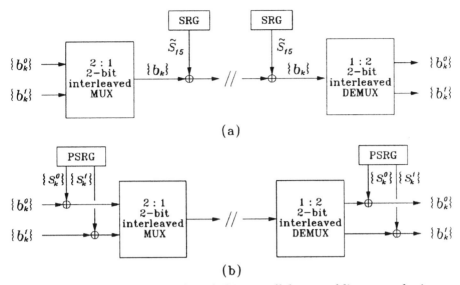

Figure 8.2. The serial and multibit-parallel scramblings employing a 2:1 2-bit interleaved multiplexer and the PRBS $\tilde{S}_{15}$ in Table 6.1. (a) Serial scrambling, (b) 2-bit parallel scrambling.

We first consider the period of the interleaved sequence in relation to the periods of the constituent decimated sequences and the periodicity of their minimal space.

THEOREM 8.2 *Let T be the periodicity of the minimal space $V[\Psi(x)]$ for the n sequences $\{ t_k^i \}$, $i = 0, 1, \cdots, n\text{-}1$, whose periods are respectively τ_i, $i = 0, 1, \cdots, n\text{-}1$. Then,*

$$T = LCM[\tau_0, \tau_1, \cdots, \tau_{n-1}]. \tag{8.5}$$

Proof : Let $V[\Psi_i(x)]$, $i = 0, 1, \cdots, n - 1$, be the minimal space for the sequence $\{ t_k^i \}$; and let $\hat{T} = LCM[\tau_0, \tau_1, \cdots, \tau_{n-1}]$. Then, by Theorem 4.37 $\Psi_i(x)$, $i = 0, 1, \cdots, n - 1$, divides $x^{\tau_i} + 1$, and hence $\Psi_i(x)$ also divides $x^{\hat{T}} + 1$.[1] Therefore, $\Psi(x)$ divides $x^{\hat{T}} + 1$, since $\Psi(x)$ is $LCM[\Psi_0(x), \Psi_1(x), \cdots, \Psi_{n-1}(x)]$ due to Theorem 4.36. Thus by the definition of periodicity we have the relation $T \leq \hat{T}$. On the other hand, by Theorem 4.29 τ_i, $i = 0, 1, \cdots, n - 1$, divides T, and hence $\hat{T}$ divides T. Therefore $\hat{T} = T$. This proves the theorem. ∎

---

[1]If a divides b, $x^a + 1$ divides $x^b + 1$. Refer to Property A.8 in Appendix A.

THEOREM 8.3 *Let W be the interleaved sequence of the n sequences $\{ t_k^i \}$, $i = 0, 1, \cdots, n$-1, whose minimal space is the sequence space $V[\Psi(x)]$ of periodicity T. Then, the period τ of the interleaved sequence W has the form*

$$\tau = \hat{n}T \tag{8.6}$$

for an integer $\hat{n}$ which is a submultiple of n and a multiple of $GCD[T, n]$.

Proof : To prove the theorem, we let $W = \{ s_k \}$. Then, $s_{i+kn} = t_k^i$. Since the period of $\{ s_k \}$ is τ, we have the relation $s_{i+kn} = s_{i+kn+\tau n} = s_{i+(k+\tau)n}$, and hence $t_k^i = t_{k+\tau}^i$, $i = 0, 1, \cdots, n-1$. So the period τ_i, $i = 0, 1, \cdots, n-1$, of $\{ t_k^i \}$ divides τ, and hence by (8.5) T divides τ. Therefore, τ takes the form in (8.6) for an integer $\hat{n}$.

Now we prove that $\hat{n}$ is a submultiple of n. By Theorem 4.29, each period τ_i, $i = 0, 1, \cdots, n-1$, of the sequence $\{ t_k^i \}$ divides T, and hence $t_k^i = t_{k+T}^i$, or $s_{i+kn} = s_{i+(k+T)n} = s_{i+kn+Tn}$, $i = 0, 1, \cdots, n-1$. Therefore, the period $\tau = \hat{n}T$ of $\{ s_k \}$ divides Tn, that is, $\hat{n}$ divides n.

Finally we prove that $\hat{n}$ is a multiple of $GCD[T, n]$. In case $GCD[T, n] = 1$, it is self-evident. So we prove it assuming that $GCD[T, n] \neq 1$, that is, n is not relatively prime to T. Then, since $\{ t_k^i \}$, $i = 0, 1, \cdots, n-1$, is the ith n-decimated sequence of $\{ s_k \}$, whose decimation factor n is not relatively prime to the period $\tau = \hat{n}T$, by Theorem 7.3, the period τ_i of $\{ t_k^i \}$ divides $LCM[\hat{n}T, n]/n$. So, by (8.5) T divides $LCM[\hat{n}T, n]/n$, which is identical to $\hat{n}Tn/nGCD[\hat{n}T, n] = \hat{n}T/GCD[\hat{n}T, n]$. Therefore, $GCD[\hat{n}T, n]$ divides $\hat{n}$, which means that $GCD[T, n]$ divides $\hat{n}$. This completes the proof.   ■

The theorem means that if the n sequences whose minimal space has the periodicity T, are interleaved, then the period τ of the interleaved sequence is $\hat{n}$ times the periodicity T for a number $\hat{n}$ which is a submultiple of n and a multiple of $GCD[T, n]$. Therefore, in case n is a submultiple of T, that is, $GCD[T, n] = n$, then the period τ of W is nT. But in other cases, it is not possible to determine the period τ from the periodicity T of the minimal space $V[\Psi(x)]$ for the n sequences $\{ t_k^i \}$'s, because there exist at least two $\hat{n}$ such that $\hat{n}$ is a submultiple of n and a multiple of $GCD[T, n]$. In fact, the period τ depends on the relations among the n sequences $\{ t_k^i \}$'s, as illustrated in the following example.

EXAMPLE 8.2 We consider the period of the interleaved sequence W of the two non-zero sequences $\{ t_k^0 \}$ and $\{ t_k^1 \}$ in the primitive space

$V[x^4 + x + 1]$ of Table 6.1. Then, by Theorem 6.9, their periods are $\tau_0 = \tau_1 = 2^L - 1 = 15$, and hence by Theorem 8.2 their minimal space $V[\Psi(x)]$ has the periodicity $T = LCM[\tau_0, \tau_1] = 15$ (Note that the minimal space is $V[x^4 + x + 1]$). Therefore, by the theorem the period τ of the interleaved sequence W is 15 or 30, since $GCD[T, n]$ for $n = 2$ is 1. In fact, if we interleave the two sequences $\tilde{S}_8$ and $\tilde{S}_9$ in Table 6.1, the interleaved sequence W has period 30, as confirmed by (8.4) ; but if we interleave the two sequences $\tilde{S}_8$ and $\tilde{S}_{10}$ in Table 6.1, the interleaved sequence W becomes the sequence $\tilde{S}_{12}$ in Table 6.1, whose period is 15. ♣

Now we consider how to determine the interleaved sequences. As a means to this, we define the interleaved polynomial.

DEFINITION 8.4 (Interleaved Polynomial) For n polynomials $P_i(x)$, $i = 0, 1, \cdots, n-1$, we define the *interleaved polynomial* $P(x)$ to be the polynomial

$$P(x) = \sum_{i=0}^{n-1} x^{n-1-i} P_i(x^n). \tag{8.7}$$

Note that the interleaved polynomial and the decimated polynomials have the inverse relation such that if $P_i(x)$'s are the decimated polynomials of $P(x)$, then $P(x)$ is the interleaved polynomial of $P_i(x)$'s, and vice versa.

EXAMPLE 8.3 We consider the four polynomials $P_0(x) = x^3 + 1$, $P_1(x) = x^3$, $P_2(x) = x^3 + x + 1$ and $P_3(x) = x^3 + x^2 + x$. Then, the interleaved polynomial $\hat{P}_0(x)$ for $P_0(x)$ and $P_1(x)$, and the interleaved polynomial $\hat{P}_1(x)$ for $P_2(x)$ and $P_3(x)$ are respectively

$$\begin{cases} \hat{P}_0(x) = x^7 + x^6 + x, \\ \hat{P}_1(x) = x^7 + x^6 + x^4 + x^3 + x^2 + x. \end{cases} ♣ \tag{8.8}$$

Since the interleaving process is the inverse operation of the decimation process, the interleaved sequence is obtained by the following theorem.

THEOREM 8.5 *Let W be the interleaved sequence of the n sequences $\{ t_k^i \}$, $i = 0, 1, \cdots, n\text{-}1$, whose polynomial expressions are $S[\Psi(x), P_i(x)]$, $i = 0, 1, \cdots, n\text{-}1$, respectively, on their minimal space $V[\Psi(x)]$. Then,*

$$W = S[\Psi(x^n), P(x)] \tag{8.9}$$

for the interleaved polynomial $P(x)$ of the n polynomials $P_i(x)$, $i = 0$, 1, $\cdots$, n-1.

   Proof : We denote by $\{\ s_k\ \}$ the sequence $S[\Psi(x^n), P(x)]$. Then, by Theorem 7.6 the sequence $\{\ t_k^i\ \}$, $i = 0, 1, \cdots, n-1$, is the ith n-decimated sequence of $\{\ s_k\ \}$, that is, $\{\ s_k\ \}$ is the interleaved sequence of $\{\ t_k^i\ \}$, $i = 0, 1, \cdots, n-1$. Therefore, we have the relation $t_k^i = s_{i+kn}$ for $i = 0, 1, \cdots, n-1$, and $k = 0, 1, \cdots$, that is, $\{\ t_0^0, t_0^1, \cdots, t_0^{n-1}\ ;\ t_1^0, t_1^1, \cdots, t_1^{n-1}\ ;\ \cdots\ \} = \{\ s_k\ \}$. This proves the theorem. ■

   The theorem indicates that if the n sequences are represented by the polynomial expressions $S[\Psi(x), P_i(x)]$, $i = 0, 1, \cdots, n-1$, on their minimal space $V[\Psi(x)]$, their interleaved sequence has the polynomial expression $S[\Psi(x^n), P(x)]$ for the interleaved polynomial $P(x)$ of the n polynomials $P_i(x)$, $i = 0, 1, \cdots, n-1$. It should be noted, however, that the polynomial expression of the interleaved sequence W in (8.9) does not necessarily imply that the minimal space for W is the sequence space $V[\Psi(x^n)]$. To obtain the minimal space for W, we need to apply Theorem 4.34 to its polynomial expression in (8.9).

EXAMPLE *8.4* We consider the interleaved sequence W_0 of the sequences $\tilde{S}_8$ and $\tilde{S}_9$ in Table 6.1, and the interleaved sequence W_1 of the sequences $\tilde{S}_{10}$ and $\tilde{S}_{15}$ also in Table 6.1. The polynomial expression of $\tilde{S}_8$, $\tilde{S}_9$, $\tilde{S}_{10}$ and $\tilde{S}_{15}$ are respectively $S[x^4 + x + 1, x^3 + 1]$, $S[x^4 + x + 1, x^3]$, $S[x^4 + x + 1, x^3 + x + 1]$ and $S[x^4 + x + 1, x^3 + x^2 + x]$ on their minimal space $V[x^4 + x + 1]$. By Theorem 8.5 and (8.8) the interleaved sequences W_0 and W_1 respectively have the polynomial expressions

$$\begin{cases} W_0 = S[x^8 + x^2 + 1, x^7 + x^6 + x], \\ W_1 = S[x^8 + x^2 + 1, x^7 + x^6 + x^4 + x^3 + x^2 + x]. \end{cases} \quad (8.10)$$

In this case, the sequence space $V[x^8 + x^2 + 1]$ is the minimal space for each interleaved sequence W_i, $i = 0, 1$ in (8.10), since $x^7 + x^6 + x$ and $x^7 + x^6 + x^4 + x^3 + x^2 + x$ are both relatively prime to $x^8 + x^2 + 1$. Applying Theorem 4.35 respectively to the sequences W_0 and W_1 of period $\tau = 30$ in (8.4), we can confirm that their polynomial expressions are represented by (8.10) on the minimal space $V[x^8 + x^2 + 1]$. ♣

EXAMPLE *8.5* We consider the interleaved sequence W of the two sequences $S[x^3+1, 1]$ and $S[x+1, 1]$. Then, by Theorem 4.36 their minimal space is $V[x^3 + 1]$, since the least common multiple of the two characteristic polynomials $x^3 + 1$ and $x + 1$ is $x^3 + 1$. Noting that the sequence $S[x + 1, 1]$ is equivalent to $S[x^3 + 1, x^2 + x + 1]$ on the minimal space $V[x^3 + 1]$, we obtain, by Theorem 8.5, $W = S[x^6 + 1, x + (x^4 + x^2 + 1)]$ $= S[x^6 + 1, x^4 + x^2 + x + 1]$. Now we apply Theorem 4.34 to this. Then, the minimal space for the interleaved sequence W becomes the sequence space $V[x^5 + x^4 + x^3 + x^2 + x + 1]$, and W takes the polynomial expression $S[x^5 + x^4 + x^3 + x^2 + x + 1, x^3 + x^2 + 1]$ on its minimal space. ♣

8.3   Minimal Space for Parallel Sequences

Combining the properties of decimation and interleaving of sequences described in the previous chapter and the previous section, we consider how to determine the minimal space for the (M, N) parallel sequences W_i's in (8.1) along with their periods.

We first consider the periods of the (M, N) parallel sequences. In case MN is relatively prime to the period of a sequence $\{ s_k \}$, the periods of the (M, N) parallel sequences for the sequence $\{ s_k \}$ is determined by the following theorem.

THEOREM 8.6 *Let W_i, $i = 0, 1, \cdots, N\text{-}1$, be the (M, N) parallel sequences for a sequence $\{ s_k \}$ of period τ ($1 < N < \tau$); and let MN be relatively prime to the period τ. Then, the period of each parallel sequence W_i, $i = 0, 1, \cdots, N\text{-}1$, is $M\tau$; and the ith parallel sequence W_i is the (imM^2)-delayed sequence of the 0th parallel sequence W_0, where m is the smallest integer that satisfies*

$$mMN = 1 \quad \text{modulo} \quad \tau. \tag{8.11}$$

Proof : Let U_i, $i = 0, 1, \cdots, MN - 1$, be the ith MN-decimated sequence of the sequence $\{ s_k \}$. Then, by Theorem 7.2, the period of U_i is τ. Noting that W_i is obtained by interleaving the M sequences U_{iM} through $U_{(i+1)M-1}$, we can easily show that the period of W_i is $M\tau$. This proves the first part of the theorem.

For the second part, it suffices to show that the sequence W_{i+1} is the mM^2-delayed sequence of the sequence W_i, $i = 0, 1, \cdots, N - 2$. By Theorem 7.2, the sequence U_{i+M} is the mM-delayed sequence of U_i, $i = 0, 1, \cdots, M(N - 1) - 1$, for m in (8.11). But W_i is the parallel sequence obtained by interleaving the M sequences U_{iM} through $U_{(i+1)M-1}$, and

W_{i+1} is obtained by interleaving the M sequences $U_{(i+1)M}$ through $U_{(i+2)M-1}$. Therefore, W_{i+1} is the $(mM)M$-delayed sequence of W_i. This completes the proof.   ∎

According to the the theorem, the (M, N) parallel sequences W_i's are closely related in case MN is relatively prime to the period τ of a sequence $\{ s_k \}$. They all have the same period $M\tau$, and they are all delayed sequences of the 0th parallel sequence W_0.

EXAMPLE 8.6 In the case of the (2,2) parallel sequences W_0 and W_1 for the PRBS $\tilde{S}_{15}$ in Table 6.1, the period of $\tilde{S}_{15}$ is $\tau = 15$, and $MN(= 4)$ is relatively prime to $\tau(= 15)$. Therefore, by the theorem the parallel sequences W_0 and W_1 have period $M\tau = 30$, and W_1 is the 16-delayed sequence of W_0 since the smallest integer meeting (8.11) is $m = 4$. We can confirm these using the element expressions of W_i's in (8.4). ♣

In case MN is not relatively prime to the period τ of $\{ s_k \}$, the periods of the parallel sequences meet the following theorem.

THEOREM 8.7 Let W_i, $i = 0, 1, \cdots, N$-1, be the (M, N) parallel sequences for a sequence $\{ s_k \}$ of period τ ; and let MN be not relatively prime to the period τ. Then, the period of each parallel sequence W_i, $i = 0, 1, \cdots, N$-1, divides $\hat{\tau}$, which is

$$\hat{\tau} = \frac{LCM[\tau, MN]}{N}. \tag{8.12}$$

Proof : Let U_i, $i = 0, 1, \cdots, MN - 1$, be the ith MN-decimated sequence of the sequence $\{ s_k \}$. Then, by Theorem 7.3, the period of U_i divides $LCM[\tau, MN]/MN$. Therefore, the period of W_i divides $\hat{\tau}$, since W_i is obtained by interleaving the M sequences U_{iM} through $U_{(i+1)M-1}$.   ∎

The theorem means that if MN is not relatively prime to the period τ of a sequence $\{ s_k \}$, the period of each parallel sequence W_i divides $\hat{\tau}$ in (8.12). However, note that it is not possible to determine the periods of the (M, N) parallel sequences W_i's directly from the period τ of the sequence $\{ s_k \}$ in case MN is not relatively prime to τ. That is, even if two sequences $\{ s_k \}$ and $\{ \hat{s}_k \}$ have the same periods (and are in the same minimal space), their (M, N) parallel sequences can have different periods, as the following example illustrates.

EXAMPLE 8.7 We consider the $(2,3)$ parallel sequences of a sequence $\{ s_k \}$ of period $\tau = 6$. In this case, $MN = 6$, which is not relatively prime to the period $\tau = 6$. So, by the theorem all the periods of the parallel sequences divide $\hat{\tau} = LCM[6,6]/3 = 2$. This implies that the periods are either 1 or 2. If we employ the power sequence $\hat{S}_1$ of period $\tau = 6$ in Table 7.3, its $(2,3)$ parallel sequences W_0, W_1 and W_2 become

$$
\begin{cases}
W_0 & = & \{\, 0,\, 0,\, 0,\, 0,\, 0,\, 0,\, \cdots \,\}, \\
W_1 & = & \{\, 0,\, 1,\, 0,\, 1,\, 0,\, 1,\, \cdots \,\}, \\
W_2 & = & \{\, 0,\, 1,\, 0,\, 1,\, 0,\, 1,\, \cdots \,\},
\end{cases}
\tag{8.13}
$$

whose periods are respectively 1, 2, and 2 ; but if we employ the power sequence $\hat{S}_3$ of period $\tau = 6$ in Table 7.3, its $(2,3)$ parallel sequences $\hat{W}_0$, $\hat{W}_1$ and $\hat{W}_2$ become

$$
\begin{cases}
\hat{W}_0 & = & \{\, 0,\, 0,\, 0,\, 0,\, 0,\, 0,\, \cdots \,\}, \\
\hat{W}_1 & = & \{\, 1,\, 1,\, 1,\, 1,\, 1,\, 1,\, \cdots \,\}, \\
\hat{W}_2 & = & \{\, 1,\, 1,\, 1,\, 1,\, 1,\, 1,\, \cdots \,\},
\end{cases}
\tag{8.14}
$$

whose periods are all 1. Therefore, we observe that even if the two sequences $\hat{S}_1$ and $\hat{S}_3$ have the same minimal spaces as well as the same periods, their $(2,3)$ parallel sequences can have different periods. ♣

In regard to the minimal space for the parallel sequences, we have the following guideline.

THEOREM 8.8 *Let W_i, $i = 0, 1, \cdots, N\text{-}1$, be the (M, N) parallel sequences for a sequence $\{ s_k \}$; and let $V[\Psi_m(x)]$ be the minimal space for the MN-decimated sequences U_i, $i = 0, 1, \cdots, MN\text{-}1$, of the sequence $\{ s_k \}$. Then, the minimal space for the (M, N) parallel sequences W_i, $i = 0, 1, \cdots, N\text{-}1$, is a subspace of the sequence space $V[\Psi_m(x^M)]$.*

Proof : Since the parallel sequence W_i, $i = 0, 1, \cdots, N - 1$, is obtained by interleaving the M decimated sequences U_{iM} through $U_{(i+1)M-1}$, by Theorem 8.5 W_i is an element of the sequence space $V[\Psi_m(x^M)]$. Therefore, $V[\Psi_m(x^M)]$ is a common space for the parallel sequences W_i's, and hence by Theorems 4.25 and 4.33 the minimal space for the parallel sequences W_i's is a subspace of the common space $V[\Psi_m(x^M)]$. ■

The theorem gives a guideline for the minimal space for the parallel sequences : *If the minimal space for the MN-decimated sequences U_i's of a sequence $\{ s_k \}$ is $V[\Psi_m(x)]$, then the minimal space for the parallel sequences W_i's for the sequence $\{ s_k \}$ is a subspace of the sequence space $V[\Psi_m(x^M)]$.* Unfortunately, it is not possible to determine the minimal space for the (M, N) parallel sequences W_i's directly from the minimal space for the MN-decimated sequences U_i's, even if the minimal space for the sequence $\{ s_k \}$ is known. Therefore, to determine the minimal space for the parallel sequences W_i's we have to apply Theorems 4.34 and 4.36 to the parallel sequences W_i's, as the following example illustrates.

EXAMPLE 8.8 We consider the $(2,3)$ parallel sequences for a sequence $\{ s_k \}$ whose minimal space is the power space $V[(x^2 + x + 1)^2]$ in Table 7.3. In this case, the minimal space for the 6-decimated sequences U_i's is $V[x + 1]$ (refer to Example 7.17). By the theorem the minimal space for the $(2,3)$ parallel sequences is a subspace of the sequence space $V[x^2 + 1]$. This implies that the minimal space is either $V[x^2 + 1]$ or $V[x + 1]$. If we employ the the power sequence $\hat{S}_1$ in Table 7.3, its $(2,3)$ parallel sequences W_i's in (8.13) have the polynomial expressions $W_0 = S[x^2 + 1, 0]$, $W_1 = S[x^2 + 1, 1]$ and $W_2 = S[x^2 + 1, 1]$ on the sequence space $V[x^2 + 1]$, which are obtained by applying (4.14) to their initial vectors $s_0 = [\, 0 \, 0 \,]^t$, $s_1 = [\, 0 \, 1 \,]^t$ and $s_2 = [\, 0 \, 1 \,]^t$ respectively. Therefore, by Theorems 4.34 and 4.36 the minimal space for the parallel sequences is the sequence space $V[x^2 + 1]$. But if we employ the power sequence $\hat{S}_3$ in Table 7.3, its $(2,3)$ parallel sequences $\hat{W}_i$'s in (8.14) have the polynomial expressions $\hat{W}_0 = S[x^2 + 1, 0]$, $\hat{W}_1 = S[x^2 + 1, x + 1]$ and $\hat{W}_2 = S[x^2 + 1, x + 1]$ on the sequence space $V[x^2 + 1]$. Therefore, by Theorems 4.34 and 4.36 the minimal space for the parallel sequences is the sequence space $V[x + 1]$. This confirms that even if two sequences $\hat{S}_1$ and $\hat{S}_3$ are in the same minimal spaces, their $(2,3)$ parallel sequences can belong to different minimal spaces. ♣

8.4   Parallel Sequences for Irreducible Sequences

In the FSS, a PRBS in a primitive space is employed for the serial scrambling, and in most practical cases MN is taken to be relatively prime to the period of the PRBS.[2] Therefore, in this section, we specifically consider the parallel sequences for the irreducible sequences.[3]

The following theorem describes how to determine the minimal space for the (M, N) parallel sequences for an irreducible sequence $\{ s_k \}$ when MN is relatively prime to the period of the irreducible sequence $\{ s_k \}$.

THEOREM 8.9 *Let W_i, $i = 0, 1, \cdots, N\text{-}1$, be the (M, N) parallel sequences for an irreducible sequence $\{ s_k \}$ in the irreducible space $V[\Psi_I(x)]$ of degree L and periodicity T_I ; and let MN, $1 < N < T_I$, be relatively prime to the periodicity T_I. Then, the minimal space for each (M, N) parallel sequence W_i, $i = 0, 1, \cdots, N\text{-}1$, is the sequence space $V[\Psi_m(x^M)]$, where $\Psi_m(x)$ is the degree-L characteristic polynomial of the minimal space $V[\Psi_m(x)]$ for the MN-decimated sequences U_i, $i = 0, 1, \cdots, MN\text{-}1$, of the irreducible sequence $\{ s_k \}$.*

Proof : By Theorem 8.6, each parallel sequence W_i, $i = 0, 1, \cdots, N - 1$, is a delayed sequence of W_0. Therefore, it suffices to show that the minimal space for W_0 is the sequence space $V[\Psi_m(x^M)]$.

Let $U_0 = S[\Psi_m(x), P(x)]$ for a polynomial $P(x)$ relatively prime to $\Psi_m(x)$; and let m be the smallest integer such that $mMN = 1$ modulo T_I. Then, by Theorem 7.2, U_i for $i = 0, 1 \cdots, MN - 1$, is the (im)-delayed sequence of U_0, and hence by Theorem 4.20, $U_i = S[\Psi_m(x), R[\Psi_m(x), x^{im}P(x)]]$. Since W_0 is the interleaved sequence of the M sequences U_0 through U_{M-1}, by Theorem 8.5, $W_0 = S[\Psi_m(x^M), \sum_{i=0}^{M-1} x^{M-1-i}R[\Psi_m(x^M), x^{imM}P(x^M)]]$. Since the degree of the remainder polynomial $R[\Psi_m(x^M), x^{imM}P(x^M)]$, $i = 0, 1, \cdots, M - 1$, is $M(L - 1)$ or less, we have the relation $\sum_{i=0}^{M-1} x^{M-1-i}R[\Psi_m(x^M), x^{imM}P(x^M)] = \sum_{i=0}^{M-1} R[\Psi_m(x^M), x^{M-1-i}x^{imM}P(x^M)] = R[\Psi_m(x^M), \sum_{i=0}^{M-1} x^{M-1-i}x^{imM}P(x^M)]$. Therefore, $W_0 = S[\Psi_m(x^M), R[\Psi_m(x^M), \sum_{i=0}^{M-1} x^{M-1-i}x^{imM}P(x^M)]]$.

Let $g(x)$ be the greatest common divisor of $\Psi_m(x^M)$ and $R[\Psi_m(x^M), \sum_{i=0}^{M-1} x^{M-1-i}x^{imM}P(x^M)]$, or the greatest common divisor of $\Psi_m(x^M)$ and $\sum_{i=0}^{M-1} x^{M-1-i}x^{imM}P(x^M) = x^{M-1}P(x^M)(x^{(mM-1)M}+1)/(x^{mM-1}+1)$.[4] Then, due to Theorem 4.34, it suffices to show that $g(x) = 1$. Since

---

[2]In the SDH/SONET system, which is a typical transmission system employing the FSS, MN is relatively prime to the period of the PRBS used for the serial scrambling. Refer to Chapter 9.

[3]Note that a PRBS in a primitive space $V[\Psi_p(x)]$ is an irreducible sequence since the primitive characteristic polynomial $\Psi_p(x)$ is irreducible.

[4]For two polynomials $a(x)$ and $b(x)$, $GCD[a(x), b(x)] = GCD[a(x), R[a(x), b(x)]]$. Refer to Property A.9 in Appendix A.

$g(x)$ does not divide x^{M-1}, $g(x)$ divides $GCD[\Psi_m(x^M), (x^{(mM-1)M} + 1)P(x^M)]$.

To prove that $g(x) = 1$, we consider the greatest common divisor of $\Psi_m(x)$ and $(x^{mM-1} + 1)P(x)$. Since $\Psi_m(x)$ is an irreducible polynomial, $GCD[\Psi_m(x), (x^{mM-1} + 1)P(x)]$ is either 1 or $\Psi_m(x)$. If $GCD[\Psi_m(x), (x^{mM-1} + 1)P(x)] = \Psi_m(x)$, $\Psi_m(x)$ divides $x^{mM-1} + 1$ since $P(x)$ is relatively prime to $\Psi_m(x)$. Hence T_I divides $mM - 1$ since the periodicity of the irreducible space $V[\Psi_m(x)]$ is T_I due to Theorems 7.2 and 4.37. But T_I also divides $mMN - N$, and hence by the relation $mMN = 1$ modulo T_I, $N = 1$ modulo T_I, which is a contradiction to the assumption $1 < N < T_I$. Therefore, $GCD[\Psi_m(x), (x^{mM-1} + 1)P(x)] = 1$, and hence $g(x) = 1$. This completes the proof. ∎

Combining this theorem with Theorem 7.19, we can determine the minimal space for the parallel sequences W_i's in the following manner : *We first find the minimal space $V[\Psi_m(x)]$ for the MN-decimated sequences U_i's using Theorem 7.19. Then, the sequence space $V[\Psi_m(x^M)]$ becomes the minimal space for the (M, N) parallel sequences W_i's.*

EXAMPLE 8.9 We consider the $(2,3)$ parallel sequences W_0, W_1 and W_2 for the sequence S_1 in Table 7.1. In this case, the sequence S_1 is an irreducible sequence in the irreducible space $V[x^4 + x^3 + x^2 + x + 1]$ of periodicity $T_I = 5$ (refer to Example 7.10), and $MN = 6$ is relatively prime to $T_I = 5$. By Theorem 7.18 the minimal space $V[\Psi_m(x)]$ for the 6-decimated sequences is identical to the minimal space for the 3-decimated sequences, which is $V[x^4 + x^3 + x^2 + x + 1]$ (refer to Example 7.14). Therefore, by Theorem 8.9 the minimal space for the $(2,3)$ parallel sequences W_i's is the sequence space $V[\Psi_m(x^2)] = V[x^8 + x^6 + x^4 + x^2 + 1]$. Since the initial vectors for W_i's are $s_0 = [00000111]^t$, $s_1 = [01111000]^t$ and $s_2 = [10000001]^t$, the $(2,3)$ parallel sequences W_i's are respectively represented by

$$
\begin{cases}
W_0 &= [0\,0\,0\,0\,0\,1\,1\,1] \cdot \mathbf{E}_{x^8+x^6+x^4+x^2+1}, \\
W_1 &= [0\,1\,1\,1\,1\,0\,0\,0] \cdot \mathbf{E}_{x^8+x^6+x^4+x^2+1}, \\
W_2 &= [1\,0\,0\,0\,0\,0\,0\,1] \cdot \mathbf{E}_{x^8+x^6+x^4+x^2+1},
\end{cases}
\tag{8.15}
$$

on their minimal space $V[x^8 + x^6 + x^4 + x^2 + 1]$. ♣

As a special case of Theorem 8.9, in case MN is a power of 2, we get the following theorem.

THEOREM 8.10 *Let* W_i, $i = 0$, 1, $\cdots$, $N\text{-}1$, *be the* (M, N) *parallel sequences for an irreducible sequence* $\{\ s_k\ \}$ *in the irreducible space* $V[\Psi_I(x)]$ *; and let* MN *be a power of 2. Then, the minimal space for the* (M, N) *parallel sequences* W_i, $i = 0$, 1, $\cdots$, $N\text{-}1$, *is the sequence space* $V[\Psi_I(x^M)]$.

Proof : This theorem is a direct outcome of Theorems 8.9 and 7.20. ∎

The theorem means that if a sequence $\{\ s_k\ \}$ is an irreducible sequence in the irreducible space $V[\Psi_I(x)]$ and if MN is a power of 2, then the irreducible polynomial $\Psi_I(x)$ *itself* becomes the polynomial $\Psi_m(x)$ whose extension $\Psi_m(x^M)$ is the characteristic polynomial of the minimal space $V[\Psi_m(x^M)]$ for the parallel sequences.

EXAMPLE 8.10 In the case of the $(2,2)$ parallel sequences W_0 and W_1 for the PRBS $\tilde{S}_{15}$ in Table 6.1, $\tilde{S}_{15}$ is an irreducible sequence in the primitive space $V[x^4 + x + 1]$ and $MN = 4$ is a power of 2. Therefore, by the theorem the minimal space for the parallel sequences W_0 and W_1 is the sequence space $V[x^8 + x^2 + 1]$. By (8.4) the initial vectors for W_0 and W_1 are $\mathbf{s}_0 = [\,1\,1\,0\,0\,0\,0\,0\,1\,]^t$ and $\mathbf{s}_1 = [\,1\,1\,0\,1\,1\,1\,0\,1\,]^t$, and hence the $(2,2)$ parallel sequences W_0 and W_1 respectively have the expressions

$$\begin{cases} W_0 = [\,1\,1\,0\,0\,0\,0\,0\,1\,] \cdot \mathbf{E}_{x^8+x^2+1}, \\ W_1 = [\,1\,1\,0\,1\,1\,1\,0\,1\,] \cdot \mathbf{E}_{x^8+x^2+1}, \end{cases} \tag{8.16}$$

on their minimal space $V[x^8 + x^2 + 1]$. Using (4.14) we can confirm that the parallel sequences W_0 and W_1 represented by (8.16) have the polynomial expressions in (8.10). ♣

8.5 Realizations of PSRGs for MPFSS

We now consider how to realize the PSRGs generating the (M, N) parallel sequences we have discussed so far. We call such a PSRG an (M, N) *PSRG*.

The (M, N) PSRG is a kind of PSRGs which generate the (M, N) parallel sequences. So, in realizing the (M, N) PSRGs, we can rely on Sections 7.8 through 7.10, which investigated how to realize the PSRGs generating any desired sequences along with their minimal and simple realizations. For the minimal realizations of (M, N) PSRGs, therefore we can employ Theorems 7.36 and 7.37 to determine the state transition matrix $\mathbf{T}$, the initial state vector $\mathbf{d}_0$, and the generating vectors $\mathbf{h}_j$'s for the PSRG.

In case the (M, N) parallel sequences W_j's are for an irreducible serial sequence $\{ s_k \}$, we can find even simpler realizations for the (M, N) PSRGs. According to Theorem 8.9, if MN is relatively prime to the period τ of an irreducible sequence $\{ s_k \}$, the minimal space for the (M, N) parallel sequences W_j's is the sequence space $V[\Psi_m(x^M)]$ for the characteristic polynomial $\Psi_m(x)$ of the minimal space for the MN-decimated sequences U_i's of $\{ s_k \}$. Further, according to Theorem 8.6 each parallel sequence W_j is the jmM^2-delayed sequence of W_0 for the smallest integer m meeting (8.11) in this case. These properties enable us to devise very simple realizations of (M, N) PSRGs.

In practice, the above irreducible case is important, since in most practical applications the multibit number M and the multiplexer number N are both even while the period τ is odd. Therefore, in this section, we concentrate on the realizations of this irreducible case.

We first consider a simple type of the PSRG engine, whose SRG maximal space has the form $V[\Psi(x^M)]$. For this, we introduce the concept of extended SRG.

DEFINITION 8.11 (Extended SRG) For an SRG of length L with the state transition matrix $\mathbf{T}$, the *M-extended SRG* is defined to be the SRG of length ML whose state transition matrix is the $ML \times ML$ matrix

$$\tilde{\mathbf{T}}_M \equiv \left[\begin{array}{cc} \mathbf{0} & \mathbf{I}_{(M-1)L \times (M-1)L} \\ \mathbf{T} & \mathbf{0} \end{array} \right], \qquad (8.17)$$

where $\mathbf{I}_{(M-1)L \times (M-1)L}$ denotes the $(M-1)L \times (M-1)L$ identity matrix. We call $\tilde{\mathbf{T}}_M$ in (8.17) the *M-extended state transition matrix* of the state transition matrix $\mathbf{T}$.

Then, the M-extended SRG of an SRG is obtained by replacing each shift register of an SRG with M shift registers, as the following example illustrates.

EXAMPLE 8.11 We consider the 2-extended SRG for the SRG in Fig.

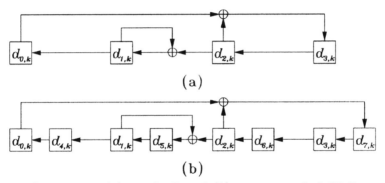

Figure 8.3. (a) An SRG and (b) its 2-extended SRG.

8.3(a), whose state transition matrix $\mathbf{T}$ is

$$\mathbf{T} = \begin{bmatrix} 0 & 1 & 0 & 0 \\ 0 & 1 & 1 & 0 \\ 0 & 0 & 0 & 1 \\ 1 & 0 & 1 & 0 \end{bmatrix}. \tag{8.18}$$

Then, by (8.17) its 2-extended state transition matrix $\tilde{\mathbf{T}}_2$ becomes

$$\tilde{\mathbf{T}}_2 = \begin{bmatrix} 0 & 0 & 0 & 0 & 1 & 0 & 0 & 0 \\ 0 & 0 & 0 & 0 & 0 & 1 & 0 & 0 \\ 0 & 0 & 0 & 0 & 0 & 0 & 1 & 0 \\ 0 & 0 & 0 & 0 & 0 & 0 & 0 & 1 \\ 0 & 1 & 0 & 0 & 0 & 0 & 0 & 0 \\ 0 & 1 & 1 & 0 & 0 & 0 & 0 & 0 \\ 0 & 0 & 0 & 1 & 0 & 0 & 0 & 0 \\ 1 & 0 & 1 & 0 & 0 & 0 & 0 & 0 \end{bmatrix}. \tag{8.19}$$

Therefore, we can depict the 2-extended SRG by rearranging the orders of the eight shift registers, as in Fig. 8.3(b). Note that the 2-extended SRG in the figure can be obtained by replacing each shift register of the SRG in Fig. 8.3(a) with a pair of shift registers. ♣

The SRG maximal space of the M-extended SRG is determined by the following theorem.

THEOREM 8.12 *For an SRG with the state transition matrix* $\mathbf{T}$, *let* $V[\Psi_{\mathbf{T}}(x)]$ *be the SRG maximal space. Then, the SRG maximal space* $V[\Psi_{\tilde{\mathbf{T}}_M}(x)]$ *of its M-extended SRG is*

$$V[\Psi_{\tilde{\mathbf{T}}_M}(x)] = V[\Psi_{\mathbf{T}}(x^M)]. \tag{8.20}$$

Proof : By Theorem 5.14 it suffices to show that the minimal polynomial $\Psi_{\tilde{\mathbf{T}}_M}(x)$ of $\tilde{\mathbf{T}}_M$ is $\Psi_{\mathbf{T}}(x^M)$ for the minimal polynomial $\Psi_{\mathbf{T}}(x)$ of $\mathbf{T}$. To prove this, we put $\Psi_{\mathbf{T}}(x) = \sum_{i=0}^{\hat{L}} \psi_i x^i$ and $\Psi_{\tilde{\mathbf{T}}_M}(x) = \sum_{i=0}^{M\hat{L}} \tilde{\psi}_i x^i$. By (8.17), we have the expression

$$
\tilde{\mathbf{T}}_M^{iM+j} =
\begin{array}{c}
\overbrace{\hspace{3cm}}^{jL} \quad \overbrace{\hspace{3cm}}^{(M-j)L} \\
\begin{bmatrix}
 & & & & \mathbf{T}^i & & \mathbf{0} \\
 & \mathbf{0} & & & & \ddots & \\
 & & & & \mathbf{0} & & \mathbf{T}^i \\
\mathbf{T}^{i+1} & & \mathbf{0} & & & & \\
 & \ddots & & & & \mathbf{0} & \\
\mathbf{0} & & \mathbf{T}^{i+1} & & & &
\end{bmatrix}
\end{array}
\tag{8.21}
$$

for $i = 0, 1, \cdots, \hat{L} - 1$ and $j = 0, 1, \cdots, M - 1$, and hence we get the relation

$$\Psi_{\tilde{\mathbf{T}}_M}(\tilde{\mathbf{T}}_M) =$$

$$
\begin{bmatrix}
\sum_{i=0}^{\hat{L}} \tilde{\psi}_{iM} \mathbf{T}^i & \sum_{i=0}^{\hat{L}-1} \tilde{\psi}_{iM+1} \mathbf{T}^i & \cdots & \sum_{i=0}^{\hat{L}-1} \tilde{\psi}_{(i+1)M-1} \mathbf{T}^i \\
\sum_{i=0}^{\hat{L}-1} \tilde{\psi}_{(i+1)M-1} \mathbf{T}^{i+1} & \sum_{i=0}^{\hat{L}} \tilde{\psi}_{iM} \mathbf{T}^i & \cdots & \sum_{i=0}^{\hat{L}-1} \tilde{\psi}_{(i+1)M-2} \mathbf{T}^i \\
\vdots & \vdots & \ddots & \vdots \\
\sum_{i=0}^{\hat{L}-1} \tilde{\psi}_{iM+1} \mathbf{T}^{i+1} & \sum_{i=0}^{\hat{L}-1} \tilde{\psi}_{iM+2} \mathbf{T}^{i+1} & \cdots & \sum_{i=0}^{\hat{L}} \tilde{\psi}_{iM} \mathbf{T}^i
\end{bmatrix}.
\tag{8.22}
$$

Since $\Psi_{\tilde{\mathbf{T}}_M}(\tilde{\mathbf{T}}_M) = 0$, $\sum_{i=0}^{\hat{L}} \tilde{\psi}_{iM} \mathbf{T}^i = 0$ and $\sum_{i=0}^{\hat{L}-1} \tilde{\psi}_{iM+j} \mathbf{T}^i = 0$, $j = 1$, $2, \cdots, M - 1$. Therefore, noting that the minimal polynomial of $\mathbf{T}$ is $\Psi_{\mathbf{T}}(x) = \sum_{i=0}^{\hat{L}} \psi_i x^i$, we obtain the relation $\tilde{\psi}_{iM} = \psi_i$, $i = 0, 1, \cdots, \hat{L}$, and $\tilde{\psi}_{iM+j} = 0$, $i = 0, 1, \cdots, \hat{L} - 1$, $j = 1, 2, \cdots, M - 1$. This proves the relation $\Psi_{\tilde{\mathbf{T}}_M}(x) = \Psi_{\mathbf{T}}(x^M)$. ∎

The theorem establishes that if the SRG maximal space of an SRG is $V[\Psi_{\mathbf{T}}(x)]$, the SRG maximal space of its M-extended SRG is $V[\Psi_{\mathbf{T}}(x^M)]$. Therefore, in case MN is relatively prime to the period of an irreducible sequence $\{ s_k \}$, if an SRG is a BSRG for the minimal space $V[\Psi_m(x)]$ for the MN-decimated sequences U_i's, its M-extended SRG is a BSRG for the sequence space $V[\Psi_m(x^M)]$. Hence we can realize a minimal (M, N) PSRG using the M-extended SRG, since the minimal space for the (M, N) parallel sequences is the sequence space $V[\Psi_m(x^M)]$ due to Theorem 8.9.

EXAMPLE 8.12 We consider a (2,3) minimal PSRG generating the (2,3) parallel sequences W_0, W_1 and W_2 in (8.15) for the sequence S_1 in Table 7.1. In this case, the minimal space for the 6-decimated sequences U_i, $i = 0, 1, \cdots, 5$, of S_1 is $V[\Psi_m(x)] = V[x^4 + x^3 + x^2 + x + 1]$, and the minimal space for the (2,3) parallel sequences W_i's is $V[\Psi_m(x^2)] = V[x^8 + x^6 + x^4 + x^2 + 1]$ (refer to Example 8.9). Since the SRG in Fig. 8.3(a) is a BSRG for the sequence space $V[\Psi_m(x)] = V[x^4 + x^3 + x^2 + x + 1]$ (refer to Example 7.30), the 2-extended SRG in Fig. 8.3(b) is a BSRG for the minimal space $V[\Psi_m(x^2)] = V[x^8 + x^6 + x^4 + x^2 + 1]$ by the theorem. Therefore, we can realize a (2,3) minimal PSRG using this 2-extended SRG. ♣

We now apply the concept of the M-extended SRG in realizing the (M, N) minimal PSRGs for MN which is relatively prime to the period of the relevant irreducible sequence.

THEOREM 8.13 *Let U_i, $i = 0, 1, \cdots, MN$-1, and W_j, $j = 0, 1, \cdots$, N-1, be respectively the ith MN-decimated sequence and the jth (M, N) parallel sequence of an irreducible sequence $\{ s_k \}$; and let MN be relatively prime to the period τ of the sequence $\{ s_k \}$. In addition, let a PSRG with the state transition matrix $\mathbf{T}$, the initial state vector $\mathbf{d}_0$ and the generating vectors $\mathbf{h}_j$, $j = 0, 1, \cdots, N$-1, be a minimal PSRG generating the N sequences U_{jM}, $j = 0, 1, \cdots, N$-1. Then, the PSRG with the state transition matrix $\hat{\mathbf{T}}$, the initial state vector $\hat{\mathbf{d}}_0$ and the generating vectors $\hat{\mathbf{h}}_j$, $j = 0, 1, \cdots, N$-1, such that*

$$
\left\{
\begin{array}{l}
\hat{\mathbf{T}} = \tilde{\mathbf{T}}_M, \\[2em]
\hat{\mathbf{d}}_0 = \begin{bmatrix} \mathbf{I} \\ \mathbf{T}^m \\ \vdots \\ \mathbf{T}^{m(M-1)} \end{bmatrix} \cdot \mathbf{d}_0, \\[3em]
\hat{\mathbf{h}}_j = \begin{bmatrix} \mathbf{h}_j \\ \mathbf{0} \\ \vdots \\ \mathbf{0} \end{bmatrix},
\end{array}
\right.
\tag{8.23}
$$

is an (M, N) minimal PSRG generating the (M, N) parallel sequences

W_j, $j = 0, 1, \cdots, N - 1$, where $\tilde{\mathbf{T}}_M$ is the M-extended state transition matrix of $\mathbf{T}$ and m is the smallest integer meeting (8.11)

Proof : We assume that $\hat{W}_j$, $j = 0, 1, \cdots, N - 1$, are the parallel sequences generated by the PSRG with the state transition matrix $\hat{\mathbf{T}}$, the initial state vector $\hat{\mathbf{d}}_0$ and the generating vectors $\hat{\mathbf{h}}_j$ in (8.23). Then, we show that $\hat{W}_j = W_j$, $j = 0, 1, \cdots, N - 1$.

Let $V[\Psi_m(x)]$ of dimension L be the minimal space for the MN-decimated sequences U_i's ; and let $U_i = \mathbf{s}_i^t \cdot \mathbf{E}_{\Psi_m(x)} = [s_0^i \ s_1^i \ \cdots \ s_{L-1}^i] \cdot \mathbf{E}_{\Psi_m(x)}$, $i = 0, 1, \cdots, MN - 1$. Then, by Theorem 8.9 the minimal space for the (M, N) parallel sequences W_j's is the sequence space $V[\Psi_m(x^M)]$. Thus we have

$$W_j = [s_0^{jM} s_0^{jM+1} \cdots s_0^{(j+1)M-1}; \cdots; s_{L-1}^{jM} s_{L-1}^{jM+1} \cdots s_{L-1}^{(j+1)M-1}] \cdot \mathbf{E}_{\Psi_m(x^M)}, \tag{8.24}$$

since W_j is obtained by interleaving the M sequences U_{jM} through $U_{(j+1)M-1}$.

By Theorem 7.36 the SRG maximal space $V[\mathbf{T}]$ is the sequence space $V[\Psi_m(x)]$, and hence by Theorem 8.12 the SRG maximal space $V[\tilde{\mathbf{T}}_M]$ of the M-extended state transition matrix $\tilde{\mathbf{T}}_M$ is the sequence space $V[\Psi_m(x^M)]$. Therefore, $\hat{W}_j$'s generated by the extended SRG are elements of the sequence space $V[\Psi_m(x^M)]$, and hence

$$\hat{W}_j = [\hat{\mathbf{h}}_j^t \cdot \hat{\mathbf{d}}_0 \ \ \hat{\mathbf{h}}_j^t \cdot \hat{\mathbf{T}} \cdot \hat{\mathbf{d}}_0 \ \ \cdots \ \ \hat{\mathbf{h}}_j^t \cdot \hat{\mathbf{T}}^{ML-1} \cdot \hat{\mathbf{d}}_0] \cdot \mathbf{E}_{\Psi_m(x^M)}. \tag{8.25}$$

Finally, comparing (8.24) and (8.25), we prove that $s_i^{jM+l} = \hat{\mathbf{h}}_j^t \cdot \hat{\mathbf{T}}^{iM+l} \cdot \hat{\mathbf{d}}_0$, $j = 0, 1, \cdots, N - 1$ for $i = 0, 1, \cdots, L - 1$, $l = 0, 1, \cdots, M - 1$. By Theorem 7.2, U_{jM+l} is the (lm)-delayed sequence of U_{jM}, and hence $s_i^{jM+l} = s_{i+lm}^{jM}$; and by (8.21), we obtain the relation $\hat{\mathbf{h}}_j^t \cdot \hat{\mathbf{T}}^{iM+l} \cdot \hat{\mathbf{d}}_0 = \mathbf{h}_j^t \cdot \mathbf{T}^i \cdot \mathbf{T}^{lm} \cdot \mathbf{d}_0 = \mathbf{h}_j^t \cdot \mathbf{T}^{i+lm} \cdot \mathbf{d}_0$, which is identical to s_{i+lm}^{jM}. Therefore, $s_i^{jM+l} = \hat{\mathbf{h}}_j^t \cdot \hat{\mathbf{T}}^{iM+l} \cdot \hat{\mathbf{d}}_0$. This completes the proof. ∎

The theorem means that if MN is relatively prime to the period of an irreducible sequence $\{ s_k \}$, the (M, N) minimal PSRG generating the (M, N) parallel sequences W_j's can be directly obtained by a minimal PSRG generating the (jM)th MN-decimated sequences U_{jM}, $j = 0, 1, \cdots, N - 1$, of the sequence $\{ s_k \}$. That is, if a PSRG with the state transition matrix $\mathbf{T}$, the initial state vector $\mathbf{d}_0$ and the generating vectors $\mathbf{h}_j$'s is a minimal PSRG generating the N sequences U_{jM}'s, then the

PSRG with the state transition matrix $\hat{\mathbf{T}}$, the initial state vector $\hat{\mathbf{d}}_0$ and the generating vectors $\hat{\mathbf{h}}_j$'s in (8.23) becomes an (M, N) minimal PSRG generating the (M, N) parallel sequences W_j's. Noting that the state transition matrix $\hat{\mathbf{T}}$ is the M-extended matrix of $\mathbf{T}$ and the generating vectors $\hat{\mathbf{h}}_j$'s have the form in (8.23), we can realize an (M, N) minimal PSRG by replacing each shift register in a minimal PSRG generating the N sequences U_{jM}'s with M shift registers and by setting the initial states $\hat{\mathbf{d}}_0$ of the extended shift registers as specified in (8.23).

EXAMPLE 8.13 We realize a $(2,3)$ minimal PSRG generating the $(2,3)$ parallel sequences W_0, W_1 and W_2 in (8.15) for the sequence S_1 in Table 7.1. The period τ of S_1 is 5, and MN is 6, which is relatively prime to τ. Since the minimal space for the 6-decimated sequences U_i, $i = 0, 1,$ $\cdots, 5$, is $V[\Psi_m(x)] = V[x^4 + x^3 + x^2 + x + 1]$ (refer to Example 8.12), the 0th, 2nd, and 4th 6-decimated sequences are represented by $U_0 = [\,0\,0\,0\,1\,]^t \cdot \mathbf{E}_{x^4+x^3+x^2+x+1}$, $U_2 = [\,0\,1\,1\,0\,]^t \cdot \mathbf{E}_{x^4+x^3+x^2+x+1}$ and $U_4 = [\,1\,0\,0\,0\,]^t \cdot \mathbf{E}_{x^4+x^3+x^2+x+1}$. To obtain a minimal PSRG generating these three sequences, we choose the nonsingular matrix

$$
\mathbf{Q} = \begin{bmatrix} 0 & 0 & 0 & 1 \\ 0 & 0 & 1 & 1 \\ 0 & 1 & 0 & 1 \\ 1 & 0 & 1 & 0 \end{bmatrix}. \tag{8.26}
$$

Then, inserting this into (7.45), we obtain the state transition matrix $\mathbf{T}$ in (8.18), the initial state vector $\mathbf{d}_0 = \mathbf{e}_3$, and the generating vectors $\mathbf{h}_0 = \mathbf{e}_0$, $\mathbf{h}_1 = \mathbf{e}_1 + \mathbf{e}_2$, $\mathbf{h}_2 = \mathbf{e}_0 + \mathbf{e}_1 + \mathbf{e}_3$. Therefore, we have the minimal PSRG generating the three sequences U_0, U_2 and U_4, shown in Fig. 8.4.

Now we insert those $\mathbf{T}$, $\mathbf{d}_0$ and $\mathbf{h}_j$'s of the minimal PSRG in Fig. 8.4 into (8.23) with $m = 1$. Then, we obtain the state transition matrix $\hat{\mathbf{T}}$ in (8.19), the initial state vector $\hat{\mathbf{d}}_0 = \mathbf{e}_3 + \mathbf{e}_6$, and the generating vectors $\hat{\mathbf{h}}_0 = \mathbf{e}_0$, $\hat{\mathbf{h}}_1 = \mathbf{e}_1 + \mathbf{e}_2$, $\hat{\mathbf{h}}_2 = \mathbf{e}_0 + \mathbf{e}_1 + \mathbf{e}_3$. Using these, we can depict the $(2,3)$ minimal PSRG generating the $(2,3)$ parallel sequences W_0, W_1 and W_2 in (8.15), as in Fig. 8.5. Note that this minimal PSRG can be directly obtained by replacing each shift register in the minimal PSRG in Fig. 8.4 with two shift registers and by setting the initial states of the extended shift registers according to the initial state vector $\hat{\mathbf{d}}_0$ in (8.23). ♣

EXAMPLE 8.14 In the case of $(2,2)$ minimal PSRGs generating the $(2,2)$ parallel sequences W_0 and W_1 in (8.16) for the sequence $\tilde{S}_{15}$ in Table

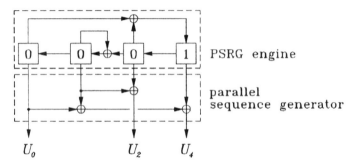

Figure 8.4. A minimal PSRG generating the 0th, the 2nd and the 4th 6-decimated sequences of S_1 in Table 7.1.

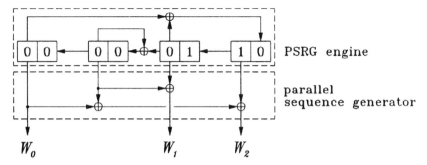

Figure 8.5. A minimal PSRG generating the (2,3) parallel sequences for S_1 in Table 7.1. This PSRG can be derived from Fig. 8.4 by replacing each shift register with two shift registers and by setting the initial states according to (8.23).

6.1, $\tau = 15$ and $MN = 4$, so they are relatively prime to each other. Since the sequence $\tilde{S}_{15}$ is an irreducible sequence in the irreducible space $V[x^4 + x + 1]$ and the decimation factor $MN = 4$ is a power of 2, we can apply Theorem 7.20 to know that the minimal space for the 4-decimated sequences U_i, $i = 0, 1, 2, 3$, is $V[\Psi_m(x)] = V[x^4 + x + 1]$. Hence the 0th and the 2nd 4-decimated sequences are represented by $U_0 = [1\,0\,0\,0]^t \cdot \mathbf{E}_{x^4+x+1}$ and $U_2 = [1\,0\,1\,0]^t \cdot \mathbf{E}_{x^4+x+1}$.

If we choose the nonsingular matrix

$$\mathbf{Q} = \begin{bmatrix} 1 & 0 & 0 & 0 \\ 0 & 0 & 0 & 1 \\ 0 & 0 & 1 & 0 \\ 0 & 1 & 0 & 0 \end{bmatrix}, \tag{8.27}$$

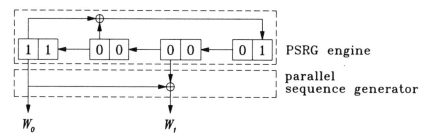

Figure 8.6. A (2,2) minimal PSRG generating the (2,2) parallel sequences for $\tilde{S}_{15}$ in Table 6.1. This is an SSRG-type realization.

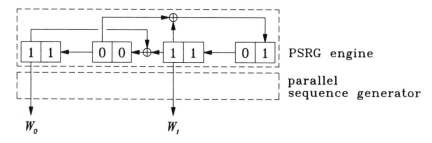

Figure 8.7. Another (2,2) minimal PSRG generating the (2,2) parallel sequences for $\tilde{S}_{15}$ in Table 6.1. This has null sequence generator part.

whose rows are $\mathbf{q}_i^t = [\,1\,0\,0\,0\,]^t \cdot \mathbf{A}_{x^4+x+1}^i$ for $i = 0, 1, 2, 3$, then by (7.45) we obtain the state transition matrix $\mathbf{T} = \mathbf{A}_{x^4+x+1}^t$, the initial state vector $\mathbf{d}_0 = \mathbf{e}_0$, and the generating vectors $\mathbf{h}_0 = \mathbf{e}_0$, $\mathbf{h}_1 = \mathbf{e}_0 + \mathbf{e}_2$ of the minimal PSRG generating the two sequences U_0 and U_2. Therefore, inserting these into (8.23) along with $m = 4$, we can obtain $\hat{\mathbf{T}}$, $\hat{\mathbf{d}}_0$ and $\hat{\mathbf{h}}_j$'s, from which we can deduce the (2,2) minimal PSRG in Fig. 8.6. Note that this is an SSRG-type realization.

To obtain a minimal PSRG with null parallel sequence generator part, we choose another nonsingular matrix

$$\mathbf{Q} = \begin{bmatrix} 1 & 0 & 0 & 0 \\ 0 & 0 & 0 & 1 \\ 1 & 0 & 1 & 0 \\ 0 & 1 & 0 & 1 \end{bmatrix}, \tag{8.28}$$

whose rows are $\mathbf{q}_i^t = [\,1\,0\,0\,0\,]^t \cdot \mathbf{A}_{x^4+x+1}^i$ for $i = 0, 1$, and $\mathbf{q}_i^t = [\,1\,0\,1\,0\,]^t \cdot \mathbf{A}_{x^4+x+1}^{i-2}$ for $i = 2, 3$. Then, in a similar manner we can obtain the minimal PSRG shown in Fig. 8.7. Note that the PSRGs in Figs. 8.6

and 8.7 do generate the parallel sequences $\{ s_k^0 \}$ and $\{ s_k^1 \}$ for use in Fig. 8.2(b). ♣

Chapter 9

Applications to Scrambling in SDH/SONET Transmission

In this chapter, we consider how to apply the serial, parallel and multibit-parallel FSS techniques developed in the previous chapters to the SDH and SONET systems, which are the most typical lightwave transmission systems employing the FSS. We first describe the FSS operations used in the SDH and SONET systems along with the frame formats of their transmission signals STM-N and STS-N. Then, we consider how to apply the parallel FSS techniques to the scrambling of the STM-1 and the STS-1 signals such that the scrambling rates can drop to one-eighth of the transmission rates. Finally, we examine how to apply the byte-parallel FSS techniques to the scrambling of the various STM-N and STS-N signals to reduce their rates to the STM-1 and the STS-1 rates respectively.

9.1 Scrambling of SDH/SONET Signals

The *synchronous digital hierarchy* (SDH) and *synchronous optical network* (SONET), standardized respectively by the ITU-T(International Telecommunication Union - Telecommunication Standardization Sector) and the T1 Committee,[1] are digital transport structures that operate by appropriately managing the payloads and transporting them through synchronous transmission network. In the SDH, the transmission signal

---

[1] For the SDH, refer to ITU-T Recommendations G.707 to 709 revised in 1992 ; and for the SONET, refer to ANSI T1.105-1991 issued in 1991.

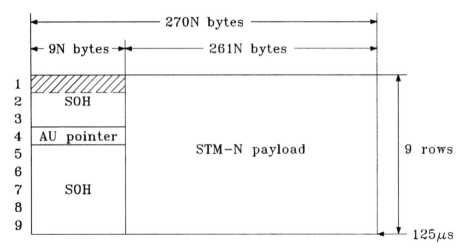

Figure 9.1. Frame structure of the SDH signal STM-N.

is called the *synchronous transport module* (STM)-N, which is formed
by byte-interleaved multiplexing N of the base-rate signal STM-1. The
STM-N signal, for which $N = 1$, 4, 16 are of primary interests, has
the 9B(bytes) $\times$ N $\times$ 270 frame structure in the 125 μs frame time, as
shown in Fig. 9.1, and thus acquires the N $\times$ 155.520 Mbps bit rate.
The 9B $\times$ N $\times$ 9 partition of the STM-N frame is assigned for the *section overhead* (SOH) and the *administrative unit* (AU) pointer, and the
remaining 9B $\times$ N $\times$ 261 is used for the payload.

In the case of the SONET, the transmission signal is called the *synchronous transport signal* (STS)-N, which is formed by byte-interleaved
multiplexing N of the base-rate signal STS-1. For the STS-N signal, N
$= 1, 3, 9, 12, 18, 24, 36, 48$, are of main interests. The STS-N signal has
the 9B $\times$ N $\times$ 90 frame structure in the 125 μs frame time, as shown
in Fig. 9.2, and thus the rate of the STS-N becomes N $\times$ 51.840 Mbps.
Note, as can be confirmed by comparing the frame formats in Figs. 9.1
and 9.2, that byte-interleaving three of the STS-N frames yields the
STM-N frame.

The STM(or STS)-N signal, which is formed by appending its overhead to the payload, is frame-synchronously scrambled before transmission, and the scrambled signal is frame-synchronously descrambled in the
receiver, as illustrated in Fig. 9.3. The overhead for frame alignment,
which is in the first row of the SDH(the shaded bytes in Figs. 9.1 and
9.2), however, are excluded from the scrambling and the descrambling

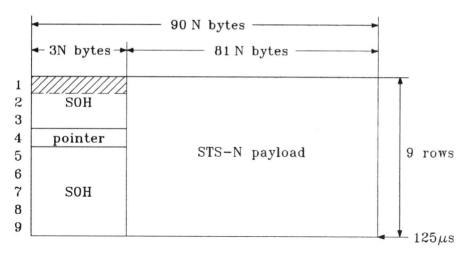

Figure 9.2. Frame structure of the SONET signal STS-N.

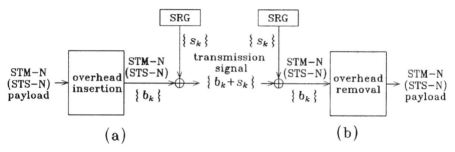

Figure 9.3. Scrambling-related blockdiagram of the SDH/SONET systems. (a) Transmitter, (b) receiver.

processes.

The SRG generating the scrambling and descrambling sequence $\{s_k\}$ in Fig. 9.3 is specified to be the terminating sequence $D_{S,0}$ of the SSRG shown in Fig. 9.4 whose characteristic polynomial is $\Psi(x) = x^7 + x + 1$.[2] The state of this SSRG is supposed to be reset to the state "1111111" at the beginning of each frame, that is, its initial state vector is $\mathbf{d}_0 = [\,1\,1\,1\,1\,1\,1\,1\,]^t$. Therefore, by (5.22a) the scrambling and descrambling

---

[2]Note that this is the characteristic polynomial $\Psi(x)$, newly introduced in this book to define the concept of the sequence space, which is the reciprocal of the "characteristic" polynomial $C(x)$ listed in Table 1.3. Refer to footnote 6 in Section 5.6.

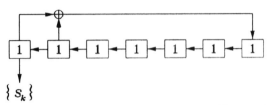

Figure 9.4. The SRG employed in the scrambling of the STM-N/ STS-N signals. The scrambling and descrambling sequence { s_k } is the PRBS $[\,1\,1\,1\,1\,1\,1\,1\,]^t \cdot \mathbf{E}_{x^7+x+1}$ of period $\tau = 127$.

Table 9.1. The elements of the PRBS { s_k } = $[\,1\,1\,1\,1\,1\,1\,1\,]^t \cdot \mathbf{E}_{x^7+x+1}$ used for scrambling of the STM-N/STS-N signals. The element in the ith row of the jth column indicates the $(8i+j)$th element s_{8i+j} in { s_k }.

$i\backslash j$	0	1	2	3	4	5	6	7
0	1	1	1	1	1	1	1	0
1	0	0	0	0	0	1	0	0
2	0	0	0	1	1	0	0	0
3	0	1	0	1	0	0	0	1
4	1	1	1	0	0	1	0	0
5	0	1	0	1	1	0	0	1
6	1	1	0	1	0	1	0	0
7	1	1	1	1	1	0	1	0
8	0	0	0	1	1	1	0	0
9	0	1	0	0	1	0	0	1
10	1	0	1	1	0	1	0	1
11	1	0	1	1	1	1	0	1
12	1	0	0	0	1	1	0	1
13	0	0	1	0	1	1	1	0
14	1	1	1	0	0	1	1	0
15	0	1	0	1	0	1	0	1
16	1	1	1	1	1	1	0	0
$\vdots$	$\vdots$	$\vdots$	$\vdots$	$\vdots$	$\vdots$	$\vdots$	$\vdots$	$\vdots$

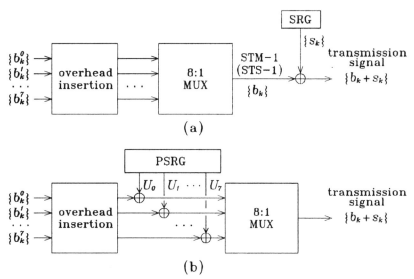

Figure 9.5. Serial scrambling and parallel scrambling of the STM-1/ STS-1 signals. (a) Serial scrambling, (b) parallel scrambling.

sequence $\{\,s_k\,\}$ becomes

$$\{\,s_k\,\} \;=\; [\,1\,1\,1\,1\,1\,1\,1\,]^t \cdot \mathbf{E}_{x^7+x+1}. \tag{9.1}$$

Since the characteristic polynomial $\Psi(x) = x^7 + x + 1$ is primitive, the sequence $\{\,s_k\,\}$ is a **PRBS** in the primitive space $V[x^7+x+1]$, and hence by Theorem 6.9 its period τ is 127. The elements of the scrambling and descrambling sequence $\{\,s_k\,\}$ in (9.1) are as listed in Table 9.1.

9.2 Bit-Parallel Scrambling of STM-1/STS-1 Signals

We first consider how to apply the parallel scrambling techniques developed in Chapter 7 to the scrambling of the STM-1 and the STS-1 signals.

In order to consider the parallel FSS, we regard the STM-1 or the STS-1 signal $\{\,b_k\,\}$ as the bit-interleaved sequence of the eight parallel sequences $\{\,b_k^j\,\}$, $j = 0, 1, \cdots, 7$, shown in Fig. 9.5(a).[3] Then, we

---

[3]In general, we can take the STM-1 or the STS-1 signal $\{\,b_k\,\}$ as the bit-interleaved sequence of the N parallel sequences $\{\,b_k^j\,\}$, $j = 0, 1, \cdots, N-1$. Among various

obtain its equivalent parallel scrambling process shown in Fig. 9.5(b), where the eight parallel scrambling sequences U_j, $j = 0, 1, \cdots, 7$, are the 8-decimated sequences of the serial scrambling sequence $\{ s_k \}$ in (9.1). If we employ this parallel scrambling for the scrambling of the STM-1 and the STS-1 signals, then their corresponding serial scrambling rates 155.520 Mbps and 51.840 Mbps drop to 19.44($= 155.520/8$) Mbps and 6.48($= 51.840/8$) Mbps respectively.

Since the serial scrambling sequence $\{ s_k \}$ in (9.1) is the PRBS in the primitive space $V[x^7 + x + 1]$ and the decimation factor $N = 8$ is a power of 2, we can apply Theorem 7.30 to find that the minimal space for the 8-decimated sequences U_j, $j = 0, 1, \cdots, 7$, is the primitive space $V[x^7 + x + 1]$. Therefore, the decimated sequences U_j, $j = 0, 1, \cdots, 7$, have the expression

$$U_j = \mathbf{s}_j^t \cdot \mathbf{E}_{x^7+x+1} \tag{9.2a}$$

for the initial vectors

$$\begin{cases}
\mathbf{s}_0 = [\,1\,0\,0\,0\,1\,0\,1\,]^t, \\
\mathbf{s}_1 = [\,1\,0\,0\,1\,1\,1\,1\,]^t, \\
\mathbf{s}_2 = [\,1\,0\,0\,0\,1\,0\,0\,]^t, \\
\mathbf{s}_3 = [\,1\,0\,1\,1\,0\,1\,1\,]^t, \\
\mathbf{s}_4 = [\,1\,0\,1\,0\,0\,1\,0\,]^t, \\
\mathbf{s}_5 = [\,1\,1\,0\,0\,1\,0\,1\,]^t, \\
\mathbf{s}_6 = [\,1\,0\,0\,0\,0\,0\,0\,]^t, \\
\mathbf{s}_7 = [\,0\,0\,0\,1\,0\,1\,0\,]^t,
\end{cases} \tag{9.2b}$$

which are taken from Table 9.1.

To realize a minimal PSRG generating the decimated sequences U_j's in (9.2), we choose the nonsingular matrix $\mathbf{Q}$

$$\mathbf{Q} = \begin{bmatrix}
\mathbf{s}_0^t \\
\mathbf{s}_0^t \cdot \mathbf{A}_{x^7+x+1} \\
\mathbf{s}_0^t \cdot \mathbf{A}^2_{x^7+x+1} \\
\mathbf{s}_0^t \cdot \mathbf{A}^3_{x^7+x+1} \\
\mathbf{s}_0^t \cdot \mathbf{A}^4_{x^7+x+1} \\
\mathbf{s}_0^t \cdot \mathbf{A}^5_{x^7+x+1} \\
\mathbf{s}_0^t \cdot \mathbf{A}^6_{x^7+x+1}
\end{bmatrix} = \begin{bmatrix}
1 & 0 & 0 & 0 & 1 & 0 & 1 \\
0 & 0 & 0 & 1 & 0 & 1 & 1 \\
0 & 0 & 1 & 0 & 1 & 1 & 0 \\
0 & 1 & 0 & 1 & 1 & 0 & 0 \\
1 & 0 & 1 & 1 & 0 & 0 & 1 \\
0 & 1 & 1 & 0 & 0 & 1 & 1 \\
1 & 1 & 0 & 0 & 1 & 1 & 1
\end{bmatrix}.$$

possible choices of N, $N = 8$ is a good choice as it matches the byte-level parallel processing.

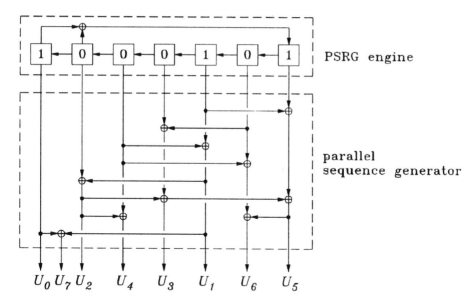

Figure 9.6. An SSRG-based minimal PSRG generating the 8-decimated sequences U_j's in (9.2). This PSRG can be used as the PSRG for parallel scrambling the STM-1 or STS-1 signal.

Inserting this into (7.45) along with (9.2b) and $\Psi_m(x) = x^7 + x + 1$, we obtain the minimal PSRG with the state transition matrix $\mathbf{T} = \mathbf{A}^t_{x^7+x+1}$, the initial state vector $\mathbf{d}_0 = \mathbf{s}_0$, and the generating vectors

$$
\left\{
\begin{array}{rl}
\mathbf{h}_0 = & [\,1\,0\,0\,0\,0\,0\,0\,]^t, \\
\mathbf{h}_1 = & [\,0\,0\,1\,0\,1\,0\,0\,]^t, \\
\mathbf{h}_2 = & [\,0\,1\,1\,0\,1\,0\,0\,]^t, \\
\mathbf{h}_3 = & [\,0\,1\,1\,1\,1\,1\,0\,]^t, \\
\mathbf{h}_4 = & [\,0\,1\,0\,0\,1\,0\,0\,]^t, \\
\mathbf{h}_5 = & [\,0\,1\,1\,1\,0\,1\,1\,]^t, \\
\mathbf{h}_6 = & [\,0\,1\,0\,1\,0\,0\,1\,]^t, \\
\mathbf{h}_7 = & [\,1\,0\,1\,0\,1\,0\,0\,]^t.
\end{array}
\right.
\tag{9.3}
$$

The circuit diagram for this minimal PSRG is as shown in Fig. 9.6, which becomes an SSRG-based realization. This PSRG can be used as the PSRG in Fig. 9.5(b) for parallel scrambling the STM-1 or STS-1 signal.

To get a simpler minimal PSRG, we choose another nonsingular ma-

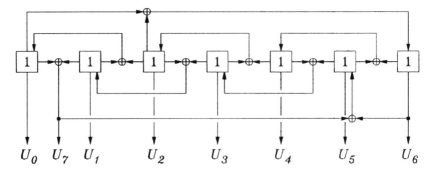

Figure 9.7. Another minimal PSRG generating the 8-decimated sequences U_j's in (9.2). This PSRG is more regularly structured and requires less number of exclusive-OR gates than the SSRG-based minimal PSRG in Fig. 9.6.

trix

$$
\mathbf{Q} = \begin{bmatrix} s_0^t \\ s_1^t \\ s_2^t \\ s_3^t \\ s_4^t \\ s_5^t \\ s_6^t \end{bmatrix} = \begin{bmatrix} 1 & 0 & 0 & 0 & 1 & 0 & 1 \\ 1 & 0 & 0 & 1 & 1 & 1 & 1 \\ 1 & 0 & 0 & 0 & 1 & 0 & 0 \\ 1 & 0 & 1 & 1 & 0 & 1 & 1 \\ 1 & 0 & 1 & 0 & 0 & 1 & 0 \\ 1 & 1 & 0 & 0 & 1 & 0 & 1 \\ 1 & 0 & 0 & 0 & 0 & 0 & 0 \end{bmatrix}.
$$

Then, by (7.45) we obtain the minimal PSRG with

$$
\mathbf{T} = \begin{bmatrix} 0 & 1 & 1 & 0 & 0 & 0 & 0 \\ 0 & 0 & 1 & 1 & 0 & 0 & 0 \\ 0 & 0 & 0 & 1 & 1 & 0 & 0 \\ 0 & 0 & 0 & 0 & 1 & 1 & 0 \\ 0 & 0 & 0 & 0 & 0 & 1 & 1 \\ 1 & 1 & 0 & 0 & 0 & 0 & 1 \\ 1 & 0 & 1 & 0 & 0 & 0 & 0 \end{bmatrix}, \tag{9.4a}
$$

$$
\mathbf{d}_0 = [1\,1\,1\,1\,1\,1\,1]^t, \tag{9.4b}
$$

$$
\mathbf{h}_i = \begin{cases} \mathbf{e}_i, & i = 0, 1, \cdots, 6, \\ \mathbf{e}_0 + \mathbf{e}_1, & i = 7. \end{cases} \tag{9.4c}
$$

The resulting circuit diagram is as shown in Fig. 9.7. This minimal PSRG can also generate the parallel sequences U_j's in (9.2), but it is better than the SSRG-based minimal PSRG in Fig. 9.6 in that it is

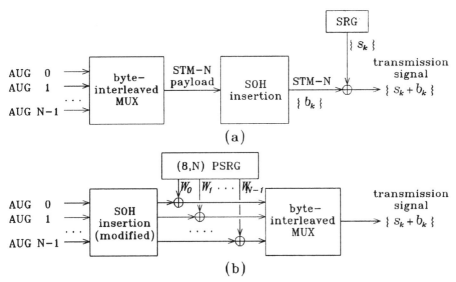

Figure 9.8. Serial scrambling and byte-parallel scrambling of STM-N signal. (a) Serial scrambling, (b) byte-parallel scrambling.

more regularly structured, while requiring less number of exclusive-OR gates.

9.3 Byte-Parallel Scrambling of STM-N Signal

We now consider how to apply the multibit-parallel scrambling techniques developed in the previous chapter to the scrambling of the STM-N signal.

In the SDH system, N *administrative unit group* (AUG) signals are byte-interleaved in the unit of eight bits to form a STM-N payload, and the SOH data are appended to it to complete an STM-N signal, which is scrambled before transmission. This is illustrated in Fig. 9.8(a). In this case, the scrambling rates are 622.080 Mbps for $N = 4$ and 2,488.320 Mbps for $N = 16$. If we employ the byte-parallel scrambling processing shown in Fig. 9.8(b) instead of the serial scrambling in Fig. 9.8(a), the scrambling processing rate could drop from the STM-N rate to the STM-1 rate of 155.520 Mbps. Since byte-interleaving is an 8-bit based operation, the byte-parallel scrambling sequences $W_j, j = 0, 1, \cdots, N-1$, in Fig. 9.8(b) are the $(8,N)$ parallel sequences for the serial scrambling sequence $\{ s_k \}$ in (9.1). Note that in the byte-parallel scrambling in

Fig. 9.8(b), the SOH insertion process should be modified accordingly. That is, byte-interleaving of the modified SOH inserted in Fig. 9.8(b) should be identical to the original SOH inserted in Fig. 9.8(a).

9.3.1   Byte-Parallel Scrambling of STM-4 Signal

For the byte-parallel scrambling of STM-4 signal, we need an $(8,4)$ minimal PSRG for the serial scrambling sequence $\{ s_k \}$ in (9.1). Since $MN = 32$ is relatively prime to the period $\tau = 127$ of the PRBS $\{ s_k \}$, we can realize this relying on Theorem 8.13.

By Theorem 7.30, the minimal space for the 32-decimated sequences U_i, $i = 0, 1, \cdots, 31$, is the sequence space $V[x^7 + x + 1]$. From Table 9.1 we find that the $(8j)$th 32-decimated sequences U_{8j}, $j = 0, 1, 2, 3$, have the expression

$$U_{8j} = s_j^t \cdot E_{x^7+x+1} \tag{9.5a}$$

for

$$\begin{cases} s_0 = [1\,1\,0\,1\,1\,1\,0]^t, \\ s_1 = [0\,0\,0\,0\,0\,1\,1]^t, \\ s_2 = [0\,1\,1\,1\,0\,1\,0]^t, \\ s_3 = [0\,1\,1\,0\,1\,1\,0]^t. \end{cases} \tag{9.5b}$$

If we choose the nonsingular matrix

$$Q = \begin{bmatrix} s_0^t \\ s_0^t \cdot A_{x^7+x+1} \\ s_1^t \\ s_1^t \cdot A_{x^7+x+1} \\ s_2^t \\ s_2^t \cdot A_{x^7+x+1} \\ s_3^t \end{bmatrix} = \begin{bmatrix} 1 & 1 & 0 & 1 & 1 & 1 & 0 \\ 1 & 0 & 1 & 1 & 1 & 0 & 0 \\ 0 & 0 & 0 & 0 & 0 & 1 & 1 \\ 0 & 0 & 0 & 0 & 1 & 1 & 0 \\ 0 & 1 & 1 & 1 & 0 & 1 & 0 \\ 1 & 1 & 1 & 0 & 1 & 0 & 1 \\ 0 & 1 & 1 & 0 & 1 & 1 & 0 \end{bmatrix},$$

we obtain, by (7.45) and $\Psi_m(x) = x^7 + x + 1$,

$$T = \begin{bmatrix} 0 & 1 & 0 & 0 & 0 & 0 & 0 \\ 0 & 0 & 1 & 0 & 1 & 0 & 0 \\ 0 & 0 & 0 & 1 & 0 & 0 & 0 \\ 0 & 0 & 0 & 0 & 1 & 0 & 1 \\ 0 & 0 & 0 & 0 & 0 & 1 & 0 \\ 0 & 1 & 0 & 0 & 0 & 0 & 1 \\ 1 & 0 & 1 & 0 & 0 & 0 & 0 \end{bmatrix}, \tag{9.6a}$$

$$\mathbf{d}_0 = [1\,1\,0\,0\,0\,1\,0]^t, \tag{9.6b}$$

$$\mathbf{h}_i = \mathbf{e}_{2i}, \quad i = 0,\,1,\,2,\,3, \tag{9.6c}$$

for the minimal PSRG generating the sequences U_{8j}'s in (9.5). Therefore, by Theorem 8.13 along with the value $m = 4$ we obtain the (8,4) minimal PSRG in Fig. 9.9, which generates the (8,4) parallel sequences W_j, $j = 0,\,1,\,2,\,3$, to be used in place of the serial scrambling sequence $\{\,s_k\,\}$ in (9.1).

9.3.2 Byte-Parallel Scrambling of STM-16 Signal

In the case of the byte-parallel scrambling of STM-16 signal, we need to realize an (8,16) minimal PSRG. We can realize this in a manner similar to the case of the (8,4) PSRG in the previous subsection, since $MN = 128$ is also relatively prime to the period $\tau = 127$ of the serial scrambling PRBS $\{\,s_k\,\}$ in (9.1).

By Theorem 7.30, the minimal space for the 128-decimated sequences U_i, $i = 0,\,1,\,\cdots,\,127$, is the sequence space $V[x^7 + x + 1]$, and by Table 9.1 the $(8j)$th 128-decimated sequences U_{8j}, $j = 0,\,1,\,\cdots,\,15$, have the expression

$$U_{8j} = \mathbf{s}_j^t \cdot \mathbf{E}_{x^7+x+1} \tag{9.7a}$$

for

$$\left\{ \begin{array}{rcl}
\mathbf{s}_0 & = & [1\,1\,1\,1\,1\,1\,1]^t, \\
\mathbf{s}_1 & = & [0\,0\,0\,0\,0\,1\,0]^t, \\
\mathbf{s}_2 & = & [0\,0\,0\,1\,1\,0\,0]^t, \\
\mathbf{s}_3 & = & [0\,1\,0\,1\,0\,0\,0]^t, \\
\mathbf{s}_4 & = & [1\,1\,1\,0\,0\,1\,0]^t, \\
\mathbf{s}_5 & = & [0\,1\,0\,1\,1\,0\,0]^t, \\
\mathbf{s}_6 & = & [1\,1\,0\,1\,0\,1\,0]^t, \\
\mathbf{s}_7 & = & [1\,1\,1\,1\,1\,0\,1]^t, \\
\mathbf{s}_8 & = & [0\,0\,0\,1\,1\,1\,0]^t, \\
\mathbf{s}_9 & = & [0\,1\,0\,0\,1\,0\,0]^t, \\
\mathbf{s}_{10} & = & [1\,0\,1\,1\,0\,1\,0]^t, \\
\mathbf{s}_{11} & = & [1\,0\,1\,1\,1\,1\,0]^t, \\
\mathbf{s}_{12} & = & [1\,0\,0\,0\,1\,1\,0]^t, \\
\mathbf{s}_{13} & = & [0\,0\,1\,0\,1\,1\,1]^t, \\
\mathbf{s}_{14} & = & [1\,1\,1\,0\,0\,1\,1]^t, \\
\mathbf{s}_{15} & = & [0\,1\,0\,1\,0\,1\,0]^t.
\end{array} \right. \tag{9.7b}$$

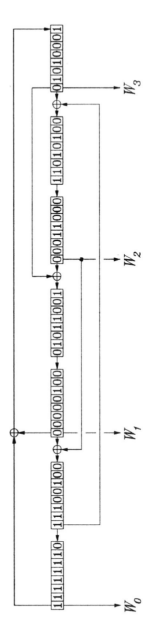

Figure 9.9. An (8,4) minimal PSRG for the serial scrambling PRBS { s_k } in (9.1). This PSRG can be used for the byte-parallel scrambling of the STM-4 in Fig. 9.8(b).

If we choose the nonsingular matrix

$$
\mathbf{Q} = \begin{bmatrix} \mathbf{s}_0^t \\ \mathbf{s}_0^t \cdot \mathbf{A}_{x^7+x+1} \\ \mathbf{s}_4^t \\ \mathbf{s}_4^t \cdot \mathbf{A}_{x^7+x+1} \\ \mathbf{s}_8^t \\ \mathbf{s}_8^t \cdot \mathbf{A}_{x^7+x+1} \\ \mathbf{s}_{12}^t \end{bmatrix} = \begin{bmatrix} 1 & 1 & 1 & 1 & 1 & 1 & 1 \\ 1 & 1 & 1 & 1 & 1 & 1 & 0 \\ 1 & 1 & 1 & 0 & 0 & 1 & 0 \\ 1 & 1 & 0 & 0 & 1 & 0 & 0 \\ 0 & 0 & 0 & 1 & 1 & 1 & 0 \\ 0 & 0 & 1 & 1 & 1 & 0 & 0 \\ 1 & 0 & 0 & 0 & 1 & 1 & 0 \end{bmatrix},
$$

then by (7.45) along with $\Psi_m(x) = x^7 + x + 1$, we obtain the state transition matrix $\mathbf{T}$ in (9.6a), the initial state vector $\mathbf{d}_0 = [\,1\,1\,1\,1\,0\,0\,1\,]^t$, and the generating vectors

$$
\begin{cases}
\mathbf{h}_0 &= \mathbf{e}_0, \\
\mathbf{h}_1 &= \mathbf{e}_1 + \mathbf{e}_2 + \mathbf{e}_4, \\
\mathbf{h}_2 &= \mathbf{e}_1 + \mathbf{e}_2, \\
\mathbf{h}_3 &= \mathbf{e}_2 + \mathbf{e}_5 + \mathbf{e}_6, \\
\mathbf{h}_4 &= \mathbf{e}_2, \\
\mathbf{h}_5 &= \mathbf{e}_3 + \mathbf{e}_4 + \mathbf{e}_6, \\
\mathbf{h}_6 &= \mathbf{e}_3 + \mathbf{e}_4, \\
\mathbf{h}_7 &= \mathbf{e}_0 + \mathbf{e}_1 + \mathbf{e}_2 + \mathbf{e}_4, \\
\mathbf{h}_8 &= \mathbf{e}_4, \\
\mathbf{h}_9 &= \mathbf{e}_1 + \mathbf{e}_5 + \mathbf{e}_6, \\
\mathbf{h}_{10} &= \mathbf{e}_5 + \mathbf{e}_6, \\
\mathbf{h}_{11} &= \mathbf{e}_2 + \mathbf{e}_3 + \mathbf{e}_4 + \mathbf{e}_6, \\
\mathbf{h}_{12} &= \mathbf{e}_6, \\
\mathbf{h}_{13} &= \mathbf{e}_0 + \mathbf{e}_1 + \mathbf{e}_2 + \mathbf{e}_3, \\
\mathbf{h}_{14} &= \mathbf{e}_0 + \mathbf{e}_1 + \mathbf{e}_2, \\
\mathbf{h}_{15} &= \mathbf{e}_1 + \mathbf{e}_4 + \mathbf{e}_5 + \mathbf{e}_6,
\end{cases}
\tag{9.8}
$$

for the minimal PSRG generating the sequences U_{8j}'s in (9.7). Therefore, by Theorem 8.13 along with the value $m = 1$ we obtain the (8,16) minimal PSRG in Fig. 9.10, which generates the (8,16) parallel sequences W_j, $j = 0, 1, \cdots, 15$, for use in place of the serial scrambling sequence $\{\,s_k\,\}$ in (9.1).

9.4　Byte-Parallel Scrambling of STS-N Signal

Finally, we consider how to apply the multibit-parallel scrambling techniques to scrambling of the SONET signal STS-N. In the SONET, the

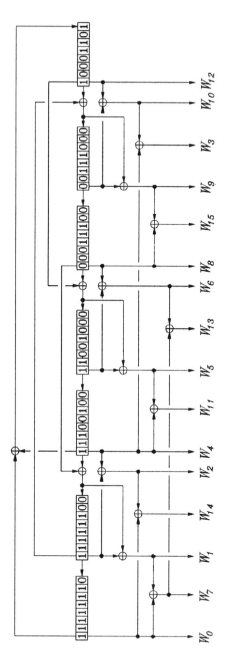

Figure 9.10. An (8,16) minimal PSRG for the serial scrambling PRBS
{ s_k } in (9.1). This PSRG can be used for the byte-parallel scrambling
of the STM-16 in Fig. 9.8(b).

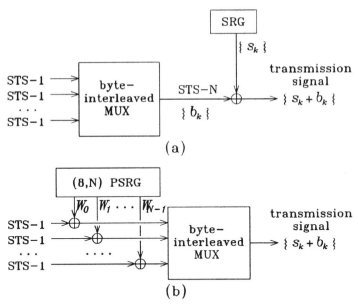

Figure 9.11. Serial scrambling and byte-parallel scrambling of STS-N signal. (a) Serial scrambling, (b) byte-parallel scrambling.

STS-1 signal is used as the base rate, and the STS-N signals for $N = 3, 9, 12, 18, 24, 36, 48$, are obtained by byte-interleaving N of STS-1 signals, as illustrated in Fig. 9.11(a).[4] Therefore, if we employ the byte-parallel scrambling as shown in Fig. 9.11(b), the serial scrambling processing rate in Fig. 9.11(a) drops from $N \times 51.840$ Mbps to the STS-1 rate of 51.840 Mbps. The byte-parallel scrambling techniques are applicable to the scramblings of all possible STS-N signals. But we limit our demonstrations in this section to the cases of the STS-3 and STS-12 signals, leaving the other cases as an exercise to the reader.

9.4.1 Byte-Parallel Scrambling of STS-3 Signal

For the byte-parallel scrambling of the STS-3 signal, we need to realize an (8,3) minimal PSRG for the serial scrambling PRBS $\{ s_k \}$ in (9.1). We can realize this employing Theorem 8.13 since $MN = 24$ is relatively prime to the period $\tau = 127$ of the sequence $\{ s_k \}$.

By Theorems 7.29 and 7.18, the minimal space for the 24-decimated

---

[4]Note that in this figure, the overhead processing which is to be located in front of the byte-interleaved-multiplexer, as for the STM-N case in Fig. 9.8, is omitted.

sequences U_i, $i = 0, 1, \cdots, 23$, is the sequence space $V[x^7+x^5+x^3+x+1]$, and by Table 9.1 the $(8j)$th 24-decimated sequences U_{8j}, $j = 0, 1, 2$, have the expression

$$U_{8j} = \mathbf{s}_j^t \cdot \mathbf{E}_{x^7+x^5+x^3+x+1} \qquad (9.9a)$$

respectively for

$$\begin{cases} \mathbf{s}_0 = [\,1\,0\,1\,0\,1\,0\,0\,]^t, \\ \mathbf{s}_1 = [\,0\,1\,1\,1\,0\,1\,1\,]^t, \\ \mathbf{s}_2 = [\,0\,0\,0\,1\,1\,0\,1\,]^t. \end{cases} \qquad (9.9b)$$

If we choose the nonsingular matrix

$$\mathbf{Q} = \begin{bmatrix} \mathbf{s}_0^t \\ \mathbf{s}_0^t \cdot \mathbf{A}_{x^7+x^5+x^3+x+1} \\ \mathbf{s}_0^t \cdot \mathbf{A}_{x^7+x^5+x^3+x+1}^2 \\ \mathbf{s}_2^t \\ \mathbf{s}_2^t \cdot \mathbf{A}_{x^7+x^5+x^3+x+1} \\ \mathbf{s}_1^t \\ \mathbf{s}_1^t \cdot \mathbf{A}_{x^7+x^5+x^3+x+1} \end{bmatrix} = \begin{bmatrix} 1 & 0 & 1 & 0 & 1 & 0 & 0 \\ 0 & 1 & 0 & 1 & 0 & 0 & 1 \\ 1 & 0 & 1 & 0 & 0 & 1 & 0 \\ 0 & 0 & 0 & 1 & 1 & 0 & 1 \\ 0 & 0 & 1 & 1 & 0 & 1 & 1 \\ 0 & 1 & 1 & 1 & 0 & 1 & 1 \\ 1 & 1 & 1 & 0 & 1 & 1 & 1 \end{bmatrix},$$

we obtain, by (7.45) and $\Psi_m(x) = x^7 + x^5 + x^3 + x + 1$,

$$\mathbf{T} = \begin{bmatrix} 0 & 1 & 0 & 0 & 0 & 0 & 0 \\ 0 & 0 & 1 & 0 & 0 & 0 & 0 \\ 0 & 1 & 0 & 1 & 0 & 0 & 0 \\ 0 & 0 & 0 & 0 & 1 & 0 & 0 \\ 0 & 0 & 0 & 1 & 0 & 1 & 0 \\ 0 & 0 & 0 & 0 & 0 & 0 & 1 \\ 1 & 0 & 0 & 0 & 0 & 1 & 0 \end{bmatrix}, \qquad (9.10a)$$

$$\mathbf{d}_0 = [\,1\,0\,1\,0\,0\,0\,1\,]^t, \qquad (9.10b)$$

$$\begin{cases} \mathbf{h}_0 = \mathbf{e}_0, \\ \mathbf{h}_1 = \mathbf{e}_5, \\ \mathbf{h}_2 = \mathbf{e}_3, \end{cases} \qquad (9.10c)$$

for the minimal PSRG generating the sequences U_{8j}'s in (9.9). There-fore, by Theorem 8.13 along with the number $m = 90$ we obtain the (8,3) minimal PSRG in Fig. 9.12, which generates the (8,3) parallel sequences W_0, W_1 and W_2.

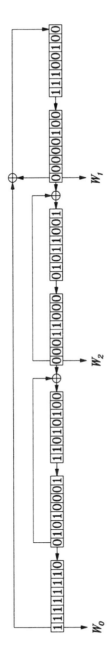

Figure 9.12. An (8,3) minimal PSRG for the serial scrambling PRBS $\{ s_k \}$ in (9.1). This PSRG can be used for the byte-parallel scrambling of the STS-3 in Fig. 9.11(b).

9.4.2   Byte-Parallel Scrambling of STS-12 Signal

In the case of the byte-parallel scrambling of STS-12 signal, we have to realize an $(8,12)$ minimal PSRG for the serial scrambling PRBS $\{\,s_k\,\}$ in (9.1). We also realize this using Theorem 8.13 since $MN = 96$ is relatively prime to the period $\tau = 127$ of the sequence $\{\,s_k\,\}$.

By Theorems 7.29 and 7.18, the minimal space for the 96-decimated sequences U_i, $i = 0, 1, \cdots, 95$, is the sequence space $V[x^7+x^5+x^3+x+1]$, and hence from Table 9.1 the $(8j)$th 96-decimated sequences U_{8j}, $j = 0, 1, \cdots, 11$, have the expression

$$U_{8j} = s_j^t \cdot E_{x^7+x^5+x^3+x+1}, \qquad (9.11a)$$

where

$$
\left\{
\begin{aligned}
s_0 &= [\,1\,1\,0\,1\,1\,0\,1\,]^t, \\
s_1 &= [\,0\,0\,1\,0\,0\,0\,1\,]^t, \\
s_2 &= [\,0\,1\,0\,0\,1\,0\,0\,]^t, \\
s_3 &= [\,0\,0\,0\,1\,1\,1\,1\,]^t, \\
s_4 &= [\,1\,1\,0\,0\,0\,1\,1\,]^t, \\
s_5 &= [\,0\,0\,0\,0\,1\,0\,1\,]^t, \\
s_6 &= [\,1\,0\,1\,1\,1\,1\,0\,]^t, \\
s_7 &= [\,1\,1\,1\,1\,1\,0\,0\,]^t, \\
s_8 &= [\,0\,1\,1\,0\,1\,0\,1\,]^t, \\
s_9 &= [\,0\,1\,0\,1\,0\,1\,1\,]^t, \\
s_{10} &= [\,1\,1\,0\,1\,1\,0\,0\,]^t, \\
s_{11} &= [\,1\,1\,0\,0\,1\,1\,0\,]^t.
\end{aligned}
\right.
\qquad (9.11b)
$$

If we choose the nonsingular matrix

$$
Q =
\begin{bmatrix}
s_{10}^t \cdot A_{x^7+x^5+x^3+x+1}^{53} \\
s_{10}^t \\
s_8^t \\
s_6^t \\
s_4^t \\
s_2^t \\
s_0^t
\end{bmatrix}
=
\begin{bmatrix}
1 & 0 & 1 & 1 & 0 & 1 & 1 \\
1 & 1 & 0 & 1 & 1 & 0 & 0 \\
0 & 1 & 1 & 0 & 1 & 0 & 1 \\
1 & 0 & 1 & 1 & 1 & 1 & 0 \\
1 & 1 & 0 & 0 & 0 & 1 & 1 \\
0 & 1 & 0 & 0 & 1 & 0 & 0 \\
1 & 1 & 0 & 1 & 1 & 0 & 1
\end{bmatrix},
$$

we obtain, by (7.45) and $\Psi_m(x) = x^7 + x^5 + x^3 + x + 1$,

$$\mathbf{T} \;=\; \begin{bmatrix} 1 & 1 & 0 & 0 & 0 & 0 & 0 \\ 0 & 1 & 1 & 0 & 0 & 0 & 0 \\ 0 & 0 & 1 & 1 & 0 & 0 & 0 \\ 0 & 0 & 0 & 1 & 1 & 0 & 0 \\ 0 & 0 & 0 & 0 & 1 & 1 & 0 \\ 0 & 0 & 0 & 0 & 0 & 1 & 1 \\ 1 & 0 & 0 & 0 & 0 & 0 & 0 \end{bmatrix}, \tag{9.12a}$$

$$\mathbf{d}_0 \;=\; [\,1\,1\,0\,1\,1\,0\,1\,]^t, \tag{9.12b}$$

$$\mathbf{h}_i \;=\; \begin{cases} \mathbf{e}_{6-j}, & i = 2j,\; j = 0, 1, \cdots, 5, \\ \mathbf{e}_{2-j} + \mathbf{e}_{5-j}, & i = 2j+1,\; j = 0, 1, 2, \\ \mathbf{e}_{5-j} + \mathbf{e}_{8-j} + \mathbf{e}_{9-j}, & i = 2j+1,\; j = 3, 4, 5, \end{cases} \tag{9.12c}$$

for the minimal PSRG generating the sequences U_{8j}'s in (9.11). Therefore, by Theorem 8.13 along with the number $m = 86$ we obtain the (8,12) minimal PSRG in Fig. 9.13, which generates the (8,12) parallel sequences W_j, $j = 0, 1, \cdots, 11$.

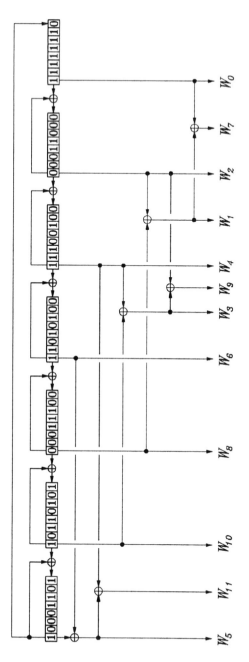

Figure 9.13. An (8,12) minimal PSRG for the serial scrambling PRBS $\{ s_k \}$ in (9.1). This PSRG can be used for the byte-parallel scrambling of the STS-12 in Fig. 9.11(b).

Part III

DISTRIBUTED SAMPLE SCRAMBLING

Chapter 10

Introduction to Distributed Sample Scrambling

Distributed sample scrambling(DSS) is a scrambling technique which employs SRGs for generating scrambling sequences and utilizes distributed samples for synchronizing the SRGs. Differently from the FSS, the DSS keeps scrambling without periodically resetting the SRG states, and therefore the DSS exhibits an excellent scrambling effect for small frame-sized signals.

In this part we will provide a full-fledged discussion on the DSS, and in this introductory chapter we will describe the operation of the DSS and its synchronization issues in relation to the relevant state transition mechanism.

10.1 Operation of DSS

A blockdiagram of the DSS is shown in Fig. 10.1. As in the case of the FSS, the input bitstream $\{ b_k \}$ is scrambled by adding the scrambler SRG sequence $\{ s_k \}$ in the scrambler, and the scrambled bitstream $\{ b_k + s_k \}$ is descrambled by adding the descrambler SRG sequence $\{ \hat{s}_k \}$ in the descrambler. In order for the descrambled bitstream $\{ b_k + s_k + \hat{s}_k \}$ to be identical to the original input bitstream $\{ b_k \}$, the descrambler SRG should be synchronized to the scrambler SRG, that is, the descrambler SRG sequence $\{ \hat{s}_k \}$ should be identical to the scrambler SRG sequence $\{ s_k \}$.

For the synchronization of the descrambler, some samples of the

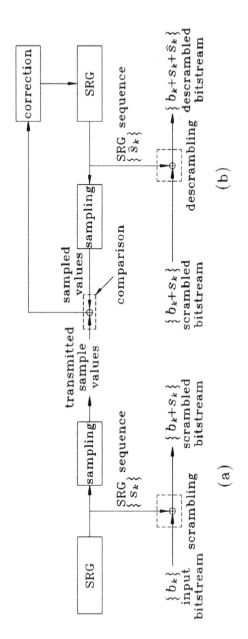

Figure 10.1. Blockdiagram of the DSS. (a) Scrambler, (b) descrambler.

scrambler SRG sequence { s_k } are taken by the sampling function in Fig. 10.1 and conveyed to the descrambler in parallel with the scrambled bitstream. The descrambler generates its own samples of SRG sequence { $\hat{s}_k$ } in the same manner and compares them to the conveyed ones. The upper exclusive-OR gates in the descrambler of Fig. 10.1 does this comparison function. If the two samples are determined to be identical, no action takes place. Otherwise, the correction logic is initiated to change the descrambler SRG state in such a manner that the descrambler SRG can eventually get synchronized to the scrambler SRG.

The DSS is distinctively different from the FSS in this synchronization process. The differences are contrasted in a number of aspects : First, the SRG states are corrected according to some predetermined rules in every reception of the samples in the DSS, while they are reset to some prespecified values in every frame start in the FSS. Second, the synchronization of the DSS solely relies on the conveyed samples, while the synchronization of the FSS relies on the a priori knowledge which was provided through some pre-communications. Third, the information to aid the synchronization, that is, the samples, is conveyed in a distributive manner in the DSS, while it is provided in bulk in the FSS. As a consequence, the synchronization in the DSS is done in a progressive manner, while it is done in one-shot in the FSS. Fourth, the data randomizing effect is much improved in the DSS since its generated SRG sequence is independent of the frame boundaries, while in the FSS the SRG sequence is synchronized to the frame boundaries.

10.2 Three-State Synchronization Mechanism for DSS

In the DSS, the scrambler and the descrambler SRGs are synchronized relying on the conveyed samples of the scrambler SRG sequence { s_k }. But the conveyed samples could be corrupted by errors during transmission, and in this case the two SRGs fail to get synchronized since in this case the synchronization processing is carried on erroneous samples. Therefore, in the DSS it is required to verify whether or not the synchronization is really acquired. Further, even if the synchronization is once acquired it can be broken by unexpected errors in the scrambler or descrambler SRGs. So it is also necessary to continuously monitor the synchronization status. As a solution to these synchronization problems

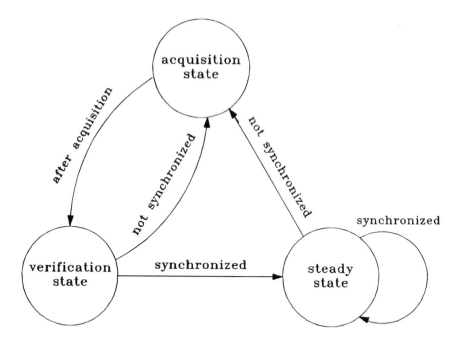

Figure 10.2. State transition diagram for the three-state synchronization mechanism.

arising in the errored environment, the DSS is additionally equipped with the *three-state synchronization mechanism*.

The three-state synchronization mechanism consists of three states of synchronization — *the acquisition state, the verification state, and the steady state* — and conditions for the state transition of the three states can be depicted as in Fig. 10.2.[1] In the acquisition state, the synchronization is acquired by changing the descrambler SRG states based on the conveyed samples of the scrambler SRG sequence. In the verification state which follows the acquisition state, the acquired synchronization is verified. Once the synchronization is verified, the DSS then moves to the steady state. Otherwise, the DSS goes back to the acquisition state and restarts the acquisition processing. Even in the steady state, the synchronization status is continuously confirmed so that the synchronization processing can be reinitiated whenever the synchronization breaks down.

---

[1] More detailed description and operation of the three-state synchronization mechanism follows in Chapter 15.

10.3   Organization of the Part

As discussed above, the DSS is identical to the FSS in the scrambling mechanism which employs SRGs for the generation of scrambling sequences, but different in the synchronization mechanism as it relies on the distributed samples instead of the prespecified state data for synchronization. So the theories on sequence spaces and SRGs developed in Chapters 4 and 5 are equally applicable to the DSS. Therefore, in this part we will concentrate more on the synchronization aspect of the DSS, investigating the sampling and correction processes in the serial, parallel, and multibit-parallel scrambling environments.

The part is organized as follows : In Chapter 11, we introduce the concept of scrambling space and discuss how to predict scrambling sequences using the samples taken out of it. In Chapter 12 we examine the operation of the serial DSS with stress put on the sampling and correction processes, and in the subsequent two chapters we extend the discussion to the cases of the parallel and the multibit-parallel DSSs. In Chapter 15, we consider the synchronization issues arising in the errored environment and discuss the three-state synchronization mechanism introduced as a means to resolve them. Finally in Chapter 16, we demonstrate how to apply the developed DSS techniques for scrambling in the cell-based ATM transmission.

Chapter 11
Prediction of Scrambling Sequences

The DSS is distinctively different from the FSS in that it takes samples out of the scrambler SRG sequences and conveys them to the descrambler for use in determining the state of the descrambler SRG. In this regard, it is important to clearly understand the properties of the scrambling sequences and the relations between the scrambling sequences and their samples.

In this chapter, we examine the behaviors of the scrambling sequences, and consider how to predict the sequences using their samples. For a given SRG, we first consider how to determine the sequence space formed by the scrambling sequences with a fixed generating vector (scrambling space), and then consider how to find the largest-dimensional sequence space that can be obtained by varying the generating vectors (scrambling maximal space). For basic SRGs, we discuss the related properties of the scrambling sequences generated by them. Finally, we consider how to choose the sampling times in sampling the scrambling sequence and how to predict the scrambling sequences using the predictable samples.

11.1 Scrambling Space

According to the discussion in Chapter 5, the configuration of an SRG is determined by its state transition matrix $\mathbf{T}$, and if its initial state vector $\mathbf{d}_0$ is additionally furnished, the SRG sequence vector $\mathbf{D}$ can be determined by (5.7). Further, if the generating vector **h** is also given, then the scrambling sequence { s_k } can be uniquely determined by (6.13). This implies that in order to determine the scrambling sequence { s_k } generated by an SRG, we have to know the state transition matrix **T**, the initial state vector **d**$_0$, and the generating vector **h**. In the DSS, however, the initial state vector **d**$_0$ is not given in an explicit form, but is implicitly provided in the form of samples. Therefore, the scrambling sequence { s_k } can be obtained only when proper processings are done on the samples to extract the correct value of **d**$_0$.

We first consider what kind of the scrambling sequences can be generated when initial state vectors of an SRG are varied. For this, we first introduce the concept of scrambling space.

DEFINITION 11.1 (Scrambling Space) For an SRG with the state transition matrix **T** and the generating vector **h**, we define the *scrambling sequence space*(in short, *scrambling space*) $V[\mathbf{T}, \mathbf{h}]$ to be the set

$$V[\mathbf{T}, \mathbf{h}] \equiv \{\{s_k\} : \{s_k\} = \mathbf{h}^t \cdot \mathbf{D} \text{for all possible initial state vectors } \mathbf{d}_0\text{'s}\}. \tag{11.1}$$

According to this definition, the scrambling space $V[\mathbf{T}, \mathbf{h}]$ of the SRG with the state transition matrix **T** and the generating vector **h** is the set composed of the scrambling sequences { s_k }'s obtained by varying initial state vectors **d**$_0$'s. Therefore, every scrambling sequence { s_k } generated by the SRG with the state transition matrix **T** and the generating vector **h** is an element of the scrambling space $V[\mathbf{T}, \mathbf{h}]$ regardless of its initial state vector **d**$_0$.

The following theorem describes how to determine the scrambling space $V[\mathbf{T}, \mathbf{h}]$.

THEOREM 11.2 *For an SRG with the state transition matrix* **T** *and the generating vector* **h**, *the scrambling space* $V[\mathbf{T}, \mathbf{h}]$ *is the sequence space* $V[\Psi_{\mathbf{T}, \mathbf{h}}(x)]$ *whose characteristic polynomial* $\Psi_{\mathbf{T}, \mathbf{h}}(x)$ *is the lowest-degree polynomial meeting the relation*

$$\mathbf{h}^t \cdot \Psi_{\mathbf{T}, \mathbf{h}}(\mathbf{T}) = 0. \tag{11.2}$$

Proof : We first prove that $V[\mathbf{T}, \mathbf{h}] \subset V[\Psi_{\mathbf{T}, \mathbf{h}}(x)]$. Let $\Psi_{\mathbf{T}, \mathbf{h}}(x) = \sum_{i=0}^{L} \psi_i x^i$, and let { s_k } be the scrambling sequence in $V[\mathbf{T}, \mathbf{h}]$ for an initial state vector **d**$_0$. Then, it suffices to show that { s_k } $\in V[\Psi_{\mathbf{T}, \mathbf{h}}(x)]$.

Since $s_k = \mathbf{h}^t \cdot \mathbf{d}_k$, $k = 0, 1, \cdots$, we have, by (5.2), $\sum_{i=0}^{L} \psi_i s_{k+i} = \sum_{i=0}^{L} \psi_i \mathbf{h}^t \cdot \mathbf{T}^i \cdot \mathbf{d}_k = \mathbf{h}^t \cdot \Psi_{\mathbf{T},\mathbf{h}}(\mathbf{T}) \cdot \mathbf{d}_k$. But this is 0 due to (11.2). Therefore by (4.1) $\{ s_k \} \in V[\Psi_{\mathbf{T},\mathbf{h}}(x)]$.

To complete the proof, we now prove that $V[\Psi_{\mathbf{T},\mathbf{h}}(x)] \subset V[\mathbf{T}, \mathbf{h}]$. Let $\{ \hat{s}_k \} = \mathbf{s}^t \cdot \mathbf{E}_{\Psi_{\mathbf{T},\mathbf{h}}(x)}$ for an initial vector $\mathbf{s}$. Then, it suffices to show that $\{ \hat{s}_k \} \in V[\mathbf{T}, \mathbf{h}]$, which is equivalent to showing that there exists an initial state vector $\mathbf{d}_0$ that makes the scrambling sequence $\{ s_k \}$ identical to $\{ \hat{s}_k \}$. Since $\Psi_{\mathbf{T},\mathbf{h}}(x)$ is the lowest-degree polynomial meeting (11.2), the L vectors $\mathbf{h}^t \cdot \mathbf{T}^i$, $i = 0, 1, \cdots, L - 1$, are linearly independent. Therefore, we can choose an initial state vector $\mathbf{d}_0$ such that $\mathbf{h}^t \cdot \mathbf{T}^i \cdot \mathbf{d}_0 = \mathbf{e}_i^t \cdot \mathbf{s}$ for $i = 0, 1, \cdots, L - 1$, and for this initial state vector the scrambling sequence $\{ s_k \}$ has the initial vector $\mathbf{s}$. Therefore, the scrambling sequence $\{ s_k \}$ is identical to $\{ \hat{s}_k \} = \mathbf{s}^t \cdot \mathbf{E}_{\Psi_{\mathbf{T},\mathbf{h}}(x)}$ since $\{ s_k \}$ is an element of the sequence space $V[\Psi_{\mathbf{T},\mathbf{h}}(x)]$. This completes the proof. ∎

The theorem describes how to determine the scrambling space $V[\mathbf{T}, \mathbf{h}]$. That is, the scrambling space $V[\mathbf{T}, \mathbf{h}]$ is identical to the sequence space $V[\Psi_{\mathbf{T},\mathbf{h}}(x)]$ whose characteristic polynomial $\Psi_{\mathbf{T},\mathbf{h}}(x)$ is the lowest-degree polynomial meeting (11.2). Therefore, any scrambling sequence $\{ s_k \}$ generated by an SRG with the state transition matrix $\mathbf{T}$ and the generating vector $\mathbf{h}$ belongs, regardless of its initial state vector $\mathbf{d}_0$, to the sequence space $V[\Psi_{\mathbf{T},\mathbf{h}}(x)]$ whose characteristic polynomial $\Psi_{\mathbf{T},\mathbf{h}}(x)$ is the lowest-degree polynomial meeting (11.2).

EXAMPLE 11.1 We consider the SRG in Fig. 11.1. The state transition matrix $\mathbf{T}$ and the generating vector $\mathbf{h}$ are respectively

$$\mathbf{T} = \begin{bmatrix} 1 & 1 & 0 & 0 & 0 \\ 0 & 0 & 1 & 0 & 1 \\ 0 & 0 & 0 & 1 & 0 \\ 0 & 0 & 0 & 0 & 1 \\ 0 & 1 & 0 & 0 & 1 \end{bmatrix}, \tag{11.3}$$

$$\mathbf{h} = [\, 0 \, 1 \, 0 \, 0 \, 0 \,]^t. \tag{11.4}$$

For these $\mathbf{T}$ and $\mathbf{h}$, the lowest-degree polynomial $\Psi_{\mathbf{T},\mathbf{h}}(x)$ meeting (11.2) is $\Psi_{\mathbf{T},\mathbf{h}}(x) = x^3 + x + 1$. Therefore, by the theorem the scrambling space $V[\mathbf{T}, \mathbf{h}]$ is the sequence space $V[x^3 + x + 1]$. In fact, the scrambling sequence $\{ s_k \}$ generated by this SRG becomes one of the 32 sequences

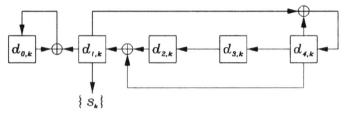

Figure 11.1. An example of SRG.

listed in Table 11.1, depending on its initial state vector $\mathbf{d}_0$. From the table we can confirm that there are only eight different sequences in the scrambling space $V[\mathbf{T}, \mathbf{h}]$(marked with $*$), which are identical to the sequences in the sequence space $V[x^3 + x + 1]$ listed in Table 4.2. ♣

11.2 Scrambling Maximal Space

We consider the largest-dimensional scrambling space among all the scrambling spaces that can be obtained by varying generating vectors.

DEFINITION 11.3 (Scrambling Maximal Space) For an SRG with the state transition matrix $\mathbf{T}$, a *scrambling maximal space* $V[\mathbf{T}]^1$ refers to the largest-dimensional scrambling space of all scrambling spaces $V[\mathbf{T}, \mathbf{h}]$'s obtained by varying the generating vectors $\mathbf{h}$'s.

We first consider the relation between the scrambling space and the SRG maximal space defined in Definition 5.13.

THEOREM 11.4 *For an SRG with the state transition matrix $\mathbf{T}$, the scrambling space $V[\mathbf{T}, \mathbf{h}]$ for a generating vector $\mathbf{h}$ is a subspace of the SRG maximal space $V[\mathbf{T}]$.*

Proof : Let $\{\, s_k \,\}$ be an element of the scrambling space $V[\mathbf{T}, \mathbf{h}]$. Then, it suffices to show that $\{\, s_k \,\} \in V[\mathbf{T}]$. By the definition of scrambling space, there exists an initial state vector $\mathbf{d}_0$ whose scrambling sequence is identical to the sequence $\{\, s_k \,\}$, and hence by the definition

---

[1]The notation $V[\mathbf{T}]$ of the scrambling maximal space is intentionally made identical to that of the SRG maximal space, defined in Definition 5.13, because the scrambling maximal space is, in fact, identical to the SRG maximal space, as will become clear in Theorem 11.5.

Table 11.1. Scrambling sequence $\{\,s_k\,\}$ generated by the SRG in Fig. 11.1 for different initial state vectors. The eight different sequences are marked with $*$.

$\mathbf{d_0}$	scrambling sequence $\{\,s_k\,\}$	
$[\,0\,0\,0\,0\,0\,]^t$	$\{\,0,0,0,0,0,0,0,0,0,0,\cdots\,\} = [\,0\,0\,0\,] \cdot \mathbf{E}_{x^3+x+1}$	$*$
$[\,0\,0\,0\,0\,1\,]^t$	$\{\,0,1,1,1,0,0,1,0,1,1,\cdots\,\} = [\,0\,1\,1\,] \cdot \mathbf{E}_{x^3+x+1}$	$*$
$[\,0\,0\,0\,1\,0\,]^t$	$\{\,0,0,1,0,1,1,1,0,0,1,\cdots\,\} = [\,0\,0\,1\,] \cdot \mathbf{E}_{x^3+x+1}$	$*$
$[\,0\,0\,0\,1\,1\,]^t$	$\{\,0,1,0,1,1,1,0,0,1,0,\cdots\,\} = [\,0\,1\,0\,] \cdot \mathbf{E}_{x^3+x+1}$	$*$
$[\,0\,0\,1\,0\,0\,]^t$	$\{\,0,1,0,1,1,1,0,0,1,0,\cdots\,\} = [\,0\,1\,0\,] \cdot \mathbf{E}_{x^3+x+1}$	
$[\,0\,0\,1\,0\,1\,]^t$	$\{\,0,0,1,0,1,1,1,0,0,1,\cdots\,\} = [\,0\,0\,1\,] \cdot \mathbf{E}_{x^3+x+1}$	
$[\,0\,0\,1\,1\,0\,]^t$	$\{\,0,1,1,1,0,0,1,0,1,1,\cdots\,\} = [\,0\,1\,1\,] \cdot \mathbf{E}_{x^3+x+1}$	
$[\,0\,0\,1\,1\,1\,]^t$	$\{\,0,0,0,0,0,0,0,0,0,0,\cdots\,\} = [\,0\,0\,0\,] \cdot \mathbf{E}_{x^3+x+1}$	
$[\,0\,1\,0\,0\,0\,]^t$	$\{\,1,0,1,1,1,0,0,1,0,1,\cdots\,\} = [\,1\,0\,1\,] \cdot \mathbf{E}_{x^3+x+1}$	$*$
$[\,0\,1\,0\,0\,1\,]^t$	$\{\,1,1,0,0,1,0,1,1,1,0,\cdots\,\} = [\,1\,1\,0\,] \cdot \mathbf{E}_{x^3+x+1}$	$*$
$[\,0\,1\,0\,1\,0\,]^t$	$\{\,1,0,0,1,0,1,1,1,0,0,\cdots\,\} = [\,1\,0\,0\,] \cdot \mathbf{E}_{x^3+x+1}$	$*$
$[\,0\,1\,0\,1\,1\,]^t$	$\{\,1,1,1,0,0,1,0,1,1,1,\cdots\,\} = [\,1\,1\,1\,] \cdot \mathbf{E}_{x^3+x+1}$	$*$
$[\,0\,1\,1\,0\,0\,]^t$	$\{\,1,1,1,0,0,1,0,1,1,1,\cdots\,\} = [\,1\,1\,1\,] \cdot \mathbf{E}_{x^3+x+1}$	
$[\,0\,1\,1\,0\,1\,]^t$	$\{\,1,0,0,1,0,1,1,1,0,0,\cdots\,\} = [\,1\,0\,0\,] \cdot \mathbf{E}_{x^3+x+1}$	
$[\,0\,1\,1\,1\,0\,]^t$	$\{\,1,1,0,0,1,0,1,1,1,0,\cdots\,\} = [\,1\,1\,0\,] \cdot \mathbf{E}_{x^3+x+1}$	
$[\,0\,1\,1\,1\,1\,]^t$	$\{\,1,0,1,1,1,0,0,1,0,1,\cdots\,\} = [\,1\,0\,1\,] \cdot \mathbf{E}_{x^3+x+1}$	
$[\,1\,0\,0\,0\,0\,]^t$	$\{\,0,0,0,0,0,0,0,0,0,0,\cdots\,\} = [\,0\,0\,0\,] \cdot \mathbf{E}_{x^3+x+1}$	
$[\,1\,0\,0\,0\,1\,]^t$	$\{\,0,1,1,1,0,0,1,0,1,1,\cdots\,\} = [\,0\,1\,1\,] \cdot \mathbf{E}_{x^3+x+1}$	
$[\,1\,0\,0\,1\,0\,]^t$	$\{\,0,0,1,0,1,1,1,0,0,1,\cdots\,\} = [\,0\,0\,1\,] \cdot \mathbf{E}_{x^3+x+1}$	
$[\,1\,0\,0\,1\,1\,]^t$	$\{\,0,1,0,1,1,1,0,0,1,0,\cdots\,\} = [\,0\,1\,0\,] \cdot \mathbf{E}_{x^3+x+1}$	
$[\,1\,0\,1\,0\,0\,]^t$	$\{\,0,1,0,1,1,1,0,0,1,0,\cdots\,\} = [\,0\,1\,0\,] \cdot \mathbf{E}_{x^3+x+1}$	
$[\,1\,0\,1\,0\,1\,]^t$	$\{\,0,0,1,0,1,1,1,0,0,1,\cdots\,\} = [\,0\,0\,1\,] \cdot \mathbf{E}_{x^3+x+1}$	
$[\,1\,0\,1\,1\,0\,]^t$	$\{\,0,1,1,1,0,0,1,0,1,1,\cdots\,\} = [\,0\,1\,1\,] \cdot \mathbf{E}_{x^3+x+1}$	
$[\,1\,0\,1\,1\,1\,]^t$	$\{\,0,0,0,0,0,0,0,0,0,0,\cdots\,\} = [\,0\,0\,0\,] \cdot \mathbf{E}_{x^3+x+1}$	
$[\,1\,1\,0\,0\,0\,]^t$	$\{\,1,0,1,1,1,0,0,1,0,1,\cdots\,\} = [\,1\,0\,1\,] \cdot \mathbf{E}_{x^3+x+1}$	
$[\,1\,1\,0\,0\,1\,]^t$	$\{\,1,1,0,0,1,0,1,1,1,0,\cdots\,\} = [\,1\,1\,0\,] \cdot \mathbf{E}_{x^3+x+1}$	
$[\,1\,1\,0\,1\,0\,]^t$	$\{\,1,0,0,1,0,1,1,1,0,0,\cdots\,\} = [\,1\,0\,0\,] \cdot \mathbf{E}_{x^3+x+1}$	
$[\,1\,1\,0\,1\,1\,]^t$	$\{\,1,1,1,0,0,1,0,1,1,1,\cdots\,\} = [\,1\,1\,1\,] \cdot \mathbf{E}_{x^3+x+1}$	
$[\,1\,1\,1\,0\,0\,]^t$	$\{\,1,1,1,0,0,1,0,1,1,1,\cdots\,\} = [\,1\,1\,1\,] \cdot \mathbf{E}_{x^3+x+1}$	
$[\,1\,1\,1\,0\,1\,]^t$	$\{\,1,0,0,1,0,1,1,1,0,0,\cdots\,\} = [\,1\,0\,0\,] \cdot \mathbf{E}_{x^3+x+1}$	
$[\,1\,1\,1\,1\,0\,]^t$	$\{\,1,1,0,0,1,0,1,1,1,0,\cdots\,\} = [\,1\,1\,0\,] \cdot \mathbf{E}_{x^3+x+1}$	
$[\,1\,1\,1\,1\,1\,]^t$	$\{\,1,0,1,1,1,0,0,1,0,1,\cdots\,\} = [\,1\,0\,1\,] \cdot \mathbf{E}_{x^3+x+1}$	

of the SRG space the SRG space $V[\mathbf{T}, \mathbf{d}_0]$ for this initial state vector $\mathbf{d}_0$ includes the sequence $\{\ s_k\ \}$. That is, $\{\ s_k\ \} \in V[\mathbf{T}, \mathbf{d}_0]$. Therefore, by Theorem 5.17 $\{\ s_k\ \} \in V[\mathbf{T}]$. This proves the theorem.   ∎

The theorem means that the scrambling space $V[\mathbf{T}, \mathbf{h}]$ for an arbitrary generating vector $\mathbf{h}$ is a subspace of the SRG maximal space $V[\mathbf{T}]$. As a consequence, the scrambling maximal space also becomes a subspace of the SRG maximal space $V[\mathbf{T}]$.

EXAMPLE 11.2 In the case of the SRG in Fig. 11.1 with the state transition matrix $\mathbf{T}$ in (11.3), the SRG maximal space is $V[\mathbf{T}] = V[x^4 + x^3 + x^2 + 1]$ (refer to Example 5.6). Therefore, by the theorem, the scrambling space $V[\mathbf{T}, \mathbf{h}]$ for an arbitrary generating vector $\mathbf{h}$ is a subspace of the sequence space $V[x^4 + x^3 + x^2 + 1]$. In fact, by Theorem 11.2, we can obtain the scrambling space $V[\mathbf{T}, \mathbf{h}]$'s listed in Table 11.2, and from the table we can confirm that all of the scrambling spaces $V[\mathbf{T}, \mathbf{h}]$'s are subspaces of the SRG maximal space $V[x^4 + x^3 + x^2 + 1]$. ♣

THEOREM 11.5 *For an SRG, its scrambling maximal space is identical to its SRG maximal space.*

Proof : For an SRG with the state transition matrix $\mathbf{T}$, let $V[\mathbf{T}]$ and $\tilde{V}[\mathbf{T}]$ respectively denote the scrambling maximal space and the SRG maximal space. Then, by Theorem 11.4, $V[\mathbf{T}] \subset \tilde{V}[\mathbf{T}]$. Therefore, it suffices to show that the dimension of $V[\mathbf{T}]$ is not less than that of $\tilde{V}[\mathbf{T}]$. Let $\Psi_{\mathbf{T}}(x)$ be the minimal polynomial of $\mathbf{T}$. Then, there exists a generating vector $\mathbf{h}$ such that the lowest-degree polynomial $\Psi_{\mathbf{T},\mathbf{h}}(x)$ satisfying the relation $\mathbf{h}^t \cdot \Psi_{\mathbf{T},\mathbf{h}}(\mathbf{T}) = 0$ is $\Psi_{\mathbf{T}}(x)$. Hence by Theorem 11.2 the scrambling space $V[\mathbf{T}, \mathbf{h}]$ for this generating vector $\mathbf{h}$ is $V[\Psi_{\mathbf{T}}(x)]$. Therefore, the dimension of $V[\mathbf{T}]$, which is identical to $V[\Psi_{\mathbf{T}}(x)]$ due to Theorem 5.14, is not less than that of $V[\Psi_{\mathbf{T}}(x)]$, since the scrambling maximal space is the largest-dimensional scrambling space. This completes the proof.   ∎

Combining the theorem with Theorem 5.14, we find that the scrambling maximal space $V[\mathbf{T}]$, the SRG maximal space, and the sequence space $V[\Psi_{\mathbf{T}}(x)]$ for the minimal polynomial $\Psi_{\mathbf{T}}(x)$ of $\mathbf{T}$ are all identical.

EXAMPLE 11.3 In the case of the SRG in Fig. 11.1 with the state transition matrix $\mathbf{T}$ in (11.3), the SRG maximal space is $V[x^4 + x^3 + x^2 + 1]$.

Table 11.2. Scrambling space $V[\mathbf{T}, \mathbf{h}]$ for the SRG in Fig. 11.1 determined for different generating vectors

generating vector $\mathbf{h}$	scrambling space $V[\mathbf{T}, \mathbf{h}]$
$[0\,0\,0\,0\,0]^t$	$V[1]$
$[0\,0\,0\,0\,1]^t$	$V[x^4 + x^3 + x^2 + 1]$
$[0\,0\,0\,1\,0]^t$	$V[x^4 + x^3 + x^2 + 1]$
$[0\,0\,0\,1\,1]^t$	$V[x^3 + x + 1]$
$[0\,0\,1\,0\,0]^t$	$V[x^4 + x^3 + x^2 + 1]$
$[0\,0\,1\,0\,1]^t$	$V[x^3 + x + 1]$
$[0\,0\,1\,1\,0]^t$	$V[x^3 + x + 1]$
$[0\,0\,1\,1\,1]^t$	$V[x^4 + x^3 + x^2 + 1]$
$[0\,1\,0\,0\,0]^t$	$V[x^3 + x + 1]$
$[0\,1\,0\,0\,1]^t$	$V[x^4 + x^3 + x^2 + 1]$
$[0\,1\,0\,1\,0]^t$	$V[x^4 + x^3 + x^2 + 1]$
$[0\,1\,0\,1\,1]^t$	$V[x^3 + x + 1]$
$[0\,1\,1\,0\,0]^t$	$V[x^4 + x^3 + x^2 + 1]$
$[0\,1\,1\,0\,1]^t$	$V[x^3 + x + 1]$
$[0\,1\,1\,1\,0]^t$	$V[x^3 + x + 1]$
$[0\,1\,1\,1\,1]^t$	$V[x + 1]$
$[1\,0\,0\,0\,0]^t$	$V[x^4 + x^3 + x^2 + 1]$
$[1\,0\,0\,0\,1]^t$	$V[x + 1]$
$[1\,0\,0\,1\,0]^t$	$V[x^4 + x^3 + x^2 + 1]$
$[1\,0\,0\,1\,1]^t$	$V[x^4 + x^3 + x^2 + 1]$
$[1\,0\,1\,0\,0]^t$	$V[x^4 + x^3 + x^2 + 1]$
$[1\,0\,1\,0\,1]^t$	$V[x^4 + x^3 + x^2 + 1]$
$[1\,0\,1\,1\,0]^t$	$V[x^4 + x^3 + x^2 + 1]$
$[1\,0\,1\,1\,1]^t$	$V[x^4 + x^3 + x^2 + 1]$
$[1\,1\,0\,0\,0]^t$	$V[x^4 + x^3 + x^2 + 1]$
$[1\,1\,0\,0\,1]^t$	$V[x^4 + x^3 + x^2 + 1]$
$[1\,1\,0\,1\,0]^t$	$V[x^4 + x^3 + x^2 + 1]$
$[1\,1\,0\,1\,1]^t$	$V[x^4 + x^3 + x^2 + 1]$
$[1\,1\,1\,0\,0]^t$	$V[x^4 + x^3 + x^2 + 1]$
$[1\,1\,1\,0\,1]^t$	$V[x^4 + x^3 + x^2 + 1]$
$[1\,1\,1\,1\,0]^t$	$V[x + 1]$
$[1\,1\,1\,1\,1]^t$	$V[x^4 + x^3 + x^2 + 1]$

Therefore, by the theorem, its scrambling maximal space $V[\mathbf{T}]$ is the sequence space $V[x^4 + x^3 + x^2 + 1]$, which we can confirm from Table 11.2. ♣

As we considered the concept of the maximal initial sate vector in relation to the SRG maximal space before, we now consider the concept of the maximal generating vector in relation to the scrambling maximal space.

DEFINITION 11.6 (Maximal Generating Vector) For an SRG with the state transition matrix $\mathbf{T}$, a *maximal generating vector* $\mathbf{h}$ refers to a generating vector that makes the scrambling space $V[\mathbf{T}, \mathbf{h}]$ the scrambling maximal space $V[\mathbf{T}]$.

THEOREM 11.7 *For an SRG with the state transition matrix $\mathbf{T}$, let $\hat{L}$ be the dimension of the scrambling maximal space $V[\mathbf{T}]$. Then, a generating vector $\mathbf{h}$ is a maximal generating vector, if and only if the discrimination matrix $\Delta_{\mathbf{T},\mathbf{h}}$ defined by*

$$\Delta_{\mathbf{T},\mathbf{h}} \equiv \begin{bmatrix} \mathbf{h}^t \\ \mathbf{h}^t \cdot \mathbf{T} \\ \vdots \\ \mathbf{h}^t \cdot \mathbf{T}^{\hat{L}-1} \end{bmatrix} \tag{11.5}$$

is of rank $\hat{L}$.

Proof : We first prove the "if" part of the theorem. Let $\mathbf{h}$ be a generating vector that makes the rank of $\Delta_{\mathbf{T},\mathbf{h}}$ $\hat{L}$. Then, the $\hat{L}$ vectors $\mathbf{h}^t \cdot \mathbf{T}^i$, $i = 0, 1, \cdots, \hat{L}-1$, are linearly independent. Therefore, the degree of the lowest-degree polynomial $\Psi_{\mathbf{T},\mathbf{h}}(x)$ that meets (11.2) is $\hat{L}$ or higher, so by Theorem 11.2, the scrambling space $V[\mathbf{T}, \mathbf{h}]$ is of dimension $\hat{L}$ or higher. However, since the dimension of the scrambling maximal space $V[\mathbf{T}]$ is $\hat{L}$, $V[\mathbf{T}, \mathbf{h}]$ should be the scrambling maximal space $V[\mathbf{T}]$. That is, $\mathbf{h}$ is a maximal generating vector.

Now we prove the "only if" part by contradiction. Suppose that there exists a maximal generating vector $\mathbf{h}$ such that the rank of $\Delta_{\mathbf{T},\mathbf{h}}$ is less than $\hat{L}$. Then, the $\hat{L}$ vectors $\mathbf{h}^t \cdot \mathbf{T}^i$, $i = 0, 1, \cdots, \hat{L} - 1$, are linearly dependent. Therefore, the degree of the lowest-degree polynomial $\Psi_{\mathbf{T},\mathbf{h}}(x)$ that meets (11.2) is less than $\hat{L}$, so by Theorem 11.2, the dimension of $V[\mathbf{T}, \mathbf{h}]$ is less than $\hat{L}$. However, it is a contradiction to the assumption that $\mathbf{h}$ is a maximal generating vector. ∎

EXAMPLE 11.4 In the case of the SRG in Fig. 11.1 with the state transition matrix **T** in (11.3), the scrambling maximal space is $V[x^4 + x^3 + x^2 + 1]$, whose dimension is 4. If we choose $\mathbf{h} = [\,1\,0\,0\,0\,0\,]^t$, by (11.5) we obtain the discrimination matrix

$$\Delta_{\mathbf{T},\mathbf{h}} = \begin{bmatrix} 1 & 0 & 0 & 0 & 0 \\ 1 & 1 & 0 & 0 & 0 \\ 1 & 1 & 1 & 0 & 1 \\ 1 & 0 & 1 & 1 & 0 \end{bmatrix}, \tag{11.6}$$

whose rank is 4. Therefore, by the theorem, this **h** is a maximal generating vector, and we can confirm this using Table 11.2. ♣

The concept of the maximal generating vector is defined in relation to the scrambling maximal space, as the concept of the maximal initial state vector was defined in relation to the SRG maximal space. But by Theorem 11.5 the scrambling maximal space is identical to the SRG maximal space. In this aspect, we can expect the maximal generating vector to be closely related to the maximal initial state vector. In fact, they turn out to possess a *dual* relation, as the following theorem describes.

THEOREM 11.8 *A maximal initial vector for an SRG with the state transition matrix* **T** *becomes a maximal generating vector for the SRG with the state transition matrix* $\mathbf{T}^t$. *Conversely, a maximal generating vector for an SRG with the state transition matrix* **T** *becomes a maximal initial vector for the SRG with the state transition matrix* $\mathbf{T}^t$.

Proof : Let $\mathbf{d}_0$ and **h** respectively be the maximal initial state vector and the maximal generating vector of an SRG with the state transition matrix **T**. Then, by (5.10) and (11.5), we have the relation $\Delta_{\mathbf{T},\mathbf{d}_0} = \Delta_{\mathbf{T}^t,\mathbf{d}_0}$ and $\Delta_{\mathbf{T}^t,\mathbf{h}} = \Delta_{\mathbf{T},\mathbf{h}}$. Therefore, by Theorems 5.16 and 11.7, we have the theorem. ∎

Noting that the state transition matrix $\mathbf{A}_{\Psi(x)}^t$ of the SSRG is the transpose of the state transition matrix $\mathbf{A}_{\Psi(x)}$ of the MSRG, we can apply Theorem 11.8 to Theorems 5.31 and 5.32 to generate the following two theorems.

THEOREM 11.9 *For the SSRG of the sequence space* $V[\Psi(x)]$ *of dimension* L, *a generating vector* **h** *is a maximal generating vector, if and*

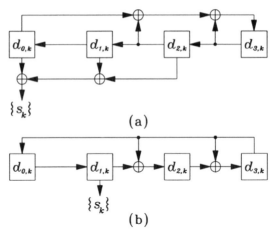

Figure 11.2. The SSRG and the MSRG of the sequence space $V[x^4 + x^3 + x^2 + 1]$. (a) The SSRG with its maximal generating vector $\mathbf{h} = [0\,1\,1\,1]^t$, (b) the MSRG with its maximal generating vector $\mathbf{h} = [0\,1\,0\,0]^t$.

only if the minimal space for the sequence $\mathbf{h}^t \cdot \mathbf{P}_{\Psi(x)}$ *is the sequence space* $V[\Psi(x)]$.

THEOREM 11.10 *For the MSRG for the sequence space* $V[\Psi(x)]$ *of dimension L, a generating vector* $\mathbf{h}$ *is a maximal generating vector, if and only if the minimal space for the sequence* $\mathbf{h}^t \cdot \mathbf{E}_{\Psi(x)}$ *is the sequence space* $V[\Psi(x)]$.

EXAMPLE 11.5 In the case of the sequence space $V[x^4 + x^3 + x^2 + 1]$, the minimal space for the sequence S_4 in Table 4.1 is $V[x^4 + x^3 + x^2 + 1]$, and its initial vectors for the elementary and primary bases are respectively $\mathbf{s} = [0\,1\,0\,0]^t$ and $\mathbf{p} = [1\,1\,1\,0]^t$ (refer to Example 5.12). Therefore, the generating vector $\mathbf{h} = \mathbf{p} = [1\,1\,1\,0]^t$ is a maximal generating vector of the SSRG for the sequence space $V[x^4 + x^3 + x^2 + 1]$, and the generating vector $\mathbf{h} = \mathbf{s} = [0\,1\,0\,0]^t$ is a maximal generating vector for the MSRG of the sequence space $V[x^4 + x^3 + x^2 + 1]$. These SRGs are shown in Figs. 11.2(a) and (b) respectively. ♣

11.3   Scrambling Maximal Spaces for BSRGs

We consider the scrambling maximal spaces and the maximal generating vectors for BSRGs.

According to Definition 5.18, a BSRG for a sequence space $V[\Psi(x)]$ is an SRG of the smallest length whose SRG maximal space is identical to the sequence space $V[\Psi(x)]$; and according to Theorem 11.5, the scrambling maximal space of an SRG is identical to its SRG maximal space. Therefore, we have the following theorem.

THEOREM 11.11 *An SRG of the smallest length whose scrambling maximal space is identical to the sequence space $V[\Psi(x)]$ is a BSRG for the sequence space $V[\Psi(x)]$.*

Combining this theorem with Theorems 5.19 and 5.20, we find that the L-dimensional scrambling space $V[\Psi(x)]$ can be generated by an SRG of length L or more, and that a length-L BSRG has a state transition matrix $\mathbf{T}$ which is similar to the companion matrix $\mathbf{A}_{\Psi(x)}$ in (4.7). The maximal generating vector for such a BSRG can be determined by the following theorem.

THEOREM 11.12 *A generating vector $\mathbf{h}$ is a maximal generating vector of the BSRG for a sequence space $V[\Psi(x)]$, if and only if the corresponding discrimination matrix $\Delta_{\mathbf{T},\mathbf{h}}$ in (11.5) is nonsingular for the state transition matrix $\mathbf{T}$ taken similar to $\mathbf{A}_{\Psi(x)}$ in (4.7).*[2]

Proof : This theorem is obtained by combining Theorems 5.19, 5.20 and 11.7. ∎

EXAMPLE 11.6 We design a smallest-length SRG whose scrambling space is the sequence space $V[x^4 + x^3 + x^2 + 1]$. By Theorem 11.11 this desired SRG is a BSRG for sequence space $V[x^4 + x^3 + x^2 + 1]$. In accordance with Theorems 5.19 and 5.20, we choose the length $L = 4$ which is identical to the dimension of $V[x^4 + x^3 + x^2 + 1]$, and the state transition matrix

$$\mathbf{T} = \begin{bmatrix} 0 & 1 & 0 & 0 \\ 0 & 0 & 1 & 0 \\ 0 & 1 & 0 & 1 \\ 1 & 1 & 0 & 1 \end{bmatrix} \tag{11.7}$$

which is similar to the companion matrix $\mathbf{A}_{x^4+x^3+x^2+1}$ in (4.8). For the BSRG thus formed, we choose the generating vector $\mathbf{h} = [0011]^t$. Then

---

[2]Note that in the case of BSRG, the discrimination matrix $\Delta_{\mathbf{T},\mathbf{h}}$ in (11.5) is a square matrix since $\hat{L}$ is identical to the length of the BSRG due to Theorem 5.19.

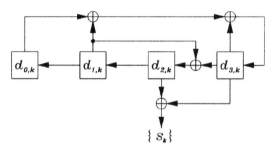

$\{ s_k \}$

Figure 11.3. A smallest-length SRG whose scrambling space is $V[x^4 + x^3 + x^2 + 1]$. This is a BSRG for the sequence space $V[x^4 + x^3 + x^2 + 1]$, and its generating vector is maximal for the BSRG.

by (11.5) we obtain the discrimination matrix

$$\Delta_{\mathbf{T},\mathbf{h}} = \begin{bmatrix} 0 & 0 & 1 & 1 \\ 1 & 0 & 0 & 0 \\ 0 & 1 & 0 & 0 \\ 0 & 0 & 1 & 0 \end{bmatrix}, \tag{11.8}$$

which is nonsingular. Therefore, by Theorem 11.12 this $\mathbf{h}$ is a maximal generating vector for the BSRG with the state transition matrix in (11.7). The resulting BSRG is as shown in Fig. 11.3. ♣

THEOREM 11.13 *For a state transition matrix $\mathbf{T}$ and a maximal generating vector $\mathbf{h}$ of a BSRG for the sequence space $V[\Psi(x)]$, the following relations hold for the relevant nonsingular discrimination matrix $\Delta_{\mathbf{T},\mathbf{h}}$:*

$$\mathbf{T} = \Delta_{\mathbf{T},\mathbf{h}}^{-1} \cdot \mathbf{A}_{\Psi(x)}^{t} \cdot \Delta_{\mathbf{T},\mathbf{h}}, \tag{11.9a}$$

$$\mathbf{h} = \Delta_{\mathbf{T},\mathbf{h}}^{t} \cdot \mathbf{e}_0. \tag{11.9b}$$

Proof : To prove (11.9a), we let $\Psi(x) = \sum_{i=0}^{L} \psi_i x^i$ and consider the matrix $\mathbf{A}_{\Psi(x)}^{t} \cdot \Delta_{\mathbf{T},\mathbf{h}}$. Then, by (11.5) and (4.7), $\mathbf{A}_{\Psi(x)}^{t} \cdot \Delta_{\mathbf{T},\mathbf{h}} = [\mathbf{T}^t \cdot \mathbf{h} \quad (\mathbf{T}^t)^2 \cdot \mathbf{h} \quad \cdots \quad (\mathbf{T}^t)^{L-1} \cdot \mathbf{h} \quad \sum_{i=0}^{L-1} \psi_i (\mathbf{T}^t)^i \cdot \mathbf{h}]^t$ (note that $\hat{L}$ in (11.5) becomes L in this case). Since $\mathbf{T}$ is similar to $\mathbf{A}_{\Psi(x)}$ due to Theorem 5.20, $\sum_{i=0}^{L} \psi_i (\mathbf{T}^t)^i \cdot \mathbf{h} = 0$, that is, $\sum_{i=0}^{L-1} \psi_i (\mathbf{T}^t)^i \cdot \mathbf{h} = (\mathbf{T}^t)^L \cdot \mathbf{h}$. Thus, $\mathbf{A}_{\Psi(x)}^{t} \cdot \Delta_{\mathbf{T},\mathbf{h}} = [\mathbf{T}^t \cdot \mathbf{h} \quad (\mathbf{T}^t)^2 \cdot \mathbf{h} \quad \cdots \quad (\mathbf{T}^t)^{L-1} \cdot \mathbf{h} \quad (\mathbf{T}^t)^L \cdot \mathbf{h}]^t = \Delta_{\mathbf{T},\mathbf{h}} \cdot \mathbf{T}$. Therefore, we have the equation (11.9a).

Equation (11.9b) is directly obtained from (11.5). ∎

The theorem indicates how closely the state transition matrix $\mathbf{T}$ and the maximal generating vector $\mathbf{h}$ of a BSRG are related with the discrimination matrix $\Delta_{\mathbf{T},\mathbf{h}}$. While it is useful to know the relations in (11.9), they are not directly applicable in designing SRGs that can generate a prespecified scrambling space because $\mathbf{T}$, $\mathbf{h}$, and $\Delta_{\mathbf{T},\mathbf{h}}$ are connected through recursive relations. Therefore, in the following we consider a more general set of expressions that enables us to determine $\mathbf{T}$ and $\mathbf{h}$ directly for the given scrambling space $V[\Psi(x)]$.

THEOREM 11.14 *For a nonsingular matrix* $\mathbf{R}$, *an SRG with the state transition matrix*

$$\mathbf{T} = \mathbf{R}^{-1} \cdot \mathbf{A}^t_{\Psi(x)} \cdot \mathbf{R} \tag{11.10a}$$

is a BSRG for the sequence space $V[\Psi(x)]$; *and the generating vector of the form*

$$\mathbf{h} = \mathbf{R}^t \cdot \mathbf{e}_0 \tag{11.10b}$$

is a maximal generating vector for this BSRG.

Proof : The first part is directly obtained by Theorem 5.20, since $\mathbf{A}^t_{\Psi(x)}$ is similar to $\mathbf{A}_{\Psi(x)}$. To prove the second part, we insert $\mathbf{T} = \mathbf{R}^{-1} \cdot \mathbf{A}^t_{\Psi(x)} \cdot \mathbf{R}$ and $\mathbf{h} = \mathbf{R}^t \cdot \mathbf{e}_0$ into (11.5). Then, we obtain $\Delta_{\mathbf{T},\mathbf{h}} = [\, \mathbf{R}^t \cdot \mathbf{e}_0 \quad \mathbf{R}^t \cdot \mathbf{A}_{\Psi(x)} \cdot \mathbf{e}_0 \quad \cdots \quad \mathbf{R}^t \cdot \mathbf{A}^{L-1}_{\Psi(x)} \cdot \mathbf{e}_0 \,]^t$, which is identical to $[\, \mathbf{R}^t \cdot \mathbf{e}_0 \quad \mathbf{R}^t \cdot \mathbf{e}_1 \quad \cdots \quad \mathbf{R}^t \cdot \mathbf{e}_{L-1} \,]^t$ due to the relation $\mathbf{A}^i_{\Psi(x)} \cdot \mathbf{e}_0 = \mathbf{e}_i$, $i = 0$, $1, \cdots, L-1$. Therefore, $\Delta_{\mathbf{T},\mathbf{h}} = \mathbf{R}$, which is nonsingular by assumption. Thus by Theorem 11.12, $\mathbf{h} = \mathbf{R}^t \cdot \mathbf{e}_0$ is a maximal generating vector. ∎

The theorem describes how to determine BSRGs for a sequence space $V[\Psi(x)]$ and their maximal generating vectors. That is, the SRG with the state transition matrix $\mathbf{T} = \mathbf{R}^{-1} \cdot \mathbf{A}^t_{\Psi(x)} \cdot \mathbf{R}$ for an arbitrary nonsingular matrix $\mathbf{R}$ is a BSRG for the sequence space $V[\Psi(x)]$, and the generating vector $\mathbf{h} = \mathbf{R}^t \cdot \mathbf{e}_0$ is maximal generating vector for this BSRG. Note that the nonsingular matrix $\mathbf{R}$ itself now becomes the discrimination matrix $\Delta_{\mathbf{T},\mathbf{h}}$ for these $\mathbf{T}$ and $\mathbf{h}$, and therefore no separate singularity test is necessary on it.

EXAMPLE 11.7 We design another BSRG for the sequence space $V[x^4 +$

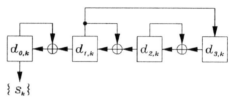

Figure 11.4. Another smallest-length SRG whose scrambling space is $V[x^4 + x^3 + x^2 + 1]$.

$x^3 + x^2 + 1]$ based on Theorem 11.14. We choose the nonsingular matrix

$$R = \begin{bmatrix} 1 & 0 & 0 & 0 \\ 1 & 1 & 0 & 0 \\ 1 & 0 & 1 & 0 \\ 1 & 1 & 1 & 1 \end{bmatrix}, \tag{11.11}$$

and insert this into (11.10). Then we obtain the following state transition matrix and generating vector :

$$T = \begin{bmatrix} 1 & 1 & 0 & 0 \\ 0 & 1 & 1 & 0 \\ 0 & 0 & 1 & 1 \\ 0 & 1 & 0 & 0 \end{bmatrix}, \tag{11.12a}$$

$$h = [1\,0\,0\,0]^t. \tag{11.12b}$$

The resulting circuit diagram for this BSRG is as shown in Fig. 11.4. Note that the BSRG in the figure is also a smallest-length SRG whose scrambling space is $V[x^4 + x^3 + x^2 + 1]$. ♣

11.4   Prediction of Scrambling Sequences

So far, we have considered the scrambling space and the scrambling maximal space formed by scrambling sequences. We now consider, assuming that the scrambling space is known, how to predict the scrambling sequences from their samples.

We consider the scrambling sequence $\{\,s_k\,\}$ generated by an SRG whose scrambling space is the sequence space $V[\Psi(x)]$ of dimension L. Then, since the scrambling sequence $\{\,s_k\,\}$ is an element of the scrambling space $V[\Psi(x)]$, it can be completely determined if its initial vector s is known. But since the initial vector s of the scrambling sequence { s_k } is an L-vector, it can not be predicted from samples less than L. Therefore, it is fundamental for a valid prediction to be furnished with a minimum of L samples of { s_k }. With this fundamental requirement in mind we consider when to take the L samples, or equivalently, how to form the sample vector.

DEFINITION 11.15 (Sampling Time) For a scrambling sequence { s_k } generated by an SRG whose scrambling space is the sequence space $V[\Psi(x)]$ of dimension L, the ith *sampling time* α_i, $i = 0, 1, \cdots, L - 1$, indicates the time when the ith *sample* z_i is taken from the scrambling sequence { s_k } such that

$$z_i = s_{\alpha_i}. \tag{11.13}$$

DEFINITION 11.16 (Sample Vector) For a scrambling sequence { s_k } generated by an SRG whose scrambling space is the sequence space $V[\Psi(x)]$ of dimension L, the *sample vector* z refers to the L-vector formed by the L samples z_i's, i.e.,

$$\mathbf{z} \equiv [\, z_0 \ \ z_1 \ \cdots \ z_{L-1} \,]^t. \tag{11.14}$$

We first examine the relation between the sample vector z and the initial vector s of the scrambling sequence.

THEOREM 11.17 *For a scrambling sequence* { s_k } *generated by an SRG whose scrambling space is the sequence space* $V[\Psi(x)]$ *of dimension L, the sample vector z formed by samples taken at the sampling times* α_i, $i = 0, 1, \cdots, L\text{-}1$, *is represented by*

$$\mathbf{z} = \Delta_\alpha \cdot \mathbf{s}, \tag{11.15}$$

where Δ_α *is the* $L \times L$ *discrimination matrix defined by*

$$\Delta_\alpha \equiv \begin{bmatrix} \mathbf{e}_0^t \cdot (\mathbf{A}_{\Psi(x)}^t)^{\alpha_0} \\ \mathbf{e}_0^t \cdot (\mathbf{A}_{\Psi(x)}^t)^{\alpha_1} \\ \vdots \\ \mathbf{e}_0^t \cdot (\mathbf{A}_{\Psi(x)}^t)^{\alpha_{L-1}} \end{bmatrix}. \tag{11.16}$$

Proof: Let s be the initial vector of the scrambling sequence { s_k } whose scrambling space is $V[\Psi(x)]$. Then, by (11.13) and (4.9) the ith sample is represented by $z_i = \mathbf{s}^t \cdot \mathbf{A}_{\Psi(x)}^{\alpha_i} \cdot \mathbf{e}_0$, $i = 0, 1, \cdots, L - 1$, which can be rewritten in the form of $z_i = \mathbf{e}_0^t \cdot (\mathbf{A}_{\Psi(x)}^t)^{\alpha_i} \cdot \mathbf{s}$, $i = 0, 1, \cdots, L - 1$. Therefore, we have the theorem. ∎

EXAMPLE 11.8 We consider the scrambling sequence $\{\, s_k \,\}$ generated by an SRG whose scrambling space is the sequence space $V[x^4 + x^3 + x^2 + 1]$. If we choose the sampling times $\alpha_i = 2i$, $i = 0, 1, 2, 3$, then by (11.16) and (4.8) we have the discrimination matrix

$$
\Delta_{\alpha,1} = \begin{bmatrix} 1 & 0 & 0 & 0 \\ 0 & 0 & 1 & 0 \\ 1 & 0 & 1 & 1 \\ 0 & 1 & 1 & 1 \end{bmatrix}. \tag{11.17}
$$

Therefore, by (11.15) the sample vector **z** becomes as listed in Table 11.3 depending on the initial vector **s**. If we choose another set of sampling times, $\alpha_0 = 1$, $\alpha_1 = 3$, $\alpha_2 = 4$ and $\alpha_3 = 5$, then we have the discrimination matrix

$$
\Delta_{\alpha,2} = \begin{bmatrix} 0 & 1 & 0 & 0 \\ 0 & 0 & 0 & 1 \\ 1 & 0 & 1 & 1 \\ 1 & 1 & 1 & 0 \end{bmatrix}. \tag{11.18}
$$

The sample vector **z** in this case becomes as listed in the right half of Table 11.3. We can confirm the sample vectors **z**'s in Table 11.3 using the sequences in Table 4.1. ♣

Even though L samples are the minimum number required for a valid prediction, it does not necessarily mean that they guarantee to determine the initial vector **s**. It is because the determination of **s** additionally depends on the sampling times as (11.15) and (11.16) indicate. Therefore, we consider the conditions to be imposed on sampling times to achieve a valid prediction.

DEFINITION 11.18 (Predictable Sampling Times) For a scrambling sequence $\{\, s_k \,\}$ generated by an SRG whose scrambling space is the sequence space $V[\Psi(x)]$ of dimension L, we call the sampling times α_i, $i = 0, 1, \cdots, L-1$, based on which the initial vector **s** of the scrambling sequence $\{\, s_k \,\}$ is predictable the *predictable sampling times*.

THEOREM 11.19 *For a scrambling sequence $\{\, s_k \,\}$ generated by an SRG whose scrambling space is the sequence space $V[\Psi(x)]$ of dimension L, the sampling times α_i, $i = 0, 1, \cdots, L\text{-}1$, are predictable sampling times, if and only if the discrimination matrix Δ_α in (11.16) is nonsingular.*

Table 11.3. The relations between the initial vector **s** and the sample vector **z** of the scrambling sequence $\{ s_k \}$ generated by an SRG whose scrambling space is the sequence space $V[x^4 + x^3 + x^2 + 1]$

initial vector	sample vector **z**	
s	(1) when $\alpha_i = 0, 2, 4, 6$	(2) when $\alpha_i = 1, 3, 4, 5$
$[0\,0\,0\,0]^t$	$[0\,0\,0\,0]^t$	$[0\,0\,0\,0]^t$
$[0\,0\,0\,1]^t$	$[0\,0\,1\,1]^t$	$[0\,1\,1\,0]^t$
$[0\,0\,1\,0]^t$	$[0\,1\,1\,1]^t$	$[0\,0\,1\,1]^t$
$[0\,0\,1\,1]^t$	$[0\,1\,0\,0]^t$	$[0\,1\,0\,1]^t$
$[0\,1\,0\,0]^t$	$[0\,0\,0\,1]^t$	$[1\,0\,0\,1]^t$
$[0\,1\,0\,1]^t$	$[0\,0\,1\,0]^t$	$[1\,1\,1\,1]^t$
$[0\,1\,1\,0]^t$	$[0\,1\,1\,0]^t$	$[1\,0\,1\,0]^t$
$[0\,1\,1\,1]^t$	$[0\,1\,0\,1]^t$	$[1\,1\,0\,0]^t$
$[1\,0\,0\,0]^t$	$[1\,0\,1\,0]^t$	$[0\,0\,1\,1]^t$
$[1\,0\,0\,1]^t$	$[1\,0\,0\,1]^t$	$[0\,1\,0\,1]^t$
$[1\,0\,1\,0]^t$	$[1\,1\,0\,1]^t$	$[0\,0\,0\,0]^t$
$[1\,0\,1\,1]^t$	$[1\,1\,1\,0]^t$	$[0\,1\,1\,0]^t$
$[1\,1\,0\,0]^t$	$[1\,0\,1\,1]^t$	$[1\,0\,1\,0]^t$
$[1\,1\,0\,1]^t$	$[1\,0\,0\,0]^t$	$[1\,1\,0\,0]^t$
$[1\,1\,1\,0]^t$	$[1\,1\,0\,0]^t$	$[1\,0\,0\,1]^t$
$[1\,1\,1\,1]^t$	$[1\,1\,1\,1]^t$	$[1\,1\,1\,1]^t$

Proof : This theorem is directly obtained from Theorem 11.17, since if Δ_α is nonsingular then it is invertible and so **s** is uniquely determined from **z**. ■

According to the theorem, for a valid prediction of the initial vector **s**, all we need to do is to choose the L sampling times α_i's such that the corresponding discrimination matrix Δ_α in (11.16) becomes nonsingular.

EXAMPLE 11.9 We consider the scrambling sequence $\{ s_k \}$ generated by an SRG whose scrambling space is the sequence space $V[x^4+x^3+x^2+1]$. If the sampling times are $\alpha_i = 2i$, $i = 0, 1, 2, 3$, then the discrimination matrix $\Delta_{\alpha,1}$ in (11.17) is nonsingular ; but if the sampling times are $\alpha_0 = 1$, $\alpha_1 = 3$, $\alpha_2 = 4$ and $\alpha_3 = 5$, then the discrimination matrix $\Delta_{\alpha,2}$ in (11.18) is singular. Therefore, it is possible to predict the sample vector **s** by using the samples taken at the predictable sampling times $\alpha_i = 2i$,

$i = 0, 1, 2, 3$, but not by using the samples whose sampling times are $\alpha_0 = 1$, $\alpha_1 = 3$, $\alpha_2 = 4$ and $\alpha_3 = 5$. In fact, we observe from Table 11.3 that in the former case the sample vector $\mathbf{z}$'s differ for different initial vectors $\mathbf{s}$'s, while in the latter case some sample vectors become identical for some different initial vectors. ♣

Based on Theorems 11.17 and 11.19, we finally state how to determine the initial vector $\mathbf{s}$ from the predictable sample vector $\mathbf{z}$.

THEOREM 11.20 *For a scrambling sequence* $\{\ s_k\ \}$ *generated by an SRG whose scrambling space is the sequence space* $V[\Psi(x)]$ *of dimension L, let* α_i, $i = 0, 1, \cdots, L\text{-}1$, *be the predictable sampling times which form the sample vector* $\mathbf{z}$. *Then, the initial vector* $\mathbf{s}$ *of the scrambling sequence is determined by*

$$\mathbf{s} = \Delta_\alpha^{-1} \cdot \mathbf{z}. \tag{11.19}$$

EXAMPLE 11.10 In the case of the scrambling sequence $\{\ s_k\ \}$ generated by an SRG whose scrambling space is the sequence space $V[x^4 + x^3 + x^2 + 1]$, by Example 11.9, $\alpha_i = 2i$, $i = 0, 1, 2, 3$, are predictable sampling times, and $\Delta_{\alpha,1}$ in (11.17) is a nonsingular discrimination matrix. Therefore, we have

$$\Delta_{\alpha,1}^{-1} = \begin{bmatrix} 1 & 0 & 0 & 0 \\ 1 & 0 & 1 & 1 \\ 0 & 1 & 0 & 0 \\ 1 & 1 & 1 & 0 \end{bmatrix}, \tag{11.20}$$

and hence by (11.19) the initial vector $\mathbf{s}$ becomes as listed in Table 11.3(1) for the 16 $\mathbf{z}_i$'s $[\,0\ 0\ 0\ 0\,]^t$ through $[\,1\ 1\ 1\ 1\,]^t$. ♣

Chapter 12

Serial Distributed Sample Scrambling

In the *distributed sample scrambling* (DSS), which is identical to the FSS in the scrambling and descrambling operations, the descrambler SRG is synchronized to the scrambler SRG using the transmitted samples of the scrambling sequence. In this chapter, we consider the synchronization process of the serial DSS and examine how to realize the scramblers and the descramblers. For this, we first mathematically model the synchronization process along with the scrambler and descrambler SRGs. Then, for the scramblers we consider how to determine the scrambler SRGs and how to sample the scrambling sequence ; and for the descramblers we consider how to determine the descrambler SRGs and how to use the samples for the synchronization, along with their efficient realization methods. Finally, we consider how to design the DSS without employing complex timing circuitry.

12.1   Mathematical Modeling

We first mathematically model the scrambler and the descrambler SRGs along with their synchronization processes.

For a scrambler SRG, we denote by $\mathbf{T}$, $\mathbf{h}$ and $\mathbf{d}_k$ respectively its state transition matrix, generating vector and state vector ; and denote by { s_k } and $V[\Psi(x)]$ of dimension L respectively its scrambling sequence and scrambling space. We denote by z_i, $i = 0, 1, \cdots, L-1$, the ith sample of the scrambling sequence { s_k } ; and denote by α_i, $i = 0, 1,$

$\cdots$, $L - 1$, its sampling time.

We put hat($\hat{}$) on the above variables and parameters to denote their counterparts in the descrambler SRG. That is, for a descrambler SRG, we denote by $\hat{T}$, $\hat{h}$ and $\hat{d}_k$ respectively its state transition matrix, generating vector and state vector ; denote by $\{\hat{s}_k\}$ and $V[\hat{\Psi}(x)]$ of dimension $\hat{L}$, respectively its descrambling sequence and descrambling space ; and denote by $\hat{z}_i$, $i = 0, 1, \cdots, L - 1$, the ith sample of the descrambling sequence $\{\hat{s}_k\}$. Since the sampling functions in the scrambler and the descrambler are identical in the DSS, the sampling time of the descrambler sample $\hat{z}_i$ becomes identical to the sampling time α_i of the scrambler sample z_i.[1]

Based on the sequence $\{s_k\}$ and $\{\hat{s}_k\}$, we can define the term synchronization of the scrambler and the descrambler SRGs as follows :

DEFINITION 12.1 (Synchronization) A descrambler SRG is said *synchronized* to the scrambler SRG at time t, if the relation

$$\hat{s}_k = s_k, \quad k = t, t + 1, \cdots, \tag{12.1}$$

holds between the scrambling sequence $\{s_k\}$ and the descrambling sequence $\{\hat{s}_k\}$.

For the synchronization of the scrambler and descrambler SRGs, the samples z_i's of the scrambling sequence $\{s_k\}$ are transmitted to the descrambler. In the descrambler, the state of the descrambler SRG is repeatedly corrected until the descrambler SRG sequence $\{\hat{s}_k\}$ becomes identical to the scrambler SRG sequence $\{s_k\}$. For each correction the transmitted sample z_i is compared to its descrambler counterpart $\hat{z}_i$, and a correction is made on the state of the descrambler SRG in case the two samples do not coincide.

For the mathematical modeling of this correction process, we define the correction time and correction vector as follows :

DEFINITION 12.2 (Correction Time) For a descrambler SRG, the ith *correction time* β_i, $i = 0, 1, \cdots, L-1$, indicates the time the descrambler SRG state is corrected in case the ith scrambler and descrambler samples z_i and $\hat{z}_i$ do not coincide.

---

[1]Note that the transmission delay of the scrambled sequences is not considered in this mathematical representation because it does not mean anything more than a translation of the whole data on the time scale.

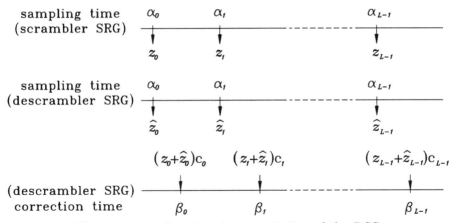

Figure 12.1. Synchronization timing of the DSS.

DEFINITION 12.3 (Correction Vector) For a descrambler SRG, the *i*th *correction vector* $\mathbf{c}_i$, $i = 0, 1, \cdots, L-1$, refers to the L-vector whose unity(i.e., non-zero) elements indicate the positions of the shift registers to be corrected at the *i*th correction time β_i in case the *i*th scrambler and descrambler samples z_i and $\hat{z}_i$ do not coincide.

Then, the correction process of the descrambler SRG is equivalent to adding a correction vector $\mathbf{c}_i$, $i = 0, 1, \cdots, L-1$, to the descrambler SRG state vector $\hat{\mathbf{d}}_k$ at the correction time β_i, $i = 0, 1, \cdots, L-1$. Therefore, the correction process can be carried out in the following manner : First, add the two samples z_i and $\hat{z}_i$, then multiply $\mathbf{c}_i$ to the sum, finally adding it at time β_i to the state vector $\hat{\mathbf{d}}_{\beta_i}$ of the descrambler SRG. The timing relations for this synchronization process are depicted in Fig. 12.1. Note that the *i*th correction is done after the *i*th sampling but no later than the $(i+1)$th sampling, that is,

$$\alpha_0 < \beta_0 \leq \alpha_1 < \beta_1 \leq \cdots \leq \alpha_{L-1} < \beta_{L-1}. \qquad (12.2)$$

After L times of correction, or at the final correction time β_{L-1}, the descrambler SRG is synchronized to the scrambling sequence $\{ s_k \}$.

12.2 Scramblers for DSS

Now we consider how to choose the state transition matrix $\mathbf{T}$ and the generating vector $\mathbf{h}$ for a DSS scrambler SRG and how to decide the sampling times α_i's when taking samples from the scrambling sequence $\{\,s_k\,\}$.

12.2.1   Scrambler SRGs

In the DSS, a PRBS is used for scrambling as in the case of the FSS. However, differently from the FSS, the scrambling sequence $\{\,s_k\,\}$ in the DSS varies depending on the initial state vector $\mathbf{d}_0$ of the scrambler SRG. According to Definition 11.1, the scrambling sequence $\{\,s_k\,\}$ is an element of the scrambling space $V[\Psi(x)]$; and according to Theorem 6.14, a non-zero sequence $\{\,s_k\,\}$ in a primitive space $V[\Psi_p(x)]$ is a PRBS. Therefore, any non-zero scrambling sequence $\{\,s_k\,\}$ generated by an SRG whose scrambling space is a primitive space $V[\Psi_p(x)]$ becomes a PRBS. In other words, *the scrambler SRG in the DSS always renders a primitive scrambling space.*

DEFINITION 12.4 (Primitive BSRG) We define a *primitive BSRG* to be a BSRG for a primitive space $V[\Psi_p(x)]$.

Then, by Theorem 11.11, a primitive BSRG for the primitive space $V[\Psi_p(x)]$ is the smallest-length SRG whose scrambling maximal space is the primitive space $V[\Psi_p(x)]$. Therefore, if a DSS uses a PRBS in the primitive space $V[\Psi_p(x)]$ for scrambling, then it is employing a primitive BSRG for the primitive space $V[\Psi_p(x)]$.

The state transition matrix and the maximal generating vector of a primitive BSRG is determined by the following two theorems.

THEOREM 12.5 *A matrix* $\mathbf{T}$ *is the state transition matrix of a primitive BSRG for the primitive space* $V[\Psi_p(x)]$, *if and only if it is similar to the companion matrix* $\mathbf{A}_{\Psi_p(x)}$.[2]

Proof : This theorem is a direct outcome of Theorem 5.20.   ∎

THEOREM 12.6 *A generating vector* $\mathbf{h}$ *is a maximal generating vector for a primitive BSRG for the primitive space* $V[\Psi_p(x)]$, *if and only if it is a non-zero vector.*

---

[2]Note that a primitive BSRG for the primitive space $V[\Psi_p(x)]$ becomes a PRBS minimal generator which generates a PRBS in the primitive space $V[\Psi_p(x)]$ described in Section 6.3. Refer to Theorem 6.17.

Proof : Let $\mathbf{T}$ be the state transition matrix of a primitive BSRG for a primitive space $V[\Psi_p(x)]$. Then, by Theorem 12.5, $\mathbf{T}$ is similar to $\mathbf{A}_{\Psi_p(x)}$, and hence by Theorems 11.5 and 5.14 the scrambling maximal space $V[\mathbf{T}]$ is $V[\Psi_p(x)]$. Therefore by Theorem 11.4, the scrambling space $V[\mathbf{T}, \mathbf{h}]$ for a generating vector $\mathbf{h}$ is a sequence subspace of $V[\Psi_p(x)]$, and hence by Theorem 4.25 the scrambling space $V[\mathbf{T}, \mathbf{h}]$ is $V[1]$ or $V[\Psi_p(x)]$ since the primitive polynomial $\Psi_p(x)$ is an irreducible polynomial.

If $\mathbf{h} \neq 0$, the lowest-degree polynomial $\Psi(x)$ meeting (11.2) is not 1, and hence by Theorem 11.2 $V[\mathbf{T}, \mathbf{h}] \neq V[1]$. Therefore, $V[\mathbf{T}, \mathbf{h}] = V[\Psi_p(x)]$, that is, $\mathbf{h}$ is a maximal generating vector. This proves the "if" part.

The "only if" part is trivial since the scrambling space $V[\mathbf{T}, \mathbf{h}]$ for a zero generating vector $\mathbf{h}$ is $V[1]$. ∎

EXAMPLE 12.1 We realize scrambler SRGs of the DSS which employs a PRBS in the primitive space $V[x^4 + x + 1]$. According to Theorem 12.5, we choose the state transition matrix $\mathbf{T} = \mathbf{A}^t_{x^4+x+1}$, and according to Theorem 12.6, we choose the non-zero generating vector $\mathbf{h} = \mathbf{e}_0$. Then, we obtain the primitive BSRG shown in Fig. 12.2(a), which is an SSRG-based realization. If we choose the state transition matrix $\mathbf{T} = \mathbf{A}_{x^4+x+1}$ and the non-zero generating vector $\mathbf{h} = \mathbf{e}_1$, we obtain the primitive BSRG shown in Fig. 12.2(b), which is an MSRG-based realization. Note that the two primitive BSRGs have the same scrambling primitive space $V[x^4 + x + 1]$. ♣

The requirements on the initial state vector $\mathbf{d}_0$ of a primitive BSRG is similar to that for the generating vector $\mathbf{h}$, as the following theorem describes.

THEOREM 12.7 *For a primitive BSRG for the primitive space $V[\Psi_p(x)]$ with the state transition matrix $\mathbf{T}$ taken similar to $\mathbf{A}_{\Psi_p(x)}$ and the non-zero generating vector $\mathbf{h}$, its scrambling sequence $\{ s_k \}$ is a PRBS in the primitive space $V[\Psi_p(x)]$, if and only if its initial state vector $\mathbf{d}_0$ is a non-zero vector.*

Proof : If $\mathbf{d}_0 \neq 0$, the scrambling sequence $\{ s_k \}$ is a non-zero sequence, and hence by Theorem 6.14 a non-zero sequence $\{ s_k \}$ in a primitive space is a PRBS. This proves the "if" part.

The "only if" part is trivial since the scrambling sequence $\{ s_k \}$ for a zero state vector is a zero sequence, which is not a PRBS. ∎

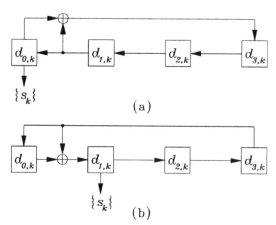

Figure 12.2. Examples of primitive BSRGs for the primitive space $V[x^4 + x + 1]$. (a) An SSRG-based realization, (b) an MSRG-based realization.

As a consequence, we find that the scrambling sequence $\{ s_k \}$ generated by a primitive BSRG is a PRBS if its initial state vector is a non-zero vector. Therefore, in the DSS all we need to do for the initial state setting is to avoid the all-zero setting.

EXAMPLE 12.2 The scrambling sequences generated by the primitive BSRGs in Fig. 12.2 become PRBSs in the primitive space $V[x^4+x+1]$ at all times except when the initial states are all set to zero simultaneously. Therefore, they can be used for the DSS. ♣

12.2.2 Sampling Times

In the DSS, the scrambling sequence $\{ s_k \}$ used in scrambling is predicted by its samples z_i's. Therefore, the sampling times α_i's of the samples z_i's should be chosen such that the scrambling sequence $\{ s_k \}$ can be predicted by the samples z_i's. Such predictable sampling times α_i's can be determined by Theorem 11.19, and the predicted scrambling sequence $\{ s_k \}$ is as described in Theorem 11.20. We combine these two into the following theorem.

THEOREM 12.8 *For a scrambling sequence $\{ s_k \}$ whose scrambling space is the sequence space $V[\Psi(x)]$ of dimension L, let α_i, $i = 0, 1, \cdots,$ L-1, be the sampling time of the ith sample z_i of the sample vector $\mathbf{z}$. Then, if the sampling times α_i's are chosen such that the discrimination* *matrix Δ_α in (11.16) is nonsingular, the scrambling sequence $\{ s_k \}$ is determined by the relation*

$$\{ s_k \} = (\Delta_\alpha^{-1} \cdot \mathbf{z})^t \cdot \mathbf{E}_{\Psi(x)} ; \qquad (12.3)$$

otherwise it is impossible to determine the scrambling sequence $\{ s_k \}$.

Therefore, in the DSS the sampling times α_i's should be chosen such that the discrimination matrix Δ_α in (11.16) is nonsingular. Only on this condition the descrambler SRG can generate the scrambler SRG sequence $\{ s_k \}$ by employing the relation in (12.3).

EXAMPLE 12.3 We consider the scrambling sequence $\{ s_k \}$ whose scrambling space is the sequence space $V[x^4+x+1]$. If we choose the sampling times $\alpha_0 = 2$, $\alpha_1 = 3$, $\alpha_2 = 6$ and $\alpha_3 = 7$, then we obtain the discrimination matrix

$$\Delta_\alpha = \begin{bmatrix} 0 & 0 & 1 & 0 \\ 0 & 0 & 0 & 1 \\ 0 & 0 & 1 & 1 \\ 1 & 1 & 0 & 1 \end{bmatrix},$$

which is singular. Therefore, by the theorem we can not determine the scrambling sequence $\{ s_k \}$ using these sampling times. However, if we choose another set of sampling times $\alpha_0 = 1$, $\alpha_1 = 3$, $\alpha_2 = 5$ and $\alpha_3 = 7$, then we obtain the discrimination matrix

$$\Delta_\alpha = \begin{bmatrix} 0 & 1 & 0 & 0 \\ 0 & 0 & 0 & 1 \\ 0 & 1 & 1 & 0 \\ 1 & 1 & 0 & 1 \end{bmatrix}, \qquad (12.4)$$

which is nonsingular. Now we can determine the scrambling sequence $\{ s_k \}$ for the DSS in the scrambling space $V[x^4 + x + 1]$. The scrambler employing these sampling times and the SRG in Fig. 12.2(b) can be depicted as Fig. 12.3. ♣

12.3   Descramblers for DSS

Assuming that the sampling times α_i's are taken to be predictable, we now consider how to choose the state transition matrix $\hat{\mathbf{T}}$ and the generating vector $\hat{\mathbf{h}}$ of a descrambler SRG and, further, how to choose the correction times β_i's and the correction vectors $\mathbf{c}_i$'s, to generate the desired sequences.

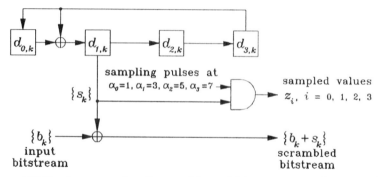

Figure 12.3. An example of scrambler which generates the scrambling sequence $\{\ s_k\ \}$ in the scrambling space $V[x^4 + x + 1]$.

12.3.1 Descrambler SRGs

Among many possible choices of the descrambler SRGs, we consider the smallest-length descrambler SRG which can be synchronized to the scrambling sequence $\{\ s_k\ \}$.

DEFINITION 12.9 (Synchronizable Descrambler SRG) A descrambler SRG is said *synchronizable* if its state matrix $\hat{T}$ and generating vector $\hat{h}$ are chosen such that it can be synchronized to the scrambler SRG.

THEOREM 12.10 *For a scrambling sequence $\{\ s_k\ \}$ whose scrambling space is the sequence space $V[\Psi(x)]$, let $\hat{T}$ and $\hat{h}$ be respectively the state transition matrix and the generating vector of the smallest-length synchronizable descrambler SRG. Then, $\hat{T}$ is similar to the companion matrix $A_{\Psi(x)}$, and $\hat{h}$ is a maximal generating vector.*

Proof : This theorem is obtained by combining Theorems 11.11 and 5.20. ∎

According to the theorem, the state transition matrix $\hat{T}$ of the smallest-length descrambler SRG for the scrambling space $V[\Psi(x)]$ of a scrambling sequence $\{\ s_k\ \}$ should be similar to the companion matrix $A_{\Psi(x)}$, and its generating vector $\hat{h}$ should be maximal in order for them to be useful in generating the scrambler SRG sequence. It should be noted that this requirement is general enough as it does not require $\hat{T}$ and $\hat{h}$ be respectively identical to T and h of the corresponding scrambler.

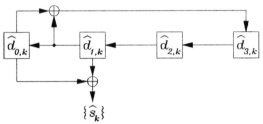

Figure 12.4. An example of the smallest-length descrambler SRG which can be synchronized to the scrambling sequence { s_k } whose scrambling space is the sequence space $V[x^4 + x + 1]$. This is an SSRG-based realization.

EXAMPLE 12.4 We realize the smallest-length synchronizable descrambler SRG which generates the scrambling sequence { s_k } whose scrambling space is the sequence space $V[x^4 + x + 1]$. We choose the state transition matrix $\hat{T} = A^t_{x^4+x+1}$, which is similar to A_{x^4+x+1} ; and the generating vector $\hat{h} = e_0 + e_1$, which is a maximal generating vector since the discrimination matrix

$$
\Delta_{\hat{T},\hat{h}} = \begin{bmatrix} 1 & 1 & 0 & 0 \\ 0 & 1 & 1 & 0 \\ 0 & 0 & 1 & 1 \\ 1 & 1 & 0 & 1 \end{bmatrix} \tag{12.5}
$$

is nonsingular.[3] Then, the resulting synchronizable descrambler SRG is as shown in Fig. 12.4, which is an SSRG-based realization. ♣

12.3.2   Correction Times and Correction Vectors

We consider how to choose the correction times β_i's and the correction vectors c_i's of the smallest-length descrambler SRG to achieve synchronization. For this, we first define the synchronization state vector as follows :

DEFINITION 12.11 (Synchronization State Vector) We call the kth state vector $\tilde{d}_k$ of the descrambler SRG the kth *synchronization state vector* if the descrambler SRG is in the synchronization state at time k.

---

[3]In fact, for a primitive BSRG, any non-zero generating vector is maximal due to Theorem 12.6.

Then, the synchronization state vector is determined by the following theorem.

THEOREM 12.12 *Let* $\hat{\mathbf{T}}$ *and* $\hat{\mathbf{h}}$ *be respectively the state transition matrix and the generating vector of a smallest-length synchronizable descrambler SRG. Then, the synchronization state vector* $\tilde{\mathbf{d}}_k$ *is*

$$\tilde{\mathbf{d}}_k = \hat{\mathbf{T}}^k \cdot \tilde{\mathbf{d}}_0, \qquad (12.6)$$

where the synchronization initial state vector is

$$\tilde{\mathbf{d}}_0 = \Delta_{\hat{\mathbf{T}},\hat{\mathbf{h}}}^{-1} \cdot \Delta_\alpha^{-1} \cdot \mathbf{z}. \qquad (12.7)$$

Proof : It suffices to prove (12.7) since $\hat{\mathbf{d}}_k = \hat{\mathbf{T}} \cdot \hat{\mathbf{d}}_{k-1}$, $k = 1, 2, \cdots$, for the state vector $\hat{\mathbf{d}}_k$ of the descrambler SRG. If the descrambler SRG is synchronized to the scrambling sequence at time 0, the scrambling sequence $\{ s_k \}$ is identical to the descrambling sequence $\{ \hat{s}_k \}$ for the synchronization initial state vector $\tilde{\mathbf{d}}_0$. By Theorem 12.8, the scrambling sequence $\{ s_k \}$ is determined by $\{ s_k \} = (\Delta_\alpha^{-1} \cdot \mathbf{z})^t \cdot \mathbf{E}_{\Psi(x)}$; and by (6.13) the initial state vector $\hat{\mathbf{s}}$ of the descrambling sequence $\{ \hat{s}_k \}$ for the synchronization initial state vector $\tilde{\mathbf{d}}_0$ is $[\, \hat{\mathbf{h}}^t \cdot \tilde{\mathbf{d}}_0 \quad \hat{\mathbf{h}}^t \cdot \hat{\mathbf{T}} \cdot \tilde{\mathbf{d}}_0 \quad \cdots \quad \hat{\mathbf{h}}^t \cdot \hat{\mathbf{T}}^{L-1} \cdot \tilde{\mathbf{d}}_0]^t = \Delta_{\hat{\mathbf{T}},\hat{\mathbf{h}}} \cdot \tilde{\mathbf{d}}_0$. Therefore, by (4.5) $\Delta_\alpha^{-1} \cdot \mathbf{z} = \Delta_{\hat{\mathbf{T}},\hat{\mathbf{h}}} \cdot \tilde{\mathbf{d}}_0$, and hence we have (12.7). ∎

If the smallest-length synchronizable descrambler SRG is synchronized to the scrambler SRG at time k, the state vector $\hat{\mathbf{d}}_k$ of this SRG at time k is the synchronization state vector $\tilde{\mathbf{d}}_k = \hat{\mathbf{T}}^k \cdot \tilde{\mathbf{d}}_0$ for $\tilde{\mathbf{d}}_0$ in (12.7). Therefore, in the DSS synchronization is achieved at the final correction time β_{L-1}, and the state vector of the descrambler SRG at time β_{L-1} becomes

$$\hat{\mathbf{d}}_{\beta_{L-1}} = \hat{\mathbf{T}}^{\beta_{L-1}} \cdot \Delta_{\hat{\mathbf{T}},\hat{\mathbf{h}}}^{-1} \cdot \Delta_\alpha^{-1} \cdot \mathbf{z}. \qquad (12.8)$$

We introduce the concept of state distance vector δ_k as a means to represent the difference between the state vector $\hat{\mathbf{d}}_k$ and the synchronization state vector $\tilde{\mathbf{d}}_k$.

DEFINITION 12.13 (State Distance Vector) For a synchronizable descrambler SRG, we define by the *state distance vector* δ_k the modulo-2 sum of the state vector $\hat{\mathbf{d}}_k$ and the synchronization state vector $\tilde{\mathbf{d}}_k$, i.e.,

$$\delta_k \equiv \hat{\mathbf{d}}_k + \tilde{\mathbf{d}}_k. \qquad (12.9)$$

In the DSS, the state distance vector $\delta_{\beta_{L-1}}$ at the final correction time β_{L-1} should become zero regardless of the initial state distance vector δ_0. To formulate this relation, we adopt a matrix that represents the relation between the final corrected state distance vector $\delta_{\beta_{L-1}}$ and the initial state distance vector δ_0. More specifically,

DEFINITION 12.14 (Correction Matrix) The *correction matrix* Λ is the matrix that relates the state distance vectors $\delta_{\beta_{L-1}}$ and δ_0 in the form

$$\delta_{\beta_{L-1}} = \Lambda \cdot \delta_0. \qquad (12.10)$$

Then, in the DSS, the correction matrix Λ should be zero to make the final corrected state distance vector $\delta_{\beta_{L-1}}$ zero regardless of the initial state distance vector δ_0.

In order to evaluate the correction matrix Λ, we consider the following lemma :

LEMMA 12.15 *Let $\hat{T}$ and $\hat{h}$ be respectively the state transition matrix and the generating vector of the smallest-length synchronizable descrambler SRG. Then, the ith sample z_i, $i = 0, 1, \cdots, L\text{-}1$, is represented by*

$$z_i = \hat{h}^t \cdot \tilde{d}_{\alpha_i} \qquad (12.11)$$

for the synchronization state vector $\tilde{d}_{\alpha_i}$.

Proof : From (12.7), $\mathbf{z} = \Delta_\alpha \cdot \Delta_{\hat{T},\hat{h}} \cdot \tilde{d}_0$, so $z_i = \mathbf{e}_i^t \cdot \Delta_\alpha \cdot \Delta_{\hat{T},\hat{h}} \cdot \tilde{d}_0$. Applying (11.16) to this, we get $z_i = \mathbf{e}_0^t \cdot (\mathbf{A}_{\Psi(x)}^t)^{\alpha_i} \cdot \Delta_{\hat{T},\hat{h}} \cdot \tilde{d}_0$. But by (11.9a), $\mathbf{A}_{\Psi(x)}^t = \Delta_{\hat{T},\hat{h}} \cdot \hat{T} \cdot \Delta_{\hat{T},\hat{h}}^{-1}$. Therefore, $z_i = \mathbf{e}_0^t \cdot \Delta_{\hat{T},\hat{h}} \cdot \hat{T}^{\alpha_i} \cdot \tilde{d}_0$. Hence, by (11.5) we have $z_i = \hat{h}^t \cdot \hat{T}^{\alpha_i} \cdot \tilde{d}_0$, and applying (12.6) to this, we finally obtain (12.11). ∎

Then, the correction matrix Λ is determined by the following theorem.

THEOREM 12.16 *Let $\hat{T}$ and $\hat{h}$ be respectively the state transition matrix and the generating vector of the smallest-length synchronizable descrambler SRG. Then, the correction matrix Λ for the correction times β_i and the correction vectors $\mathbf{c}_i$, $i = 0, 1, \cdots, L\text{-}1$, is*

$$\begin{aligned}
\Lambda = \; &(\hat{T}^{\beta_{L-1}-\beta_{L-2}} + \mathbf{c}_{L-1} \cdot \hat{h}^t \cdot \hat{T}^{\alpha_{L-1}-\beta_{L-2}}) \cdot \\
&(\hat{T}^{\beta_{L-2}-\beta_{L-3}} + \mathbf{c}_{L-2} \cdot \hat{h}^t \cdot \hat{T}^{\alpha_{L-2}-\beta_{L-3}}) \cdot \\
&\cdots (\hat{T}^{\beta_1-\beta_0} + \mathbf{c}_1 \cdot \hat{h}^t \cdot \hat{T}^{\alpha_1-\beta_0}) \cdot (\hat{T}^{\beta_0} + \mathbf{c}_0 \cdot \hat{h}^t \cdot \hat{T}^{\alpha_0}). \quad (12.12)
\end{aligned}$$

Proof : For the synchronization state vector $\tilde{\mathbf{d}}_k$, we have, by (12.6), the relations

$$
\tilde{\mathbf{d}}_{\alpha_i} = \begin{cases} \hat{\mathbf{T}}^{\alpha_0} \cdot \tilde{\mathbf{d}}_0, & i = 0, \\ \hat{\mathbf{T}}^{\alpha_i - \beta_{i-1}} \cdot \tilde{\mathbf{d}}_{\beta_{i-1}}, & i = 1, 2, \cdots, L-1, \end{cases} \tag{12.13}
$$

$$
\tilde{\mathbf{d}}_{\beta_i} = \begin{cases} \hat{\mathbf{T}}^{\beta_0} \cdot \tilde{\mathbf{d}}_0, & i = 0, \\ \hat{\mathbf{T}}^{\beta_i - \beta_{i-1}} \cdot \tilde{\mathbf{d}}_{\beta_{i-1}}, & i = 1, 2, \cdots, L-1, \end{cases} \tag{12.14}
$$

and inserting (12.13) into (12.11), we obtain

$$
z_i = \begin{cases} \hat{\mathbf{h}}^t \cdot \hat{\mathbf{T}}^{\alpha_0} \cdot \tilde{\mathbf{d}}_0, & i = 0, \\ \hat{\mathbf{h}}^t \cdot \hat{\mathbf{T}}^{\alpha_i - \beta_{i-1}} \cdot \tilde{\mathbf{d}}_{\beta_{i-1}}, & i = 1, 2, \cdots, L-1. \end{cases} \tag{12.15}
$$

Since the correction process corresponds to a modulo-2 sum of $(z_i + \hat{z}_i)\mathbf{c}_i$, we have the relations

$$
\hat{\mathbf{d}}_{\alpha_i} = \begin{cases} \hat{\mathbf{T}}^{\alpha_0} \cdot \hat{\mathbf{d}}_0, & i = 0, \\ \hat{\mathbf{T}}^{\alpha_i - \beta_{i-1}} \cdot \hat{\mathbf{d}}_{\beta_{i-1}}, & i = 1, 2, \cdots, L-1, \end{cases} \tag{12.16}
$$

$$
\hat{\mathbf{d}}_{\beta_i} = \begin{cases} \hat{\mathbf{T}}^{\beta_0} \cdot \hat{\mathbf{d}}_0 + (z_0 + \hat{z}_0)\mathbf{c}_0, & i = 0, \\ \hat{\mathbf{T}}^{\beta_i - \beta_{i-1}} \cdot \hat{\mathbf{d}}_{\beta_{i-1}} + (z_i + \hat{z}_i)\mathbf{c}_i, & i = 1, 2, \cdots, L-1, \end{cases} \tag{12.17}
$$

and from (12.16) we obtain

$$
\hat{z}_i = \begin{cases} \hat{\mathbf{h}}^t \cdot \hat{\mathbf{T}}^{\alpha_0} \cdot \hat{\mathbf{d}}_0, & i = 0, \\ \hat{\mathbf{h}}^t \cdot \hat{\mathbf{T}}^{\alpha_i - \beta_{i-1}} \cdot \hat{\mathbf{d}}_{\beta_{i-1}}, & i = 1, 2, \cdots, L-1. \end{cases} \tag{12.18}
$$

By (12.15), (12.18) and (12.9), we have

$$
z_i + \hat{z}_i = \begin{cases} \hat{\mathbf{h}}^t \cdot \hat{\mathbf{T}}^{\alpha_0} \cdot \delta_0, & i = 0, \\ \hat{\mathbf{h}}^t \cdot \hat{\mathbf{T}}^{\alpha_i - \beta_{i-1}} \cdot \delta_{\beta_{i-1}}, & i = 1, 2, \cdots, L-1, \end{cases}
$$

and inserting this into (12.17) and then summing the result to (12.14), we obtain the relation

$$
\delta_{\beta_i} = \begin{cases} \hat{\mathbf{T}}^{\beta_0} \cdot \delta_0 + (\hat{\mathbf{h}}^t \cdot \hat{\mathbf{T}}^{\alpha_0} \cdot \delta_0)\mathbf{c}_0, & i = 0, \\ \hat{\mathbf{T}}^{\beta_i - \beta_{i-1}} \cdot \delta_{\beta_{i-1}} + (\hat{\mathbf{h}}^t \cdot \hat{\mathbf{T}}^{\alpha_i - \beta_{i-1}} \cdot \delta_{\beta_{i-1}})\mathbf{c}_i, \\ \qquad\qquad\qquad i = 1, 2, \cdots, L-1, \end{cases}
$$

which can be rewritten in the form

$$
\delta_{\beta_i} = \begin{cases} (\hat{\mathbf{T}}^{\beta_0} + \mathbf{c}_0 \cdot \hat{\mathbf{h}}^t \cdot \hat{\mathbf{T}}^{\alpha_0}) \cdot \delta_0, & i = 0, \\ (\hat{\mathbf{T}}^{\beta_i - \beta_{i-1}} + \mathbf{c}_i \cdot \hat{\mathbf{h}}^t \cdot \hat{\mathbf{T}}^{\alpha_i - \beta_{i-1}}) \cdot \delta_{\beta_{i-1}}, & i = 1, 2, \cdots, L-1. \end{cases}
$$

By applying this repeatedly to (12.10), we obtain the correction matrix Λ in (12.12). ∎

As an aid to determining the correction times β_i's and the correction vectors c_i's that make the correction matrix Λ null, we introduce the following two lemmas.

LEMMA 12.17 *For the smallest-length synchronizable descrambler SRG with the state transition matrix $\hat{T}$ and the generating vector $\hat{h}$, let $\tilde{H}$ denote the $L \times L$ matrix*

$$
\tilde{H} \equiv \begin{bmatrix} \hat{h}^t \cdot \hat{T}^{\alpha_0} \\ \hat{h}^t \cdot \hat{T}^{\alpha_1} \\ \vdots \\ \hat{h}^t \cdot \hat{T}^{\alpha_{L-1}} \end{bmatrix}. \tag{12.19}
$$

Then, we have the relations

$$
\tilde{H} = \Delta_\alpha \cdot \Delta_{\hat{T},\hat{h}}, \tag{12.20}
$$

$$
\hat{h}^t \cdot \hat{T}^{\alpha_i} \cdot \tilde{H}^{-1} = e_i^t, \quad i = 0, 1, \cdots, L-1. \tag{12.21}
$$

Proof : By (11.9a), we have the relation $\hat{T} = \Delta_{\hat{T},\hat{h}}^{-1} \cdot A_{\Psi(x)}^t \cdot \Delta_{\hat{T},\hat{h}}$. Inserting this into (12.19) and then applying the relation $\hat{h}^t \cdot \Delta_{\hat{T},\hat{h}}^{-1} = e_0^t$ obtained by (11.5), we get the equation (12.20).

Equation (12.21) is obvious due to (12.19) and the relation $\tilde{H} \cdot \tilde{H}^{-1} = I$. ∎

LEMMA 12.18 *Let Λ_i, $i = 1, 2, \cdots, L-1$, denote the $L \times L$ matrix*

$$
\Lambda_i \equiv (\hat{T}^{\beta_{L-1}-\beta_{L-2}} + c_{L-1} \cdot \hat{h}^t \cdot \hat{T}^{\alpha_{L-1}-\beta_{L-2}}) \cdots
$$
$$
(\hat{T}^{\beta_i-\beta_{i-1}} + c_i \cdot \hat{h}^t \cdot \hat{T}^{\alpha_i-\beta_{i-1}}). \tag{12.22}
$$

Then, we have the relations

$$
\begin{cases} \Lambda = \Lambda_1 \cdot (\hat{T}^{\beta_0} + c_0 \cdot \hat{h}^t \cdot \hat{T}^{\alpha_0}), \\ \Lambda_i = \Lambda_{i+1} \cdot (\hat{T}^{\beta_i-\beta_{i-1}} + c_i \cdot \hat{h}^t \cdot \hat{T}^{\alpha_i-\beta_{i-1}}), \quad i = 1, 2, \cdots, L-2, \end{cases} \tag{12.23}
$$

$$
\Lambda_i \cdot \hat{T}^{\beta_{i-1}} \cdot \tilde{H}^{-1} \cdot e_j = \begin{cases} \hat{T}^{\beta_{L-1}} \cdot \tilde{H}^{-1} \cdot e_j, & 0 \le j < i \le L-1, \\ \Lambda \cdot \tilde{H}^{-1} \cdot e_j, & 1 \le i \le j \le L-1, \end{cases} \tag{12.24}
$$

$$
\Lambda \cdot \tilde{H}^{-1} \cdot e_i = \begin{cases} \Lambda_{i+1} \cdot (\hat{T}^{\beta_i} \cdot \tilde{H}^{-1} \cdot e_i + c_i), & i = 0, 1, \cdots, L-2, \\ \hat{T}^{\beta_{L-1}} \cdot \tilde{H}^{-1} \cdot e_{L-1} + c_{L-1}, & i = L-1. \end{cases} \tag{12.25}
$$

Proof : Proof of (12.23) is trivial.

We first show (12.24) for the case $0 \le j < i \le L-1$. By (12.23) and (12.21) we have the relation $\Lambda_i \cdot \hat{\mathbf{T}}^{\beta_i-1} \cdot \tilde{\mathbf{H}}^{-1} \cdot \mathbf{e}_j = \Lambda_{i+1} \cdot (\hat{\mathbf{T}}^{\beta_i} \cdot \tilde{\mathbf{H}}^{-1} \cdot \mathbf{e}_j + \mathbf{c}_i \cdot \hat{\mathbf{h}}^t \cdot \hat{\mathbf{T}}^{\alpha_i} \cdot \tilde{\mathbf{H}}^{-1} \cdot \mathbf{e}_j) = \Lambda_{i+1} \cdot (\hat{\mathbf{T}}^{\beta_i} \cdot \tilde{\mathbf{H}}^{-1} \cdot \mathbf{e}_j + \mathbf{c}_i \cdot \mathbf{e}_i^t \cdot \mathbf{e}_j) = \Lambda_{i+1} \cdot \hat{\mathbf{T}}^{\beta_i} \cdot \tilde{\mathbf{H}}^{-1} \cdot \mathbf{e}_j$. Applying this procedure repeatedly, we finally obtain $\Lambda_i \cdot \hat{\mathbf{T}}^{\beta_i-1} \cdot \tilde{\mathbf{H}}^{-1} \cdot \mathbf{e}_j = \Lambda_{L-1} \cdot \hat{\mathbf{T}}^{\beta_{L-2}} \cdot \tilde{\mathbf{H}}^{-1} \cdot \mathbf{e}_j$. For Λ_{L-1} on the right hand side, we apply (12.22) to obtain the relation $\Lambda_i \cdot \hat{\mathbf{T}}^{\beta_i-1} \cdot \tilde{\mathbf{H}}^{-1} \cdot \mathbf{e}_j = \hat{\mathbf{T}}^{\beta_{L-1}} \cdot \tilde{\mathbf{H}}^{-1} \cdot \mathbf{e}_j + \mathbf{c}_{L-1} \cdot \hat{\mathbf{h}}^t \cdot \hat{\mathbf{T}}^{\alpha_{L-1}} \cdot \tilde{\mathbf{H}}^{-1} \cdot \mathbf{e}_j$. Therefore, by (12.21), $\Lambda_i \cdot \hat{\mathbf{T}}^{\beta_i-1} \cdot \tilde{\mathbf{H}}^{-1} \cdot \mathbf{e}_j = \hat{\mathbf{T}}^{\beta_{L-1}} \cdot \tilde{\mathbf{H}}^{-1} \cdot \mathbf{e}_j + \mathbf{c}_{L-1} \cdot \mathbf{e}_{L-1}^t \cdot \mathbf{e}_j = \hat{\mathbf{T}}^{\beta_{L-1}} \cdot \tilde{\mathbf{H}}^{-1} \cdot \mathbf{e}_j$.

Next we prove (12.24) for the case $1 \le i \le j \le L-1$ by induction. For each $j = 1, 2, \cdots, L-1$, we have, by (12.23), the relation $\Lambda \cdot \tilde{\mathbf{H}}^{-1} \cdot \mathbf{e}_j = \Lambda_1 \cdot (\hat{\mathbf{T}}^{\beta_0} \cdot \tilde{\mathbf{H}}^{-1} \cdot \mathbf{e}_j + \mathbf{c}_0 \cdot \hat{\mathbf{h}}^t \cdot \hat{\mathbf{T}}^{\alpha_0} \cdot \tilde{\mathbf{H}}^{-1} \cdot \mathbf{e}_j)$. Inserting (12.21) into this, we obtain $\Lambda \cdot \tilde{\mathbf{H}}^{-1} \cdot \mathbf{e}_j = \Lambda_1 \cdot (\hat{\mathbf{T}}^{\beta_0} \cdot \tilde{\mathbf{H}}^{-1} \cdot \mathbf{e}_j + \mathbf{c}_0 \cdot \mathbf{e}_0^t \cdot \mathbf{e}_j) = \Lambda_1 \cdot \hat{\mathbf{T}}^{\beta_0} \cdot \tilde{\mathbf{H}}^{-1} \cdot \mathbf{e}_j$. This proves the equation for $i = 1$. Now we assume that the relation (12.24) holds for $i = 1, 2, \cdots, i < j$. Then, by (12.23) and (12.21) we obtain $\Lambda \cdot \tilde{\mathbf{H}}^{-1} \cdot \mathbf{e}_j = \Lambda_{i+1} \cdot (\hat{\mathbf{T}}^{\beta_i} \cdot \tilde{\mathbf{H}}^{-1} \cdot \mathbf{e}_j + \mathbf{c}_i \cdot \hat{\mathbf{h}}^t \cdot \hat{\mathbf{T}}^{\alpha_i} \cdot \tilde{\mathbf{H}}^{-1} \cdot \mathbf{e}_j) = \Lambda_{i+1} \cdot (\hat{\mathbf{T}}^{\beta_i} \cdot \tilde{\mathbf{H}}^{-1} \cdot \mathbf{e}_j + \mathbf{c}_i \cdot \mathbf{e}_i^t \cdot \mathbf{e}_j) = \Lambda_{i+1} \cdot \hat{\mathbf{T}}^{\beta_i} \cdot \tilde{\mathbf{H}}^{-1} \cdot \mathbf{e}_j$. This implies that (12.24) also holds for $i + 1$, and it completes the proof of (12.24) for the case $1 \le i \le j \le L-1$.

Finally, we prove (12.25). We first consider the case when $i = 0$. By (12.23) and (12.21), we obtain the relation $\Lambda \cdot \tilde{\mathbf{H}}^{-1} \cdot \mathbf{e}_0 = \Lambda_1 \cdot (\hat{\mathbf{T}}^{\beta_0} \cdot \tilde{\mathbf{H}}^{-1} \cdot \mathbf{e}_0 + \mathbf{c}_0 \cdot \hat{\mathbf{h}}^t \cdot \hat{\mathbf{T}}^{\alpha_0} \cdot \tilde{\mathbf{H}}^{-1} \cdot \mathbf{e}_0) = \Lambda_1 \cdot (\hat{\mathbf{T}}^{\beta_0} \cdot \tilde{\mathbf{H}}^{-1} \cdot \mathbf{e}_0 + \mathbf{c}_0 \cdot \mathbf{e}_0^t \cdot \mathbf{e}_0) = \Lambda_1 \cdot (\hat{\mathbf{T}}^{\beta_0} \cdot \tilde{\mathbf{H}}^{-1} \cdot \mathbf{e}_0 + \mathbf{c}_0)$. This proves (12.25) for $i = 0$. Now we consider the cases when $i = 1, 2, \cdots, L-2$. Putting $j = i$ in (12.24), we get $\Lambda \cdot \tilde{\mathbf{H}}^{-1} \cdot \mathbf{e}_i = \Lambda_i \cdot \hat{\mathbf{T}}^{\beta_i-1} \cdot \tilde{\mathbf{H}}^{-1} \cdot \mathbf{e}_i$, and hence by (12.23) and (12.21) we obtain $\Lambda \cdot \tilde{\mathbf{H}}^{-1} \cdot \mathbf{e}_i = \Lambda_{i+1} \cdot (\hat{\mathbf{T}}^{\beta_i} \cdot \tilde{\mathbf{H}}^{-1} \cdot \mathbf{e}_i + \mathbf{c}_i \cdot \hat{\mathbf{h}}^t \cdot \hat{\mathbf{T}}^{\alpha_i} \cdot \tilde{\mathbf{H}}^{-1} \cdot \mathbf{e}_i) = \Lambda_{i+1} \cdot (\hat{\mathbf{T}}^{\beta_i} \cdot \tilde{\mathbf{H}}^{-1} \cdot \mathbf{e}_i + \mathbf{c}_i)$. To prove (12.25) for the case $i = L-1$, we put $i = j = L-1$ in (12.24). Then we get $\Lambda \cdot \tilde{\mathbf{H}}^{-1} \cdot \mathbf{e}_{L-1} = \Lambda_{L-1} \cdot \hat{\mathbf{T}}^{\beta_{L-2}} \cdot \tilde{\mathbf{H}}^{-1} \cdot \mathbf{e}_{L-1}$. Inserting (12.22) and (12.21) into this, we obtain $\Lambda \cdot \tilde{\mathbf{H}}^{-1} \cdot \mathbf{e}_{L-1} = \hat{\mathbf{T}}^{\beta_{L-1}} \cdot \tilde{\mathbf{H}}^{-1} \cdot \mathbf{e}_{L-1} + \mathbf{c}_{L-1} \cdot \hat{\mathbf{h}}^t \cdot \hat{\mathbf{T}}^{\alpha_{L-1}} \cdot \tilde{\mathbf{H}}^{-1} \cdot \mathbf{e}_{L-1} = \hat{\mathbf{T}}^{\beta_{L-1}} \cdot \tilde{\mathbf{H}}^{-1} \cdot \mathbf{e}_{L-1} + \mathbf{c}_{L-1}$. This completes the proof of (12.25). ∎

Then, we get the following theorem for choosing the correction times β_i's and the correction vectors $\mathbf{c}_i$'s that make the correction matrix Λ in (12.12) zero.

THEOREM 12.19 *Let $\hat{\mathbf{T}}$ and $\hat{\mathbf{h}}$ be respectively the state transition matrix and the generating vector of the smallest-length synchronizable descrambler SRG. Then, the correction matrix Λ in (12.12) becomes a zero* *matrix if and only if the correction vector $\mathbf{c}_i$ has, for the correction time β_i, the expression*

$$\mathbf{c}_i = \begin{cases} \hat{\mathbf{T}}^{\beta_i} \cdot \tilde{\mathbf{H}}^{-1} \cdot (\mathbf{e}_i + \sum_{j=i+1}^{L-1} u_{i,j}\mathbf{e}_j), & i = 0, 1, \cdots, L\text{-}2, \\ \hat{\mathbf{T}}^{\beta_{L-1}} \cdot \tilde{\mathbf{H}}^{-1} \cdot \mathbf{e}_{L-1}, & i = L\text{-}1, \end{cases} \qquad (12.26)$$

where $u_{i,j}$ is either 0 or 1 for $i = 0, 1, \cdots, L\text{-}2$, and $j = i+1, i+2, \cdots, L\text{-}1$.

Proof : To prove the "if" part of the theorem we show that $\Lambda \cdot \tilde{\mathbf{H}}^{-1} \cdot \mathbf{e}_i = 0$, $i = 0, 1, \cdots, L - 1$ for $\mathbf{c}_i$'s chosen as in (12.26). We show this by induction. From (12.25), we have the relation $\Lambda \cdot \tilde{\mathbf{H}}^{-1} \cdot \mathbf{e}_{L-1} = \hat{\mathbf{T}}^{\beta_{L-1}} \cdot \tilde{\mathbf{H}}^{-1} \cdot \mathbf{e}_{L-1} + \mathbf{c}_{L-1}$, and inserting (12.26) into this, we obtain $\Lambda \cdot \tilde{\mathbf{H}}^{-1} \cdot \mathbf{e}_{L-1} = \mathbf{c}_{L-1} + \mathbf{c}_{L-1} = 0$. This proves the equation for $i = L - 1$. Now we assume $\Lambda \cdot \tilde{\mathbf{H}}^{-1} \cdot \mathbf{e}_j = 0$ for $j = L - 1, L - 2, \cdots, i$. Then, from (12.25), we have the relation $\Lambda \cdot \tilde{\mathbf{H}}^{-1} \cdot \mathbf{e}_{i-1} = \Lambda_i \cdot (\hat{\mathbf{T}}^{\beta_{i-1}} \cdot \tilde{\mathbf{H}}^{-1} \cdot \mathbf{e}_{i-1} + \mathbf{c}_{i-1})$. Inserting (12.26) into this, we obtain $\Lambda \cdot \tilde{\mathbf{H}}^{-1} \cdot \mathbf{e}_{i-1} = \sum_{j=i}^{L-1} u_{i-1,j} \cdot \Lambda_i \cdot \hat{\mathbf{T}}^{\beta_{i-1}} \cdot \tilde{\mathbf{H}}^{-1} \cdot \mathbf{e}_j$, which can be rewritten, by (12.24), as $\Lambda \cdot \tilde{\mathbf{H}}^{-1} \cdot \mathbf{e}_{i-1} = \sum_{j=i}^{L-1} u_{i-1,j} \cdot \Lambda \cdot \tilde{\mathbf{H}}^{-1} \cdot \mathbf{e}_j$. But it is zero, since $\Lambda \cdot \tilde{\mathbf{H}}^{-1} \cdot \mathbf{e}_j = 0$ for $j = L - 1, L - 2, \cdots, i$, by assumption. This completes the proof of the "if" part.

Next we prove the "only if" part of the theorem. Since $\hat{\mathbf{T}}$ and $\tilde{\mathbf{H}}$ are both nonsingular, a vector $\mathbf{c}_i$ can be uniquely expressed as

$$\mathbf{c}_i = \hat{\mathbf{T}}^{\beta_i} \cdot \tilde{\mathbf{H}}^{-1} \cdot \sum_{j=0}^{L-1} v_{i,j}\mathbf{e}_j, \quad i = 0, 1, \cdots, L - 1, \qquad (12.27)$$

for the basis vectors $\mathbf{e}_j$, $j = 0, 1, \cdots, L - 1$. To prove the theorem, it suffices to show that $v_{i,i} = 1$ and $v_{i,j} = 0$ for $i = 0, 1, \cdots, L - 1$, and $j = 0, 1, \cdots, i - 1$ when Λ is a zero matrix.

For $i = 0, 1, \cdots, L-2$, we insert (12.27) into (12.25). Then we obtain $\Lambda \cdot \tilde{\mathbf{H}}^{-1} \cdot \mathbf{e}_i = \Lambda_{i+1} \cdot (\hat{\mathbf{T}}^{\beta_i} \cdot \tilde{\mathbf{H}}^{-1} \cdot \mathbf{e}_i + \hat{\mathbf{T}}^{\beta_i} \cdot \tilde{\mathbf{H}}^{-1} \cdot \sum_{j=0}^{L-1} v_{i,j} \cdot \mathbf{e}_j)$. This can be rewritten, by (12.24), in the form $\Lambda \cdot \tilde{\mathbf{H}}^{-1} \cdot \mathbf{e}_i = \sum_{j=0}^{i-1} v_{i,j} \cdot \hat{\mathbf{T}}^{\beta_{L-1}} \cdot \tilde{\mathbf{H}}^{-1} \cdot \mathbf{e}_j + (1 + v_{i,i}) \cdot \hat{\mathbf{T}}^{\beta_{L-1}} \cdot \tilde{\mathbf{H}}^{-1} \cdot \mathbf{e}_i + \sum_{j=i+1}^{L-1} v_{i,j} \cdot \Lambda \cdot \tilde{\mathbf{H}}^{-1} \cdot \mathbf{e}_j$. Since Λ is a zero matrix, we have the equality $\hat{\mathbf{T}}^{\beta_{L-1}} \cdot \tilde{\mathbf{H}}^{-1} \cdot \{\sum_{j=0}^{i-1} v_{i,j} \cdot \mathbf{e}_j + (1 + v_{i,i}) \cdot \mathbf{e}_i\} = 0$, and hence $v_{i,i} = 1$ and $v_{i,j} = 0$ for $j = 0, 1, \cdots, i - 1$.

For the case $i = L - 1$, we obtain, from (12.25), the equality $\mathbf{c}_{L-1} = \hat{\mathbf{T}}^{\beta_{L-1}} \cdot \tilde{\mathbf{H}}^{-1} \cdot \mathbf{e}_{L-1}$. This implies that $v_{L-1,L-1} = 1$ and $v_{L-1,j} = 0$ for $j = 0, 1, \cdots, L - 2$. This completes the proof. ∎

This theorem implies that the correction vectors c_i's in (12.26) make the correction matrix Λ in (12.12) zero for arbitrarily chosen correction times β_i's. That is, we can arbitrarily select correction times β_i's, and we can select correction vectors c_i's according to the expression in (12.26). We observe from the correction vector expression in (12.26) the following two facts : First, the ith correction vector c_i to make the correction matrix Λ a zero matrix depends only on the ith correction time β_i ; and, secondly the correction vector c_{L-1} is unique, but other correction vectors c_{L-1-i} for $i = 1, 2, \cdots, L-1$, could be as many as 2^i depending on the choice of $u_{i,j}$. Once correction vectors are all determined we can choose some simple combination of them by appropriately selecting the correction times, as the following example illustrates.

EXAMPLE 12.5 For the scrambling sequence $\{ s_k \}$ in the scrambling space $V[x^4 + x + 1]$ whose sampling times are $\alpha_0 = 1$, $\alpha_1 = 3$, $\alpha_2 = 5$ and $\alpha_3 = 7$, we consider the correction vectors c_i's of the descrambler SRG in Fig. 12.4. By (12.2) the feasible correction times are $\beta_0 = 2, 3$; $\beta_1 = 4, 5$; $\beta_2 = 6, 7$; and $\beta_3 = 8, 9, \cdots$, in this case. Inserting (12.4) and (12.5) into (12.20), we obtain the matrix

$$\tilde{H} = \begin{bmatrix} 0 & 1 & 1 & 0 \\ 1 & 1 & 0 & 1 \\ 0 & 1 & 0 & 1 \\ 0 & 1 & 1 & 1 \end{bmatrix}.$$

Therefore inserting this and $\hat{T} = A^t_{x^4+x+1}$ into (12.26), we get the correction vectors c_i's in Table 12.1 for each feasible correction time. If we choose the correction times $\beta_0 = 3$, $\beta_1 = 5$, $\beta_2 = 7$ and $\beta_3 = 9$, it is possible to select the common correction vector $c_i = [\,0\,0\,1\,1\,]^t$, $i = 0$, 1, 2, 3, as marked with $*$ in the table. Using these, we can design the descrambler shown in Fig. 12.5. This is one of many possible descramblers that can be used in conjunction with the scrambler in Fig. 12.3.
♣

As a special case, we consider the uniform sampling and uniform correction case in which the sampling and the correction times are uniformly spaced. The following theorem describes how we can obtain a common correction vector in this case.

THEOREM 12.20 *Let $\hat{T}$ and $\hat{h}$ be respectively the state transition matrix and the generating vector of the smallest-length synchronizable descrambler SRG ; and let the sampling times α_i's be uniformly spaced,*

Table 12.1. Correction vectors c_i's of the descrambler SRG in Fig. 12.4 for different feasible correction times. The scrambling space of the scrambling sequence $\{\, s_k \,\}$ is $V[x^4 + x + 1]$, and the sampling times are $\alpha_0 = 1$, $\alpha_1 = 3$, $\alpha_2 = 5$ and $\alpha_3 = 7$. The asterisks(*) indicate the case with common correction vectors $c_0 = c_1 = c_2 = c_3$.

c_0	$\beta_0 = 2$	$[\,0\,1\,1\,1\,]^t$	
		$[\,0\,1\,0\,1\,]^t$	
		$[\,1\,1\,1\,1\,]^t$	
		$[\,1\,0\,0\,1\,]^t$	
		$[\,1\,1\,0\,1\,]^t$	
		$[\,1\,0\,1\,1\,]^t$	
		$[\,0\,0\,0\,1\,]^t$	
		$[\,0\,0\,1\,1\,]^t$	
	$\beta_0 = 3$	$[\,1\,1\,1\,1\,]^t$	
		$[\,1\,0\,1\,1\,]^t$	
		$[\,1\,1\,1\,0\,]^t$	
		$[\,0\,0\,1\,1\,]^t$	*
		$[\,1\,0\,1\,0\,]^t$	
		$[\,0\,1\,1\,1\,]^t$	
		$[\,0\,0\,1\,0\,]^t$	
		$[\,0\,1\,1\,0\,]^t$	
c_1	$\beta_1 = 4$	$[\,1\,0\,0\,1\,]^t$	
		$[\,1\,0\,1\,1\,]^t$	
		$[\,0\,0\,0\,1\,]^t$	
		$[\,0\,0\,1\,1\,]^t$	
	$\beta_1 = 5$	$[\,0\,0\,1\,1\,]^t$	*
		$[\,0\,1\,1\,1\,]^t$	
		$[\,0\,0\,1\,0\,]^t$	
		$[\,0\,1\,1\,0\,]^t$	
c_2	$\beta_2 = 6$	$[\,1\,0\,0\,1\,]^t$	
		$[\,1\,0\,1\,1\,]^t$	
	$\beta_2 = 7$	$[\,0\,0\,1\,1\,]^t$	*
		$[\,0\,1\,1\,1\,]^t$	
c_3	$\beta_3 = 8$	$[\,1\,0\,0\,1\,]^t$	
	$\beta_3 = 9$	$[\,0\,0\,1\,1\,]^t$	*
	$\vdots$	$\vdots$	

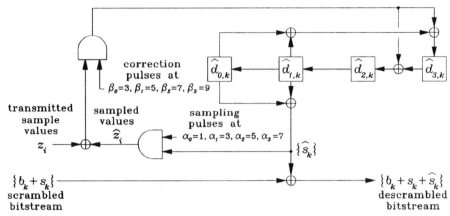

Figure 12.5. An example of descrambler which can be synchronized to the scrambler in Fig. 12.3.

that is, $\alpha_i = i\alpha$, $i = 0, 1, \cdots, L\text{-}1$. *Then, the correction matrix* Λ *in (12.12) becomes a zero matrix for the uniform correction times* $\beta_i = \beta + i\alpha$, $i = 0, 1, \cdots, L\text{-}1$, $\beta \leq \alpha$, *and for the correction vectors*

$$\mathbf{c}_i = \hat{\mathbf{T}}^{\beta+(L-1)\alpha} \cdot \tilde{\mathbf{H}}^{-1} \cdot \mathbf{e}_{L-1}, \quad i = 0, 1, \cdots L-1. \qquad (12.28)$$

Proof : In case of $i = L - 1$, (12.28) is obvious due to (12.26). To prove (12.28) for $i \neq L-1$, we insert $u_{i,j} = \hat{\mathbf{h}}^t \cdot \hat{\mathbf{T}}^{(L-1+j-i)\alpha} \cdot \tilde{\mathbf{H}}^{-1} \cdot \mathbf{e}_{L-1}$, $i = 0, 1, \cdots, L-2$, $j = i+1, i+2, \cdots, L-1$, to (12.26). Then, we have $\mathbf{c}_i = \hat{\mathbf{T}}^{\beta+i\alpha} \cdot \tilde{\mathbf{H}}^{-1} \cdot \{\mathbf{e}_i + \sum_{j=i+1}^{L-1}(\hat{\mathbf{h}}^t \cdot \hat{\mathbf{T}}^{(L-1+j-i)\alpha} \cdot \tilde{\mathbf{H}}^{-1} \cdot \mathbf{e}_{L-1})\mathbf{e}_j\}$. But $\sum_{j=0}^{i}(\hat{\mathbf{h}}^t \cdot \hat{\mathbf{T}}^{(L-1+j-i)\alpha} \cdot \tilde{\mathbf{H}}^{-1} \cdot \mathbf{e}_{L-1})\mathbf{e}_j = \mathbf{e}_i$ by (12.21). Therefore $\mathbf{c}_i$ can be rewritten as $\mathbf{c}_i = \hat{\mathbf{T}}^{\beta+i\alpha} \cdot \tilde{\mathbf{H}}^{-1} \cdot \sum_{j=0}^{L-1}(\hat{\mathbf{h}}^t \cdot \hat{\mathbf{T}}^{(L-1+j-i)\alpha} \cdot \tilde{\mathbf{H}}^{-1} \cdot \mathbf{e}_{L-1})\mathbf{e}_j$. For the summation part of $\mathbf{c}_i$, it is not difficult to apply (12.19) to derive the relation $\sum_{j=0}^{L-1}(\hat{\mathbf{h}}^t \cdot \hat{\mathbf{T}}^{(L-1+j-i)\alpha} \cdot \tilde{\mathbf{H}}^{-1} \cdot \mathbf{e}_{L-1})\mathbf{e}_j = \tilde{\mathbf{H}} \cdot \hat{\mathbf{T}}^{(L-1-i)\alpha} \cdot \tilde{\mathbf{H}}^{-1} \cdot \mathbf{e}_{L-1}$. Therefore $\mathbf{c}_i = \hat{\mathbf{T}}^{\beta+i\alpha} \cdot \tilde{\mathbf{H}}^{-1} \cdot \tilde{\mathbf{H}} \cdot \hat{\mathbf{T}}^{(L-1-i)\alpha} \cdot \tilde{\mathbf{H}}^{-1} \cdot \mathbf{e}_{L-1} = \hat{\mathbf{T}}^{\beta+(L-1)\alpha} \cdot \tilde{\mathbf{H}}^{-1} \cdot \mathbf{e}_{L-1}$. ∎

This theorem means that in the case of uniform sampling and uniform correction it is always possible to choose common correction vectors, i.e., $\mathbf{c}_0 = \mathbf{c}_1 = \cdots = \mathbf{c}_{L-1}$, to make the correction matrix Λ null. Equation (12.28) describes how to choose such a correction vector. Note that such a common correction vector is unique since the correction vector $\mathbf{c}_{L-1}$ is unique.

EXAMPLE 12.6 For the scrambling sequence $\{\, s_k \,\}$ in the scrambling space $V[x^4 + x + 1]$ with uniform sampling times $\alpha_i = 3i$, $i = 0, 1, 2,$ 3, we consider correction times β_i's and correction vectors c_i's of the descrambler SRG in Fig. 12.4. For these sampling times, the discrimination matrix becomes

$$\Delta_\alpha = \begin{bmatrix} 1 & 0 & 0 & 0 \\ 0 & 0 & 0 & 1 \\ 0 & 0 & 1 & 1 \\ 0 & 1 & 0 & 1 \end{bmatrix}, \qquad (12.29)$$

which is nonsingular. Inserting (12.5) and (12.29) into (12.20), we obtain the matrix

$$\tilde{\mathbf{H}} = \begin{bmatrix} 1 & 1 & 0 & 0 \\ 1 & 1 & 0 & 1 \\ 1 & 1 & 1 & 0 \\ 1 & 0 & 1 & 1 \end{bmatrix}.$$

Therefore, if we select the uniform correction times $\beta_i = 1 + 3i$, $i = 0,$ 1, 2, 3, then by (12.28) we obtain the common correction vector $c_0 = c_1 = c_2 = c_3 = [\,0\,1\,0\,1\,]^t$. ♣

12.3.3 Efficient Realization Methods

We finally consider how to choose the state transition matrix $\hat{\mathbf{T}}$, the generating vector $\hat{\mathbf{h}}$, the correction times β_i's and the correction vectors c_i's to efficiently realize the descramblers.

THEOREM 12.21 *For the scrambling sequence $\{\, s_k \,\}$ whose scrambling space is the sequence space $V[\Psi(x)]$ of dimension L, let the sampling times α_i, $i = 0, 1, \cdots, L\text{-}1$, be the predictable sampling times of the scrambling sequence $\{\, s_k \,\}$. Then, for a nonsingular matrix $\mathbf{R}$, the descrambler with the state transition matrix $\hat{\mathbf{T}}$, the generating vector $\hat{\mathbf{h}}$ and the correction vectors c_i's of the form*

$$\hat{\mathbf{T}} = \mathbf{R}^{-1} \cdot \mathbf{A}_{\Psi(x)}^t \cdot \mathbf{R}, \qquad (12.30a)$$

$$\hat{\mathbf{h}} = \mathbf{R}^t \cdot \mathbf{e}_0, \qquad (12.30b)$$

$$c_i = \begin{cases} \mathbf{R}^{-1} \cdot (\mathbf{A}_{\Psi(x)}^t)^{\beta_i} \cdot \Delta_\alpha^{-1} \cdot (\mathbf{e}_i + \sum_{j=i+1}^{L-1} u_{i,j}\mathbf{e}_j), \\ \qquad i = 0, 1, \cdots, L\text{-}2, \\ \mathbf{R}^{-1} \cdot (\mathbf{A}_{\Psi(x)}^t)^{\beta_{L-1}} \cdot \Delta_\alpha^{-1} \cdot \mathbf{e}_{L-1}, \quad i = L\text{-}1, \end{cases} \qquad (12.30c)$$

for the correction times β_i's and the arbitrary binary parameters $u_{i,j}$'s is a smallest-length synchronizable descrambler.

Proof : The state transition matrix $\hat{\mathbf{T}}$ in (12.30a) is similar to $\mathbf{A}_{\Psi(x)}$, so it meets Theorem 12.10. Inserting (12.30a) and (12.30b) into (11.5), and applying the relations $\mathbf{e}_0^t \cdot (\mathbf{A}_{\Psi(x)}^t)^i = \mathbf{e}_i$, $i = 0, 1, \cdots, L-1$, we obtain $\Delta_{\hat{\mathbf{T}},\hat{\mathbf{h}}} = \mathbf{R}$, which is nonsingular. Therefore, $\hat{\mathbf{h}}$ in (12.30b) is a maximal generating vector which meets Theorem 12.10. Inserting $\Delta_{\hat{\mathbf{T}},\hat{\mathbf{h}}} = \mathbf{R}$ into (12.20), we get $\tilde{\mathbf{H}} = \Delta_\alpha \cdot \mathbf{R}$, and inserting this and (12.30a) into (12.26) we obtain the correction vectors $\mathbf{c}_i$'s in (12.30c).   ■

The theorem states that for an arbitrary nonsingular matrix $\mathbf{R}$ the descrambler with the state transition matrix $\hat{\mathbf{T}}$ in (12.30a), the generating vector $\hat{\mathbf{h}}$ in (12.30b) and the correction vectors $\mathbf{c}_i$'s in (12.30c) for correction times β_i's becomes a smallest-length synchronizable descrambler. Therefore, it is easy to determine the three parameters $\hat{\mathbf{T}}$, $\hat{\mathbf{h}}$ and $\mathbf{c}_i$'s by inserting an arbitrary nonsingular matrix $\mathbf{R}$ along with arbitrary correction times β_i's. The most efficient descrambler, however, can be obtained out of repeated trials on a number of different parameters.

To further investigate the correction vector expression in (12.30c), we insert $\mathbf{R} = \mathbf{I}$ into (12.30a), (12.30b) and (12.30c) respectively. Then, we obtain the descrambler with the parameters

$$\hat{\mathbf{T}}_S = \mathbf{A}_{\Psi(x)}^t, \tag{12.31a}$$

$$\hat{\mathbf{h}}_S = \mathbf{e}_0, \tag{12.31b}$$

$$\mathbf{c}_i^S = \begin{cases} (\mathbf{A}_{\Psi(x)}^t)^{\beta_i} \cdot \Delta_\alpha^{-1} \cdot (\mathbf{e}_i + \sum_{j=i+1}^{L-1} u_{i,j}\mathbf{e}_j), \\ \qquad i = 0, 1, \cdots, L\text{-}2, \\ (\mathbf{A}_{\Psi(x)}^t)^{\beta_{L-1}} \cdot \Delta_\alpha^{-1} \cdot \mathbf{e}_{L-1}, \quad i = L\text{-}1, \end{cases} \tag{12.31c}$$

which is an SSRG-based realization. If expressed in terms of $\mathbf{c}_i^S$ the correction vector $\mathbf{c}_i$ in (12.30c) has the expression

$$\mathbf{c}_i = \mathbf{R}^{-1} \cdot \mathbf{c}_i^S, \quad i = 0, 1, \cdots, L-1. \tag{12.32}$$

Note that $\mathbf{c}_i^S$'s in (12.31c) depend only on the correction times β_i's. So it is possible to choose the correction times β_i's to make all the correction vectors $\mathbf{c}_i^S$'s identical, and further for such correction times the common correction vectors can be determined by (12.32).

EXAMPLE 12.7 For the scrambling sequence $\{\, s_k \,\}$ in the scrambling space $V[x^4 + x + 1]$ with the sampling times $\alpha_0 = 1$, $\alpha_1 = 3$, $\alpha_2 = 5$ and $\alpha_3 = 7$, we realize its counterpart descramblers. If we insert $\mathbf{R} = \mathbf{I}$ along with Δ_α in (12.4) into (12.30), we obtain the SSRG with the state transition matrix $\hat{\mathbf{T}} = \mathbf{A}^t_{x^4+x+1}$, the generating vector $\hat{\mathbf{h}} = \mathbf{e}_0$, and the correction vectors $\mathbf{c}_i^S$'s in Table 12.2 for all feasible correction times β_i's. If we choose the correction times $\beta_0 = 3$, $\beta_1 = 5$, $\beta_2 = 7$ and $\beta_3 = 9$, it is possible to select the common correction vector $\mathbf{c}_i^S = [\, 0\,1\,0\,1\,]^t$, $i = 0, 1, 2, 3$, as marked with $*$ in the table. Using this, we can realize the descrambler shown in Fig. 12.6(a). For the same correction times β_i's, we now choose the nonsingular matrix

$$
\mathbf{R} = \begin{bmatrix} 0 & 1 & 0 & 0 \\ 1 & 0 & 1 & 0 \\ 0 & 1 & 0 & 1 \\ 1 & 1 & 0 & 0 \end{bmatrix}.
$$

Then, by (12.30a), (12.30b) and (12.32) we obtain another descrambler with

$$
\hat{\mathbf{T}} = \begin{bmatrix} 0 & 1 & 0 & 0 \\ 1 & 0 & 1 & 0 \\ 0 & 0 & 0 & 1 \\ 0 & 1 & 1 & 0 \end{bmatrix},
$$

$$
\hat{\mathbf{h}} = [\, 0\,1\,0\,0\,]^t,
$$

$$
\mathbf{c}_i = [\, 1\,0\,0\,0\,]^t, \quad i = 0, 1, 2, 3,
$$

whose circuit diagram is shown in Fig. 12.6(b). ♣

12.4 DSS with Minimized Timing Circuitry

In the DSS, timing circuits are necessary to generate clocks for the sampling and the correction times. Moreover, additional timing circuits are needed to generate the clocks that indicate when the samples are actually transmitted. We define such a time the sample transmission time, or more specifically,

DEFINITION 12.22 (Sample Transmission Time) We call the time γ_i, $i = 0, 1, \cdots, L - 1$, at which the sample z_i is transmitted, the ith *sample transmission time.*

Table 12.2. Correction vectors c_i^S's in (12.31c) for different feasible correction times. The scrambling space of the scrambling sequence $\{ s_k \}$ is $V[x^4 + x + 1]$, and the sampling times are $\alpha_0 = 1$, $\alpha_1 = 3$, $\alpha_2 = 5$ and $\alpha_3 = 7$. The asterisks(*) indicate the case with common correction vectors $c_0^S = c_1^S = c_2^S = c_3^S$.

c_0^S	$\beta_0 = 2$	$[\,1\,0\,0\,0\,]^t$	
		$[\,1\,1\,1\,0\,]^t$	
		$[\,0\,0\,0\,1\,]^t$	
		$[\,1\,0\,1\,0\,]^t$	
		$[\,0\,1\,1\,1\,]^t$	
		$[\,1\,1\,0\,0\,]^t$	
		$[\,0\,0\,1\,1\,]^t$	
		$[\,0\,1\,0\,1\,]^t$	
	$\beta_0 = 3$	$[\,0\,0\,0\,1\,]^t$	
		$[\,1\,1\,0\,0\,]^t$	
		$[\,0\,0\,1\,0\,]^t$	
		$[\,0\,1\,0\,1\,]^t$	*
		$[\,1\,1\,1\,1\,]^t$	
		$[\,1\,0\,0\,0\,]^t$	
		$[\,0\,1\,1\,0\,]^t$	
		$[\,1\,0\,1\,1\,]^t$	
c_1^S	$\beta_1 = 4$	$[\,1\,0\,1\,0\,]^t$	
		$[\,1\,1\,0\,0\,]^t$	
		$[\,0\,0\,1\,1\,]^t$	
		$[\,0\,1\,0\,1\,]^t$	
	$\beta_1 = 5$	$[\,0\,1\,0\,1\,]^t$	*
		$[\,1\,0\,0\,0\,]^t$	
		$[\,0\,1\,1\,0\,]^t$	
		$[\,1\,0\,1\,1\,]^t$	
c_2^S	$\beta_2 = 6$	$[\,1\,0\,1\,0\,]^t$	
		$[\,1\,1\,0\,0\,]^t$	
	$\beta_2 = 7$	$[\,0\,1\,0\,1\,]^t$	*
		$[\,1\,0\,0\,0\,]^t$	
c_3^S	$\beta_3 = 8$	$[\,1\,0\,1\,0\,]^t$	
	$\beta_3 = 9$	$[\,0\,1\,0\,1\,]^t$	*
	$\vdots$	$\vdots$	

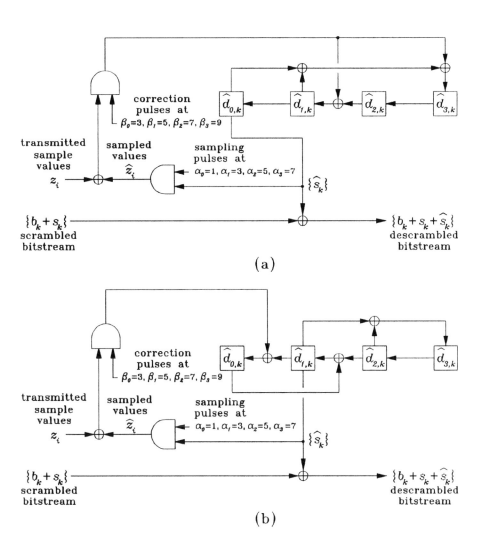

Figure 12.6. Examples of the smallest-length descrambler SRG which can be synchronized to the scrambler in Fig. 12.3. (a) An SSRG-based realization, (b) another realization.

If samples are transmitted at the moment they are sampled, the sampling time itself becomes the sample transmission time. But, in general, the sample transmission time is decided depending on the available slots in the frame format, so in practice it is impossible to change the sample transmission times arbitrarily. Consequently, timing circuitry is required to generate various kinds of clocks for the relevant storing and retrieval of samples, and such a timing circuit is usually complicated. Therefore, it is worthwhile to consider methods to minimize the complexity of the related timing circuit.

12.4.1   Concurrent Sampling

Since the available slots for sample transmission are not changeable, we rather consider changing the sampling times to be identical to the sample transmission times. Then, additional timing circuitry for the generation of the sampling times will not be necessary because other timing circuits that readily exist to generate clocks for frame-formatting can be utilized for this. However, this does not always work because the sampling times fitted to the available transmission slots may not be the predictable sampling times. Therefore, it does not provide a complete solution.

For a complete solution to the timing circuitry problem, we need to allow for independence between the sampling time and the sample transmission time, but, instead, take the samples at the sample transmission times, yet keeping the effect of sampling at the desired predictable sampling times. We call this *concurrent sampling* as the sampling occurs concurrently with the sample insertion for transmission. To better describe this solution, we introduce the concept of sampling vector.

DEFINITION 12.23 (Sampling Vector) For a sample z taken from an SRG at sampling time α, we define the *sampling vector* $\mathbf{v}$ to be the vector describing the relation of the sample z to the corresponding state vector $\mathbf{d}_\alpha$, i.e.,

$$z = \mathbf{v}^t \cdot \mathbf{d}_\alpha. \tag{12.33}$$

In the DSSs we have discussed so far, all the samples have been taken directly from the scrambling sequence at the sampling times, and consequently all the sampling vectors have been identical to the generating vector $\mathbf{h}$.

We consider how much the sampling vector changes as the sampling times move to the sample transmission times.

THEOREM 12.24 *For a sample z taken from an SRG with the state transition matrix $\mathbf{T}$ and the generating vector $\mathbf{h}$, let $\mathbf{v}$ be its sampling vector taken at sampling time α. Then, the sample z is identical to the sample taken at a sampling time $\hat{\alpha}$ using the sampling vector*

$$\hat{\mathbf{v}} = (\mathbf{T}^t)^{\alpha - \hat{\alpha}} \cdot \mathbf{v}. \tag{12.34}$$

Proof : Since the sample z has the sampling time α and the sampling vector $\mathbf{v}$, we have the relation $z = \mathbf{v}^t \cdot \mathbf{d}_\alpha$, which can be rewritten in the form $z = \mathbf{v}^t \cdot \mathbf{T}^{\alpha - \hat{\alpha}} \cdot \mathbf{d}_{\hat{\alpha}}$. Therefore, by (12.33) the sampling vector of z at the sampling time $\hat{\alpha}$ is $\hat{\mathbf{v}} = (\mathbf{T}^t)^{\alpha - \hat{\alpha}} \cdot \mathbf{v}$. ∎

The theorem means that the sample z taken at the sampling time α using the sampling vector $\mathbf{v}$ is identical to the sample taken at the sampling time $\hat{\alpha}$ using the sampling vector $\hat{\mathbf{v}}$ in (12.34). Therefore, in the DSS the sample z_i to be sampled at sampling time α_i can be identically obtained at the sample transmission time γ_i if the sampling vector is switched to

$$\hat{\mathbf{v}} = (\mathbf{T}^t)^{\alpha_i - \gamma_i} \cdot \mathbf{h}, \tag{12.35}$$

since the original sampling vector used to be the generating vector $\mathbf{h}$.

EXAMPLE 12.8 For the scrambling sequence $\{ s_k \}$ in the scrambling space $V[x^4 + x + 1]$, we consider the case when the available transmission times are $\gamma_0 = 2$, $\gamma_1 = 3$, $\gamma_2 = 6$ and $\gamma_3 = 7$. In this case, we can not take the sampling times α_i's identical to the sample transmission times γ_i's because they are not predictable sampling times, that is, the corresponding discrimination matrix in this case becomes singular (refer to Example 12.3). So we choose the predictable sampling times $\alpha_0 = 1$, $\alpha_1 = 3$, $\alpha_2 = 5$ and $\alpha_3 = 7$, as in the scrambler shown in Fig. 12.3, and realize the concurrent sampling for this. In this case, we have $\mathbf{T} = \mathbf{A}_{x^4 + x + 1}$ and $\mathbf{h} = \mathbf{e}_1$, so from (12.35) we obtain the sampling vectors

$$\begin{cases} \hat{\mathbf{v}}_0 &= \mathbf{A}_{x^4 + x + 1}^{t^{1-2}} \cdot \mathbf{e}_1 = \mathbf{e}_2, \\ \hat{\mathbf{v}}_1 &= \mathbf{A}_{x^4 + x + 1}^{t^{3-3}} \cdot \mathbf{e}_1 = \mathbf{e}_1, \\ \hat{\mathbf{v}}_2 &= \mathbf{A}_{x^4 + x + 1}^{t^{5-6}} \cdot \mathbf{e}_1 = \mathbf{e}_2, \\ \hat{\mathbf{v}}_3 &= \mathbf{A}_{x^4 + x + 1}^{t^{7-7}} \cdot \mathbf{e}_1 = \mathbf{e}_1. \end{cases}$$

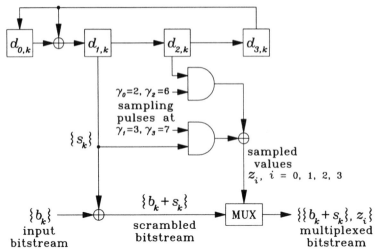

Figure 12.7. An example of concurrent sampling based scrambler which takes samples at the sample transmission times 2, 3, 6, 7 with the effect of sampling at the sampling times 1, 3, 5, 7.

The resulting scrambler is as shown in Fig. 12.7. As it employs the concurrent sampling, the taken samples can now be directly multiplexed with the scrambled bitstream for transmission. That is, the samples are inserted at the sample transmission time γ_i's to the preassigned slots in the transmission frame format. So we need to generate the sample transmission times γ_i's but not the sampling times α_i's, and the sample transmission times are readily generated elsewhere in the system for the multiplexing or frame formatting purpose. If we employ the scrambler in Fig. 12.3, however, we need timing circuits for generating the sampling times α_i's and additional circuits to store and retrieve the samples for transmission. ♣

12.4.2 Immediate Correction

We know that, differently from the case of the sampling times, the correction times can be chosen in an arbitrary manner. Therefore, we can arrange them to come immediately after the sampling times, thus eliminating the additional timing circuit needed to generate the correction times. We call such a correction scheme an *immediate correction*.

In fact, the correction made at the correction time β_i in $\gamma_i < \beta_i \leq \gamma_{i+1}$ using the correction vector $\mathbf{c}_i$ in (12.32) has the same effect as the correction made at the correction time $\tilde{\beta}_i$ in $\alpha_i < \tilde{\beta}_i \leq \alpha_{i+1}$ using the correction vector $\tilde{\mathbf{c}}_i = \mathbf{T}^{\tilde{\beta}_i - \beta_i} \cdot \mathbf{c}_i$. But this $\tilde{\mathbf{c}}_i$ preserves the form in (12.32), and therefore we can use (12.32) without change even in the case of immediate correction.

If we combine the immediate correction with the concurrent sampling, we can design descramblers that do not require any additional circuitry for the sampling time and correction time generation, or for the relevant sample storing and retrieval. In this arrangement, we can move the sampling times α_i's to the sample transmission times γ_i's at the cost of the sampling vector generating circuit, and we can choose the correction times β_i's to be $\beta_i = \gamma_i + 1$ at no additional cost.

EXAMPLE 12.9 For the scrambling sequence $\{ s_k \}$ in the scrambling space $V[x^4 + x + 1]$ with the sampling times $\alpha_0 = 1$, $\alpha_1 = 3$, $\alpha_2 = 5$, $\alpha_3 = 7$, and the sample transmission times $\gamma_0 = 2$, $\gamma_1 = 3$, $\gamma_2 = 6$, $\gamma_3 = 7$, we realize its counterpart descrambler which does not require sampling time and correction time generation. We employ the descrambler SRG embedded in Fig. 12.6(a), whose state transition matrix is $\hat{\mathbf{T}} = \mathbf{A}_{x^4+x+1}^t$ and the generating vector is $\hat{\mathbf{h}} = \mathbf{e}_0$. We first move the sampling times α_i's to the concurrent sampling times γ_i's. Then, by (12.35) we get the sampling vectors

$$\begin{cases} \hat{\mathbf{v}}_0 = \mathbf{A}_{x^4+x+1}^{1-2} \cdot \mathbf{e}_0 = \mathbf{e}_0 + \mathbf{e}_3, \\ \hat{\mathbf{v}}_1 = \mathbf{A}_{x^4+x+1}^{3-3} \cdot \mathbf{e}_0 = \mathbf{e}_0, \\ \hat{\mathbf{v}}_2 = \mathbf{A}_{x^4+x+1}^{5-6} \cdot \mathbf{e}_0 = \mathbf{e}_0 + \mathbf{e}_3, \\ \hat{\mathbf{v}}_3 = \mathbf{A}_{x^4+x+1}^{7-7} \cdot \mathbf{e}_0 = \mathbf{e}_0. \end{cases}$$

Now we choose the immediate correction times $\beta_0 = \gamma_0 + 1 = 3$, $\beta_1 = \gamma_1 + 1 = 4$, $\beta_2 = \gamma_2 + 1 = 7$ and $\beta_3 = \gamma_3 + 1 = 8$. Then, by (12.31c) we can select the correction vectors $\mathbf{c}_0^S = \mathbf{c}_2^S = [\,0\,1\,0\,1\,]^t$ and $\mathbf{c}_1^S = \mathbf{c}_3^S = [\,1\,0\,1\,0\,]^t$. Using these parameters, we obtain the descrambler shown in Fig. 12.8. This is the descrambler which can be used in conjunction with the scrambler in Fig. 12.7. We observe from the figure that the only timing signal necessary in the descrambler is the sample transmission time signal, which is readily generated elsewhere in the system for frame deformatting. ♣

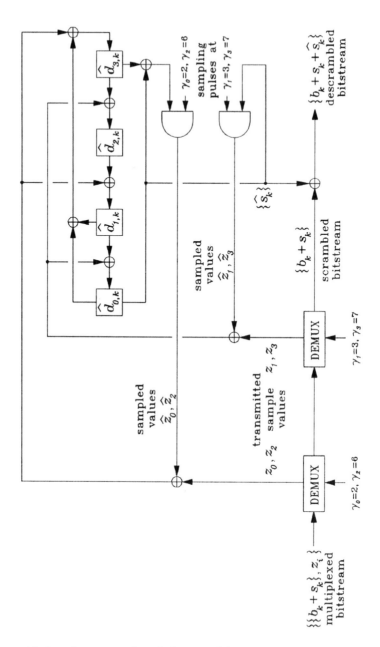

Figure 12.8. An example of descrambler employing the concurrent sampling and the immediate correction.

Chapter 13

Parallel Distributed Sample Scrambling

In the *parallel distributed sample scrambling* (PDSS), the parallel input data bitstreams are scrambled before multiplexing, and the scrambled data bitstream is descrambled after demultiplexing. In this chapter, we discuss issues involved with the parallel realizations of DSSs. We first consider how to realize the PSRGs for PDSS along with the minimal realizations of PSRGs. Then, we examine how to achieve the parallel sampling along with the concurrent parallel sampling. Finally, we consider the parallel correction, with single, double and multiple corrections occurring in the form of immediate corrections.

13.1   Considerations for Parallel DSS

The serial and the parallel DSSs can be contrasted as shown in Fig. 13.1(a) and (b). In the serial scrambling, N parallel input data bitstreams $\{ b_k^j \}$, $j = 0, 1, \cdots, N - 1$, are multiplexed, and then the multiplexed bitstream $\{ b_k \}$ is scrambled by adding a scrambling sequence $\{ s_k \}$ to it. In the receiving part, the scrambled bitstream $\{ b_k + s_k \}$ is descrambled by adding a descrambling sequence $\{ \hat{s}_k \}$ to it, and then the descrambled data $\{ b_k + s_k + \hat{s}_k \}$ is demultiplexed. To make the demultiplexed bitstreams $\{ b_k^j + s_k^j + \hat{s}_k^j \}$, $j = 0, 1, \cdots, N - 1$, identical to the parallel input data bitstreams $\{ b_k^j \}$, the samples z_i's of the scrambling sequence $\{ s_k \}$ are taken by the sampling function and conveyed to the descrambler over the available bit slots of the scrambled

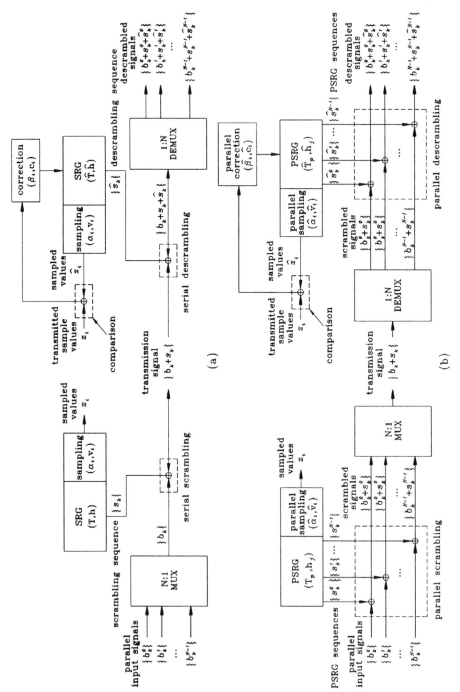

Figure 13.1. Serial and parallel DSS. (a) Serial DSS, (b) parallel DSS.

bitstream $\{ b_k + s_k \}$. The descrambler generates its own samples $\hat{z}_i$'s of the descrambling sequence $\{ \hat{s}_k \}$ in the same manner and compares them to their conveyed counterparts. If the two sets of samples are different, the correction logic is initiated to synchronize the descrambler SRG states to the scrambler SRG states, thus eventually making the descrambling sequence $\{ \hat{s}_k \}$ identical to the scrambling sequence $\{ s_k \}$.

In the parallel DSS, N parallel input data bitstreams $\{ b_k^j \}$, $j = 0$, $1, \cdots, N - 1$, are scrambled before multiplexing by adding the parallel scrambling sequences $\{ s_k^j \}$, $j = 0, 1, \cdots, N - 1$, respectively to the parallel input bitstreams $\{ b_k^j \}$, $j = 0, 1, \cdots, N - 1$, and the scrambled bitstream $\{ b_k + s_k \}$ is descrambled after demultiplexing by adding the parallel descrambling sequences $\{ \hat{s}_k^j \}$, $j = 0, 1, \cdots, N - 1$, to the demultiplexed bitstreams respectively. To make the scrambled bitstream $\{ b_k + s_k \}$ in Fig. 13.1(b) identical to the serial-scrambled bitstream $\{ b_k + s_k \}$ in Fig. 13.1(a) for the same parallel input data bitstreams $\{ b_k^j \}$'s, the parallel scrambling sequences $\{ s_k^j \}$, $j = 0, 1, \cdots, N - 1$, should be the N-decimated sequences of the serial scrambling sequence $\{ s_k \}$,[1] and the samples z_i's generated by the serial and the parallel scramblers should be the same. In addition, the correction logic should also be changed accordingly.

13.2 PSRGs for PDSS

We consider how to realize PSRGs generating the parallel scrambling sequences, which are the decimated sequences of the serial scrambling sequence.

As described in Chapter 12, in the DSS the scrambling space is a primitive space, and the scrambling sequence is a PRBS in the primitive scrambling space. Therefore, we concentrate on PSRGs for an irreducible scrambling sequence. Note that a PRBS in a primitive space is an irreducible sequence. The following theorem describes how to determine the minimal PSRGs for an irreducible scrambling sequence $\{ s_k \}$.

THEOREM 13.1 *For an irreducible scrambling sequence* $\{ s_k \} = \mathbf{s}^t \cdot \mathbf{E}_{\Psi_I(x)}$ *in the irreducible space* $V[\Psi_I(x)]$ *of dimension* L, *let* $\{ s_k^j \}$, j

---

[1]The relation between the serial scrambling sequence and the parallel scrambling sequences is described in Chapter 7.

$= 0, 1, \cdots, N\text{-}1$, *be its jth N-decimated sequence for an N relatively prime to the periodicity T_I of $V[\Psi_I(x)]$. Then, for a nonsingular matrix* $\mathbf{R}$, *the PSRG with the state transition matrix* $\mathbf{T}$, *the initial state vector* $\mathbf{d}_0$ *and the generating vectors* $\mathbf{h}_j$ *'s of the form*

$$\mathbf{T} = \mathbf{R} \cdot (\mathbf{A}^t_{\Psi_I(x)})^N \cdot \mathbf{R}^{-1}, \tag{13.1a}$$

$$\mathbf{d}_0 = \mathbf{R} \cdot \mathbf{s}, \tag{13.1b}$$

$$\mathbf{h}_j = (\mathbf{R}^t)^{-1} \cdot \mathbf{A}^j_{\Psi_I(x)} \cdot \mathbf{e}_0, \quad j = 0, 1, \cdots, N-1, \tag{13.1c}$$

is a minimal PSRG generating the N-decimated sequences $\{ s^j_k \}$'s.

Proof : By Theorem 7.16 the period τ of the irreducible scrambling sequence $\{ s_k \}$ is the periodicity T_I, so the decimation factor N is relatively prime to τ. Therefore, by Theorem 7.19 the minimal space of the N-decimated sequences $\{ s^j_k \}$'s is the irreducible space $V[\Psi_m(x)]$ of dimension L, whose characteristic polynomial $\Psi_m(x)$ divides $x^{2^L-1}+1$ and its extension $\Psi_m(x^N)$ is divided by $\Psi_I(x)$. By Theorem 4.10, the initial vector $\mathbf{s}_j$, $j = 0, 1, \cdots, N-1$, of the jth decimated sequence $\{ s^j_k \}$ becomes $\mathbf{s}_j = [\, \mathbf{s}^t \cdot \mathbf{A}^j_{\Psi_I(x)} \cdot \mathbf{e}_0 \quad \mathbf{s}^t \cdot \mathbf{A}^{N+j}_{\Psi_I(x)} \cdot \mathbf{e}_0 \quad \cdots \quad \mathbf{s}^t \cdot \mathbf{A}^{N(L-1)+j}_{\Psi_I(x)} \cdot \mathbf{e}_0 \,]^t$, which can be rewritten in the form $\mathbf{s}_j = [\, \mathbf{s} \, (\mathbf{A}^t_{\Psi_I(x)})^N \cdot \mathbf{s} \cdots (\mathbf{A}^t_{\Psi_I(x)})^{N(L-1)} \cdot \mathbf{s} \,]^t \cdot \mathbf{A}^j_{\Psi_I(x)} \cdot \mathbf{e}_0$. Since $\Psi_m(x)$ is a degree-L irreducible polynomial such that $\Psi_I(x)$ divides $\Psi_m(x^N)$, the minimal polynomial of $(\mathbf{A}^t_{\Psi_I(x)})^N$ is $\Psi_m(x)$. So the matrix $[\, \mathbf{s} \quad (\mathbf{A}^t_{\Psi_I(x)})^N \cdot \mathbf{s} \quad \cdots \quad (\mathbf{A}^t_{\Psi_I(x)})^{N(L-1)} \cdot \mathbf{s} \,]$ is nonsingular. Inserting $\mathbf{Q} = \mathbf{R} \cdot [\, \mathbf{s} \quad (\mathbf{A}^t_{\Psi_I(x)})^N \cdot \mathbf{s} \quad \cdots \quad (\mathbf{A}^t_{\Psi_I(x)})^{N(L-1)} \cdot \mathbf{s} \,]$ into (7.45), we obtain $\mathbf{T} = \mathbf{R} \cdot (\mathbf{A}^t_{\Psi_I(x)})^N \cdot \mathbf{R}^{-1}$,[2] $\mathbf{d}_0 = \mathbf{R} \cdot \mathbf{s}$ and $\mathbf{h}_j = (\mathbf{R}^t)^{-1} \cdot \mathbf{A}^j_{\Psi_I(x)} \cdot \mathbf{e}_0$. ∎

According to the theorem, the PSRG with the state transition matrix $\mathbf{T}$ in (13.1a), the initial state vector $\mathbf{d}_0$ in (13.1b) and the generating vectors $\mathbf{h}_j$'s in (13.1c) for an arbitrary nonsingular matrix $\mathbf{R}$ is a minimal PSRG generating the N-decimated sequence $\{ s^i_k \}$'s, if the scrambling sequence $\{ s_k \} = \mathbf{s}^t \cdot \mathbf{E}_{\Psi_I(x)}$ is an irreducible sequence in the irreducible space $V[\Psi_I(x)]$ and if its decimation factor N is relatively

---

[2]If the minimal polynomial of an $L \times L$ matrix $\mathbf{M}$ is a degree-L polynomial $\Psi(x)$, then for a nonsingular matrix $\mathbf{Q} = [\, \mathbf{v} \quad \mathbf{M} \cdot \mathbf{v} \quad \cdots \quad \mathbf{M}^{L-1} \cdot \mathbf{v} \,]$ formed by the matrix $\mathbf{M}$ and an L-vector $\mathbf{v}$ we have the relation $\mathbf{A}_{\Psi(x)} = \mathbf{Q}^{-1} \cdot \mathbf{M} \cdot \mathbf{Q}$. Refer to Property B.18 in Appendix B.

prime to the periodicity T_I of $V[\Psi_I(x)]$. From the equations in (13.1), we find that the initial state vector $\mathbf{d}_0$ depends on the initial vector s of the scrambling sequence $\{ s_k \}$ but the state transition matrix $\mathbf{T}$ and the generating vectors $\mathbf{h}_j$'s *do not*. This implies that the PSRG with the state transition matrix $\mathbf{T}$ in (13.1a) and the generating vectors $\mathbf{h}_j$'s in (13.1c) can generate the N-decimated sequences of an arbitrary scrambling sequence in the irreducible space $V[\Psi_I(x)]$ by setting its initial state vector to $\mathbf{d}_0$ in (13.1b). Therefore, we can determine the state transition matrix $\mathbf{T}$ and the generating vectors $\mathbf{h}_j$'s of a minimal PSRG for parallel DSS by inserting an arbitrary nonsingular matrix $\mathbf{R}$ into (13.1a) and (13.1c) respectively.

EXAMPLE 13.1 For the serial scrambling sequence $\{ s_k \}$ whose scrambling space is the irreducible space $V[x^4 + x + 1]$, we realize minimal PSRGs generating the 2-decimated sequences $\{ s_k^j \}$, $j = 0, 1$. Then, by inserting $\mathbf{R} = \mathbf{I}$ into (13.1a) and (13.1c) along with $\Psi_I(x) = x^4 + x + 1$ and $N = 2$, we obtain

$$\mathbf{T} = \begin{bmatrix} 0 & 0 & 1 & 0 \\ 0 & 0 & 0 & 1 \\ 1 & 1 & 0 & 0 \\ 0 & 1 & 1 & 0 \end{bmatrix},$$

$$\begin{cases} \mathbf{h}_0 = \mathbf{e}_0, \\ \mathbf{h}_1 = \mathbf{e}_1. \end{cases}$$

The resulting minimal PSRG is as shown in Fig. 13.2(a). If we choose another nonsingular matrix

$$\mathbf{R} = \begin{bmatrix} 1 & 0 & 0 & 0 \\ 0 & 0 & 1 & 0 \\ 1 & 1 & 0 & 0 \\ 0 & 0 & 1 & 1 \end{bmatrix}, \tag{13.2}$$

then we obtain the minimal PSRG with

$$\mathbf{T} = \mathbf{A}_{x^4+x+1}^t, \tag{13.3a}$$

$$\begin{cases} \mathbf{h}_0 = \mathbf{e}_0, \\ \mathbf{h}_1 = \mathbf{e}_0 + \mathbf{e}_2. \end{cases} \tag{13.3b}$$

This minimal PSRG is an SSRG-based realization depicted in Fig. 13.2(b). ♣

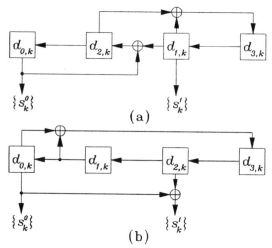

Figure 13.2. Examples of minimal PSRGs generating the 2-decimated sequences $\{ s_k^0 \}$ and $\{ s_k^1 \}$ of the scrambling sequence $\{ s_k \}$ in the irreducible sequence space $V[x^4 + x + 1]$. (a) The minimal PSRG for $\mathbf{R} = \mathbf{I}$, (b) the SSRG-based minimal PSRG.

13.3   Parallel Sampling for PDSS

We consider the sampling in the parallel DSS, which we call a *parallel sampling*. The following theorem describes this.

THEOREM 13.2 *For a serial scrambling sequence $\{ s_k \}$, let $\{ s_k^j \}$, $j = 0, 1, \cdots, N$-1, be its jth N-decimated sequence. Then, a sample z taken from the serial scrambling sequence $\{ s_k \}$ at the sampling time α corresponds to the sample taken from the lth decimated sequence $\{ s_k^l \}$ at the parallel sampling time $\hat{\alpha}$, where l and $\hat{\alpha}$ are respectively the remainder and the quotient of α divided by N.*

    Proof : Let the sample z be taken from the scrambling sequence $\{ s_k \}$ at the sampling time α. Then, $z = s_\alpha$, or $z = s_{l+N\hat{\alpha}}$. Therefore, by (7.2) $z = s_{\hat{\alpha}}^l$, that is, z is identical to the sample taken from the lth decimated sequence $\{ s_k^l \}$ at the parallel sampling time $\hat{\alpha}$.   ■

EXAMPLE 13.2 For the serial scrambling sequence $\{ s_k \}$ whose scrambling space is the irreducible space $V[x^4 + x + 1]$ and its sampling times are $\alpha_0 = 1$, $\alpha_1 = 3$, $\alpha_2 = 5$, $\alpha_3 = 7$, we consider the parallel sampling for the case $N = 2$. Then, for the 0th sample z_0 the remainder and the

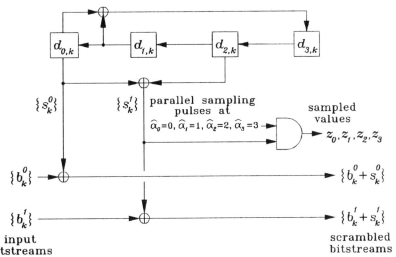

Figure 13.3. An example of a parallel scrambler for the scrambling sequence { s_k } whose scrambling space is the irreducible sequence space $V[x^4 + x + 1]$ and its sampling times are $\alpha_0 = 1$, $\alpha_1 = 3$, $\alpha_2 = 5$, $\alpha_3 = 7$.

quotient of $\alpha_0 = 1$ divided by $N = 2$ are $l_0 = 1$ and $\hat{\alpha}_0 = 0$ respectively. Therefore, by the theorem z_0 is taken from the 1st 2-decimated sequence { s_k^1 } at the parallel sampling time $\hat{\alpha}_0 = 0$. In a similar manner, for the 1st, the 2nd and the 3rd samples z_1, z_2 and z_3, the remainders and quotients of $\alpha_1 = 3$, $\alpha_2 = 5$, $\alpha_3 = 7$ divided by $N = 2$, are respectively $l_1 = 1$, $\hat{\alpha}_1 = 1$, $l_2 = 1$, $\hat{\alpha}_2 = 2$, $l_3 = 1$, $\hat{\alpha}_3 = 3$. Therefore, z_1, z_2 and z_3 are all taken from { s_k^1 } at the parallel sampling times $\hat{\alpha}_1 = 1$, $\hat{\alpha}_2 = 2$ and $\hat{\alpha}_3 = 3$ respectively. If we employ the minimal PSRG in Fig. 13.2(b), we can implement this parallel sampling as depicted in Fig. 13.3. ♣

Now we consider how to realize concurrent sampling in the parallel DSS. For this, we first consider how the sample transmission time changes in the parallel DSS, which we call the *parallel sample transmission time*.

THEOREM 13.3 *For a serial scrambling sequence* { s_k }, *let* { s_k^j }, *j* = 0, 1, $\cdots$, *N-1, be its jth N-decimated sequence. Then, a sample z transmitted over the serial-scrambled signal at the sample transmission time γ is transmitted over the mth parallel-scrambled signal at the parallel sample transmission time $\hat{\gamma}$, where m and $\hat{\gamma}$ are respectively the remainder and the quotient of γ divided by N.*

Proof : This theorem can be proved in a way similar to that of Theorem 13.2. ∎

EXAMPLE 13.3 For the serial scrambling sequence $\{ s_k \}$ whose scrambling space is the irreducible space $V[x^4 + x + 1]$ and whose sample transmission times are $\gamma_0 = 1$, $\gamma_1 = 2$, $\gamma_2 = 5$, $\gamma_3 = 6$, we consider the parallel sample transmission for the case $N = 2$. Then, for the 0th sample z_0 the remainder and the quotient of $\gamma_0 = 1$ divided by $N = 2$ are $m_0 = 1$ and $\hat{\gamma}_0 = 0$ respectively. Therefore, by the theorem z_0 is transmitted over the 1st parallel-scrambled signal at the parallel sample transmission time $\hat{\gamma}_0 = 0$. In a similar manner, we obtain, for the 1st, the 2nd and the 3rd samples z_1, z_2 and z_3, the remainders $m_1 = 0$, $m_2 = 1$, $m_3 = 0$, and the quotients $\hat{\gamma}_1 = 1$, $\hat{\gamma}_2 = 2$, $\hat{\gamma}_3 = 3$ respectively. Therefore, z_1, z_2 and z_3 are to be transmitted over the 0th, the 1st and the 0th parallel-scrambled signals respectively at the parallel sample transmission times $\hat{\gamma}_1 = 1$, $\hat{\gamma}_2 = 2$ and $\hat{\gamma}_3 = 3$. ♣

Then, the concurrent sampling can be realized in the PSRG as the following theorem describes.

THEOREM 13.4 *For a sample z taken from the lth parallel sequence $\{ s_k^l \}$ generated by a PSRG with the state transition matrix $\mathbf{T}$ and the generating vectors $\mathbf{h}_j$, $j = 0, 1, \cdots, N\text{-}1$, let $\hat{\alpha}$ and $\hat{\gamma}$ be its parallel sampling time and parallel sample transmission time respectively. Then, the sample z is identical to the sample taken at the parallel sample transmission time $\hat{\gamma}$ using the sampling vector*

$$\mathbf{v} = (\mathbf{T}^t)^{\hat{\alpha}-\hat{\gamma}} \cdot \mathbf{h}_l. \tag{13.4}$$

Proof : Since the sample z has the sampling time $\hat{\alpha}$ and the sampling vector $\mathbf{h}_l$, inserting $\alpha = \hat{\alpha}$ and $\mathbf{v} = \mathbf{h}_l$ into (12.33) we have the relation $z = \mathbf{h}_l^t \cdot \mathbf{d}_{\hat{\alpha}}$, which can be rewritten in the form $z = \mathbf{h}_l^t \cdot \mathbf{T}^{\hat{\alpha}-\hat{\gamma}} \cdot \mathbf{d}_{\hat{\gamma}}$. Therefore, by (12.33), the sampling vector of z at the parallel sample transmission time $\hat{\gamma}$ is $\mathbf{v} = (\mathbf{T}^t)^{\hat{\alpha}-\hat{\gamma}} \cdot \mathbf{h}_l$. ∎

The theorem implies that the sample z taken from the lth parallel sequence $\{ s_k^l \}$ at the parallel sampling time $\hat{\alpha}$ is identical to the sample taken at the parallel sample transmission time $\hat{\gamma}$ using the sampling vector $\mathbf{v}$ in (13.4). Therefore, we can realize the concurrent parallel sampling by applying (13.4) to all the samples z_i's.

EXAMPLE 13.4 For the serial scrambling sequence $\{ s_k \}$ whose scrambling space is the irreducible space $V[x^4 + x + 1]$, we consider the case when its sampling times are $\alpha_0 = 1$, $\alpha_1 = 3$, $\alpha_2 = 5$, $\alpha_3 = 7$, and its sample transmission times are $\gamma_0 = 1$, $\gamma_1 = 2$, $\gamma_2 = 5$, $\gamma_3 = 6$. Then, its parallel sampling times are $\hat{\alpha}_0 = 0$, $\hat{\alpha}_1 = 1$, $\hat{\alpha}_2 = 2$, $\hat{\alpha}_3 = 3$ (refer to Example 13.2) ; the parallel sequence numbers are $l_0 = 1$, $l_1 = 1$, $l_2 = 1$, $l_3 = 1$ (refer to Example 13.2) ; and the parallel sample transmission times are $\hat{\gamma}_0 = 0$, $\hat{\gamma}_1 = 1$, $\hat{\gamma}_2 = 2$, $\hat{\gamma}_3 = 3$ (refer to Example 13.3). If we employ the minimal PSRG in Fig. 13.2(b), by (13.4) we obtain the sampling vectors

$$
\begin{aligned}
\mathbf{v}_0 &= \mathbf{A}^{0-0}_{x^4+x+1} \cdot \mathbf{h}_1 = \mathbf{h}_1, \\
\mathbf{v}_1 &= \mathbf{A}^{1-1}_{x^4+x+1} \cdot \mathbf{h}_1 = \mathbf{h}_1, \\
\mathbf{v}_2 &= \mathbf{A}^{2-2}_{x^4+x+1} \cdot \mathbf{h}_1 = \mathbf{h}_1, \\
\mathbf{v}_3 &= \mathbf{A}^{3-3}_{x^4+x+1} \cdot \mathbf{h}_1 = \mathbf{h}_1
\end{aligned}
$$

for the concurrent sampling. Using these along with the transmitted signal numbers $m_0 = 1$, $m_1 = 0$, $m_2 = 1$, $m_3 = 0$ (refer to Example 13.3), we obtain the parallel scrambler shown in Fig. 13.4. ♣

13.4 Parallel Correction for PDSS

Now, we consider how to realize parallel corrections to achieve synchronization in the parallel DSS. As an aid to describing the relevant correction times and correction vectors in the descrambler PSRGs, we consider the following three lemmas.

LEMMA 13.5 *For the serial scrambling sequence* $\{ s_k \}$ *whose scrambling space is the irreducible sequence space* $V[\Psi_I(x)]$ *of dimension L, let* $\{ s_k^j \}$, $j = 0, 1, \cdots, N\text{-}1$, *be its jth N-decimated sequence for an N relatively prime to the periodicity T_I of $V[\Psi_I(x)]$. Then, the synchronization state vector $\tilde{\mathbf{d}}_k$ of the descrambler PSRG with the state transition matrix $\mathbf{T}$ in (13.1a) and the generating vectors $\mathbf{h}_j$, $j = 0, 1, \cdots, N\text{-}1$, in (13.1c) generating the decimated sequences* $\{ s_k^j \}$*'s is*

$$\tilde{\mathbf{d}}_k = \mathbf{T}^k \cdot \tilde{\mathbf{d}}_0, \tag{13.5}$$

where the synchronization initial state vector is

$$\tilde{\mathbf{d}}_0 = \mathbf{R} \cdot \Delta_\alpha^{-1} \cdot \mathbf{z} \tag{13.6}$$

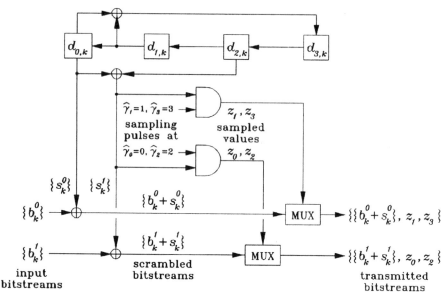

Figure 13.4. An example of parallel scrambler employing the concurrent parallel sampling.

for the relevant nonsingular matrix **R**.

Proof : It suffices to prove (13.6) since $\tilde{\mathbf{d}}_k = \mathbf{T} \cdot \tilde{\mathbf{d}}_{k-1}, k = 1, 2, \cdots$. By Theorem 12.8, the scrambling sequence $\{ s_k \}$ is predicted by $\{ s_k \}$ $= (\Delta_\alpha^{-1} \cdot \mathbf{z})^t \cdot \mathbf{E}_{\Psi_I(x)}$. Therefore, inserting $\mathbf{s} = \Delta_\alpha^{-1} \cdot \mathbf{z}$ into (13.1b), we obtain the synchronization initial state vector in (13.6). ∎

LEMMA 13.6 *For the descrambler PSRG with the state transition matrix* **T** *in (13.1a) and the generating vectors* $\mathbf{h}_j$, $j = 0, 1, \cdots, N\text{-}1$, *in (13.1c), let* $\tilde{\mathbf{V}}$ *denote the* $L \times L$ *matrix*

$$\tilde{\mathbf{V}} \equiv \left[\begin{array}{c} \mathbf{v}_0^t \cdot \mathbf{T}^{\hat{\gamma}_0} \\ \mathbf{v}_1^t \cdot \mathbf{T}^{\hat{\gamma}_1} \\ \vdots \\ \mathbf{v}_{L-1}^t \cdot \mathbf{T}^{\hat{\gamma}_{L-1}} \end{array} \right], \tag{13.7}$$

where $\hat{\gamma}_i$ *and* $\mathbf{v}_i$, $i = 0, 1, \cdots, L\text{-}1$, *are respectively the parallel sample transmission time and the sampling vector of the ith sample* z_i. *Then,*

$$\tilde{\mathbf{V}} = \Delta_\alpha \cdot \mathbf{R}^{-1}, \tag{13.8}$$

$$\mathbf{v}_i^t \cdot \mathbf{T}^{\hat{\gamma}_i} \cdot \tilde{\mathbf{V}}^{-1} = \mathbf{e}_i^t, \quad i = 0, 1, \cdots, L - 1. \tag{13.9}$$

Proof : By (13.4), we have the relation $\mathbf{v}_i^t \cdot \mathbf{T}^{\hat{\gamma}_i} = \mathbf{h}_{l_i}^t \cdot \mathbf{T}^{\hat{\alpha}_i}$ for i = 0, 1, $\cdots$, $L-1$. Inserting (13.1a) and (13.1c) into this, we obtain $\mathbf{v}_i^t \cdot \mathbf{T}^{\hat{\gamma}_i} = \mathbf{e}_0^t \cdot (\mathbf{A}_{\Psi_I(x)}^t)^{l_i + N\hat{\alpha}_i} \cdot \mathbf{R}^{-1}$, which can be rewritten in the form of $\mathbf{v}_i^t \cdot \mathbf{T}^{\hat{\gamma}_i} = \mathbf{e}_0^t \cdot (\mathbf{A}_{\Psi_I(x)}^t)^{\alpha_i} \cdot \mathbf{R}^{-1}$. Therefore, by (11.16) we get $\mathbf{v}_i^t \cdot \mathbf{T}^{\hat{\gamma}_i} = \mathbf{e}_i^t \cdot \Delta_\alpha \cdot \mathbf{R}^{-1}$, which is the ith row of (13.8).

Equation (13.9) is obvious from (13.7) and the relation $\tilde{\mathbf{V}} \cdot \tilde{\mathbf{V}}^{-1} = \mathbf{I}$. ∎

LEMMA 13.7 *For the descrambler PSRG with the state transition matrix* $\mathbf{T}$ *in (13.1a) and the generating vectors* $\mathbf{h}_j$, $j = 0, 1, \cdots, N\text{-}1$, *in (13.1c), the* ith *sample* z_i, $i = 0, 1, \cdots, L\text{-}1$, *is represented by*

$$z_i = \mathbf{v}_i^t \cdot \tilde{\mathbf{d}}_{\hat{\gamma}_i} \tag{13.10}$$

for the synchronization state vector $\tilde{\mathbf{d}}_{\hat{\gamma}_i}$.

Proof : By (13.6), we have the relation $\mathbf{z} = \Delta_\alpha \cdot \mathbf{R}^{-1} \cdot \tilde{\mathbf{d}}_0$, and inserting (13.8) into this, we obtain $\mathbf{z} = \tilde{\mathbf{V}} \cdot \tilde{\mathbf{d}}_0$, or equivalently, $z_i = \mathbf{e}_i^t \cdot \tilde{\mathbf{V}} \cdot \tilde{\mathbf{d}}_0$, $i = 0, 1, \cdots, L-1$. Inserting (13.7) into this and then applying (13.5), we obtain $z_i = \mathbf{v}_i^t \cdot \mathbf{T}^{\hat{\gamma}_i} \cdot \tilde{\mathbf{d}}_0 = \mathbf{v}_i^t \cdot \tilde{\mathbf{d}}_{\hat{\gamma}_i}$. ∎

Now we consider how to choose the correction times $\hat{\beta}_i$'s and the correction vectors $\mathbf{c}_i$'s of the descrambler PSRGs. For the correction times $\hat{\beta}_i$'s, it is efficient to choose the time slots immediately following the sample transmission times, since immediate corrections do not require additional timing circuitry generating the correction time pulse. In realizing the immediate correction in the parallel DSS, the correction process could vary depending on the parallel sample transmission times $\hat{\gamma}_i$'s. For example, if a parallel sample transmission time is different from others, the corresponding correction occurs at a correction time which is different from others, while if a multiple number of sample transmission times are identical, multiple corrections can occur at the same correction time. Therefore, we consider the correction process of the parallel DSS for three different cases -- the single correction, the double correction, and the multiple correction cases.

13.4.1 Single Correction

In the single correction case, all sample transmission times $\hat{\gamma}_i$'s are different, and the correction times $\hat{\beta}_i$'s are located between two adjacent

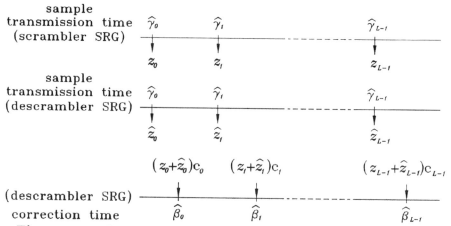

Figure 13.5. Synchronization timing for the single correction case.

sample transmission times, that is,

$$\hat{\gamma}_0 < \hat{\beta}_0 \leq \hat{\gamma}_1 < \hat{\beta}_1 \leq \cdots \leq \hat{\gamma}_{L-1} < \hat{\beta}_{L-1}. \qquad (13.11)$$

The synchronization timing employing the single correction along with the concurrent sampling can be depicted as shown in Fig. 13.5. In the figure, $\hat{\gamma}_i$, $i = 0, 1, \cdots, L-1$, is the ith sample transmission time, at which the ith samples z_i and $\hat{z}_i$ are taken respectively from the scrambler and the descrambler PSRGs. The ith correction occurs at the ith correction time $\hat{\beta}_i$, which comes later than the ith sample transmission time $\hat{\gamma}_i$ but not later than the $(i+1)$th sample transmission time $\hat{\gamma}_{i+1}$.

We first consider the correction matrix Λ which describes the relation between the initial state distance vector and the finally corrected state distance vector.

THEOREM 13.8 *For the descrambler PSRG with the state transition matrix $\mathbf{T}$ in (13.1a) and the generating vectors $\mathbf{h}_j$, $j = 0, 1, \cdots, N\text{-}1$, in (13.1c), let $\hat{\gamma}_i$ and $\mathbf{v}_i$, $i = 0, 1, \cdots, L\text{-}1$, be respectively the parallel sample transmission time and the sampling vector of the ith sample z_i. Then, the correction matrix Λ for the single correction case with the correction times $\hat{\beta}_i$ and the correction vectors $\mathbf{c}_i$, $i = 0, 1, \cdots, L\text{-}1$, is*

$$
\begin{aligned}
\Lambda = {}& (\mathbf{T}^{\hat{\beta}_{L-1}-\hat{\beta}_{L-2}} + \mathbf{c}_{L-1} \cdot \mathbf{v}_{L-1}^t \cdot \mathbf{T}^{\hat{\gamma}_{L-1}-\hat{\beta}_{L-2}}) \cdot \\
& (\mathbf{T}^{\hat{\beta}_{L-2}-\hat{\beta}_{L-3}} + \mathbf{c}_{L-2} \cdot \mathbf{v}_{L-2}^t \cdot \mathbf{T}^{\hat{\gamma}_{L-2}-\hat{\beta}_{L-3}}) \cdot \\
& \cdots (\mathbf{T}^{\hat{\beta}_1-\hat{\beta}_0} + \mathbf{c}_1 \cdot \mathbf{v}_1^t \cdot \mathbf{T}^{\hat{\gamma}_1-\hat{\beta}_0}) \cdot (\mathbf{T}^{\hat{\beta}_0} + \mathbf{c}_0 \cdot \mathbf{v}_0^t \cdot \mathbf{T}^{\hat{\gamma}_0}). \quad (13.12)
\end{aligned}
$$

Proof : For the synchronization state vector $\tilde{\mathbf{d}}_k$, we have, by (13.5), the relations

$$
\tilde{\mathbf{d}}_{\hat{\gamma}_i} = \begin{cases} \mathbf{T}^{\hat{\gamma}_0} \cdot \tilde{\mathbf{d}}_0, & i = 0, \\ \mathbf{T}^{\hat{\gamma}_i - \hat{\beta}_{i-1}} \cdot \tilde{\mathbf{d}}_{\hat{\beta}_{i-1}}, & i = 1, 2, \cdots, L-1, \end{cases} \tag{13.13}
$$

$$
\tilde{\mathbf{d}}_{\hat{\beta}_i} = \begin{cases} \mathbf{T}^{\hat{\beta}_0} \cdot \tilde{\mathbf{d}}_0, & i = 0, \\ \mathbf{T}^{\hat{\beta}_i - \hat{\beta}_{i-1}} \cdot \tilde{\mathbf{d}}_{\hat{\beta}_{i-1}}, & i = 1, 2, \cdots, L-1, \end{cases} \tag{13.14}
$$

and inserting (13.13) into (13.10), we obtain

$$
z_i = \begin{cases} \mathbf{v}_0^t \cdot \mathbf{T}^{\hat{\gamma}_0} \cdot \tilde{\mathbf{d}}_0, & i = 0, \\ \mathbf{v}_i^t \cdot \mathbf{T}^{\hat{\gamma}_i - \hat{\beta}_{i-1}} \cdot \tilde{\mathbf{d}}_{\hat{\beta}_{i-1}}, & i = 1, 2, \cdots, L-1. \end{cases} \tag{13.15}
$$

For the state vector $\hat{\mathbf{d}}_k$ of the descrambler SRG, noting that the single correction process corresponds to a modulo-2 sum of $(z_i + \hat{z}_i)\mathbf{c}_i$, we have the relations

$$
\hat{\mathbf{d}}_{\hat{\gamma}_i} = \begin{cases} \mathbf{T}^{\hat{\gamma}_0} \cdot \hat{\mathbf{d}}_0, & i = 0, \\ \mathbf{T}^{\hat{\gamma}_i - \hat{\beta}_{i-1}} \cdot \hat{\mathbf{d}}_{\hat{\beta}_{i-1}}, & i = 1, 2, \cdots, L-1, \end{cases} \tag{13.16}
$$

$$
\hat{\mathbf{d}}_{\hat{\beta}_i} = \begin{cases} \mathbf{T}^{\hat{\beta}_0} \cdot \hat{\mathbf{d}}_0 + (z_0 + \hat{z}_0)\mathbf{c}_0, & i = 0, \\ \mathbf{T}^{\hat{\beta}_i - \hat{\beta}_{i-1}} \cdot \hat{\mathbf{d}}_{\hat{\beta}_{i-1}} + (z_i + \hat{z}_i)\mathbf{c}_i, & i = 1, 2, \cdots, L-1, \end{cases} \tag{13.17}
$$

and from (13.16) we obtain

$$
\hat{z}_i = \begin{cases} \mathbf{v}_0^t \cdot \mathbf{T}^{\hat{\gamma}_0} \cdot \hat{\mathbf{d}}_0, & i = 0, \\ \mathbf{v}_i^t \cdot \mathbf{T}^{\hat{\gamma}_i - \hat{\beta}_{i-1}} \cdot \hat{\mathbf{d}}_{\hat{\beta}_{i-1}}, & i = 1, 2, \cdots, L-1. \end{cases} \tag{13.18}
$$

By (13.15), (13.18) and (12.9), we have

$$
z_i + \hat{z}_i = \begin{cases} \mathbf{v}_0^t \cdot \mathbf{T}^{\hat{\gamma}_0} \cdot \delta_0, & i = 0, \\ \mathbf{v}_i^t \cdot \mathbf{T}^{\hat{\gamma}_i - \hat{\beta}_{i-1}} \cdot \delta_{\hat{\beta}_{i-1}}, & i = 1, 2, \cdots, L-1, \end{cases}
$$

and inserting this into (13.17), then summing the result with (13.14), we obtain the relation

$$
\delta_{\hat{\beta}_i} = \begin{cases} \mathbf{T}^{\hat{\beta}_0} \cdot \delta_0 + (\mathbf{v}_0^t \cdot \mathbf{T}^{\hat{\gamma}_0} \cdot \delta_0)\mathbf{c}_0, & i = 0, \\ \mathbf{T}^{\hat{\beta}_i - \hat{\beta}_{i-1}} \cdot \delta_{\hat{\beta}_{i-1}} + (\mathbf{v}_i^t \cdot \mathbf{T}^{\hat{\gamma}_i - \hat{\beta}_{i-1}} \cdot \delta_{\hat{\beta}_{i-1}})\mathbf{c}_i, \\ \qquad\qquad i = 1, 2, \cdots, L-1, \end{cases}
$$

which can be rewritten in the form

$$
\delta_{\hat{\beta}_i} = \begin{cases} (\mathbf{T}^{\hat{\beta}_0} + \mathbf{c}_0 \cdot \mathbf{v}_0^t \cdot \mathbf{T}^{\hat{\gamma}_0}) \cdot \delta_0, & i = 0, \\ (\mathbf{T}^{\hat{\beta}_i - \hat{\beta}_{i-1}} + \mathbf{c}_i \cdot \mathbf{v}_i^t \cdot \mathbf{T}^{\hat{\gamma}_i - \hat{\beta}_{i-1}}) \cdot \delta_{\hat{\beta}_{i-1}}, & i = 1, 2, \cdots, L-1. \end{cases}
$$

Applying this repeatedly to the basic relation in (12.10), we obtain the correction matrix Λ in (13.12). ∎

For the synchronization, the correction matrix Λ in (13.12) should be a zero matrix. The following lemma and the theorem that follows describe how to choose the correction vectors $\mathbf{c}_i$'s to make the correction matrix zero.

LEMMA 13.9 *Let* Λ_i, $i = 1, 2, \cdots, L-1$, *denote the* $L \times L$ *matrix*

$$
\Lambda_i \equiv (\mathbf{T}^{\hat{\beta}_{L-1} - \hat{\beta}_{L-2}} + \mathbf{c}_{L-1} \cdot \mathbf{v}_{L-1}^t \cdot \mathbf{T}^{\hat{\gamma}_{L-1} - \hat{\beta}_{L-2}}) \cdots
$$
$$
\cdot (\mathbf{T}^{\hat{\beta}_i - \hat{\beta}_{i-1}} + \mathbf{c}_i \cdot \mathbf{v}_i^t \cdot \mathbf{T}^{\hat{\gamma}_i - \hat{\beta}_{i-1}}). \tag{13.19}
$$

Then, we have the relations

$$
\begin{cases} \Lambda = \Lambda_1 \cdot (\mathbf{T}^{\hat{\beta}_0} + \mathbf{c}_0 \cdot \mathbf{v}_0^t \cdot \mathbf{T}^{\hat{\gamma}_0}), \\ \Lambda_i = \Lambda_{i+1} \cdot (\mathbf{T}^{\hat{\beta}_i - \hat{\beta}_{i-1}} + \mathbf{c}_i \cdot \mathbf{v}_i^t \cdot \mathbf{T}^{\hat{\gamma}_i - \hat{\beta}_{i-1}}), & i = 1, 2, \cdots, L-2, \end{cases}
$$
$$
\tag{13.20}
$$

$$
\Lambda_i \cdot \mathbf{T}^{\hat{\beta}_{i-1}} \cdot \tilde{\mathbf{V}}^{-1} \cdot \mathbf{e}_j = \begin{cases} \mathbf{T}^{\hat{\beta}_{L-1}} \cdot \tilde{\mathbf{V}}^{-1} \cdot \mathbf{e}_j, & 0 \le j < i \le L-1, \\ \Lambda \cdot \tilde{\mathbf{V}}^{-1} \cdot \mathbf{e}_j, & 1 \le i \le j \le L-1, \end{cases} \tag{13.21}
$$

$$
\Lambda \cdot \tilde{\mathbf{V}}^{-1} \cdot \mathbf{e}_i = \begin{cases} \Lambda_{i+1} \cdot (\mathbf{T}^{\hat{\beta}_i} \cdot \tilde{\mathbf{V}}^{-1} \cdot \mathbf{e}_i + \mathbf{c}_i), & i = 0, 1, \cdots, L-2, \\ \mathbf{T}^{\hat{\beta}_{L-1}} \cdot \tilde{\mathbf{V}}^{-1} \cdot \mathbf{e}_{L-1} + \mathbf{c}_{L-1}, & i = L-1. \end{cases}
$$
$$
\tag{13.22}
$$

Proof : This lemma can be proved in a manner similar to the proof of Lemma 12.18. ∎

THEOREM 13.10 *For the descrambler PSRG with the state transition matrix* $\mathbf{T}$ *in (13.1a) and the generating vectors* $\mathbf{h}_j$, $j = 0, 1, \cdots, N\text{-}1$, *in (13.1c), the correction matrix* Λ *in (13.12) becomes a zero matrix, if and only if the correction vector* $\mathbf{c}_i$ *has, for the correction times* $\hat{\beta}_i$'s, *the expression*

$$
\mathbf{c}_i = \begin{cases} \mathbf{T}^{\hat{\beta}_i} \cdot \tilde{\mathbf{V}}^{-1} \cdot (\mathbf{e}_i + \sum_{j=i+1}^{L-1} u_{i,j} \mathbf{e}_j), & i = 0, 1, \cdots, L\text{-}2, \\ \mathbf{T}^{\hat{\beta}_{L-1}} \cdot \tilde{\mathbf{V}}^{-1} \cdot \mathbf{e}_{L-1}, & i = L\text{-}1, \end{cases} \tag{13.23}
$$

where $u_{i,j}$ is either 0 or 1 for $i = 0, 1, \cdots, L\text{-}2$ and $j = i+1, i+2, \cdots, L\text{-}1$.

Proof : This theorem can be proved by employing Lemma 13.9 in a similar manner to the proof of Theorem 12.19. ∎

According to the theorem, the correction vectors c_i's in (13.23) make the correction matrix Λ in (13.12) zero for arbitrarily chosen correction times $\hat{\beta}_i$'s. Therefore, to realize an immediate correction, we can select correction times $\hat{\beta}_i$'s such that $\hat{\beta}_i = \hat{\gamma}_i + 1$, and then select correction vectors c_i's using (13.23).

EXAMPLE 13.5 For the serial scrambling sequence $\{ s_k \}$ in the scrambling space $V[x^4 + x + 1]$ whose sampling times are $\alpha_0 = 1$, $\alpha_1 = 3$, $\alpha_2 = 5$ and $\alpha_3 = 7$, and whose sample transmission times are $\gamma_0 = 1$, $\gamma_1 = 2$, $\gamma_2 = 5$ and $\gamma_3 = 6$, we consider the correction vectors c_i's of the PSRG in Fig. 13.4 for an immediate correction. Since the parallel sample transmission times are $\hat{\gamma}_0 = 0$, $\hat{\gamma}_1 = 1$, $\hat{\gamma}_2 = 2$ and $\hat{\gamma}_3 = 3$ (refer to Example 13.3), this corresponds to the single correction case, and the correction times become $\hat{\beta}_0 = 1$, $\hat{\beta}_1 = 2$, $\hat{\beta}_2 = 3$ and $\hat{\beta}_3 = 4$. Inserting (12.4) and (13.2) into (13.8), we get

$$
\tilde{V} = \begin{bmatrix} 1 & 0 & 1 & 0 \\ 0 & 1 & 0 & 1 \\ 1 & 1 & 1 & 0 \\ 0 & 1 & 1 & 1 \end{bmatrix},
\tag{13.24}
$$

and inserting this, along with (13.3a) and the above correction times $\hat{\beta}_i$'s, to (13.23), we obtain the correction vectors

$$
\begin{aligned}
c_0 &= [\,1\,0\,1\,0\,]^t, [\,1\,1\,0\,1\,]^t, [\,0\,0\,0\,1\,]^t, [\,1\,1\,1\,1\,]^t, \\
&\quad [\,0\,1\,1\,0\,]^t, [\,1\,0\,0\,0\,]^t, [\,0\,1\,0\,0\,]^t, [\,0\,0\,1\,1\,]^t, \\
c_1 &= [\,1\,1\,1\,1\,]^t, [\,1\,0\,0\,0\,]^t, [\,0\,1\,0\,0\,]^t, [\,0\,0\,1\,1\,]^t, \\
c_2 &= [\,1\,1\,1\,1\,]^t, [\,1\,0\,0\,0\,]^t, \\
c_3 &= [\,1\,1\,1\,1\,]^t,
\end{aligned}
$$

From these, we can select the common correction vectors $c_0 = c_1 = c_2 = c_3 = [\,1\,1\,1\,1\,]^t$, thus obtaining the parallel descrambler depicted in Fig. 13.6. ♣

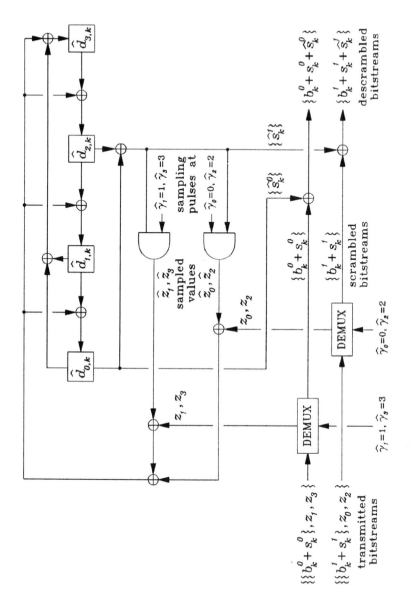

Figure 13.6. An example of parallel descrambler employing the concurrent parallel sampling and the immediate single correction. This parallel descrambler can be used in conjunction with the parallel scrambler in Fig. 13.4.

13.4.2 Double Correction

In the double correction case, two samples are transmitted at the same parallel sample transmission time. In general, the time slots for sample conveyance are repeated frame by frame, so we can assume that the sample transmission times are uniformly distributed. The even-indexed and the odd-indexed sampling vectors are identical in this case, that is,

$$\hat{\gamma}_{2i} = \hat{\gamma}_{2i+1} = i\hat{\gamma}, \quad i = 0, 1, \cdots, J-1, \tag{13.25a}$$

$$\begin{cases} \mathbf{v}_0 = \mathbf{v}_2 = \cdots = \mathbf{v}_{2J-2}, \\ \mathbf{v}_1 = \mathbf{v}_3 = \cdots = \mathbf{v}_{2J-1}, \end{cases} \tag{13.25b}$$

where J is $L/2$ for an even L and $(L+1)/2$ for an odd L. We also assume for an efficient correction that the correction times are uniformly distributed, and the even-indexed and the odd-indexed correction vectors are identical, that is,

$$\hat{\beta}_{2i} = \hat{\beta}_{2i+1} = i\hat{\gamma} + \hat{\beta}, \quad i = 0, 1, \cdots, J-1, \tag{13.26a}$$

$$\begin{cases} \mathbf{c}_0 = \mathbf{c}_2 = \cdots = \mathbf{c}_{2J-2}, \\ \mathbf{c}_1 = \mathbf{c}_3 = \cdots = \mathbf{c}_{2J-1}. \end{cases} \tag{13.26b}$$

Using these parameters, we can depict the synchronization timing for the double correction case as shown in Fig. 13.7. It should be noted that in case L is even, L times of sampling and correction occur for synchronization, while in case L is odd, $(L+1)$ times sampling and correction occur.

The correction matrix Λ of the double correction case is determined by the following theorem.

THEOREM 13.11 *For the descrambler PSRG with the state transition matrix* $\mathbf{T}$ *in (13.1a) and the generating vector* $\mathbf{h}_j$, $j = 0, 1, \cdots, N\text{-}1$, *in (13.1c), the correction matrix* Λ *of the double correction is*

$$\Lambda = (\mathbf{T}^{\hat{\gamma}} + \mathbf{c}_0 \cdot \mathbf{v}_0^t \cdot \mathbf{T}^{\hat{\gamma}-\hat{\beta}} + \mathbf{c}_1 \cdot \mathbf{v}_1^t \cdot \mathbf{T}^{\hat{\gamma}-\hat{\beta}})^{J-1} \cdot (\mathbf{T}^{\hat{\beta}} + \mathbf{c}_0 \cdot \mathbf{v}_0^t + \mathbf{c}_1 \cdot \mathbf{v}_1^t). \tag{13.27}$$

Proof : For the synchronization state vector $\tilde{\mathbf{d}}_k$, we have, by (13.5), the relations

$$\tilde{\mathbf{d}}_{i\hat{\gamma}} = \begin{cases} \tilde{\mathbf{d}}_0, & i = 0, \\ \mathbf{T}^{\hat{\gamma}-\hat{\beta}} \cdot \tilde{\mathbf{d}}_{(i-1)\hat{\gamma}+\hat{\beta}}, & i = 1, 2, \cdots, J-1, \end{cases} \tag{13.28}$$

$$\tilde{\mathbf{d}}_{i\hat{\gamma}+\hat{\beta}} = \begin{cases} \mathbf{T}^{\hat{\beta}} \cdot \tilde{\mathbf{d}}_0, & i = 0, \\ \mathbf{T}^{\hat{\gamma}} \cdot \tilde{\mathbf{d}}_{(i-1)\hat{\gamma}+\hat{\beta}}, & i = 1, 2, \cdots, J-1, \end{cases} \tag{13.29}$$

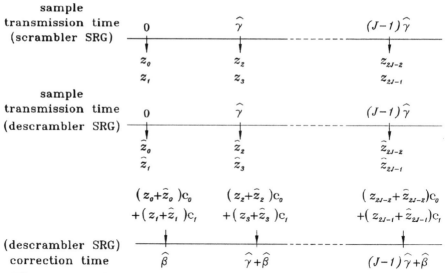

Figure 13.7. Synchronization timing for the double correction case.

and inserting (13.28) and (13.25) into (13.10), we obtain

$$
z_{2i} = \begin{cases} \mathbf{v}_0^t \cdot \tilde{\mathbf{d}}_0, & i = 0, \\ \mathbf{v}_0^t \cdot \mathbf{T}^{\hat{\gamma}-\hat{\beta}} \cdot \tilde{\mathbf{d}}_{(i-1)\hat{\gamma}+\hat{\beta}}, & i = 1, 2, \cdots, J-1, \end{cases} \quad (13.30a)
$$

$$
z_{2i+1} = \begin{cases} \mathbf{v}_1^t \cdot \tilde{\mathbf{d}}_0, & i = 0, \\ \mathbf{v}_1^t \cdot \mathbf{T}^{\hat{\gamma}-\hat{\beta}} \cdot \tilde{\mathbf{d}}_{(i-1)\hat{\gamma}+\hat{\beta}}, & i = 1, 2, \cdots, J-1. \end{cases} \quad (13.30b)
$$

For the state vector $\hat{\mathbf{d}}_k$ of the descrambler PSRG, noting that the double correction process corresponds to the sum $(z_{2i} + \hat{z}_{2i})\mathbf{c}_0 + (z_{2i+1} + \hat{z}_{2i+1})\mathbf{c}_1$, we have the relations

$$
\hat{\mathbf{d}}_{i\hat{\gamma}} = \begin{cases} \hat{\mathbf{d}}_0, & i = 0, \\ \mathbf{T}^{\hat{\gamma}-\hat{\beta}} \cdot \hat{\mathbf{d}}_{(i-1)\hat{\gamma}+\hat{\beta}}, & i = 1, 2, \cdots, J-1, \end{cases} \quad (13.31)
$$

$$
\hat{\mathbf{d}}_{i\hat{\gamma}+\hat{\beta}} = \begin{cases} \mathbf{T}^{\hat{\beta}} \cdot \hat{\mathbf{d}}_0 + (z_0 + \hat{z}_0)\mathbf{c}_0 + (z_1 + \hat{z}_1)\mathbf{c}_1, & i = 0, \\ \mathbf{T}^{\hat{\gamma}} \cdot \hat{\mathbf{d}}_{(i-1)\hat{\gamma}+\hat{\beta}} + (z_{2i} + \hat{z}_{2i})\mathbf{c}_0 + (z_{2i+1} + \hat{z}_{2i+1})\mathbf{c}_1, \\ \quad i = 1, 2, \cdots, J-1, \end{cases} \quad (13.32)
$$

and from (13.31) and (13.25) we obtain

$$
\hat{z}_{2i} = \begin{cases} \mathbf{v}_0^t \cdot \hat{\mathbf{d}}_0, & i = 0, \\ \mathbf{v}_0^t \cdot \mathbf{T}^{\hat{\gamma}-\hat{\beta}} \cdot \hat{\mathbf{d}}_{(i-1)\hat{\gamma}+\hat{\beta}}, & i = 1, 2, \cdots, J-1, \end{cases} \quad (13.33a)
$$

$$\hat{z}_{2i+1} = \begin{cases} \mathbf{v}_1^t \cdot \hat{\mathbf{d}}_0, & i = 0, \\ \mathbf{v}_1^t \cdot \mathbf{T}^{\hat{\gamma}-\hat{\beta}} \cdot \hat{\mathbf{d}}_{(i-1)\hat{\gamma}+\hat{\beta}}, & i = 1, 2, \cdots, J-1. \end{cases} \quad (13.33b)$$

By (13.30), (13.33) and (12.9), we have

$$z_{2i} + \hat{z}_{2i} = \begin{cases} \mathbf{v}_0^t \cdot \delta_0, & i = 0, \\ \mathbf{v}_0^t \cdot \mathbf{T}^{\hat{\gamma}-\hat{\beta}} \cdot \delta_{(i-1)\hat{\gamma}+\hat{\beta}}, & i = 1, 2, \cdots, J-1, \end{cases}$$

$$z_{2i+1} + \hat{z}_{2i+1} = \begin{cases} \mathbf{v}_1^t \cdot \delta_0, & i = 0, \\ \mathbf{v}_1^t \cdot \mathbf{T}^{\hat{\gamma}-\hat{\beta}} \cdot \delta_{(i-1)\hat{\gamma}+\hat{\beta}}, & i = 1, 2, \cdots, J-1, \end{cases}$$

and inserting these into (13.32), then summing the result with (13.29), we get

$$\delta_{i\hat{\gamma}+\hat{\beta}} = \begin{cases} (\mathbf{T}^{\hat{\beta}} + \mathbf{c}_0 \cdot \mathbf{v}_0^t + \mathbf{c}_1 \cdot \mathbf{v}_1^t) \cdot \delta_0, & i = 0, \\ (\mathbf{T}^{\hat{\gamma}} + \mathbf{c}_0 \cdot \mathbf{v}_0^t \cdot \mathbf{T}^{\hat{\gamma}-\hat{\beta}} + \mathbf{c}_1 \cdot \mathbf{v}_1^t \cdot \mathbf{T}^{\hat{\gamma}-\hat{\beta}}) \cdot \delta_{(i-1)\hat{\gamma}+\hat{\beta}}, \\ \qquad i = 1, 2, \cdots, J-1. \end{cases}$$

Applying this relation to the basic relation in (12.10), we obtain the correction matrix Λ in (13.27). ■

The following two theorems describe how to choose the correction vectors $\mathbf{c}_0$ and $\mathbf{c}_1$ to make the correction matrix Λ in (13.27) zero respectively for the even-lengthed and the odd-lengthed PSRGs.

THEOREM 13.12A *For the descrambler PSRG with an even length L, the state transition matrix $\mathbf{T}$ in (13.1a) and the generating vector $\mathbf{h}_j$, $j = 0, 1, \cdots, N\text{-}1$, in (13.1c), the correction matrix Λ in (13.27) becomes a zero matrix, if and only if the correction vectors $\mathbf{c}_0$ and $\mathbf{c}_1$ have, for a correction delay $\hat{\beta}$, the expression*

$$\begin{cases} \mathbf{c}_0 = \mathbf{T}^{(J-1)\hat{\gamma}+\hat{\beta}} \cdot \tilde{\mathbf{V}}^{-1} \cdot \mathbf{e}_{L-2}, \\ \mathbf{c}_1 = \mathbf{T}^{(J-1)\hat{\gamma}+\hat{\beta}} \cdot \tilde{\mathbf{V}}^{-1} \cdot \mathbf{e}_{L-1}. \end{cases} \quad (13.34)$$

Proof : We first prove the "if" part of the theorem. Since $\mathbf{T}$ is nonsingular, we can rewrite (13.27) as $\Lambda = \Lambda_*^J \cdot \mathbf{T}^{\hat{\beta}-\hat{\gamma}}$, where

$$\Lambda_* = \mathbf{T}^{\hat{\gamma}} + \mathbf{c}_0 \cdot \mathbf{v}_0^t \cdot \mathbf{T}^{\hat{\gamma}-\hat{\beta}} + \mathbf{c}_1 \cdot \mathbf{v}_1^t \cdot \mathbf{T}^{\hat{\gamma}-\hat{\beta}}. \quad (13.35)$$

Inserting (13.34) to (13.35), we obtain

$$\Lambda_* = \mathbf{T}^{(J-1)\hat{\gamma}+\hat{\beta}} \cdot \tilde{\mathbf{V}}^{-1} \cdot \mathbf{A}_e \cdot \tilde{\mathbf{V}} \cdot \mathbf{T}^{-(J-1)\hat{\gamma}-\hat{\beta}}, \quad (13.36)$$

where

$$\mathbf{A}_e = \tilde{\mathbf{V}} \cdot \mathbf{T}^{\hat{\gamma}} \cdot \tilde{\mathbf{V}}^{-1} + \mathbf{e}_{L-2} \cdot \mathbf{v}_0^t \cdot \mathbf{T}^{J\hat{\gamma}} \cdot \tilde{\mathbf{V}}^{-1} + \mathbf{e}_{L-1} \cdot \mathbf{v}_1^t \cdot \mathbf{T}^{J\hat{\gamma}} \cdot \tilde{\mathbf{V}}^{-1}. \quad (13.37)$$

But by (13.9) along with (13.25) and an even L, we have the relations

$$\begin{cases} \mathbf{v}_0^t \cdot \mathbf{T}^{i\hat{\gamma}} \cdot \tilde{\mathbf{V}}^{-1} = \mathbf{e}_{2i}^t, & i = 0, 1, \cdots, J-1, \\ \mathbf{v}_1^t \cdot \mathbf{T}^{i\hat{\gamma}} \cdot \tilde{\mathbf{V}}^{-1} = \mathbf{e}_{2i+1}^t, & i = 0, 1, \cdots, J-1. \end{cases} \quad (13.38)$$

So, (13.37) reduces to $\mathbf{A}_e = [\,0\ 0\ \mathbf{e}_0\ \mathbf{e}_1\ \cdots\ \mathbf{e}_{L-3}\,]$ due to this. Therefore, $\mathbf{A}_e^J = 0$, and $\Lambda_*^J = 0$ by (13.36), and hence $\Lambda = 0$.

Now we prove the "only if" part. We evaluate $\Lambda \cdot \tilde{\mathbf{V}}^{-1} \cdot \mathbf{e}_{L-2}$ by (13.27) and (13.38) repeatedly. Then, we finally obtain the relation $\Lambda \cdot \tilde{\mathbf{V}}^{-1} \cdot \mathbf{e}_{L-2} = \mathbf{T}^{(J-1)\hat{\gamma}+\hat{\beta}} \cdot \tilde{\mathbf{V}}^{-1} \cdot \mathbf{e}_{L-2} + \mathbf{c}_0$. In a similar manner, applying (13.27) and (13.38) repeatedly to $\Lambda \cdot \tilde{\mathbf{V}}^{-1} \cdot \mathbf{e}_{L-1}$, we obtain $\Lambda \cdot \tilde{\mathbf{V}}^{-1} \cdot \mathbf{e}_{L-1} = \mathbf{T}^{(J-1)\hat{\gamma}+\hat{\beta}} \cdot \tilde{\mathbf{V}}^{-1} \cdot \mathbf{e}_{L-1} + \mathbf{c}_1$. Therefore, if $\Lambda = 0$, then $\mathbf{c}_0 = \mathbf{T}^{(J-1)\hat{\gamma}+\hat{\beta}} \cdot \tilde{\mathbf{V}}^{-1} \cdot \mathbf{e}_{L-2}$ and $\mathbf{c}_1 = \mathbf{T}^{(J-1)\hat{\gamma}+\hat{\beta}} \cdot \tilde{\mathbf{V}}^{-1} \cdot \mathbf{e}_{L-1}$. ∎

THEOREM 13.12B *For the descrambler PSRG with an odd length L, the state transition matrix $\mathbf{T}$ in (13.1a) and the generating vector $\mathbf{h}_j$, $j = 0, 1, \cdots, N\text{-}1$, in (13.1c), the correction matrix Λ in (13.27) becomes a zero matrix, if and only if the correction vectors $\mathbf{c}_0$ and $\mathbf{c}_1$ have, for a correction delay $\hat{\beta}$, the expression*

$$\begin{cases} \mathbf{c}_0 = \mathbf{T}^{(J-1)\hat{\gamma}+\hat{\beta}} \cdot \tilde{\mathbf{V}}^{-1} \cdot \mathbf{e}_{L-1} + a_{L-1}\mathbf{c}_1, \\ \mathbf{c}_1 = \mathbf{T}^{(J-1)\hat{\gamma}+\hat{\beta}} \cdot \tilde{\mathbf{V}}^{-1} \cdot \{\, \mathbf{e}_0 + \mathbf{e}_1 + a_0\mathbf{e}_1 + u(a_0\mathbf{e}_1 + a_1\mathbf{e}_0)\,\}, \end{cases} \quad (13.39)$$

where

$$[\,a_0\ a_1\ \cdots\ a_{L-1}\,] \equiv \mathbf{v}_1^t \cdot \mathbf{T}^{(J-1)\hat{\gamma}} \cdot \tilde{\mathbf{V}}^{-1}, \quad (13.40)$$

and u is either 0 or 1.

Proof : To prove the theorem, we first show that both of a_0 and a_1 can not be 0, that is, $(a_0, a_1) = (1,0)$, $(0,1)$ or $(1,1)$. From (13.40), we have the relation $\mathbf{v}_1^t \cdot \mathbf{T}^{(J-1)\hat{\gamma}} = [\,a_0\ a_1\ \cdots\ a_{L-1}\,] \cdot \tilde{\mathbf{V}}$, and inserting (13.7) and (13.25) for an odd L into this, we obtain

$$\mathbf{v}_1^t \cdot \mathbf{T}^{(J-1)\hat{\gamma}} = \sum_{i=0}^{J-1} a_{2i}\mathbf{v}_0^t \cdot \mathbf{T}^{i\hat{\gamma}} + \sum_{i=0}^{J-2} a_{2i+1}\mathbf{v}_1^t \cdot \mathbf{T}^{i\hat{\gamma}}.$$

If $a_0 = a_1 = 0$, the $L-2$ vectors $\mathbf{v}_0^t \cdot \mathbf{T}^{i\hat{\gamma}}$, $\mathbf{v}_1^t \cdot \mathbf{T}^{i\hat{\gamma}}$, $i = 1, 2, \cdots, J-1$ are linearly dependent, which means that $\tilde{\mathbf{V}}$ is singular. Therefore, (a_0, a_1) $\neq (0,0)$.

Now we prove the "if" part of the theorem. Since $\mathbf{T}$ is nonsingular, we can rewrite (13.27) as $\Lambda = \Lambda_*^J \cdot \mathbf{T}^{\hat{\beta}-\hat{\gamma}}$ for the Λ_* in (13.35). Due to the nonsingular $\tilde{\mathbf{V}}$ and (13.39), Λ_* can be rewritten as

$$\Lambda_* = \mathbf{T}^{(J-1)\hat{\gamma}+\hat{\beta}} \cdot \tilde{\mathbf{V}}^{-1} \cdot \mathbf{A}_o \cdot \tilde{\mathbf{V}} \cdot \mathbf{T}^{-(J-1)\hat{\gamma}-\hat{\beta}} \tag{13.41}$$

for

$$\mathbf{A}_o = \tilde{\mathbf{V}} \cdot \mathbf{T}^{\hat{\gamma}} \cdot \tilde{\mathbf{V}}^{-1} + \mathbf{e}_{L-1} \cdot \mathbf{v}_0^t \cdot \mathbf{T}^{J\hat{\gamma}} \cdot \tilde{\mathbf{V}}^{-1} + $$
$$\{ \mathbf{e}_0 + \mathbf{e}_1 + a_0\mathbf{e}_1 + u(a_0\mathbf{e}_1 + a_1\mathbf{e}_0) \} \cdot$$
$$(a_{L-1}\mathbf{v}_0^t \cdot \mathbf{T}^{J\hat{\gamma}} \cdot \tilde{\mathbf{V}}^{-1} + \mathbf{v}_1^t \cdot \mathbf{T}^{J\hat{\gamma}} \cdot \tilde{\mathbf{V}}^{-1}). \tag{13.42}$$

Since $a_{L-1} = \mathbf{v}_1 \cdot \mathbf{T}^{(J-1)\hat{\gamma}} \cdot \tilde{\mathbf{V}}^{-1} \cdot \mathbf{e}_{L-1}$ due to (13.40), $\mathbf{A}_o$ in (13.42) reduces to

$$\mathbf{A}_o = [\mathbf{I} + \{ \mathbf{e}_0 + \mathbf{e}_1 + a_0\mathbf{e}_1 + u(a_0\mathbf{e}_1 + a_1\mathbf{e}_0) \} \cdot \mathbf{v}_1^t \cdot \mathbf{T}^{(J-1)\hat{\gamma}} \cdot \tilde{\mathbf{V}}^{-1}] \cdot$$
$$[\tilde{\mathbf{V}} \cdot \mathbf{T}^{\hat{\gamma}} \cdot \tilde{\mathbf{V}}^{-1} + \mathbf{e}_{L-1} \cdot \mathbf{v}_0^t \cdot \mathbf{T}^{J\hat{\gamma}} \cdot \tilde{\mathbf{V}}^{-1}]. \tag{13.43}$$

By (13.41), $\mathbf{A}_o^J = 0$ implies $\Lambda_*^J = 0$, which again implies $\Lambda = 0$. Therefore, to complete the proof, it suffices to show that $\mathbf{A}_o^J = 0$.

We denote by $\mathbf{A}_1$ and $\mathbf{A}_2$ respectively the first and the second terms of $\mathbf{A}_o$ in (13.43). By (13.9) along with (13.25), we have the relations

$$\begin{cases} \mathbf{v}_0^t \cdot \mathbf{T}^{i\hat{\gamma}} \cdot \tilde{\mathbf{V}}^{-1} = \mathbf{e}_{2i}^t, & i = 0, 1, \cdots, J-1, \\ \mathbf{v}_1^t \cdot \mathbf{T}^{i\hat{\gamma}} \cdot \tilde{\mathbf{V}}^{-1} = \mathbf{e}_{2i+1}^t, & i = 0, 1, \cdots, J-2, \end{cases} \tag{13.44}$$

for an odd L. So $\mathbf{A}_2$ reduces to

$$\mathbf{A}_2 = [0 \; 0 \; \mathbf{e}_0 \; \mathbf{e}_1 \; \cdots \; \mathbf{e}_{L-3}] + \mathbf{e}_{L-2} \cdot \mathbf{v}_1^t \cdot \mathbf{T}^{(J-1)\hat{\gamma}} \cdot \tilde{\mathbf{V}}^{-1}. \tag{13.45}$$

But $\mathbf{e}_{L-2} \cdot \mathbf{v}_1^t \cdot \mathbf{T}^{(J-1)\hat{\gamma}} \cdot \tilde{\mathbf{V}}^{-1} \cdot \{ \mathbf{e}_0 + \mathbf{e}_1 + a_0\mathbf{e}_1 + u(a_0\mathbf{e}_1 + a_1\mathbf{e}_0) \} = \mathbf{e}_{L-2}\{a_0 + a_1 + a_0a_1 + u(a_0a_1 + a_1a_0)\} = \mathbf{e}_{L-2}$ due to (13.40) and the relation $a_0 + a_1 + a_0a_1 = 1$. So we have $\mathbf{A}_2 \cdot \mathbf{A}_1 = [0 \; 0 \; \mathbf{e}_0 \; \mathbf{e}_1 \; \cdots \; \mathbf{e}_{L-3}]$, and hence $\mathbf{A}_o^J = \mathbf{A}_1 \cdot (\mathbf{A}_2 \cdot \mathbf{A}_1)^{J-1} \cdot \mathbf{A}_2 = \mathbf{A}_1 \cdot [0 \; 0 \; \cdots \; 0 \; \mathbf{e}_0] \cdot \mathbf{A}_2$, which become 0 due to (13.45).

Finally we prove the "only if" part of the theorem. To do this, we evaluate $\Lambda \cdot \tilde{\mathbf{V}}^{-1} \cdot \mathbf{e}_{L-1}$, $\Lambda \cdot \tilde{\mathbf{V}}^{-1} \cdot \mathbf{e}_0$ and $\Lambda \cdot \tilde{\mathbf{V}}^{-1} \cdot \mathbf{e}_1$ by applying (13.27) and (13.44) repeatedly. Then, we finally get

$$
\begin{aligned}
\Lambda \cdot \tilde{\mathbf{V}}^{-1} \cdot \mathbf{e}_{L-1} &= \mathbf{T}^{(J-1)\hat{\gamma}+\hat{\beta}} \cdot \tilde{\mathbf{V}}^{-1} \cdot \mathbf{e}_{L-1} \\
&\quad + \mathbf{c}_0 + \mathbf{c}_1(\mathbf{v}_1^t \cdot \mathbf{T}^{(J-1)\hat{\gamma}} \cdot \tilde{\mathbf{V}}^{-1} \cdot \mathbf{e}_{L-1}), \quad (13.46a) \\
\Lambda \cdot \tilde{\mathbf{V}}^{-1} \cdot \mathbf{e}_0 &= \mathbf{T}^{(J-1)\hat{\gamma}+\hat{\beta}} \cdot \tilde{\mathbf{V}}^{-1} \cdot \mathbf{e}_0 \\
&\quad + \mathbf{c}_1(\mathbf{v}_1^t \cdot \mathbf{T}^{(J-1)\hat{\gamma}} \cdot \tilde{\mathbf{V}}^{-1} \cdot \mathbf{e}_0) + \Lambda_*^{J-1} \cdot \mathbf{c}_0, \quad (13.46b) \\
\Lambda \cdot \tilde{\mathbf{V}}^{-1} \cdot \mathbf{e}_1 &= \mathbf{T}^{(J-1)\hat{\gamma}+\hat{\beta}} \cdot \tilde{\mathbf{V}}^{-1} \cdot \mathbf{e}_1 \\
&\quad + \mathbf{c}_1(\mathbf{v}_1^t \cdot \mathbf{T}^{(J-1)\hat{\gamma}} \cdot \tilde{\mathbf{V}}^{-1} \cdot \mathbf{e}_1) + \Lambda_*^{J-1} \cdot \mathbf{c}_1. \quad (13.46c)
\end{aligned}
$$

If $\Lambda = 0$, by (13.40) we have the relations

$$
\begin{aligned}
\mathbf{c}_0 &= \mathbf{T}^{(J-1)\hat{\gamma}+\hat{\beta}} \cdot \tilde{\mathbf{V}}^{-1} \cdot \mathbf{e}_{L-1} + a_{L-1}\mathbf{c}_1, & (13.47a) \\
a_0\mathbf{c}_1 &= \mathbf{T}^{(J-1)\hat{\gamma}+\hat{\beta}} \cdot \tilde{\mathbf{V}}^{-1} \cdot \mathbf{e}_0 + \Lambda_*^{J-1} \cdot \mathbf{c}_0, & (13.47b) \\
a_1\mathbf{c}_1 &= \mathbf{T}^{(J-1)\hat{\gamma}+\hat{\beta}} \cdot \tilde{\mathbf{V}}^{-1} \cdot \mathbf{e}_1 + \Lambda_*^{J-1} \cdot \mathbf{c}_1. & (13.47c)
\end{aligned}
$$

These two sets of equations yield (13.39) due to the following lemma. ■

LEMMA 13.13 *If $\Lambda = 0$,*

$$
\Lambda_*^{J-1} \cdot \mathbf{c}_0 = u\mathbf{T}^{(J-1)\hat{\gamma}+\hat{\beta}} \cdot \tilde{\mathbf{V}}^{-1} \cdot (a_1\mathbf{e}_0 + \mathbf{e}_1) \tag{13.48a}
$$

for $a_0 = 1$, and

$$
\Lambda_*^{J-1} \cdot \mathbf{c}_1 = u\mathbf{T}^{(J-1)\hat{\gamma}+\hat{\beta}} \cdot \tilde{\mathbf{V}}^{-1} \cdot (\mathbf{e}_0 + a_0\mathbf{e}_1) \tag{13.48b}
$$

for $a_1 = 1$, where u is either 0 or 1.

Proof : We will prove (13.48a) only, as (13.48b) can be proved in a similar manner.

We define $\tilde{\mathbf{V}}_0$ to be the $L \times L$ matrix

$$
\tilde{\mathbf{V}}_0 =
\begin{bmatrix}
\mathbf{v}_1^t \\
\mathbf{v}_0^t \cdot \mathbf{T}^{\hat{\gamma}} \\
\mathbf{v}_1^t \cdot \mathbf{T}^{\hat{\gamma}} \\
\vdots \\
\mathbf{v}_0^t \cdot \mathbf{T}^{(J-1)\hat{\gamma}} \\
\mathbf{v}_1^t \cdot \mathbf{T}^{(J-1)\hat{\gamma}}
\end{bmatrix}.
\tag{13.49}
$$

Then, by (13.40) we have the relation

$$
\tilde{\mathbf{V}}_0 = \begin{bmatrix} 0 & 1 & 0 & \cdots & 0 \\ 0 & 0 & 1 & \cdots & 0 \\ \vdots & \vdots & \vdots & \ddots & \vdots \\ 0 & 0 & 0 & \cdots & 1 \\ 1 & a_1 & a_2 & \cdots & a_{L-1} \end{bmatrix} \cdot \tilde{\mathbf{V}}. \tag{13.50}
$$

Therefore, to prove (13.48a), it suffices to show that $\Lambda_*^{J-1} \cdot \mathbf{c}_0 = u\mathbf{T}^{(J-1)\hat{\gamma}+\hat{\beta}} \cdot \tilde{\mathbf{V}}_0^{-1} \cdot \mathbf{e}_0$.

From the relation $\tilde{\mathbf{V}}_0 \cdot \tilde{\mathbf{V}}_0^{-1} = \mathbf{I}$, we derive the expressions

$$
\begin{cases} \mathbf{v}_0^t \cdot \mathbf{T}^{i\hat{\gamma}} \cdot \tilde{\mathbf{V}}_0^{-1} = \mathbf{e}_{2i-1}^t, & i = 1, 2, \cdots, J-1, \\ \mathbf{v}_1^t \cdot \mathbf{T}^{i\hat{\gamma}} \cdot \tilde{\mathbf{V}}_0^{-1} = \mathbf{e}_{2i}^t, & i = 0, 1, \cdots, J-1, \end{cases} \tag{13.51}
$$

and applying these to $\Lambda \cdot \tilde{\mathbf{V}}_0^{-1} \cdot \mathbf{e}_{2i}$ and $\Lambda \cdot \tilde{\mathbf{V}}_0^{-1} \cdot \mathbf{e}_{2i-1}$ repeatedly, we finally get

$$
\begin{cases} \Lambda \cdot \tilde{\mathbf{V}}_0^{-1} \cdot \mathbf{e}_{2i} = \Lambda_*^{J-1-i} \cdot \mathbf{c}_1 + \mathbf{T}^{(J-1)\hat{\gamma}+\hat{\beta}} \cdot \tilde{\mathbf{V}}_0^{-1} \cdot \mathbf{e}_{2i} \\ \quad +(\mathbf{v}_0^t \cdot \tilde{\mathbf{V}}_0^{-1} \cdot \mathbf{e}_{2i})\Lambda_*^{J-1} \cdot \mathbf{c}_0, \ i = 0, 1, \cdots, J-1, \\ \Lambda \cdot \tilde{\mathbf{V}}_0^{-1} \cdot \mathbf{e}_{2i-1} = \Lambda_*^{J-1-i} \cdot \mathbf{c}_0 + \mathbf{T}^{(J-1)\hat{\gamma}+\hat{\beta}} \cdot \tilde{\mathbf{V}}_0^{-1} \cdot \mathbf{e}_{2i-1} \\ \quad +(\mathbf{v}_0^t \cdot \tilde{\mathbf{V}}_0^{-1} \cdot \mathbf{e}_{2i-1})\Lambda_*^{J-1} \cdot \mathbf{c}_0, \ i = 1, 2, \cdots, J-1. \end{cases} \tag{13.52}
$$

So, if $\Lambda = 0$, or equivalently, $\Lambda_*^J = 0$, equation (13.52) yields

$$
\begin{cases} \Lambda_*^{J-1} \cdot \mathbf{c}_1 = \Lambda_*^i \cdot \mathbf{T}^{(J-1)\hat{\gamma}+\hat{\beta}} \cdot \tilde{\mathbf{V}}_0^{-1} \cdot \mathbf{e}_{2i}, & i = 1, 2, \cdots, J-1, \\ \Lambda_*^{J-1} \cdot \mathbf{c}_0 = \Lambda_*^i \cdot \mathbf{T}^{(J-1)\hat{\gamma}+\hat{\beta}} \cdot \tilde{\mathbf{V}}_0^{-1} \cdot \mathbf{e}_{2i-1}, & i = 1, 2, \cdots, J-1, \end{cases} \tag{13.53}
$$

$$
\Lambda_*^{i+1} \cdot \mathbf{T}^{(J-1)\hat{\gamma}+\hat{\beta}} \cdot \tilde{\mathbf{V}}_0^{-1} \cdot \mathbf{e}_j = 0, \ 0 \le j < 2i+1 \le L-1, \tag{13.54}
$$

when premultiplied by Λ_*^i and Λ_*^{i+1} respectively ; and reduces to

$$
\Lambda_*^{J-1} \cdot \mathbf{c}_1 = \mathbf{T}^{(J-1)\hat{\gamma}+\hat{\beta}} \cdot \tilde{\mathbf{V}}_0^{-1} \cdot \mathbf{e}_0 + (\mathbf{v}_0^t \cdot \tilde{\mathbf{V}}_0^{-1} \cdot \mathbf{e}_0)\Lambda_*^{J-1} \cdot \mathbf{c}_0, \tag{13.55}
$$

when i is set to 0. Since $\mathbf{T}$ and $\tilde{\mathbf{V}}_0$ are both nonsingular, the vector $\Lambda_*^{J-1} \cdot \mathbf{c}_0$ can be uniquely expressed as

$$
\Lambda_*^{J-1} \cdot \mathbf{c}_0 = \sum_{j=0}^{L-1} v_j \mathbf{T}^{(J-1)\hat{\gamma}+\hat{\beta}} \cdot \tilde{\mathbf{V}}_0^{-1} \cdot \mathbf{e}_j, \tag{13.56}
$$

for the basis vectors $\mathbf{e}_j$, $j = 0, 1, \cdots, L - 1$. Therefore, it suffices to show that $v_j = 0$ for $j = 1, 2, \cdots, L - 1$, when $\Lambda = 0$. But it is obvious when $J = 1$, so we consider the case $J \geq 2$. We show this by induction.

To prove $v_{L-1} = v_{L-2} = 0$, we multiply (13.56) by Λ_*^{J-1}. Then we obtain $\Lambda_*^{2J-2} \cdot \mathbf{c}_0 = \sum_{j=0}^{L-3} v_j \Lambda_*^{J-1} \cdot \mathbf{T}^{(J-1)\hat{\gamma}+\hat{\beta}} \cdot \tilde{\mathbf{V}}_0^{-1} \cdot \mathbf{e}_j + v_{L-2}\Lambda_*^{J-1} \cdot \mathbf{T}^{(J-1)\hat{\gamma}+\hat{\beta}} \cdot \tilde{\mathbf{V}}_0^{-1} \cdot \mathbf{e}_{L-2} + v_{L-1}\Lambda_*^{J-1} \cdot \mathbf{T}^{(J-1)\hat{\gamma}+\hat{\beta}} \cdot \tilde{\mathbf{V}}_0^{-1} \cdot \mathbf{e}_{L-1}$. Applying (13.53), (13.54) and the relation $\Lambda_*^J = 0$ to this, we get the relation

$$v_{L-2}\Lambda_*^{J-1} \cdot \mathbf{c}_0 + v_{L-1}\Lambda_*^{J-1} \cdot \mathbf{c}_1 = 0. \tag{13.57}$$

Inserting (13.55) into this, we obtain $\{v_{L-1}(\mathbf{v}_0^t \cdot \tilde{\mathbf{V}}_0^{-1} \cdot \mathbf{e}_0) + v_{L-2}\}\Lambda_*^{J-1} \cdot \mathbf{c}_0 = v_{L-1}\mathbf{T}^{(J-1)\hat{\gamma}+\hat{\beta}} \cdot \tilde{\mathbf{V}}_0^{-1} \cdot \mathbf{e}_0$. If we had $v_{L-1} = 1$, then we would have the relation $\Lambda_*^{J-1} \cdot \mathbf{c}_0 = \mathbf{T}^{(J-1)\hat{\gamma}+\hat{\beta}} \cdot \tilde{\mathbf{V}}_0^{-1} \cdot \mathbf{e}_0$. But, in view of (13.56), it is a contradiction to $v_{L-1} = 1$. Therefore, $v_{L-1} = 0$, and by (13.57) $v_{L-2} = 0$. Now we assume $v_{L-1} = v_{L-2} = \cdots = v_{2k+2} = v_{2k+1} = 0$ for a $k = J - 1, J - 2, \cdots, 1$. To prove $v_{2k} = v_{2k-1} = 0$, we multiply (13.56) by Λ_*^k, and then apply (13.53), (13.54), and the relation $\Lambda_*^J = 0$. Then, we obtain $v_{2k-1}\Lambda_*^{J-1} \cdot \mathbf{c}_0 + v_{2k}\Lambda_*^{J-1} \cdot \mathbf{c}_1 = 0$. This leads to $v_{2k} = v_{2k-1} = 0$ by the same reasoning that was applied to (13.57). This completes the proof for (13.48a). ∎

EXAMPLE 13.6 For the serial scrambling sequence $\{ s_k \}$ in the scrambling space $V[x^4 + x + 1]$ (with $L = 4$) whose sampling times and sample transmission times are respectively $\alpha_0 = 1$, $\alpha_1 = 3$, $\alpha_2 = 5$, $\alpha_3 = 7$, and $\gamma_0 = 0$, $\gamma_1 = 1$, $\gamma_2 = 4$, $\gamma_3 = 5$, we realize a parallel descrambler using the PSRG in Fig. 13.2(b). By Theorem 13.2, its parallel sampling times are $\hat{\alpha}_0 = 0$, $\hat{\alpha}_1 = 1$, $\hat{\alpha}_2 = 2$, $\hat{\alpha}_3 = 3$; by Theorem 13.3, its parallel sample transmission times are $\hat{\gamma}_0 = 0$, $\hat{\gamma}_1 = 0$, $\hat{\gamma}_2 = 2$, $\hat{\gamma}_3 = 2$; and by Theorem 13.4, its sampling vectors at the concurrent sampling times $\hat{\gamma}_i$'s are

$$\begin{aligned}
\mathbf{v}_0 &= \mathbf{A}_{x^4+x+1}^{0-0} \cdot \mathbf{h}_1 = [\,1\,0\,1\,0\,]^t, \\
\mathbf{v}_1 &= \mathbf{A}_{x^4+x+1}^{1-0} \cdot \mathbf{h}_1 = [\,0\,1\,0\,1\,]^t, \\
\mathbf{v}_2 &= \mathbf{A}_{x^4+x+1}^{2-2} \cdot \mathbf{h}_1 = [\,1\,0\,1\,0\,]^t, \\
\mathbf{v}_3 &= \mathbf{A}_{x^4+x+1}^{3-2} \cdot \mathbf{h}_1 = [\,0\,1\,0\,1\,]^t,
\end{aligned}$$

since the remainder of α_i divided by $N = 2$ is $l_i = 1$ for all $i = 0, 1, 2, 3$. From the parallel sample transmission times $\hat{\gamma}_i$'s, we observe that this case corresponds to the double correction case with even L ($= 4$) and the sampling interval $\hat{\gamma} = 2$. To realize the immediate correction, we insert the correction delay $\hat{\beta} = 1$ into (13.34) along with the values $J = 2$, $\mathbf{T} = \mathbf{A}^t_{x^4+x+1}$ and $\tilde{\mathbf{V}}$ in (13.24). Then, we obtain the correction vectors

$$\begin{cases} \mathbf{c}_0 = [\,1\,1\,1\,1\,]^t, \\ \mathbf{c}_1 = [\,0\,1\,1\,1\,]^t, \end{cases}$$

From these, we can draw the resulting parallel descrambler as shown in Fig. 13.8. ♣

EXAMPLE 13.7 For the serial scrambling sequence $\{\,s_k\,\}$ in the irreducible scrambling space $V[x^5 + x^3 + 1]$ (with $L = 5$) whose sampling times and sample transmission times are $\alpha_0 = 0$, $\alpha_1 = 2$, $\alpha_2 = 4$, $\alpha_3 = 6$, $\alpha_4 = 8$, and $\gamma_0 = 0$, $\gamma_1 = 1$, $\gamma_2 = 4$, $\gamma_3 = 5$, $\gamma_4 = 8$, we realize a parallel descrambler with $N = 2$. By (11.16), we have the discrimination matrix

$$\Delta_\alpha = \begin{bmatrix} 1 & 0 & 0 & 0 & 0 \\ 0 & 0 & 1 & 0 & 0 \\ 0 & 0 & 0 & 0 & 1 \\ 0 & 1 & 0 & 0 & 1 \\ 0 & 1 & 0 & 1 & 1 \end{bmatrix},$$

which is nonsingular. Therefore, the sampling times α_i's are predictable. By Theorem 13.2, its parallel sampling times are $\hat{\alpha}_0 = 0$, $\hat{\alpha}_1 = 1$, $\hat{\alpha}_2 = 2$, $\hat{\alpha}_3 = 3$, $\hat{\alpha}_4 = 4$; and by Theorem 13.3, its parallel sample transmission times are $\hat{\gamma}_0 = 0$, $\hat{\gamma}_1 = 0$, $\hat{\gamma}_2 = 2$, $\hat{\gamma}_3 = 2$, $\hat{\gamma}_4 = 4$. We insert $\mathbf{R} = \mathbf{I}$ into (13.1a) and (13.1c) along with $N = 2$ and $\Psi_I(x) = x^5 + x^3 + 1$. Then, we obtain the PSRG with the state transition matrix $\mathbf{T}$ and the generating vectors $\mathbf{h}_0$ and $\mathbf{h}_1$ which are

$$\mathbf{T} = \begin{bmatrix} 0 & 0 & 1 & 0 & 0 \\ 0 & 0 & 0 & 1 & 0 \\ 0 & 0 & 0 & 0 & 1 \\ 1 & 0 & 0 & 1 & 0 \\ 0 & 1 & 0 & 0 & 1 \end{bmatrix},$$

$$\mathbf{h}_0 = \mathbf{e}_0,$$
$$\mathbf{h}_1 = \mathbf{e}_1,$$

By Theorem 13.4, the sampling vectors at the concurrent sampling times $\hat{\gamma}_i$'s are

$$\mathbf{v}_0 = \mathbf{T}^{\hat{\gamma}_0 - 0} \cdot \mathbf{h}_0 = [\,1\,0\,0\,0\,0\,]^t,$$

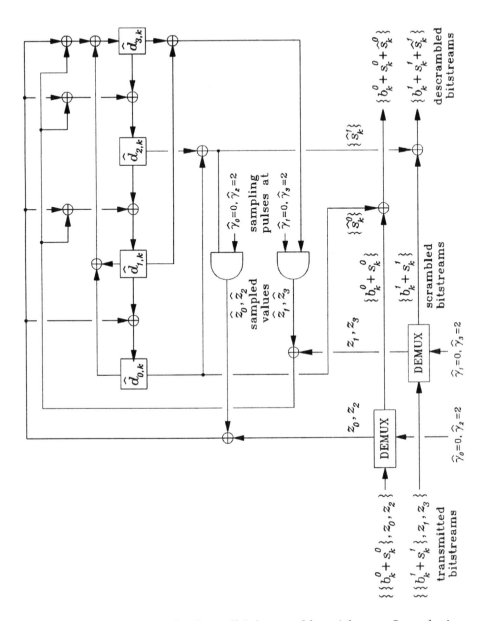

Figure 13.8. An example of parallel descrambler with even L employing the concurrent parallel sampling and the immediate double correction.

$$\mathbf{v}_1 = \mathbf{T}^{t^{1-0}} \cdot \mathbf{h}_0 = [\,0\,0\,1\,0\,0\,]^t,$$
$$\mathbf{v}_2 = \mathbf{T}^{t^{2-2}} \cdot \mathbf{h}_0 = [\,1\,0\,0\,0\,0\,]^t,$$
$$\mathbf{v}_3 = \mathbf{T}^{t^{3-2}} \cdot \mathbf{h}_0 = [\,0\,0\,1\,0\,0\,]^t,$$
$$\mathbf{v}_4 = \mathbf{T}^{t^{4-4}} \cdot \mathbf{h}_0 = [\,1\,0\,0\,0\,0\,]^t,$$

since the remainder of α_i divided by $N = 2$ is $l_i = 0$ for all $i = 0, 1, 2,$ 3, 4. From the parallel sample transmission times $\hat{\gamma}_i$'s, we observe that this case corresponds to the double correction case with odd $L\ (= 5)$ and the sampling interval $\hat{\gamma} = 2$. By (13.8), $\tilde{\mathbf{V}} = \Delta_\alpha$, and by (13.40) $a_0 = 1, a_1 = 0, a_2 = 0, a_3 = 1, a_4 = 0$. To realize the immediate correction, we insert these values and the correction delay $\hat{\beta} = 1$ into (13.39) along with the value $J = 3$. Then, we obtain the correction vectors

$$\begin{cases} \mathbf{c}_0 = [\,0\,1\,1\,0\,0\,]^t, \\ \mathbf{c}_1 = [\,1\,1\,0\,1\,1\,]^t, \end{cases} \tag{13.58a}$$

$$\begin{cases} \mathbf{c}_0 = [\,0\,1\,1\,0\,0\,]^t, \\ \mathbf{c}_1 = [\,1\,0\,1\,0\,1\,]^t. \end{cases} \tag{13.58b}$$

Taking the correction vectors in (13.58b), we can draw the resulting parallel descrambler as shown in Fig. 13.9. ♣

13.4.3 Multiple Correction

In the multiple correction case, K samples are transmitted at the same parallel sample transmission time. As in the case of the double correction, we can assume that the sample transmission times are uniformly distributed, and the sampling vectors at these sample transmission times are all identical, that is,

$$\hat{\gamma}_{Ki} = \hat{\gamma}_{Ki+1} = \cdots = \hat{\gamma}_{Ki+K-1} = i\hat{\gamma}, \ i = 0, 1, \cdots, J - 1, \tag{13.59a}$$

$$\mathbf{v}_j = \mathbf{v}_{K+j} = \cdots = \mathbf{v}_{K(J-1)+j}, \ j = 0, 1, \cdots, K - 1. \tag{13.59b}$$

For convenience, we consider the case when K is a submultiple of the length L of the PSRG only, as the other cases can be treated in a similar manner, referring to the double correction case with odd L. In this case. $J = L/K$, which is an integer. We also assume for an efficient correction that the correction times are uniformly distributed, and the correction vectors are all identical, that is,

$$\hat{\beta}_{Ki} = \hat{\beta}_{Ki+1} = \cdots = \hat{\beta}_{Ki+K-1} = i\hat{\gamma} + \hat{\beta}, \ i = 0, 1, \cdots, J - 1, \tag{13.60a}$$

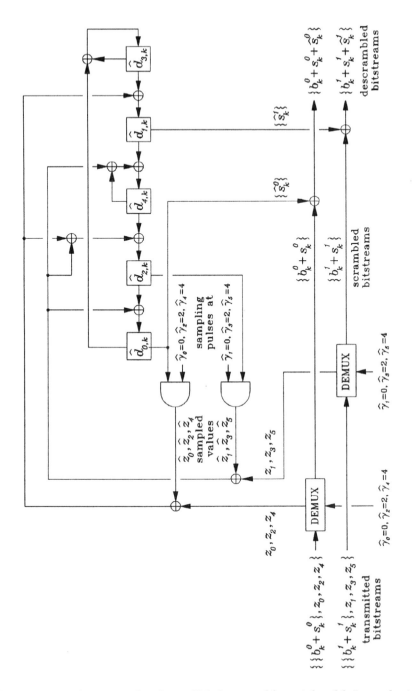

Figure 13.9. An example of parallel descrambler with odd L employing the concurrent parallel sampling and the immediate double correction.

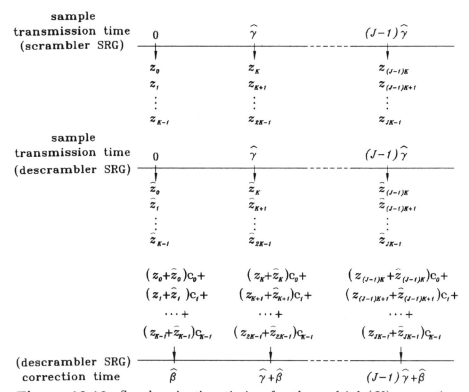

Figure 13.10. Synchronization timing for the multiple(K) correction case.

$$\mathbf{c}_j = \mathbf{c}_{K+j} = \cdots = \mathbf{c}_{K(J-1)+j}, \quad j = 0, 1, \cdots, K - 1. \quad (13.60b)$$

Then, the corresponding synchronization timing for this K-correction case appears as drawn in Fig. 13.10.

The following two theorems respectively describe how to determine the correction matrix Λ and the correction vector $\mathbf{c}_i$'s that make the correction matrix null for the synchronization in the K-correction case.

THEOREM 13.14 *For the descrambler PSRG with the state transition matrix* $\mathbf{T}$ *in (13.1a) and the generating vector* $\mathbf{h}_j$, $j = 0, 1, \cdots, N\text{-}1$, *in (13.1c), the correction matrix* Λ *of the K-correction is*

$$\Lambda = (\mathbf{T}^{\hat{\gamma}} + \mathbf{c}_0 \cdot \mathbf{v}_0^t \cdot \mathbf{T}^{\hat{\gamma}-\hat{\beta}} + \mathbf{c}_1 \cdot \mathbf{v}_1^t \cdot \mathbf{T}^{\hat{\gamma}-\hat{\beta}} + \cdots + \mathbf{c}_{K-1} \cdot \mathbf{v}_{K-1}^t \cdot \mathbf{T}^{\hat{\gamma}-\hat{\beta}})^{J-1} \cdot$$

$$(\mathbf{T}^{\hat{\beta}} + \mathbf{c}_0 \cdot \mathbf{v}_0^t + \mathbf{c}_1 \cdot \mathbf{v}_1^t + \cdots + \mathbf{c}_{K-1} \cdot \mathbf{v}_{K-1}^t). \quad (13.61)$$

Proof : This theorem can be proved in a manner similar to the proof of Theorem 13.11. ∎

THEOREM 13.15 *For the descrambler PSRG with the state transition matrix* $\mathbf{T}$ *in (13.1a) and the generating vector* $\mathbf{h}_j$, $j = 0, 1, \cdots, N\text{-}1$, *in (13.1c), the correction matrix* Λ *in (13.61) becomes a zero matrix, if and only if the correction vectors* $\mathbf{c}_i$, $i = 0, 1, \cdots, K\text{-}1$, *have, for a correction delay* $\hat{\beta}$, *the expression*

$$\mathbf{c}_i = \mathbf{T}^{(J-1)\hat{\gamma}+\hat{\beta}} \cdot \tilde{\mathbf{V}}^{-1} \cdot \mathbf{e}_{L-K+i}. \tag{13.62}$$

Proof : We first prove the "if" part of the theorem. Since $\mathbf{T}$ is nonsingular, we can rewrite (13.61) as $\Lambda = \Lambda_*^J \cdot \mathbf{T}^{\hat{\beta}-\hat{\gamma}}$, where

$$\Lambda_* = \mathbf{T}^{\hat{\gamma}} + \mathbf{c}_0 \cdot \mathbf{v}_0^t \cdot \mathbf{T}^{\hat{\gamma}-\hat{\beta}} + \mathbf{c}_1 \cdot \mathbf{v}_1^t \cdot \mathbf{T}^{\hat{\gamma}-\hat{\beta}} + \cdots + \mathbf{c}_{K-1} \cdot \mathbf{v}_{K-1} \cdot \mathbf{T}^{\hat{\gamma}-\hat{\beta}}. \tag{13.63}$$

Due to (13.7) and (13.59), this can be rewritten as

$$\Lambda_* = \mathbf{T}^{(J-1)\hat{\gamma}+\hat{\beta}} \cdot \tilde{\mathbf{V}}^{-1} \cdot \mathbf{A} \cdot \tilde{\mathbf{V}} \cdot \mathbf{T}^{-(J-1)\hat{\gamma}-\hat{\beta}} \tag{13.64}$$

for

$$\begin{aligned}
\mathbf{A} &= \tilde{\mathbf{V}} \cdot \mathbf{T}^{\hat{\gamma}} \cdot \tilde{\mathbf{V}}^{-1} + \mathbf{e}_{L-K} \cdot \mathbf{v}_0^t \cdot \mathbf{T}^{J\hat{\gamma}} \cdot \tilde{\mathbf{V}}^{-1} \\
&\quad + \mathbf{e}_{L-K+1} \cdot \mathbf{v}_1^t \cdot \mathbf{T}^{J\hat{\gamma}} \cdot \tilde{\mathbf{V}}^{-1} + \cdots \\
&\quad + \mathbf{e}_{L-1} \cdot \mathbf{v}_{K-1}^t \cdot \mathbf{T}^{J\hat{\gamma}} \cdot \tilde{\mathbf{V}}^{-1}.
\end{aligned} \tag{13.65}$$

But by (13.9) along with (13.59), we have the relations

$$\mathbf{v}_j^t \cdot \mathbf{T}^{i\hat{\gamma}} \cdot \tilde{\mathbf{V}}^{-1} = \mathbf{e}_{iK+j}^t, \; i = 0, 1, \cdots, J-1, \; j = 0, 1, \cdots, K-1, \tag{13.66}$$

and this makes (13.65) reduce to $\mathbf{A} = [\, 0 \, \cdots \, 0 \; \mathbf{e}_0 \; \mathbf{e}_1 \; \cdots \; \mathbf{e}_{L-K-1} \,]$. Therefore, $\mathbf{A}^J = 0$, and $\Lambda_*^J = 0$ by (13.64), and hence $\Lambda = 0$.

Now we prove the "only if" part. We evaluate $\Lambda \cdot \tilde{\mathbf{V}}^{-1} \cdot \mathbf{e}_{L-K+j}$, $j = 0, 1, \cdots, K-1$, by applying (13.61) and (13.66) repeatedly. Then, we finally obtain the relation $\Lambda \cdot \tilde{\mathbf{V}}^{-1} \cdot \mathbf{e}_{L-K+j} = \mathbf{T}^{(J-1)\hat{\gamma}+\hat{\beta}} \cdot \tilde{\mathbf{V}}^{-1} \cdot \mathbf{e}_{L-K+j} + \mathbf{c}_j$. Therefore, if $\Lambda = 0$, then $\mathbf{c}_j = \mathbf{T}^{(J-1)\hat{\gamma}+\hat{\beta}} \cdot \tilde{\mathbf{V}}^{-1} \cdot \mathbf{e}_{L-K+j}$. ∎

EXAMPLE 13.8 For the serial scrambling sequence $\{\, s_k \,\}$ in the irreducible scrambling space $V[x^6 + x + 1]$ whose sampling and sample transmission times are $\alpha_0 = \gamma_0 = 0$, $\alpha_1 = \gamma_1 = 1$, $\alpha_2 = \gamma_2 = 2$, α_3

$= \gamma_3 = 9$, $\alpha_4 = \gamma_4 = 10$, $\alpha_5 = \gamma_5 = 11$, we realize a parallel descrambler with $N = 3$. By (11.16), we have the discrimination matrix

$$\Delta_\alpha = \begin{bmatrix} 1 & 0 & 0 & 0 & 0 & 0 \\ 0 & 1 & 0 & 0 & 0 & 0 \\ 0 & 0 & 1 & 0 & 0 & 0 \\ 0 & 0 & 0 & 1 & 1 & 0 \\ 0 & 0 & 0 & 0 & 1 & 1 \\ 1 & 1 & 0 & 0 & 0 & 1 \end{bmatrix},$$

which is nonsingular. Therefore, the sampling times α_i's are predictable. By Theorems 13.2 and 13.3, its parallel sampling times and parallel sample transmission times are $\hat{\alpha}_0 = \hat{\gamma}_0 = 0$, $\hat{\alpha}_1 = \hat{\gamma}_1 = 0$, $\hat{\alpha}_2 = \hat{\gamma}_2 = 0$, $\hat{\alpha}_3 = \hat{\gamma}_3 = 3$, $\hat{\alpha}_4 = \hat{\gamma}_4 = 3$, $\hat{\alpha}_5 = \hat{\gamma}_5 = 3$. We insert $\mathbf{R} = \mathbf{I}$ into (13.1a) and (13.1c) along with $N = 3$ and $\Psi_I(x) = x^6 + x + 1$. Then, we obtain the PSRG with the state transition matrix $\mathbf{T}$ and the generating vectors $\mathbf{h}_0$, $\mathbf{h}_1$ and $\mathbf{h}_2$ which are

$$\mathbf{T} = \begin{bmatrix} 0 & 0 & 0 & 1 & 0 & 0 \\ 0 & 0 & 0 & 0 & 1 & 0 \\ 0 & 0 & 0 & 0 & 0 & 1 \\ 1 & 1 & 0 & 0 & 0 & 0 \\ 0 & 1 & 1 & 0 & 0 & 0 \\ 0 & 0 & 1 & 1 & 0 & 0 \end{bmatrix},$$

$$\mathbf{h}_0 = \mathbf{e}_0,$$
$$\mathbf{h}_1 = \mathbf{e}_1,$$
$$\mathbf{h}_2 = \mathbf{e}_2.$$

From the parallel sample transmission times $\hat{\gamma}_i$'s, we observe that this case corresponds to the triple ($K = 3$) correction case with $J = 2$ and the sampling interval $\hat{\gamma} = 3$. By (13.8), we get $\tilde{\mathbf{V}} = \Delta_\alpha$, and for the immediate correction, we have $\hat{\beta} = 1$. Inserting these parameters into (13.62), we obtain the correction vectors

$$\begin{cases} \mathbf{c}_0 = [\,0\,1\,0\,1\,0\,0\,]^t, \\ \mathbf{c}_1 = [\,0\,1\,1\,1\,1\,0\,]^t, \\ \mathbf{c}_2 = [\,0\,1\,1\,0\,1\,1\,]^t. \end{cases}$$

The resulting triple correction based parallel descrambler is as shown in Fig. 13.11. ♣

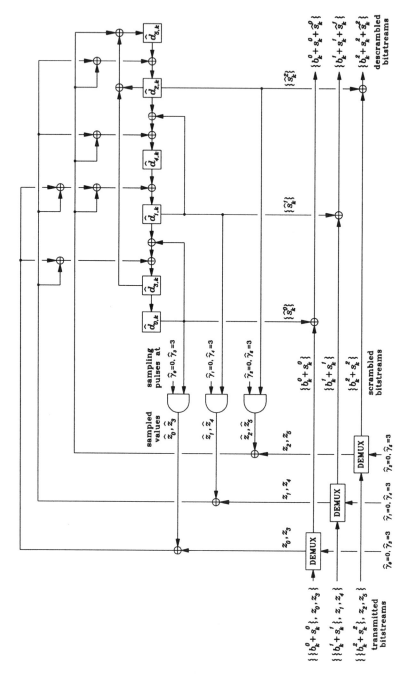

Figure 13.11. An example of parallel descrambler employing the concurrent parallel sampling and the immediate triple correction.

Chapter 14
Multibit-Parallel Distributed Sample Scrambling

The *multibit-parallel distributed sample scrambling* (MPDSS) is an extension of the PDSS in which parallel sequences are generated to match the multibit-interleaved multiplexed bitstream. So, in the MPDSS, the parallel input data bitstreams are scrambled before multibit-interleaved multiplexing, and the scrambled data bitstream is descrambled after multibit-interleaved demultiplexing. In this chapter, we discuss issues involved with the multibit-parallel realizations of DSSs. We first consider how to realize the PSRGs for MPDSS along with the minimal realizations of PSRGs. Then, we examine how to achieve the multibit-parallel sampling for MPDSS along with the concurrent multibit-parallel sampling. Finally, we consider the multibit-parallel correction for MPDSS.

14.1 Considerations for MPDSS

The serial and the multibit-parallel DSSs can be contrasted as shown in Fig. 14.1(a) and (b). In the MPDSS, N parallel input data bitstream $\{ b_k^j \}$, $j = 0, 1, \cdots, N - 1$, are scrambled before M-bit interleaved multiplexing by adding the parallel sequences $\{ s_k^j \}$, $j = 0, 1, \cdots,$ $N - 1$, respectively to the parallel input data bitstreams $\{ b_k^j \}$, $j = 0,$ $1, \cdots, N - 1$, and the scrambled bitstream $\{ b_k + s_k \}$ is descrambled after demultiplexing by adding the parallel descrambling sequences $\{ \hat{s}_k^j \}$, $j = 0, 1, \cdots, N - 1$, to the demultiplexed bitstreams respectively.

To make the scrambled bitstream $\{ b_k + s_k \}$ in Fig. 14.1(b) identical

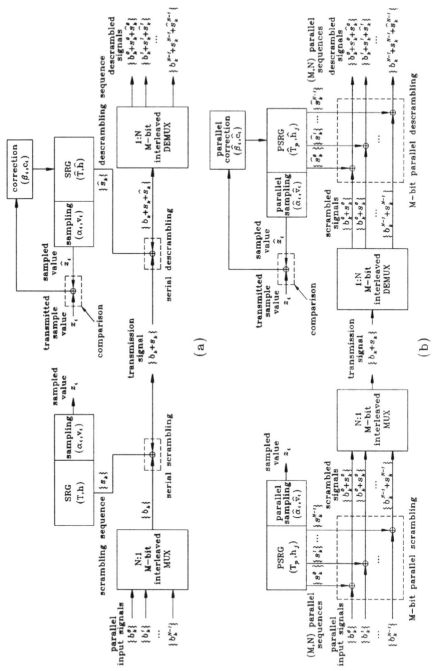

Figure 14.1. Serial and multibit-parallel DSS. (a) Serial DSS, (b) multibit-parallel DSS.

to the serial-scrambled bitstream $\{\, b_k + s_k \,\}$ in Fig. 14.1(a) for the same parallel input data bitstreams $\{\, b_k^j \,\}$'s, the parallel scrambling sequences $\{\, s_k^j \,\}$, $j = 0, 1, \cdots, N - 1$, should be the (M, N) parallel sequences of the serial scrambling sequence $\{\, s_k \,\}$.[1] The samples z_i's generated by the serial and the parallel scramblers should be the same, and the correction logic should be also changed accordingly.

14.2   PSRGs for MPDSS

We consider how to realize (M, N) PSRGs generating the (M, N) parallel scrambling sequences of the serial scrambling sequence.

As described in Chapter 12, in the DSS the scrambling space is a primitive space, and the scrambling sequence is a PRBS in the primitive scrambling space. Therefore, we concentrate on (M, N) PSRGs for an irreducible scrambling sequence. Note that a PRBS in a primitive space is an irreducible sequence. The following theorem describes how to determine the minimal (M, N) PSRGs for an irreducible scrambling sequence $\{\, s_k \,\}$.

THEOREM 14.1 *For the irreducible scrambling sequence* $\{\, s_k \,\} = \mathbf{s}^t \cdot \mathbf{E}_{\Psi_I(x)}$ *in the irreducible space* $V[\Psi_I(x)]$ *of dimension* L, *let* $\{\, s_k^j \,\}$, $j = 0, 1, \cdots, N\text{-}1$, *be the* jth (M, N) *parallel sequence of* $\{\, s_k \,\}$ *for an* MN *relatively prime to the periodicity* T_I *of* $V[\Psi_I(x)]$. *Then, for an* $L \times L$ *nonsingular matrix* $\mathbf{R}$ *and its* $ML \times L$ *augmentation* $\tilde{\mathbf{R}}_M$ *defined by*

$$\tilde{\mathbf{R}}_M = \begin{bmatrix} \mathbf{R} \\ \mathbf{R} \cdot \mathbf{A}_{\Psi_I(x)}^t \\ \vdots \\ \mathbf{R} \cdot (\mathbf{A}_{\Psi_I(x)}^t)^{M-1} \end{bmatrix}, \qquad (14.1)$$

the PSRG with the state transition matrix $\mathbf{T}$, *the initial state vector* $\mathbf{d}_0$ *and the generation vectors* $\mathbf{h}_j$'s *of the form*

$$\mathbf{T} = \begin{bmatrix} \mathbf{0} & \mathbf{I}_{(M-1)L \times (M-1)L} \\ \mathbf{R} \cdot (\mathbf{A}_{\Psi_I(x)}^t)^{MN} \cdot \mathbf{R}^{-1} & \mathbf{0} \end{bmatrix}, \qquad (14.2a)$$

$$\mathbf{d}_0 = \tilde{\mathbf{R}}_M \cdot \mathbf{s}, \qquad (14.2b)$$

---

[1]The relation between the serial scrambling sequence and the (M, N) parallel scrambling sequences is described in Chapter 8.

$$\mathbf{h}_j = \begin{bmatrix} (\mathbf{R}^t)^{-1} \cdot \mathbf{A}_{\Psi_I(x)}^{jM} \cdot \mathbf{e}_0 \\ 0 \\ \vdots \\ 0 \end{bmatrix}, \quad j = 0, 1, \cdots, N-1, \quad (14.2c)$$

is a minimal (M, N) *PSRG generating the* (M, N) *parallel sequences* { s_k^j } *'s.*

Proof: We prove the theorem using Theorems 13.1 and 8.13. Let U_i, $i = 0, 1, \cdots, MN-1$, be the ith MN-decimated sequence of the scrambling sequence { s_k } = $\mathbf{s}^t \cdot \mathbf{E}_{\Psi_I(x)}$. Then, since the decimation factor MN is relatively prime to the periodicity T_I of $V[\Psi_I(x)]$, by Theorem 13.1, the PSRG with the state transition matrix $\mathbf{T} = \mathbf{R} \cdot (\mathbf{A}_{\Psi_I(x)}^t)^{MN} \cdot \mathbf{R}^{-1}$, the initial state vector $\mathbf{d}_0 = \mathbf{R} \cdot \mathbf{s}$ and the generating vectors $\mathbf{h}_j = (\mathbf{R}^t)^{-1} \cdot \mathbf{A}_{\Psi_I(x)}^{jM} \cdot \mathbf{e}_0$, $j = 0, 1, \cdots, N-1$, is a minimal PSRG generating the (jM)th decimated sequences U_{jM}, $j = 0, 1, \cdots, N-1$. Inserting these parameters into (8.23), we obtain the minimal (M, N) PSRG with the state transition matrix $\mathbf{T}$ in (14.2a), the generating vectors $\mathbf{h}_j$'s in (14.2c), and the initial state vector $\mathbf{d}_0 = [\ \mathbf{R}^t\ (\mathbf{R} \cdot (\mathbf{A}_{\Psi_I(x)}^t)^{mMN})^t \cdots (\mathbf{R} \cdot (\mathbf{A}_{\Psi_I(x)}^t)^{m(M-1)MN})^t\ ]^t \cdot \mathbf{s}$, where m is the smallest integer that makes $mMN = 1$ modulo T_I. Therefore, it suffices to show that this $\mathbf{d}_0$ is identical to the $\mathbf{d}_0$ in (14.2b). Since the periodicity of $V[\Psi_I(x)]$ is T_I, $\Psi_I(x)$ divides $x^{T_I} + 1$, and since $\Psi_I(\mathbf{A}_{\Psi_I(x)}^t) = 0$, we get $(\mathbf{A}_{\Psi_I(x)}^t)^{T_I} + \mathbf{I} = 0$, or $(\mathbf{A}_{\Psi_I(x)}^t)^{T_I} = \mathbf{I}$. Thus, we have $(\mathbf{A}_{\Psi_I(x)}^t)^{mMN} = \mathbf{A}_{\Psi_I(x)}^t$, and therefore $\mathbf{d}_0 = [\ \mathbf{R}^t\ (\mathbf{R} \cdot \mathbf{A}_{\Psi_I(x)}^t)^t \cdots (\mathbf{R} \cdot (\mathbf{A}_{\Psi_I(x)}^t)^{M-1})^t\ ]^t \cdot \mathbf{s}$. This completes the proof. ∎

The theorem discusses how to build a minimal (M, N) PSRG generating (M, N) parallel sequences. As a consequence, we know that if the scrambling sequence { s_k } = $\mathbf{s}^t \cdot \mathbf{E}_{\Psi_I(x)}$ is an irreducible sequence in the irreducible space $V[\Psi_I(x)]$ and if MN is relatively prime to the periodicity T_I of $V[\Psi_I(x)]$, then the PSRG with the state transition matrix $\mathbf{T}$ in (14.2a), the initial state transition matrix $\mathbf{d}_0$ in (14.2b) and the generating vectors $\mathbf{h}_j$'s in (14.2c) for an arbitrary nonsingular matrix $\mathbf{R}$ becomes a minimal (M, N) PSRG generating the (M, N) parallel sequences { s_k^j }'s. From (14.2), we observe that the initial state vector $\mathbf{d}_0$ depends on the initial vector $\mathbf{s}$ of the scrambling sequence { s_k } but

the state transition matrix $\mathbf{T}$ and the generating vectors $\mathbf{h}_j$'s *do not.*
This implies that the PSRG with the transition matrix $\mathbf{T}$ in (14.2a) and
the generating vectors $\mathbf{h}_j$'s in (14.2c) can generate the (M, N) parallel
sequences for an arbitrary scrambling sequence in the irreducible space
$V[\Psi_I(x)]$ with the initial state vector setting done by (14.2b). In this
case, the state transition matrix and the generating vectors of a minimal
(M, N) PSRG for the MPDSS are determined by inserting an arbitrary
nonsingular matrix $\mathbf{R}$ into (14.2a) and (14.2c) respectively. Note that
the state transition matrix $\mathbf{T}$ in (14.2a) is the M-extended state tran-
sition matrix of $\mathbf{R} \cdot (\mathbf{A}_{\Psi_I(x)}^t)^{MN} \cdot \mathbf{R}^{-1}$. Also, note that the initial state
vector $\mathbf{d}_0$ can not be an arbitrary ML-vector as it should be a vector de-
termined through (14.2b) for the 2^L possible values of the initial vector
s.

EXAMPLE 14.1 For the serial scrambling sequence $\{ s_k \}$ whose scrambling space is the irreducible space $V[x^4 + x + 1]$, we realize minimal
(2,2) PSRGs generating its (2,2) parallel sequences $\{ s_k^j \}$, $j = 0, 1$. If
we insert

$$
\begin{aligned}
\mathbf{R} &= \mathbf{I} \\
&= \begin{bmatrix} 1 & 0 & 0 & 0 \\ 0 & 1 & 0 & 0 \\ 0 & 0 & 1 & 0 \\ 0 & 0 & 0 & 1 \end{bmatrix}
\end{aligned}
\tag{14.3a}
$$

into (14.1) along with $\Psi_I(x) = x^4 + x + 1$, $L = 4$, $M = 2$ and $N = 2$,
we obtain

$$
\begin{aligned}
\tilde{\mathbf{R}}_2 &= \begin{bmatrix} \mathbf{I} \\ \mathbf{A}_{x^4+x+1}^t \end{bmatrix} \\
&= \begin{bmatrix} 1 & 0 & 0 & 0 \\ 0 & 1 & 0 & 0 \\ 0 & 0 & 1 & 0 \\ 0 & 0 & 0 & 1 \\ 0 & 1 & 0 & 0 \\ 0 & 0 & 1 & 0 \\ 0 & 0 & 0 & 1 \\ 1 & 1 & 0 & 0 \end{bmatrix},
\end{aligned}
\tag{14.3b}
$$

and from (14.2a) and (14.2c) we obtain

$$\mathbf{T} = \begin{bmatrix} 0 & 0 & 0 & 0 & 1 & 0 & 0 & 0 \\ 0 & 0 & 0 & 0 & 0 & 1 & 0 & 0 \\ 0 & 0 & 0 & 0 & 0 & 0 & 1 & 0 \\ 0 & 0 & 0 & 0 & 0 & 0 & 0 & 1 \\ 1 & 1 & 0 & 0 & 0 & 0 & 0 & 0 \\ 0 & 1 & 1 & 0 & 0 & 0 & 0 & 0 \\ 0 & 0 & 1 & 1 & 0 & 0 & 0 & 0 \\ 1 & 1 & 0 & 1 & 0 & 0 & 0 & 0 \end{bmatrix}, \tag{14.3c}$$

$$\begin{cases} \mathbf{h}_0 = \mathbf{e}_0, \\ \mathbf{h}_1 = \mathbf{e}_2. \end{cases} \tag{14.3d}$$

The resulting minimal (2,2) PSRG is as shown in Fig. 14.2(a). Inserting (14.3b) and the 16 possible initial vectors of s into (14.2b), we obtain the initial state vectors $\mathbf{d}_0$'s listed in Table 14.1. Thererfore, the initial state vector $\mathbf{d}_0$ for the (2,2) PSRG should be one of these 16 vectors.

If we choose another nonsingular matrix

$$\mathbf{R} = [\, \mathbf{e}_0 \quad \mathbf{A}^4_{x^4+x+1} \cdot \mathbf{e}_0 \quad \mathbf{A}^8_{x^4+x+1} \cdot \mathbf{e}_0 \quad \mathbf{A}^{12}_{x^4+x+1} \cdot \mathbf{e}_0 \,]^t$$

$$= \begin{bmatrix} 1 & 0 & 0 & 0 \\ 1 & 1 & 0 & 0 \\ 1 & 0 & 1 & 0 \\ 1 & 1 & 1 & 1 \end{bmatrix}, \tag{14.4a}$$

then we obtain

$$\tilde{\mathbf{R}}_2 = \begin{bmatrix} 1 & 0 & 0 & 0 \\ 1 & 1 & 0 & 0 \\ 1 & 0 & 1 & 0 \\ 1 & 1 & 1 & 1 \\ 0 & 1 & 0 & 0 \\ 0 & 1 & 1 & 0 \\ 0 & 1 & 0 & 1 \\ 1 & 0 & 1 & 1 \end{bmatrix}, \tag{14.4b}$$

and the minimal (2,2) PSRG with

$$\mathbf{T} = \begin{bmatrix} \mathbf{0} & \mathbf{I}_{4\times4} \\ \mathbf{A}^t_{x^4+x+1} & \mathbf{0} \end{bmatrix}, \tag{14.4c}$$

$$\begin{cases} \mathbf{h}_0 = \mathbf{e}_0, \\ \mathbf{h}_1 = \mathbf{e}_0 + \mathbf{e}_2. \end{cases} \tag{14.4d}$$

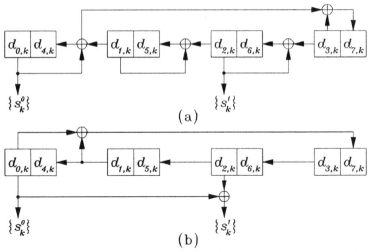

Figure 14.2. Examples of minimal (2,2) PSRGs generating the (2,2) parallel sequences $\{ s_k^0 \}$ and $\{ s_k^1 \}$ of the scrambling sequence $\{ s_k \}$ in the irreducible space $V[x^4 + x + 1]$. (a) The minimal (2,2) PSRG for $\mathbf{R} = \mathbf{I}$, (b) the SSRG-based minimal (2,2) PSRG.

This PSRG is an SSRG-based realization depicted in Fig. 14.2(b). For this PSRG, the initial state vector $\mathbf{d}_0$ should be one of the 16 vectors listed in Table 14.1. ♣

14.3 Parallel Sampling for MPDSS

We consider the sampling in the MPDSS, which we call a *multibit-parallel sampling*.

THEOREM 14.2 *For the serial scrambling sequence $\{ s_k \}$, let $\{ s_k^j \}$, $j = 0, 1, \cdots, N\text{-}1$, be the jth (M, N) parallel sequence of $\{ s_k \}$. Then, a sample z taken from the serial scrambling sequence $\{ s_k \}$ at the sampling time α corresponds to the sample taken from the lth (M, N) parallel sequence $\{ s_k^l \}$ at the multibit-parallel sampling time $\hat{\alpha}$, where l is the quotient of the integer r divided by M (i.e., $r = lM + \hat{r}$), and $\hat{\alpha}$ is M times the integer q plus the remainder of the integer r divided by M (i.e., $\hat{\alpha} = qM + \hat{r}$), where q and r are respectively the quotient and the remainder of α divided by MN (i.e., $\alpha = qMN + r$).*

Table 14.1. The relations between the initial vector of the scrambling sequence and the initial state vectors of the PSRGs in Fig. 14.2.

initial vector	initial state vector $\mathbf{d}_0$ of the PSRGs	
$\mathbf{s}$	for $\tilde{\mathbf{R}}_2$ in (14.3b)	for $\tilde{\mathbf{R}}_2$ in (14.4b)
$[0\,0\,0\,0]^t$	$[0\,0\,0\,0\,0\,0\,0\,0]^t$	$[0\,0\,0\,0\,0\,0\,0\,0]^t$
$[0\,0\,0\,1]^t$	$[0\,0\,0\,1\,0\,0\,1\,0]^t$	$[0\,0\,0\,1\,0\,0\,1\,1]^t$
$[0\,0\,1\,0]^t$	$[0\,0\,1\,0\,0\,1\,0\,0]^t$	$[0\,0\,1\,1\,0\,1\,0\,1]^t$
$[0\,0\,1\,1]^t$	$[0\,0\,1\,1\,0\,1\,1\,0]^t$	$[0\,0\,1\,0\,0\,1\,1\,0]^t$
$[0\,1\,0\,0]^t$	$[0\,1\,0\,0\,1\,0\,0\,1]^t$	$[0\,1\,0\,1\,1\,1\,1\,0]^t$
$[0\,1\,0\,1]^t$	$[0\,1\,0\,1\,1\,0\,1\,1]^t$	$[0\,1\,0\,0\,1\,1\,0\,1]^t$
$[0\,1\,1\,0]^t$	$[0\,1\,1\,0\,1\,1\,0\,1]^t$	$[0\,1\,1\,0\,1\,0\,1\,1]^t$
$[0\,1\,1\,1]^t$	$[0\,1\,1\,1\,1\,1\,1\,1]^t$	$[0\,1\,1\,1\,1\,0\,0\,0]^t$
$[1\,0\,0\,0]^t$	$[1\,0\,0\,0\,0\,0\,0\,1]^t$	$[1\,1\,1\,1\,0\,0\,0\,1]^t$
$[1\,0\,0\,1]^t$	$[1\,0\,0\,1\,0\,0\,1\,1]^t$	$[1\,1\,1\,0\,0\,0\,1\,0]^t$
$[1\,0\,1\,0]^t$	$[1\,0\,1\,0\,0\,1\,0\,1]^t$	$[1\,1\,0\,0\,0\,1\,0\,0]^t$
$[1\,0\,1\,1]^t$	$[1\,0\,1\,1\,0\,1\,1\,1]^t$	$[1\,1\,0\,1\,0\,1\,1\,1]^t$
$[1\,1\,0\,0]^t$	$[1\,1\,0\,0\,1\,0\,0\,0]^t$	$[1\,0\,1\,0\,1\,1\,1\,1]^t$
$[1\,1\,0\,1]^t$	$[1\,1\,0\,1\,1\,0\,1\,0]^t$	$[1\,0\,1\,1\,1\,1\,0\,0]^t$
$[1\,1\,1\,0]^t$	$[1\,1\,1\,0\,1\,1\,0\,0]^t$	$[1\,0\,0\,1\,1\,0\,1\,0]^t$
$[1\,1\,1\,1]^t$	$[1\,1\,1\,1\,1\,1\,1\,0]^t$	$[1\,0\,0\,0\,1\,0\,0\,1]^t$

Proof : Since the sample z is taken from the scrambling sequence $\{\,s_k\,\}$ at the sampling time $\alpha = qMN + r = qMN + lM + \hat{r}$, we have the relation $z = s_\alpha = s_{qMN+lM+\hat{r}}$. Therefore, by (8.1), $z = s^l_{qM+\hat{r}} = s^l_{\hat{\alpha}}$, that is, z is identical to the sample taken from the lth (M,N) parallel sequence $\{\,s^l_k\,\}$ at the sampling time $\hat{\alpha}$. ∎

EXAMPLE 14.2 For the serial scrambling sequence $\{\,s_k\,\}$ whose scrambling space is the irreducible space $V[x^4 + x + 1]$ and the sampling times are $\alpha_0 = 1$, $\alpha_1 = 3$, $\alpha_2 = 5$, $\alpha_3 = 7$, we consider the multibit-parallel sampling for the case $M = N = 2$. Then, for the 0th sample z_0, the quotient and the remainder of $\alpha_0 = 1$ divided by $MN = 4$ are $q_0 = 0$ and $r_0 = 1$ respectively, and hence we get the values $l_0 = 0$ and $\hat{\alpha}_0 = 1$. Therefore, by the theorem the sample z_0 is to be taken from the 0th $(2,2)$ parallel sequence $\{\,s^0_k\,\}$ at the multibit-parallel sampling time $\hat{\alpha}_0 = 1$. In a similar manner, for the 1st, the 2nd and the 3rd samples z_1, z_2 and z_3, we get the integers $l_1 = 1$, $\hat{\alpha}_1 = 1$, $l_2 = 0$, $\hat{\alpha}_2 = 3$, $l_3 = 1$, $\hat{\alpha}_3$

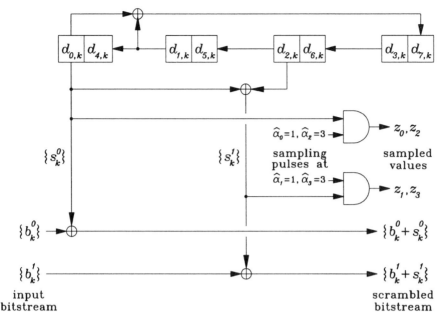

Figure 14.3. An example of a multibit-parallel scrambler for the scrambling sequence $\{\,s_k\,\}$ whose scrambling space is the irreducible space $V[x^4 + x + 1]$ and its sampling times are $\alpha_0 = 1$, $\alpha_1 = 3$, $\alpha_2 = 5$, $\alpha_3 = 7$.

$= 3$. Therefore, z_1, z_2 and z_3 are to be taken from the $\{\,s_k^1\,\}$, $\{\,s_k^0\,\}$ and $\{\,s_k^1\,\}$ respectively at the multibit-parallel sampling times $\hat{\alpha}_1 = 1$, $\hat{\alpha}_2 = 3$ and $\hat{\alpha}_3 = 3$. If we employ the minimal PSRG in Fig. 14.2(b), we can implement this multibit-parallel sampling as depicted in Fig. 14.3. ♣

Now we consider how to realize concurrent sampling in the MPDSS. For this, we first consider how the sample transmission time changes in the MPDSS, which we call the *multibit-parallel sample transmission time*.

THEOREM 14.3 *For the serial scrambling sequence $\{\,s_k\,\}$, let $\{\,s_k^j\,\}$, $j = 0, 1, \cdots, N$-1, be the jth (M, N) parallel sequence of $\{\,s_k\,\}$. Then, a sample z transmitted over the serial-scrambled signal $\{\,s_k\,\}$ at the sample transmission time γ is transmitted over the mth (M, N) parallel sequence $\{\,s_k^m\,\}$ at the multibit-parallel sample transmission time $\hat{\gamma}$, where m is the quotient of the integer r divided by M (i.e., $r = mM + \hat{r}$), and $\hat{\gamma}$ is M times the integer q plus the remainder of the integer r divided by*

M (i.e., $\hat{\gamma} = qM + \hat{r}$), where q and r are respectively the quotient and the remainder of γ divided by MN (i.e., $\gamma = qMN + r$).

Proof : This theorem can be proved in a way similar to that of Theorem 14.2. ∎

EXAMPLE 14.3 For the serial scrambling sequence { s_k } whose scrambling space is the irreducible space $V[x^4 + x + 1]$ and its sample transmission times are $\gamma_0 = 1$, $\gamma_1 = 2$, $\gamma_2 = 5$, $\gamma_3 = 6$, we consider the multibit-parallel sample transmission for the case $M = N = 2$. Then, for the 0th sample z_0, the quotient and the remainder of $\gamma_0 = 1$ divided by $MN = 4$ are $q_0 = 0$ and $r_0 = 1$ respectively, and hence we get the values $m_0 = 0$ and $\hat{\gamma}_0 = 1$. Therefore, by the theorem the sample z_0 is to be transmitted over the 0th multibit-parallel scrambled signal at the multibit-parallel sample transmission time $\hat{\gamma}_0 = 1$. In a similar manner, for the 1st, the 2nd and the 3rd samples z_1, z_2 and z_3, we get the integers $m_1 = 1$, $\hat{\gamma}_1 = 0$, $m_2 = 0$, $\hat{\gamma}_2 = 3$, $m_3 = 1$, $\hat{\gamma}_3 = 2$. Therefore, z_1, z_2 and z_3 are to be transmitted over the 1st, the 0th, and the 1st multibit-parallel scrambled signals respectively at the multibit-parallel sample transmission times $\hat{\gamma}_1 = 0$, $\hat{\gamma}_2 = 3$ and $\hat{\gamma}_3 = 2$. ♣

Then, the concurrent sampling can be realized in the (M, N) PSRG as the following theorem describes.

THEOREM 14.4 *Let a sample z be taken at the multibit-parallel sampling time $\hat{\alpha}$ from the lth (M, N) parallel sequence { s_k^l } generated by an (M, N) PSRG with the state transition matrix $\mathbf{T}$ and the generating vectors $\mathbf{h}_j$, $j = 0, 1, \cdots, N\text{-}1$. Then, the sample z is identical to the sample taken at the multibit-parallel sample transmission time $\hat{\gamma}$ using the sampling vector*

$$\mathbf{v} = (\mathbf{T}^t)^{\hat{\alpha}-\hat{\gamma}} \cdot \mathbf{h}_l. \tag{14.5}$$

Proof : This theorem can be proved in a manner similar to that of Theorem 13.4. ∎

Now that the sample taken from the lth (M, N) parallel sequence { s_k^l } at the multibit-parallel sampling time $\hat{\alpha}$ is identical to the sample taken at the multibit-parallel sample transmission time $\hat{\gamma}$ using the sampling vector $\mathbf{v}$ in (14.5), we can realize the concurrent multibit-parallel sampling by applying (14.5) for all the samples z_i's.

EXAMPLE 14.4 For the serial scrambling sequence $\{\, s_k \,\}$ whose scrambling space is the irreducible space $V[x^4 + x + 1]$, we consider the case when its sampling times are $\alpha_0 = 1$, $\alpha_1 = 3$, $\alpha_2 = 5$, $\alpha_3 = 7$, and the sample transmission times are $\gamma_0 = 1$, $\gamma_1 = 2$, $\gamma_2 = 5$, $\gamma_3 = 6$. Then, its multibit-parallel sampling times are $\hat{\alpha}_0 = 1$, $\hat{\alpha}_1 = 1$, $\hat{\alpha}_2 = 3$, $\hat{\alpha}_3 = 3$ (refer to Example 14.2) ; the sampled (2,2) parallel sequence numbers are $l_0 = 0$, $l_1 = 1$, $l_2 = 0$, $l_3 = 1$ (refer to Example 14.2) ; and its multibit-parallel sample transmission times are $\hat{\gamma}_0 = 1$, $\hat{\gamma}_1 = 0$, $\hat{\gamma}_2 = 3$, $\hat{\gamma}_3 = 2$ (refer to Example 14.3). If we employ the minimal (2,2) PSRG in Fig. 14.2(b), by (14.5) and (14.4) we obtain the sampling vectors

$$
\begin{cases}
\mathbf{v}_0 = (\mathbf{T}^t)^{1-1} \cdot \mathbf{h}_0 = \mathbf{e}_0, \\
\mathbf{v}_1 = (\mathbf{T}^t)^{1-0} \cdot \mathbf{h}_1 = \mathbf{e}_4 + \mathbf{e}_6, \\
\mathbf{v}_2 = (\mathbf{T}^t)^{3-3} \cdot \mathbf{h}_0 = \mathbf{e}_0, \\
\mathbf{v}_3 = (\mathbf{T}^t)^{3-2} \cdot \mathbf{h}_1 = \mathbf{e}_4 + \mathbf{e}_6,
\end{cases}
$$

for the concurrent sampling. Using these along with the transmitted signal numbers $m_0 = 0$, $m_1 = 1$, $m_2 = 0$, $m_3 = 1$ (refer to Example 14.3), we obtain the multibit-parallel scrambler shown in Fig. 14.4. ♣

14.4 Parallel Correction for MPDSS

As a preliminary step toward the parallel correction for MPDSS, we first consider the synchronization states of (M, N) PSRGs.

THEOREM 14.5 *For the serial scrambling sequence $\{\, s_k \,\}$ in the irreducible space $V[\Psi_I(x)]$ of dimension L, let $\{\, s_k^j \,\}$, $j = 0, 1, \cdots, N\text{-}1$, be the jth (M, N) parallel sequence of $\{\, s_k \,\}$ for an MN relatively prime to the periodicity T_I of $V[\Psi_I(x)]$. Then, the synchronization state vector $\tilde{\mathbf{d}}_k$ of the descrambler (M, N) PSRG with the state transition matrix $\mathbf{T}$ in (14.2a) and the generating vectors $\mathbf{h}_j$, $j = 0, 1, \cdots, N\text{-}1$, in (14.2c) generating the (M, N) parallel sequences $\{\, s_k^j \,\}$'s is*

$$
\tilde{\mathbf{d}}_k = \mathbf{T}^k \cdot \tilde{\mathbf{d}}_0, \tag{14.6}
$$

where the synchronization initial state vector $\tilde{\mathbf{d}}_0$ is

$$
\tilde{\mathbf{d}}_0 = \tilde{\mathbf{R}}_M \cdot \Delta_\alpha^{-1} \cdot \mathbf{z} \tag{14.7}
$$

for the $ML \times L$ matrix $\tilde{\mathbf{R}}_M$ in (14.1).

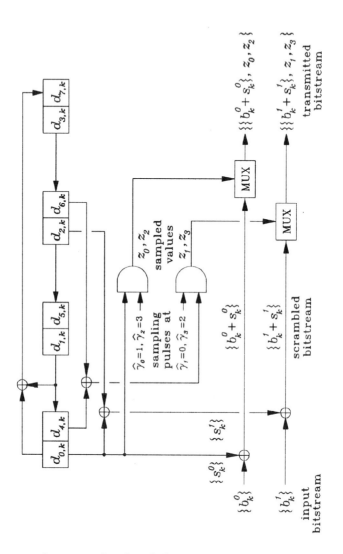

Figure 14.4. An example of multibit-parallel scrambler employing the concurrent multibit-parallel sampling.

Proof : It suffices to prove (14.7) since $\tilde{\mathbf{d}}_k = \mathbf{T} \cdot \tilde{\mathbf{d}}_{k-1}$, $k = 0, 1, \cdots$.
By Theorem 12.8, the scrambling sequence $\{ s_k \}$ is predicted by $\{ s_k \}$
$= (\Delta_\alpha^{-1} \cdot \mathbf{z})^t \cdot \mathbf{E}_{\Psi_I(x)}$. Therefore, inserting $\mathbf{s} = \Delta_\alpha^{-1} \cdot \mathbf{z}$ into (14.2b), we
obtain the synchronization initial state vector $\tilde{\mathbf{d}}_0$ in (14.7). ∎

According to (14.2a), the state transition matrix $\mathbf{T}$ is the M-extended state transition matrix of $\mathbf{R} \cdot (\mathbf{A}_{\Psi_I(x)}^t)^{MN} \cdot \mathbf{R}^{-1}$, so the length of
the relevant PSRG is M times the dimension L of the irreducible scrambling space $V[\Psi_I(x)]$. Therefore, in the MPDSS, the (M, N) PSRG of
length ML must be synchronized using only L samples, since L samples
z_i, $i = 0, 1, \cdots, L - 1$ are transmitted in the original serial DSS. It is
impossible, in general, to achieve synchronization of length-ML PSRG
using L samples, but it is possible in the (M, N) PSRG case since the
synchronization initial state vector of the (M, N) PSRG, which is an
ML-vector, is related to an L-vector in the following manner.

THEOREM 14.6 *For the descrambler (M, N) PSRG with the state transition matrix $\mathbf{T}$ in (14.2a) and the generating vectors $\mathbf{h}_j$, $j = 0, 1, \cdots$,
N-1, in (14.2c), let $\tilde{\mathbf{d}}_0^i = 0, 1, \cdots, M$-1, denote a subvector of the
synchronization initial state vector $\tilde{\mathbf{d}}_0$ such that*

$$\tilde{\mathbf{d}}_0 = \begin{bmatrix} \tilde{\mathbf{d}}_0^0 \\ \tilde{\mathbf{d}}_0^1 \\ \vdots \\ \tilde{\mathbf{d}}_0^{M-1} \end{bmatrix}. \tag{14.8}$$

Then,

$$\tilde{\mathbf{d}}_0 = \tilde{\mathbf{R}}_M \cdot \mathbf{R}^{-1} \cdot \tilde{\mathbf{d}}_0^0. \tag{14.9}$$

Proof : By (14.7) and (14.8), we have the relations $\tilde{\mathbf{d}}_0^i = \mathbf{R} \cdot (\mathbf{A}_{\Psi_I(x)}^t)^i \cdot \Delta_\alpha^{-1} \cdot \mathbf{z}$ for $i = 0, 1, \cdots, M - 1$. Therefore, we get, for $i = 0$, $\mathbf{z} = \Delta_\alpha \cdot \mathbf{R}^{-1} \cdot \tilde{\mathbf{d}}_0^0$, and inserting this to (14.7) we obtain (14.9). ∎

As for the case of PDSS, we can consider the correction process of
the MPDSS in three different cases $--$ the single, the double, and the
multiple corrections $--$ according to the sample transmission times. In
this section, we concentrate on the single correction case in which
a sample transmission time is different from others, as one can easily
extend it to the other cases in a similar manner.

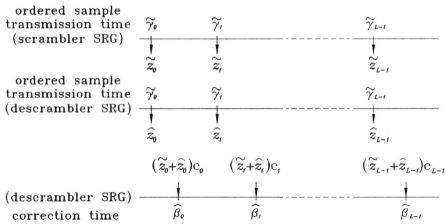

Figure 14.5. Synchronization timing for the MPDSS.

In the MPDSS, the multibit-parallel sample transmission times $\hat{\gamma}_i$'s could be out of order with respect to the original serial sample transmission times γ_i's (refer to Example 14.4). So, we rearrange the samples z_i's in the order of the multibit-parallel sample transmission times $\hat{\gamma}_i$'s, and denote the *ordered samples* as $\tilde{z}_i$; the corresponding *ordered multibit-parallel sample transmission times* as $\tilde{\gamma}_i$; the corresponding *ordered sampling vectors* as $\tilde{\mathbf{v}}_i$; and the *rearranged serial sampling times* that match the ordered samples as $\tilde{\alpha}_i$, for $i = 0, 1, \cdots, L-1$.

For the single correction case, the synchronization timing for the MPDSS with the concurrent sampling can be depicted as in Fig. 14.5. In the figure, $\tilde{\gamma}_i$, $i = 0, 1, \cdots, L-1$, is the ith ordered multibit-parallel sample transmission time, at which the ith ordered samples $\tilde{z}_i$ and $\hat{z}_i$ are taken respectively from the scrambler and the descrambler (M, N) PSRGs.[2] The ith correction time $\hat{\beta}_i$ comes later than the ith ordered sample transmission time $\tilde{\gamma}_i$ but not later than the $(i+1)$th ordered sample transmission time $\tilde{\gamma}_{i+1}$, that is,

$$\tilde{\gamma}_0 < \hat{\beta}_0 \leq \tilde{\gamma}_1 < \hat{\beta}_1 \leq \cdots \leq \tilde{\gamma}_{L-1} < \hat{\beta}_{L-1}. \tag{14.10}$$

For the ordered variables $\tilde{\gamma}_i$ and $\tilde{\mathbf{v}}_i$, the correction matrix Λ which describes the relation between the initial state distance vector and the finally corrected state distance vector takes the expression in the following theorem.

---

[2]For convenience, we keep $\hat{z}_i$ to denote the ordered sample in the descrambler.

THEOREM 14.7 *For the descrambler* (M, N) *PSRG with the state transition matrix* $\mathbf{T}$ *in (14.2a) and the generating vectors* $\mathbf{h}_j$, $j = 0, 1, \cdots,$ $N\text{-}1$, *in (14.2c), let* $\tilde{\gamma}_i$ *and* $\tilde{\mathbf{v}}_i$, $i = 0, 1, \cdots, L\text{-}1$, *be respectively the ordered multibit-parallel sample transmission time and the ordered sampling vector of the samples* z_i*'s. Then, the correction matrix* Λ *for the correction times* $\hat{\beta}_i$ *and the correction vectors* $\mathbf{c}_i$, $i = 0, 1, \cdots, L\text{-}1$, *is*

$$
\begin{aligned}
\Lambda \;=\; & (\mathbf{T}^{\hat{\beta}_{L-1}-\hat{\beta}_{L-2}} + \mathbf{c}_{L-1} \cdot \tilde{\mathbf{v}}_{L-1}^t \cdot \mathbf{T}^{\tilde{\gamma}_{L-1}-\hat{\beta}_{L-2}}) \cdot \\
& (\mathbf{T}^{\hat{\beta}_{L-2}-\hat{\beta}_{L-3}} + \mathbf{c}_{L-2} \cdot \tilde{\mathbf{v}}_{L-2}^t \cdot \mathbf{T}^{\tilde{\gamma}_{L-2}-\hat{\beta}_{L-3}}) \cdot \\
& \cdots (\mathbf{T}^{\hat{\beta}_1-\hat{\beta}_0} + \mathbf{c}_1 \cdot \tilde{\mathbf{v}}_1^t \cdot \mathbf{T}^{\tilde{\gamma}_1-\hat{\beta}_0}) \cdot (\mathbf{T}^{\hat{\beta}_0} + \mathbf{c}_0 \cdot \tilde{\mathbf{v}}_0^t \cdot \mathbf{T}^{\tilde{\gamma}_0}). \quad (14.11)
\end{aligned}
$$

Proof : The proof can be done in a way similar to that of Theorem 13.8, based on the following two lemmas. ∎

LEMMA 14.8 *For the descrambler* (M, N) *PSRG with the state transition matrix* $\mathbf{T}$ *in (14.2a) and the generating vectors* $\mathbf{h}_j$, $j = 0, 1,$ $\cdots, N\text{-}1$, *in (14.2c), let* $\tilde{\gamma}_i$ *and* $\tilde{\mathbf{v}}_i$, $i = 0, 1, \cdots, L\text{-}1$, *be respectively the ordered multibit-parallel sample transmission time and the ordered sampling vector of the samples* z_i*'s. Then, the relation*

$$
\tilde{\mathbf{v}}_i^t \cdot \mathbf{T}^{\tilde{\gamma}_i} \cdot \tilde{\mathbf{R}}_M \cdot \Delta_{\tilde{\alpha}}^{-1} \;=\; \mathbf{e}_i^t, \quad i = 0, 1, \cdots, L - 1, \quad (14.12)
$$

holds for the rearranged discrimination matrix $\Delta_{\tilde{\alpha}}$ *which is defined by*

$$
\Delta_{\tilde{\alpha}} \;\equiv\; \begin{bmatrix} \mathbf{e}_0^t \cdot (\mathbf{A}_{\Psi_I(x)}^t)^{\tilde{\alpha}_0} \\ \mathbf{e}_0^t \cdot (\mathbf{A}_{\Psi_I(x)}^t)^{\tilde{\alpha}_1} \\ \vdots \\ \mathbf{e}_0^t \cdot (\mathbf{A}_{\Psi_I(x)}^t)^{\tilde{\alpha}_{L-1}} \end{bmatrix} \quad (14.13)
$$

for the rearranged serial sampling time $\tilde{\alpha}_i$*'s.*

Proof : To prove the lemma, we show the relation $\tilde{\mathbf{v}}_i^t \cdot \mathbf{T}^{\tilde{\gamma}_i} \cdot \tilde{\mathbf{R}}_M = \mathbf{e}_i^t \cdot \Delta_{\tilde{\alpha}}$, $i = 0, 1, \cdots, L-1$. By (14.5), we get the relation $\tilde{\mathbf{v}}_i^t \cdot \mathbf{T}^{\tilde{\gamma}_i} \cdot \tilde{\mathbf{R}}_M = \hat{\mathbf{h}}_{l_i}^t \cdot \mathbf{T}^{\hat{\alpha}_i^o} \cdot \tilde{\mathbf{R}}_M$ for the ordered original multibit-parallel sampling time $\hat{\alpha}_i^o$ corresponding to the ordered multibit-parallel sample transmission time $\tilde{\gamma}_i$. Let q_i and $\hat{r}_i$ be respectively the quotient and the remainder of $\hat{\alpha}_i^o$ divided by M. Then, $\tilde{\mathbf{v}}_i^t \cdot \mathbf{T}^{\tilde{\gamma}_i} \cdot \tilde{\mathbf{R}}_M = \hat{\mathbf{h}}_{l_i}^t \cdot \mathbf{T}^{q_i M + \hat{r}_i} \cdot \tilde{\mathbf{R}}_M$. Inserting (14.2a) and (14.2c) to this, we can easily obtain the relation $\tilde{\mathbf{v}}_i^t \cdot \mathbf{T}^{\tilde{\gamma}_i} \cdot \tilde{\mathbf{R}}_M = \mathbf{e}_0^t \cdot (\mathbf{A}_{\Psi_I(x)}^t)^{q_i M N + l_i M + \hat{r}_i}$. But, since $q_i M N + l_i M + \hat{r}_i = \tilde{\alpha}_i$ by Theorem 14.2, we get $\tilde{\mathbf{v}}_i^t \cdot \mathbf{T}^{\tilde{\gamma}_i} \cdot \tilde{\mathbf{R}}_M = \mathbf{e}_0^t \cdot (\mathbf{A}_{\Psi_I(x)}^t)^{\tilde{\alpha}_i} = \mathbf{e}_i^t \cdot \Delta_{\tilde{\alpha}}$, where the last equality is due to (14.13). ∎

LEMMA 14.9 *For the descrambler* (M, N) *PSRG with the state transition matrix* $\mathbf{T}$ *in (14.2a) and the generating vectors* $\mathbf{h}_j$, $j = 0, 1, \cdots$, $N\text{-}1$, *in (14.2c), the ith ordered transmitted sample* $\tilde{z}_i$, $i = 0, 1, \cdots$, $L\text{-}1$, *is represented by*

$$\tilde{z}_i = \tilde{\mathbf{v}}_i^t \cdot \tilde{\mathbf{d}}_{\tilde{\gamma}_i} \tag{14.14}$$

for the synchronization state vector $\tilde{\mathbf{d}}_k$.

Proof : By (14.6) and (14.7), we obtain the relation $\tilde{\mathbf{v}}_i^t \cdot \tilde{\mathbf{d}}_{\tilde{\gamma}_i} = \tilde{\mathbf{v}}_i^t \cdot \mathbf{T}^{\tilde{\gamma}_i} \cdot \tilde{\mathbf{d}}_0 = \tilde{\mathbf{v}}_i^t \cdot \mathbf{T}^{\tilde{\gamma}_i} \cdot \tilde{\mathbf{R}}_M \cdot \Delta_\alpha^{-1} \cdot \mathbf{z}$. Applying (14.12) to this, we get $\tilde{\mathbf{v}}_i^t \cdot \tilde{\mathbf{d}}_{\tilde{\gamma}_i} = \mathbf{e}_i^t \cdot \Delta_{\tilde{\alpha}} \cdot \Delta_\alpha^{-1} \cdot \mathbf{z}$. Noting that the rearranged sampling times $\tilde{\alpha}_0, \tilde{\alpha}_1, \cdots, \tilde{\alpha}_{L-1}$ are obtained by reordering the original sampling times $\alpha_0, \alpha_1, \cdots, \alpha_{L-1}$ such that $[\, \tilde{\alpha}_0 \, \tilde{\alpha}_1 \, \cdots \, \tilde{\alpha}_{L-1} \,]^t = \mathbf{P} \cdot [\, \alpha_0 \, \alpha_1 \, \cdots \, \alpha_{L-1} \,]^t$ for a permuting matrix $\mathbf{P}$, we can deduce the relation $\Delta_{\tilde{\alpha}} \cdot \Delta_\alpha^{-1} = \mathbf{P}$. Therefore, $\tilde{\mathbf{v}}_i^t \cdot \tilde{\mathbf{d}}_{\tilde{\gamma}_i} = \mathbf{P} \cdot \mathbf{z} = \tilde{z}_i$. ∎

Finally, we consider how to determine the correction times and correction vectors for the descrambler (M, N) PSRG. To achieve synchronization in the descrambler (M, N) PSRG, the finally corrected state distance vector $\delta_{\hat{\beta}_{L-1}}$ should be zero regardless of the initial state distance vector δ_0. In general it is impossible to synchronize a length-ML PSRG using only L samples, but in the (M, N) PSRG case, it becomes possible due to the relation in (14.9). We assume that *the initial state vector* $\hat{\mathbf{d}}_0$ *of the descrambler* (M, N) *PSRG is set to* 0. Then, the initial state distance vector δ_0 becomes the synchronization initial state vector $\tilde{\mathbf{d}}_0$, and hence the finally corrected state distance vector $\delta_{\hat{\beta}_{L-1}}$ is $\Lambda \cdot \tilde{\mathbf{d}}_0$ for the correction matrix Λ in (14.11). Inserting (14.9) into this, we obtain the relation

$$\delta_{\hat{\beta}_{L-1}} = \Lambda \cdot \tilde{\mathbf{R}}_M \cdot \mathbf{R}^{-1} \cdot \tilde{\mathbf{d}}_0^0. \tag{14.15}$$

Therefore, to achieve synchronization in the MPDSS, the correction times $\hat{\beta}_i$'s and the correction vectors $\mathbf{c}_i$'s should be chosen such that $\delta_{\hat{\beta}_{L-1}}$ in (14.15) becomes zero for an arbitrary L-vector $\tilde{\mathbf{d}}_0^0$, or equivalently, *the correction times* $\hat{\beta}_i$'s *and the correction vectors* $\mathbf{c}_i$'s *should be chosen such that the* $ML \times L$ *matrix* $\Lambda \cdot \tilde{\mathbf{R}}_M$ *is a zero matrix*.[3]

As an aid to determining such correction times and correction vectors we consider the following lemma.

---

[3]In fact, it is easy to show that the same result yields as long as the initial state vector $\hat{\mathbf{d}}_0$ is chosen to be one of the synchronization initial state vectors $\tilde{\mathbf{d}}_0$'s.

LEMMA 14.10 *Let* Λ_i, $i = 1, 2, \cdots, L\text{-}1$, *denote the* $ML \times ML$ *matrix*

$$\Lambda_i \equiv (\mathbf{T}^{\hat{\beta}_{L-1}-\hat{\beta}_{L-2}} + \mathbf{c}_{L-1} \cdot \tilde{\mathbf{v}}_{L-1}^t \cdot \mathbf{T}^{\tilde{\gamma}_{L-1}-\hat{\beta}_{L-2}}) \cdots$$
$$(\mathbf{T}^{\hat{\beta}_i-\hat{\beta}_{i-1}} + \mathbf{c}_i \cdot \tilde{\mathbf{v}}_i^t \cdot \mathbf{T}^{\tilde{\gamma}_i-\hat{\beta}_{i-1}}). \tag{14.16}$$

Then, we have the relations

$$\begin{cases} \Lambda &= \Lambda_1 \cdot (\mathbf{T}^{\hat{\beta}_0} + \mathbf{c}_0 \cdot \tilde{\mathbf{v}}_0^t \cdot \mathbf{T}^{\tilde{\gamma}_0}), \\ \Lambda_i &= \Lambda_{i+1} \cdot (\mathbf{T}^{\hat{\beta}_i-\hat{\beta}_{i-1}} + \mathbf{c}_i \cdot \tilde{\mathbf{v}}_i^t \cdot \mathbf{T}^{\tilde{\gamma}_i-\hat{\beta}_{i-1}}), \quad i = 1, 2, \cdots, L\text{-}2, \end{cases} \tag{14.17}$$

$$\Lambda_i \cdot \mathbf{T}^{\hat{\beta}_i-1} \cdot \tilde{\mathbf{R}}_M \cdot \Delta_{\tilde{\alpha}}^{-1} \cdot \mathbf{e}_j = \begin{cases} \mathbf{T}^{\hat{\beta}_{L-1}} \cdot \tilde{\mathbf{R}}_M \cdot \Delta_{\tilde{\alpha}}^{-1} \cdot \mathbf{e}_j, & 0 \le j < i \le L\text{-}1, \\ \Lambda \cdot \tilde{\mathbf{R}}_M \cdot \Delta_{\tilde{\alpha}}^{-1} \cdot \mathbf{e}_j, & 1 \le i \le j \le L\text{-}1, \end{cases} \tag{14.18}$$

$$\Lambda \cdot \tilde{\mathbf{R}}_M \cdot \Delta_{\tilde{\alpha}}^{-1} \cdot \mathbf{e}_i = \begin{cases} \Lambda_{i+1} \cdot (\mathbf{T}^{\hat{\beta}_i} \cdot \tilde{\mathbf{R}}_M \cdot \Delta_{\tilde{\alpha}}^{-1} \cdot \mathbf{e}_i + \mathbf{c}_i), \\ \qquad i = 0, 1, \cdots, L\text{-}2, \\ \mathbf{T}^{\hat{\beta}_{L-1}} \cdot \tilde{\mathbf{R}}_M \cdot \Delta_{\tilde{\alpha}}^{-1} \cdot \mathbf{e}_{L-1} + \mathbf{c}_{L-1}, \\ \qquad i = L\text{-}1. \end{cases} \tag{14.19}$$

Proof : This lemma can be proved in a manner similar to the proof of Lemma 12.18. ∎

Then, we obtain the following theorem for choosing the correction times $\hat{\beta}_i$'s and the correction vectors $\mathbf{c}_i$'s that make $\Lambda \cdot \tilde{\mathbf{R}}_M$ zero.

THEOREM 14.11 *For the descrambler* (M, N) *PSRG with the state transition matrix* $\mathbf{T}$ *in (14.2a) and the generating vectors* $\mathbf{h}_j$, $j = 0,$ $1, \cdots, N\text{-}1,$ *in (14.2c), the* $ML \times L$ *matrix* $\Lambda \cdot \tilde{\mathbf{R}}_M$ *becomes a zero matrix, if and only if the correction vector* $\mathbf{c}_i$ *has, for the correction times* $\hat{\beta}_i$'s, *the expression*

$$\mathbf{c}_i = \begin{cases} \mathbf{T}^{\hat{\beta}_i} \cdot \tilde{\mathbf{R}}_M \cdot \Delta_{\tilde{\alpha}}^{-1} \cdot (\mathbf{e}_i + \sum_{j=i+1}^{L-1} u_{i,j}\mathbf{e}_j), & i = 0, 1, \cdots, L\text{-}2, \\ \mathbf{T}^{\hat{\beta}_{L-1}} \cdot \tilde{\mathbf{R}}_M \cdot \Delta_{\tilde{\alpha}}^{-1} \cdot \mathbf{e}_{L-1}, & i = L\text{-}1, \end{cases} \tag{14.20}$$

where $u_{i,j}$ *is either 0 or 1 for* $i = 0, 1, \cdots, L\text{-}2$ *and* $j = i+1, i+2, \cdots,$ $L\text{-}1.$

Proof : This theorem can be proved in a manner similar to the proof of Theorem 12.19, based on Lemma 14.10. ∎

According to the theorem, the correction vectors c_i's in (14.20) make the $ML \times L$ matrix $\Lambda \cdot \tilde{\mathbf{R}}_M$ a zero matrix for any arbitrarily chosen correction times $\hat{\beta}_i$'s. Therefore, to realize an immediate correction, we can select correction times $\hat{\beta}_i$'s such that $\hat{\beta}_i = \tilde{\gamma}_i + 1$, and select correction vectors c_i's using (14.20).

EXAMPLE 14.5 For the serial scrambling sequence $\{ s_k \}$ in the scrambling space $V[x^4 + x + 1]$ whose sampling times are $\alpha_0 = 1$, $\alpha_1 = 3$, $\alpha_2 = 5$, $\alpha_3 = 7$, and whose sample transmission times are $\gamma_0 = 1$, $\gamma_1 = 2$, $\gamma_2 = 5$, $\gamma_3 = 6$, we determine the correction vectors c_i's of the (2,2) PSRG in Fig. 14.4 for the immediate correction. Since the multibit-parallel sample transmission times are $\hat{\gamma}_0 = 1$, $\hat{\gamma}_1 = 0$, $\hat{\gamma}_2 = 3$, $\hat{\gamma}_3 = 2$ in this case (refer to Example 14.3), it corresponds to the single correction case. For these parallel sample transmission times, we obtain the ordered sample transmission times $\tilde{\gamma}_0 = 0$, $\tilde{\gamma}_1 = 1$, $\tilde{\gamma}_2 = 2$, $\tilde{\gamma}_3 = 3$, and the rearranged serial sampling times $\tilde{\alpha}_0 = 3$, $\tilde{\alpha}_1 = 1$, $\tilde{\alpha}_2 = 7$, $\tilde{\alpha}_3 = 5$. So by (14.13) the rearranged discrimination matrix $\Delta_{\tilde{\alpha}}$ becomes

$$\Delta_{\tilde{\alpha}} = \begin{bmatrix} 0 & 0 & 0 & 1 \\ 0 & 1 & 0 & 0 \\ 1 & 1 & 0 & 1 \\ 0 & 1 & 1 & 0 \end{bmatrix}.$$

Employing the matrix $\mathbf{R}$ in (14.4a), we get the matrices $\tilde{\mathbf{R}}_2$ and $\mathbf{T}$ respectively in (14.4b) and (14.4c), and inserting these along with the immediate correction times $\hat{\beta}_0 = 1$, $\hat{\beta}_1 = 2$, $\hat{\beta}_2 = 3$, $\hat{\beta}_3 = 4$ to (14.20), we obtain the correction vectors

$$
\begin{aligned}
\mathbf{c}_0 &= [0\,0\,1\,0\,1\,1\,0\,0]^t, \; [1\,0\,0\,0\,1\,1\,1\,1]^t, \\
&\quad [0\,0\,1\,1\,0\,0\,1\,0]^t, \; [0\,1\,1\,1\,1\,0\,1\,0]^t, \\
&\quad [1\,0\,0\,1\,0\,0\,0\,1]^t, \; [1\,1\,0\,1\,1\,0\,0\,1]^t, \\
&\quad [0\,1\,1\,0\,0\,1\,0\,0]^t, \; [1\,1\,0\,0\,0\,1\,1\,1]^t, \\
\mathbf{c}_1 &= [0\,0\,1\,1\,0\,1\,0\,1]^t, \; [1\,1\,0\,1\,0\,1\,1\,1]^t, \\
&\quad [0\,1\,0\,1\,1\,1\,1\,0]^t, \; [1\,0\,1\,1\,1\,1\,0\,0]^t, \\
\mathbf{c}_2 &= [0\,0\,1\,0\,1\,1\,0\,0]^t, \; [1\,0\,0\,1\,0\,0\,0\,1]^t, \\
\mathbf{c}_3 &= [1\,1\,0\,1\,0\,1\,1\,1]^t.
\end{aligned}
$$

If we select the correction vectors $\mathbf{c}_0 = \mathbf{c}_2 = [0\,0\,1\,0\,1\,1\,0\,0]^t$ and $\mathbf{c}_1 = \mathbf{c}_3 = [1\,1\,0\,1\,0\,1\,1\,1]^t$, we finally obtain the multibit-parallel descrambler depicted in Fig. 14.6. Note that this descrambler can be used in conjunction with the multibit-parallel scrambler in Fig. 14.4. ♣

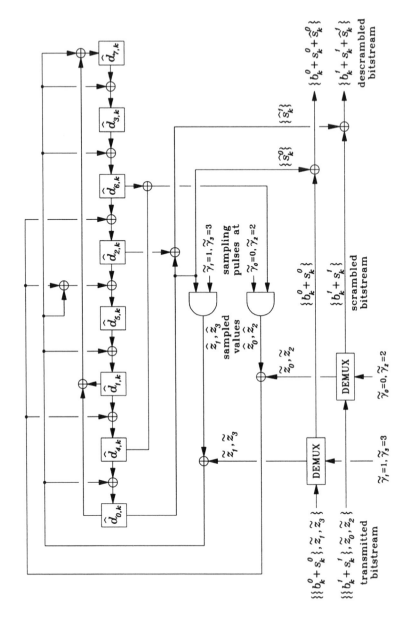

Figure 14.6. An example of multibit-parallel descrambler employing the concurrent sampling and the immediate single correction. This descrambler can be used in conjunction with the multibit-parallel scrambler in Fig. 14.4.

Chapter 15

Three-State Synchronization Mechanism under Sample Errors

Synchronization is a critical issue especially in the DSS since it solely relies on the transmitted samples for synchronization as opposed to the FSS case in which synchronization is achieved by resetting the shift registers to some prespecified states. The transmitted samples, however, could get corrupted by errors during transmission, while the prespecified initial state vector never gets errored. If guided by such errored samples, the DSS descrambler can not reach the synchronization state, and therefore synchronization becomes a very critical issue in the DSS.

As a means to handle the synchronization problem in a systematic manner we can employ the three-state synchronization mechanism introduced in Chapter 10 which consists of acquisition, verification, and steady states. In this chapter we will examine the synchronization behavior of the DSS based on this three-state synchronization mechanism. For this, we will first describe the synchronization mechanism itself, then consider the effects of sample errors on synchronization, and finally examine the synchronization behaviors in the constituent verification and steady states.

15.1 Three-State Synchronization Mechanism

With the possible transmission errors taken into account, the blockdiagram of the DSS in Fig. 10.1 can be redrawn as in Fig. 15.1. Due to the *sample error* ε_i, the *received sample* $\tilde{z}_i$ does not coincide with the

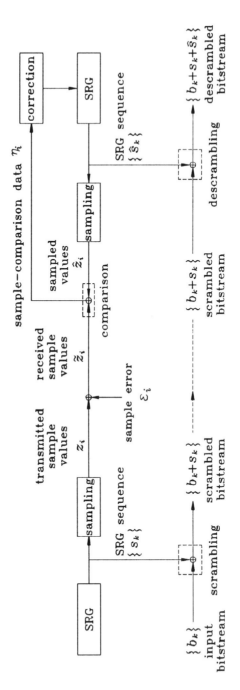

Figure 15.1. Blockdiagram of the DSS

Transmitted sample z_i any more.[1] It rather takes the relation

$$\tilde{z}_i = z_i + \varepsilon_i, \tag{15.1}$$

and, accordingly, the *sample-comparison data* η_i changes into

$$\eta_i = \tilde{z}_i + \hat{z}_i = z_i + \hat{z}_i + \varepsilon_i. \tag{15.2}$$

This errored sample-comparison data drives the correction block in such a manner that the descrambler SRG state gets modified to a wrong direction. It accompanies generation of a wrong SRG sequence { $\hat{s}_k$ } as well as wrong sampled values $\hat{z}_i$'s after completing the correction process. If such malfunctioning repeats, the DSS may never reach the synchronization state. This demonstrates how critical a sample error could be, even though the degree of impact may differ depending on the stage the sample error comes in.

As briefly discussed in Chapter 10, the DSS descrambler is equipped with the three-state synchronization mechanism shown in Fig. 10.2 for its synchronization.[2] In the acquisition state the descrambler acquires the synchronization state by repeating the comparison and correction processings based on the transmitted samples. Once the acquisition process is completed, the descrambler moves into the verification state in which the acquired synchronization is verified according to some predetermined criteria. If the acquired synchronization is confirmed to be true, the descrambler then enters the steady state and stays there as long as it meets some predetermined criteria. The descrambler returns back to the acquisition state if the predetermined criteria are not met in either state.

There are a number of different methods in deciding the details within the three-state synchronization mechanism, such as the predetermined criteria for the verification and the steady states. In examining the synchronization behaviors of the DSS in this chapter, we assume the following rather simple but general criteria : In the acquisition

---

[1]In fact, the transmission error affects both the samples z_i's and the scrambled sequence { $b_k + s_k$ }. The error appearing in the scrambled sequence, however, has no influence on the synchronization of the descrambler SRG. Therefore, we consider the error appearing on the samples only, which we call the *sample error*.

[2]The three-state synchronization mechanism is rooted in the ITU-T Recommendation I.432. It is also possible to consider other types of synchronization mechanisms based on two states, four states or others. However, we take this three-state based one as it exhibits good performance while keeping the structure comparatively simple.

state, the descrambler repeats the comparison and correction processing for the first L transmitted samples $\tilde{z}_0$ through $\tilde{z}_{L-1}$, then moves to the verification state without query. In the verification state, it compares the next L_V transmitted samples $\tilde{z}_L$ through $\tilde{z}_{L+L_V-1}$ with their descrambler-generated counterparts $\hat{z}_L$ through $\hat{z}_{L+L_V-1}$. If the number of 1's in the resulting L_V sample-comparison data η_L through η_{L+L_V-1} is not greater than a predetermined threshold value N_V, then is enters the steady state ; and otherwise it returns to the acquisition state. In the steady state, it repeats the comparison processing in units of L_S transmitted and descrambler-generated sample pairs, and stays there as long as the number of 1's in the resulting L_S sample-comparison data η_0^S through $\eta_{L_S-1}^S$ is not greater than a predetermined threshold value N_S, returning back to the acquisition state otherwise.[3]

15.2 Effects of Errored Samples in Acquisition State

According to the synchronization process we have discussed in the previous chapters, a properly designed DSS descrambler, which can be any of a serial, a parallel and a multibit-parallel descrambler, achieves synchronization after the reception of L correct samples regardless of the initial states of the descrambler SRG. If interpreted in the errored environment, this implies that the descrambler gets synchronized to the scrambler on receiving L consecutive unerrored samples. Unfortunately, the descrambler has no reference to use in judging the correctness of the samples received during the acquisition state, so it admits all the received samples to modify its own SRG states in compliance with them. The sample error occurred during the acquisition state therefore resides in the descrambler SRG and thus impacts all the sequences generated afterwards. As such, it is critical to be free from sample errors in the acquisition state. In this section we will consider such synchronization behaviors in the acquisition state, and examine how the sample errors occurred in the acquisition state affect the sample-comparison data after

---

[3]The comparison in units of L_S sample pairs is an on-going process applied to each consecutive L_S pairs of transmitted and descrambler-generated samples from the moment the descrambler enters the steady state. As such, we can not assign the absolute time index with reference to the acquisition-starting time. Therefore we reinitialize the time-index in units of L_S samples, affixing a superscript S to indicate this.

acquisition.

We first consider the synchronization behaviors for the case when sample error occurs during the acquisition state. According to Theorem 11.20, a scrambling sequence $\{ s_k \} = \mathbf{s}^t \cdot \mathbf{E}_{\Psi(x)}$ in the scrambling space $V[\Psi(x)]$ of dimension L is determined by

$$\{ s_k \} = (\Delta_\alpha^{-1} \cdot \mathbf{z})^t \cdot \mathbf{E}_{\Psi(x)} \tag{15.3a}$$

for the *transmitted sample vector* $\mathbf{z} = [\, z_0 \quad z_1 \quad \cdots \quad z_{L-1} \,]^t$ formed by predictable sampling times α_i's ; and according to Chapter 12(or 13, 14), a properly designed DSS(or PDSS, MPDSS) descrambler gets synchronized to generate the scrambling sequence $\{ s_k \}$ in (15.3a) after L times of corrections. However, in reality, the descrambler receives the samples $\tilde{z}_i$'s, not the transmitted samples z_i's, so the scrambling sequence $\{ s_k \}$ is modified to

$$\{ \tilde{s}_k \} = (\Delta_\alpha^{-1} \cdot \tilde{\mathbf{z}})^t \cdot \mathbf{E}_{\Psi(x)} \tag{15.3b}$$

for the *received sample vector* $\tilde{\mathbf{z}} \equiv [\, \tilde{z}_0 \quad \tilde{z}_1 \quad \cdots \quad \tilde{z}_{L-1} \,]^t$, which we call an *estimated scrambling sequence*. This implies that the descrambler, after completing the comparison and correction process using the received samples $\tilde{z}_i$'s, generates the estimated scrambling sequence $\{ \tilde{s}_k \}$ in (15.3b), not the transmitted scrambling sequence $\{ s_k \}$ in (15.3a).

To examine how much the estimated scrambling sequence $\{ \tilde{s}_k \}$ differs from the original transmitted scrambling sequence $\{ s_k \}$, we define the *estimation-error sequence* $\{ \sigma_k \}$ to represent the difference between them, i.e.,

$$\{ \sigma_k \} \equiv \{ s_k \} + \{ \tilde{s}_k \}. \tag{15.4}$$

Noting that the descrambling sequence $\{ \hat{s}_k \}$ generated by a properly designed DSS descrambler becomes identical to the estimated scrambling sequence $\{ \tilde{s}_k \}$ at the final correction using the $(L-1)$th sample, we find that the estimation-error sequence $\{ \sigma_k \}$ becomes the difference between the scrambling sequence $\{ s_k \}$ and the descrambling sequence $\{ \hat{s}_k \}$, i.e., $\{ \sigma_k \} = \{ s_k \} + \{ \hat{s}_k \}$ at the end of the acquisition state and beyond. The estimation-error sequence $\{ \sigma_k \}$ is represented by the sample errors as the following theorem describes.

THEOREM 15.1 *Let the acquisition state sample error vector $\mathcal{E}_A$ denote the L-vector*

$$\mathcal{E}_A \equiv [\, \varepsilon_0 \quad \varepsilon_1 \quad \cdots \quad \varepsilon_{L-1} \,]^t, \tag{15.5}$$

where ε_i, $i = 0, 1, \cdots, L\text{-}1$, is the ith sample error occurred in the acquisition state. Then, the estimation-error sequence $\{ \sigma_k \}$ is determined by

$$\{ \sigma_k \} = (\Delta_\alpha^{-1} \cdot \mathcal{E}_A)^t \cdot \mathbf{E}_{\Psi(x)}. \tag{15.6}$$

Proof: Inserting (15.3a) and (15.3b) into (15.4), we get the relation $\{ \sigma_k \} = (\Delta_\alpha^{-1} \cdot (\mathbf{z} + \tilde{\mathbf{z}}))^t \cdot \mathbf{E}_{\Psi(x)}$. But, by (15.1) $\mathbf{z} + \tilde{\mathbf{z}} = \mathcal{E}_A$. Therefore, $\{ \sigma_k \} = (\Delta_\alpha^{-1} \cdot \mathcal{E}_A)^t \cdot \mathbf{E}_{\Psi(x)}$. ■

The theorem means that the estimation-error sequence $\{ \sigma_k \}$ is determined by the sample error vector $\mathcal{E}_A$ in a deterministic manner, regardless of the original scrambling sequence $\{ s_k \}$. Comparing (15.3a) and (15.6), we observe that the estimation-error sequence $\{ \sigma_k \}$ corresponds to the scrambling sequence whose sample vector is the sample error vector $\mathcal{E}_A$. Further, we find, from (15.6), that the estimation-error sequence $\{ \sigma_k \}$ becomes zero, if and only if the sample vector $\mathcal{E}_A$ is zero. This implies that if an error occurs in the transmitted samples, the descrambling sequence $\{ \hat{s}_k \}$ can never be the same as the scrambling sequence $\{ s_k \}$ even if the correction process is completed.

EXAMPLE 15.1 We consider the scrambling sequence $\{ s_k \}$ in the scrambling space $V[x^4 + x^3 + x^2 + 1]$ whose sampling times are $\alpha_i = 2i$, $i = 0, 1, 2, 3$. If the third sample z_3 is the only erroneous one, or $\mathcal{E}_A = \mathbf{e}_3$, then by (15.6) and Table 11.3 the estimation-error sequence $\{ \sigma_k \}$ becomes $[0\,1\,0\,0]^t \cdot \mathbf{E}_{x^4+x^3+x^2+1}$, which is the sequence S_4 in Table 4.1. For the other error patterns, the corresponding estimation-error sequences become as shown in Table 15.1. ♣

We now consider how the sample errors occurred in the acquisition state affect the sample-comparison data after acquisition.

THEOREM 15.2 *Let ζ_i, $i = 0, 1, \cdots$, be the sampled value taken from the estimation-error sequence $\{ \sigma_k \}$ at the ith sampling time α_i. Then, the sample-comparison data η_i, $i = L, L + 1, \cdots$, after acquisition is represented by*

$$\eta_i = \zeta_i + \varepsilon_i, \tag{15.7}$$

where ε_i, $i = L, L+1, \cdots$, denotes the ith sample error after acquisition.

Proof : Noting that the descrambling sequence $\{ \hat{s}_k \}$ becomes identical to the estimated scrambling sequence $\{ \tilde{s}_k \}$ after acquisition,

Table 15.1. Estimation-error sequence $\{ \sigma_k \}$ for various acquisition state sample error vectors

sample error vector $\mathcal{E}_A$	estimation-error sequence $\{ \sigma_k \}$
$[\,0\,0\,0\,0\,]^t$	$\{\, 0, 0, 0, 0, 0, 0, 0, 0, 0, 0, 0, 0, 0, 0, \cdots \}$
$[\,0\,0\,0\,1\,]^t$	$\{\, 0, 1, 0, 0, 0, 1, 1, 0, 1, 0, 0, 0, 1, 1, \cdots \}$
$[\,0\,0\,1\,0\,]^t$	$\{\, 0, 1, 0, 1, 1, 1, 0, 0, 1, 0, 1, 1, 1, 0, \cdots \}$
$[\,0\,0\,1\,1\,]^t$	$\{\, 0, 0, 0, 1, 1, 0, 1, 0, 0, 0, 1, 1, 0, 1, \cdots \}$
$[\,0\,1\,0\,0\,]^t$	$\{\, 0, 0, 1, 1, 0, 1, 0, 0, 0, 1, 1, 0, 1, 0, \cdots \}$
$[\,0\,1\,0\,1\,]^t$	$\{\, 0, 1, 1, 1, 0, 0, 1, 0, 1, 1, 1, 0, 0, 1, \cdots \}$
$[\,0\,1\,1\,0\,]^t$	$\{\, 0, 1, 1, 0, 1, 0, 0, 0, 1, 1, 0, 1, 0, 0, \cdots \}$
$[\,0\,1\,1\,1\,]^t$	$\{\, 0, 0, 1, 0, 1, 1, 1, 0, 0, 1, 0, 1, 1, 1, \cdots \}$
$[\,1\,0\,0\,0\,]^t$	$\{\, 1, 1, 0, 1, 0, 0, 0, 1, 1, 0, 1, 0, 0, 0, \cdots \}$
$[\,1\,0\,0\,1\,]^t$	$\{\, 1, 0, 0, 1, 0, 1, 1, 1, 0, 0, 1, 0, 1, 1, \cdots \}$
$[\,1\,0\,1\,0\,]^t$	$\{\, 1, 0, 0, 0, 1, 1, 0, 1, 0, 0, 0, 1, 1, 0, \cdots \}$
$[\,1\,0\,1\,1\,]^t$	$\{\, 1, 1, 0, 0, 1, 0, 1, 1, 1, 0, 0, 1, 0, 1, \cdots \}$
$[\,1\,1\,0\,0\,]^t$	$\{\, 1, 1, 1, 0, 0, 1, 0, 1, 1, 1, 0, 0, 1, 0, \cdots \}$
$[\,1\,1\,0\,1\,]^t$	$\{\, 1, 0, 1, 0, 0, 0, 1, 1, 0, 1, 0, 0, 0, 1, \cdots \}$
$[\,1\,1\,1\,0\,]^t$	$\{\, 1, 0, 1, 1, 1, 0, 0, 1, 0, 1, 1, 1, 0, 0, \cdots \}$
$[\,1\,1\,1\,1\,]^t$	$\{\, 1, 1, 1, 1, 1, 1, 1, 1, 1, 1, 1, 1, 1, 1, \cdots \}$

we insert $z_i = s_{\alpha_i}$ and $\hat{z}_i = \hat{s}_{\alpha_i} = \tilde{s}_{\alpha_i}$ into (15.2). Then we obtain $\eta_i = s_{\alpha_i} + \tilde{s}_{\alpha_i} + \varepsilon_i$ for $i = L, L+1, \cdots$. Therefore, by (15.4), $\eta_i = \sigma_{\alpha_i} + \varepsilon_i = \zeta_i + \varepsilon_i$. ∎

This theorem means that the sample-comparison data η_i after acquisition is affected by two terms, ζ_i and ε_i. The first term ζ_i, which is directly obtained by sampling the estimation-error sequence $\{ \sigma_k \}$, stems from the sample errors occurred in the acquisition state ; and the second term ε_i is the sample error randomly occurred to the sample z_i after acquisition. Therefore, the sample-comparison data η_i after acquisition is determined by the sampled value ζ_i of the estimation-error sequence $\{ \sigma_k \}$ added to the random error ε_i.

EXAMPLE 15.2 In the case of the scrambling sequence $\{ s_k \}$ in the scrambling space $V[x^4 + x^3 + x^2 + 1]$ whose sampling times are $\alpha_i = 2i$, $i = 0, 1, \cdots$, the sampled values ζ_i's of the estimation-error sequence $\{ \sigma_k \}$ in Table 15.1 are as listed in Table 15.2 for various error-patterns

Table 15.2. Sampled values ζ_i's of estimation-error sequence $\{\sigma_k\}$ for various acquisition state sample error vectors

sample error vector $\mathcal{E}_A$	sampled values of estimation-error sequence			
	$\zeta_4 \cdot\cdot\ \zeta_7$	$\zeta_8 \cdot \zeta_{10}$	$\zeta_{11} \cdot\cdot\cdot\ \zeta_{16}$	$\zeta_{17} \cdots$
$[0\,0\,0\,0]^t$	0 0 0 0	0 0 0	0 0 0 0 0 0	0 $\cdots$
$[0\,0\,0\,1]^t$	1 0 1 0	0 0 1	1 0 1 0 0 0	1 $\cdots$
$[0\,0\,1\,0]^t$	1 1 1 0	0 1 0	1 1 1 0 1 0	0 $\cdots$
$[0\,0\,1\,1]^t$	0 1 0 0	0 1 1	0 1 0 0 0 1	1 $\cdots$
$[0\,1\,0\,0]^t$	0 1 1 0	1 0 0	0 1 1 0 1 0	0 $\cdots$
$[0\,1\,0\,1]^t$	1 1 0 0	1 0 1	1 1 0 0 1 0	1 $\cdots$
$[0\,1\,1\,0]^t$	1 0 0 0	1 1 0	1 0 0 0 1 1	0 $\cdots$
$[0\,1\,1\,1]^t$	0 0 1 0	1 1 1	0 0 1 0 1 1	1 $\cdots$
$[1\,0\,0\,0]^t$	1 1 0 1	0 0 0	1 1 0 1 0 0	0 $\cdots$
$[1\,0\,0\,1]^t$	0 1 1 1	0 0 1	0 1 1 1 0 0	1 $\cdots$
$[1\,0\,1\,0]^t$	0 0 1 1	0 1 0	0 0 1 1 0 1	0 $\cdots$
$[1\,0\,1\,1]^t$	1 0 0 1	0 1 1	1 0 0 1 0 1	1 $\cdots$
$[1\,1\,0\,0]^t$	1 0 1 1	1 0 0	1 0 1 1 1 0	0 $\cdots$
$[1\,1\,0\,1]^t$	0 0 0 1	1 0 1	0 0 0 1 1 0	1 $\cdots$
$[1\,1\,1\,0]^t$	0 1 0 1	1 1 0	0 1 0 1 1 1	0 $\cdots$
$[1\,1\,1\,1]^t$	1 1 1 1	1 1 1	1 1 1 1 1 1	1 $\cdots$

occurred in the acquisition state. The sample-comparison data η_i for $i = 4, 5, \cdots$, therefore become the sampled values in Table 15.2 added to the random error ε_i that may occur at the corresponding indexed time.

♣

15.3   Synchronization Verification in Verification State

In this section, we consider how to verify that the descrambler is synchronized to the scrambler after acquisition, or equivalently, how to decide whether or not the sample vector $\mathcal{E}_A$ is zero.

In the verification state, we observe the L_V sample-comparison data η_i, $i = L, L+1, \cdots, L+L_V - 1$, and based on this observation we verify whether or not the error vector $\mathcal{E}_A$ is zero. To describe this relation, we first define the L_V-vectors for the sample comparison data η_i, the estimation-error sample ζ_i, and the verification state sample error ε_i as follows :

DEFINITION 15.3 (Sample Vectors in Verification State) We define the *sample-comparison vector* $\mathcal{Y}_V$, the *sample vector* $\mathcal{Z}_V$ *of the estimation-error sequence*, and the *sample error vector* $\mathcal{E}_V$ *in the verification state*, respectively, to be

$$\mathcal{Y}_V \equiv [\, \eta_L \;\; \eta_{L+1} \;\; \cdots \;\; \eta_{L+L_V-1}\,]^t, \tag{15.8a}$$
$$\mathcal{Z}_V \equiv [\, \zeta_L \;\; \zeta_{L+1} \;\; \cdots \;\; \zeta_{L+L_V-1}\,]^t, \tag{15.8b}$$
$$\mathcal{E}_V \equiv [\, \varepsilon_L \;\; \varepsilon_{L+1} \;\; \cdots \;\; \varepsilon_{L+L_V-1}\,]^t. \tag{15.8c}$$

Then, by Theorem 15.2, we have the relation

$$\mathcal{Y}_V = \mathcal{Z}_V + \mathcal{E}_V. \tag{15.9}$$

Note that the sample vector $\mathcal{Z}_V$ is formed out of the sampled values of the estimation error sequence $\{\,\sigma_k\,\}$ which is determined by the acquisition state sample error vector $\mathcal{E}_A$ through (15.6).

We first assume that no error occurs in the verification state, or $\mathcal{E}_V = \mathbf{0}$, and consider the guidelines for choosing the sample number L_V to be observed for synchronization verification.

DEFINITION 15.4 (Weight of Vector) For an L-vector $\mathbf{v}$, we define by the *weight* $W[\mathbf{v}]$ the number of unity elements in the vector $\mathbf{v}$.

THEOREM 15.5 *If the sample number L_V to be observed for verification of synchronization is chosen such that*

$$W[\mathcal{Z}_V] \geq 1 \;\; \text{for } \mathcal{E}_A \neq \mathbf{0}, \tag{15.10}$$

then the relation

$$\begin{cases} W[\mathcal{Y}_V] = 0, & \text{if } \mathcal{E}_A = \mathbf{0}, \\ W[\mathcal{Y}_V] \geq 1, & \text{if } \mathcal{E}_A \neq \mathbf{0}, \end{cases} \tag{15.11}$$

holds for $\mathcal{E}_V = \mathbf{0}$.

Proof : By (15.9), if $\mathcal{E}_V = \mathbf{0}$, $\mathcal{Y}_V = \mathcal{Z}_V$, and hence $W[\mathcal{Y}_V] = W[\mathcal{Z}_V]$. Therefore, if $\mathcal{E}_A = \mathbf{0}$, then $W[\mathcal{Y}_V] = W[\mathcal{Z}_V] = 0$; and if $\mathcal{E}_A \neq \mathbf{0}$, then by (15.10), $W[\mathcal{Y}_V] \geq 1$. ∎

According to the theorem, if we choose the observation sample number L_V to meet the relation (15.10), the weight of the sample-comparison vector $\mathcal{Y}_V$ becomes $W[\mathcal{Y}_V] = 0$ for $\mathcal{E}_A = \mathbf{0}$ and $W[\mathcal{Y}_V] \geq 1$ for $\mathcal{E}_A \neq \mathbf{0}$, in case $\mathcal{E}_V = \mathbf{0}$. Therefore, as long as there occur no error in the verification state, we can always correctly decide whether or not the descrambler is truly synchronized to the scrambler by observing the L_V sample-comparison data.

EXAMPLE 15.3 In the case of the scrambling sequence $\{ s_k \}$ in the scrambling space $V[x^4 + x^3 + x^2 + 1]$ whose sampling times are $\alpha_i = 2i$, $i = 0, 1, \cdots$, we consider the sampled values of the estimation-error sequence in Table 15.2. We observe from the table that $W[\mathcal{Z}_V] \geq 1$ for all non-zero sample error vectors $\mathcal{E}_A$'s if L_V is chosen to be 4 or larger. So, we take $L_V = 4$, and make decision on the synchronization in such a manner that synchronization is acquired if the sample-comparison vector $\mathcal{Y}_V$ is $\mathbf{0}$, and not acquired otherwise. Then, according to the theorem, this decision is correct as long as there occurs no error in the verification state. ♣

Even if the observation sample number L_V is chosen to meet (15.10), (15.11) leads to a wrong decision if $\mathcal{E}_V$ is non-zero. In fact, the probability of wrong decision could become quite high in this case. So we need a robust decision which can make a correct decision under multiple errors in the verification state. The following theorem describes how to extend Theorem 15.5 in this regards.

THEOREM 15.6 *If the sample number L_V to be observed for verification of synchronization is chosen such that*

$$W[\mathcal{Z}_V] \geq 2N_V + 1 \quad for \ \mathcal{E}_A \neq \mathbf{0}, \tag{15.12}$$

for a number N_V, then the relation

$$\begin{cases} W[\mathcal{Y}_V] \leq N_V, & if \ \mathcal{E}_A = \mathbf{0}, \\ W[\mathcal{Y}_V] \geq N_V + 1, & if \ \mathcal{E}_A \neq \mathbf{0}, \end{cases} \tag{15.13}$$

holds for $W[\mathcal{E}_V] \leq N_V$.

Proof : If $\mathcal{E}_A = \mathbf{0}$, by (15.9) $\mathcal{Y}_V = \mathcal{E}_V$, and hence $W[\mathcal{Y}_V] = W[\mathcal{E}_V]$. Therefore, $W[\mathcal{Y}_V] \leq N_V$ since $W[\mathcal{E}_V] \leq N_V$ by assumption. On the other hand, if $\mathcal{E}_A \neq \mathbf{0}$, by (15.9) we have the relation $W[\mathcal{Y}_V] \geq W[\mathcal{Z}_V] - W[\mathcal{E}_V]$. But $W[\mathcal{Z}_V] \geq 2N_V + 1$ by (15.12), and $W[\mathcal{E}_V] \leq N_V$ by the assumption. Therefore, $W[\mathcal{Y}_V] \geq N_V + 1$. ∎

The theorem states that in case $W[\mathcal{E}_V] \leq N_V$, if we choose the observation sample number L_V to meet the relation (15.12), the weight of the sample-comparison vector $\mathcal{Y}_V$ is $W[\mathcal{Y}_V] \leq N_V$ for $\mathcal{E}_A = \mathbf{0}$ and $W[\mathcal{Y}_V] \geq N_V + 1$ for $\mathcal{E}_A \neq \mathbf{0}$. Therefore, if we choose the sample-comparison data number L_V meeting (15.12), we can correctly verify the synchronization status as long as the number of errors occurred in the verification state does not exceed N_V. The observation sample number L_V will become larger as the tolerable number N_V of verification state errors increases. This implies that a longer decision time is necessary in order to avoid wrong decisions under more errors, and the probability of wrong decision becomes lower in this case. Note that Theorem 15.5 is a special case of Theorem 15.6, in which $N_V = 0$.

EXAMPLE 15.4 We continue the previous example, which corresponds to an example of Theorem 15.5 for $N_V = 0$. In case $N_V = 1$, we observe from Table 15.2 that $W[\mathcal{Z}_V] \geq 3$ for all non-zero sample error vectors $\mathcal{E}_A$'s if L_V is 7 or larger. Similarly, in case $N_V = 2$, we find that $W[\mathcal{Z}_V] \geq 5$ for all $\mathcal{E}_A \neq \mathbf{0}$ if L_V is 13 or larger. Therefore, if we choose $L_V = 7$, and make decision whether or not the descrambler is synchronized to the scrambler based on (15.13), then the decision is correct as long as there occurs a single or no error in the verification state. Likewise, if we choose $L_V = 13$, and make decision on the synchronization relying on (15.13), then the decision is correct as long as two or less errors occur in the verification state. ♣

We now develop Theorem 15.6 into a form in which the acquisition state error $\mathcal{E}_A$ is incorporated in a more explicit manner.

THEOREM 15.7 *If the sample number L_V to be observed for verification of synchronization is chosen such that*

$$W[\mathcal{Z}_V] \geq 2N_V + 1 \quad for \ 0 \leq W[\mathcal{E}_A] \leq N_A, \tag{15.14}$$

for two numbers N_V and N_A, then the relation

$$\begin{cases} W[\mathcal{Y}_V] \leq N_V, & if \ \mathcal{E}_A = \mathbf{0}, \\ W[\mathcal{Y}_V] \geq N_V + 1, & if \ 1 \leq W[\mathcal{E}_A] \leq N_A, \end{cases} \tag{15.15}$$

holds for $W[\mathcal{E}_V] \leq N_V$.

Proof : This theorem can be proved in a manner similar to that of Theorem 15.6. ∎

The theorem states that in case $W[\mathcal{E}_A] \leq N_A$ and $W[\mathcal{E}_V] \leq N_V$, if we choose the observed sample number L_V to meet (15.14), the weight of the sample-comparison vector $\mathcal{Y}_V$ is $W[\mathcal{Y}_V] \leq N_V$ for $\mathcal{E}_A = \mathbf{0}$ and $W[\mathcal{Y}_V] \geq N_V + 1$ for $\mathcal{E}_A \neq \mathbf{0}$. Therefore, if we choose the sample-comparison data number L_V meeting (15.14), we can correctly verify the synchronization status as long as a maximum of N_A errors occur in the acquisition state and a maximum of N_V errors occur in the verification state. Comparing (15.12) and (15.14), we observe that the observation sample number L_V meeting (15.14) is smaller than that meeting (15.12). So the required decision time decreases as the number N_A of the acquisition state sample errors gets more restricted, but the probability to make a wrong decision increases accordingly.

EXAMPLE 15.5 In the previous example, we consider the case of $N_A = 1$ and $N_V = 2$. Then, from Table 15.2 we observe that $W[\mathcal{Z}_V] \geq 5$ for all $\mathcal{E}_A \neq \mathbf{0}$ if L_V is 10 or larger. Therefore, if we choose $L_V = 10$, and make decision on the synchronization based on (15.15), then the decision is correct as long as there occurs a single error in the acquisition state and two errors in the verification state. ♣

15.4   Synchronization Confirmation in Steady State

We finally consider the operation in the steady state. The objective function of the steady state is to confirm that the descrambler stays in the synchronization state.

For the scrambling and the descrambling sequences $\{ s_k \}$ and $\{ \hat{s}_k \}$ in the scrambling space $V[\Psi(x)]$, we denote by $\{ \sigma_k^S \}$ the *error sequence* representing their difference, that is, $\{ \sigma_k^S \} \equiv \{ s_k \} + \{ \hat{s}_k \}$. Then, the error sequence $\{ \sigma_k^S \}$ is represented by

$$\{ \sigma_k^S \} = \sigma^t \cdot \mathbf{E}_{\Psi(x)}, \qquad (15.16)$$

where the *initial error vector* σ is the difference between the initial vectors $\mathbf{s}$ and $\hat{\mathbf{s}}$ respectively of the scrambling sequence $\{ s_k \}$ and the descrambling sequence $\{ \hat{s}_k \}$, that is, $\sigma \equiv \mathbf{s} + \hat{\mathbf{s}}$. Then, the initial error vector σ is zero if the descrambler is synchronized to the scrambler, and non-zero otherwise.

In the steady state, we observe the L_S sample-comparison data η_i^S, $i = 0, 1, \cdots, L_S - 1$, and based on this observation we confirm whether or not the initial error vector σ is zero. For this, we define the L_S-vectors for the sample-comparison data η_i^S, the estimation-error sample ζ_i^S in the steady state, and the sample error ε_i^S in the steady state as follows :

DEFINITION 15.8 (Sample Vectors in Steady State) We define the *sample-comparison vector* $\mathcal{Y}_S$, the *sample vector* $\mathcal{Z}_S$ *of the error sequence*, and the *sample error vector* $\mathcal{E}_S$ *in the steady state* respectively to be

$$\mathcal{Y}_S \equiv [\,\eta_0^S \ \ \eta_1^S \ \ \cdots \ \ \eta_{L_S-1}^S\,]^t, \tag{15.17a}$$

$$\mathcal{Z}_S \equiv [\,\zeta_0^S \ \ \zeta_1^S \ \ \cdots \ \ \zeta_{L_S-1}^S\,]^t, \tag{15.17b}$$

$$\mathcal{E}_S \equiv [\,\varepsilon_0^S \ \ \varepsilon_1^S \ \ \cdots \ \ \varepsilon_{L_S-1}^S\,]^t. \tag{15.17c}$$

Then, by (15.7), the sample-comparison vector $\mathcal{Y}_S$ is represented by

$$\mathcal{Y}_S = \mathcal{Z}_S + \mathcal{E}_S. \tag{15.18}$$

Note that the sample vector $\mathcal{Z}_S$ is formed out of the sampled values of the error sequence $\{\,\sigma_k^S\,\}$, which depends on the initial error vector σ.

The following theorem describes how to confirm whether or not the descrambler stays in synchronization with the scrambler.

THEOREM 15.9 *If the sample number L_S to be observed for confirmation of synchronization is chosen such that*

$$W[\mathcal{Z}_S] \geq 2N_S + 1 \quad for \ \sigma \neq \mathbf{0}, \tag{15.19}$$

for a number N_S, then the relation

$$\begin{cases} W[\mathcal{Y}_S] \leq N_S, & if \ \sigma = \mathbf{0}, \\ W[\mathcal{Y}_S] \geq N_S + 1, & if \ \sigma \neq \mathbf{0}, \end{cases} \tag{15.20}$$

holds for $W[\mathcal{E}_S] \leq N_S$.

Proof : If $\sigma = \mathbf{0}$, $\mathcal{Z}_S = \mathbf{0}$, and by (15.18) $\mathcal{Y}_S = \mathcal{E}_S$. Therefore, $W[\mathcal{Y}_S] \leq N_S$ since $W[\mathcal{E}_S] \leq N_S$ by assumption. On the other hand, if $\sigma \neq \mathbf{0}$, by (15.18) we have the relation $W[\mathcal{Y}_S] \geq W[\mathcal{Z}_S] - W[\mathcal{E}_S]$. But, $W[\mathcal{Z}_S] \geq 2N_S + 1$ by (15.19), and $W[\mathcal{E}_S] \leq N_S$ by assumption. Therefore, $W[\mathcal{Y}_S] \geq N_S + 1$. ∎

Table 15.3. Sampled values ζ_i^S's of error sequence $\{ \sigma_k^S \}$ for various initial error vectors

error vector σ	sampled values of error sequence			
	$\zeta_0^S \cdots \zeta_3^S$	$\zeta_4^S \cdot \zeta_6^S$	$\zeta_7^S \cdots \zeta_{12}^S$	$\zeta_{13}^S \cdots$
$[0\,0\,0\,0]^t$	0 0 0 0	0 0 0	0 0 0 0 0 0	0 $\cdots$
$[0\,0\,0\,1]^t$	0 0 1 1	0 1 0	0 0 1 1 0 1	0 $\cdots$
$[0\,0\,1\,0]^t$	0 1 1 1	0 0 1	0 1 1 1 0 0	1 $\cdots$
$[0\,0\,1\,1]^t$	0 1 0 0	0 1 1	0 1 0 0 0 1	1 $\cdots$
$[0\,1\,0\,0]^t$	0 0 0 1	1 0 1	0 0 0 1 1 0	1 $\cdots$
$[0\,1\,0\,1]^t$	0 0 1 0	1 1 1	0 0 1 0 1 1	1 $\cdots$
$[0\,1\,1\,0]^t$	0 1 1 0	1 0 0	0 1 1 0 1 0	0 $\cdots$
$[0\,1\,1\,1]^t$	0 1 0 1	1 1 0	0 1 0 1 1 1	0 $\cdots$
$[1\,0\,0\,0]^t$	1 0 1 0	0 0 1	1 0 1 0 0 0	1 $\cdots$
$[1\,0\,0\,1]^t$	1 0 0 1	0 1 1	1 0 0 1 0 1	1 $\cdots$
$[1\,0\,1\,0]^t$	1 1 0 1	0 0 0	1 1 0 1 0 0	0 $\cdots$
$[1\,0\,1\,1]^t$	1 1 1 0	0 1 0	1 1 1 0 1 0	0 $\cdots$
$[1\,1\,0\,0]^t$	1 0 1 1	1 0 0	1 0 1 1 1 0	0 $\cdots$
$[1\,1\,0\,1]^t$	1 0 0 0	1 1 0	1 0 0 0 1 1	0 $\cdots$
$[1\,1\,1\,0]^t$	1 1 0 0	1 0 1	1 1 0 0 1 0	1 $\cdots$
$[1\,1\,1\,1]^t$	1 1 1 1	1 1 1	1 1 1 1 1 1	1 $\cdots$

The theorem states that in case $W[\mathcal{E}_S] \le N_S$, if we choose the observed sample number L_S to meet (15.19), the weight of the sample-comparison vector $\mathcal{Y}_S$ is $W[\mathcal{Y}_S] \le N_S$ for $\sigma = \mathbf{0}$ and $W[\mathcal{Y}_S] \ge N_S + 1$ for $\sigma \ne \mathbf{0}$. Therefore, if we choose the sample-comparison number L_S meeting (15.19), we can correctly confirm the synchronization status as long as the number of errors occurred in the steady state does not exceed N_S.

EXAMPLE 15.6 In the case of the scrambling sequence $\{ s_k \}$ in the scrambling space $V[x^4 + x^3 + x^2 + 1]$ whose sampling times are $\alpha_i = 2i$, $i = 0, 1, \cdots$, we obtain, from Table 4.1, the sampled values ζ_i^S's of the error sequence $\{ \sigma_k \}$ shown in Table 15.3. We can find from the table that the observation sample number L_S meeting (15.19) for $N_S = 0$(or 1, 2) is 4(or 7, 13) or larger. Therefore, if we choose $L_S = 4$(or 7, 13), and make decision on the synchronization based on (15.20), then the decision is correct as long as no error(or 1, 2 errors) occurs in the steady state. ♣

Chapter 16

Applications to Cell-Based ATM and High-Speed Data Networks

We now consider how to apply the serial, parallel and multibit-parallel DSS techniques along with synchronization mechanisms we have developed in the previous chapters to scrambling in practical transmission networks. We first describe the DSS operation used in the cell-based ATM transmission for BISDN, which is the most typical lightwave transmission system employing the DSS. Then, we consider how to achieve concurrent sampling and immediate correction for eliminating additional timing circuits in this application. Also we consider how to apply the parallel DSS techniques to the cell-based ATM scrambling in conjunction with concurrent parallel sampling and immediate parallel correction. Further, we examine the synchronization mechanisms of the ATM scrambling and analyze their performances. Finally, as an extension to the ATM applications, we discuss how to design DSS scrambler and descrambler for use in future high-speed data networks.

16.1 Scrambling of Cell-Based ATM Signals

The *asynchronous transfer mode*(ATM) is a communication mode standardized by ITU-T as a means to service integration in the *broadband integrated services digital network*(BISDN). In the BISDN, the information data are carried in the form of ATM cells, which consist of 5 header bytes and 48 payload bytes. The ATM cells can be either carried over the payload space of the STM-1(or STM-4) signal or self-carried in a contigu-

ous stream, both at the rate of 155.520(or 622.080) Mbps. The former
is called the *SDH-based ATM transmission*, and the latter is called the
cell-based ATM transmission. In the case of the SDH-based ATM transmission, the ATM-cell stream at 155.520(or 622.080) Mbps gets first
self-synchronously scrambled,[1] and then mapped to the payload space
of STM-1(or STM-4) signal, which is frame-synchronously scrambled
before transmission as for usual STM-N signals described in Chapter 9.
In the case of the cell-based ATM transmission, the ATM-cell stream is
distributed-sample scrambled once, and then directly transmitted.

In the scrambling of the cell-based ATM signal, the DSS employs the
scrambling sequence $\{ s_k \}$ in the primitive scrambling space $V[x^{31} + x^3 + 1]$ of dimension $L = 31$, which is generated by the SRG shown in Fig.
16.1.[2] This SRG is the SSRG of the sequence space $V[x^{31} + x^3 + 1]$, whose
state transition matrix $\mathbf{T}$ and the generating vector $\mathbf{h}$ are respectively

$$\mathbf{T} = \mathbf{A}_{x^{31}+x^3+1}^t, \tag{16.1a}$$

$$\mathbf{h} = \mathbf{e}_0 + \mathbf{e}_3. \tag{16.1b}$$

The 31 samples z_0 through z_{30} taken from the scrambling sequence $\{ s_k \}$ are conveyed to the descrambler over the *header error control* (HEC)
field in the ATM cell header in unit of two samples per ATM cell, as
illustrated in Fig. 16.2. This implies that the ith sample transmission
times γ_i, $i = 0, 1, \cdots, 30$ are[3]

$$\left\{ \begin{array}{ll} \gamma_{2i} = 424i, & i = 0, 1, \cdots, 15, \\ \gamma_{2i+1} = 424i + 1, & i = 0, 1, \cdots, 14, \end{array} \right. \tag{16.2}$$

since the ATM cell is composed of 53 bytes or 424 bits. The sampling
times α_i, $i = 0, 1, \cdots, 30$, of the samples z_i's are supposed to be uniformly
distributed, i.e.,

$$\alpha_i = 212i - 211, \ i = 0, 1, \cdots, 30. \tag{16.3}$$

Comparing the sample transmission times γ_i's and the sampling times
α_i's, we find that the sample transmission time and the sampling time

---

[1]A more detailed description on self-synchronous scrambling of the SDH-based
ATM signal can be found in Chapter 20.

[2]Note that the polynomial $x^{31} + x^3 + 1$ is the characteristic polynomial $\Psi(x)$ which
is the reciprocal of the "characteristic" polynomial $C(x)$ listed in Table 1.3. Refer to
footnote 6 in Section 5.6.

[3]Note that the reference time 0 can be chosen arbitrarily. For convenience, we
take the sample transmission time of the 0th sample z_0 as the reference time 0.

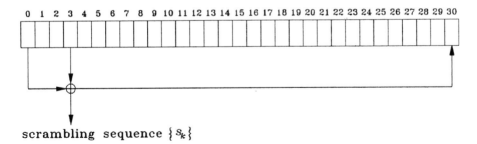

scrambling sequence $\{s_k\}$

Figure 16.1. The SRG employed in the scrambling of the cell-based ATM transmission signal.

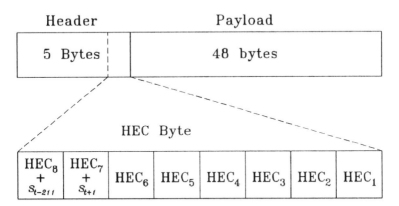

Figure 16.2. Data structure for sample conveyance within the ATM cell.

for an even-indexed sample z_{2i}, $i = 0, 1, \cdots, 15$, do not coincide, while those for an odd-indexed sample z_{2i+1}, $i = 0, 1, \cdots, 14$, do. That is, each even-indexed sample is transmitted after being stored for 211 bit-times. In Fig. 16.2, the sample s_{t-211} indicates such an even-indexed sample which is transmitted at the sample transmission time t but is taken at the sampling time $t - 211$; and s_{t+1} indicates the odd-indexed sample which is sampled and transmitted at time $t + 1$. Therefore, the scrambler for distributed-sample scrambling of the cell-based ATM transmission signal can be drawn as in Fig. 16.3. In the figure, we observe that two kinds of timing signals are required -- one for generating the sample transmission times γ_i's in (16.2), and the other for generating the sampling times α_i's in (16.3).

In order to check if the sampling times α_i's in (16.3) are predictable, we insert (16.3) and $\Psi(x) = x^{31} + x^3 + 1$ to (11.16). Then, we obtain the discrimination matrix

$$
\Delta_\alpha =
\left[
\begin{array}{l}
1\,0\,1\,1\,0\,0\,0\,1\,0\,1\,0\,0\,1\,1\,1\,1\,1\,1\,0\,0\,0\,1\,1\,1\,0\,1\,1\,0\,0\,0 \\
0\,1\,0\,0\,0\,0\,0\,0\,0\,0\,0\,0\,0\,0\,0\,0\,0\,0\,0\,0\,0\,0\,0\,0\,0\,0\,0\,0\,0\,0 \\
0\,0\,1\,0\,0\,1\,0\,0\,1\,0\,0\,1\,0\,0\,1\,0\,0\,1\,0\,0\,0\,0\,0\,0\,1\,0\,0\,0 \\
0\,0\,1\,1\,0\,1\,0\,0\,0\,1\,0\,0\,0\,0\,0\,1\,0\,0\,0\,0\,1\,1\,0\,0\,1\,0\,1\,0\,0\,0 \\
0\,0\,0\,1\,0\,0\,0\,0\,1\,0\,0\,0\,0\,0\,1\,0\,1\,0\,0\,0\,1\,0\,0\,0\,0\,0\,0\,0\,1\,0 \\
0\,0\,0\,1\,0\,1\,0\,0\,0\,1\,1\,0\,1\,1\,0\,1\,0\,1\,1\,0\,0\,1\,1\,0\,0\,1\,0\,0\,0\,1\,0 \\
0\,0\,0\,0\,0\,0\,0\,1\,0\,0\,1\,0\,0\,0\,0\,0\,0\,0\,0\,0\,1\,0\,0\,1\,0\,0\,0\,0\,0 \\
0\,0\,1\,0\,0\,0\,0\,0\,0\,0\,1\,0\,0\,1\,0\,0\,0\,1\,0\,0\,0\,0\,0\,0\,0\,1\,0\,0\,0\,0 \\
0\,0\,0\,1\,0\,0\,1\,0\,0\,1\,0\,0\,0\,0\,1\,0\,0\,0\,0\,0\,0\,0\,0\,0\,0\,0\,1\,1\,0 \\
0\,0\,1\,0\,1\,0\,0\,1\,0\,1\,1\,0\,1\,1\,0\,1\,0\,1\,1\,1\,0\,1\,0\,1\,0\,1\,1\,0\,0\,0\,0 \\
1\,0\,0\,0\,1\,0\,0\,1\,0\,0\,1\,1\,0\,1\,1\,1\,1\,0\,1\,1\,1\,1\,1\,0\,1\,1\,1\,0 \\
0\,0\,0\,0\,0\,0\,0\,0\,1\,1\,0\,1\,0\,0\,1\,1\,0\,1\,0\,0\,0\,0\,0\,0\,0\,0 \\
1\,1\,0\,1\,0\,0\,1\,0\,1\,1\,0\,1\,0\,1\,0\,0\,0\,0\,0\,0\,0\,0\,0\,0\,0\,0\,0 \\
0\,0\,1\,1\,0\,1\,0\,0\,0\,0\,0\,0\,0\,0\,0\,1\,1\,0\,0\,1\,0\,1\,0\,0\,0\,0\,0 \\
1\,0\,0\,0\,1\,0\,0\,0\,1\,0\,1\,1\,0\,0\,1\,1\,1\,1\,0\,1\,0\,1\,0\,0\,0\,0\,0\,1\,1\,0 \\
0\,0\,0\,0\,0\,1\,0\,0\,0\,0\,0\,1\,0\,0\,0\,0\,1\,0\,0\,0\,0\,0\,1\,0\,1\,0\,0\,1\,0 \\
0\,0\,1\,1\,0\,1\,0\,0\,0\,0\,0\,0\,0\,1\,1\,0\,1\,1\,1\,0\,0\,1\,0\,0\,1\,0\,0\,1\,0\,1 \\
0\,0\,1\,1\,1\,1\,1\,0\,0\,1\,1\,0\,0\,0\,1\,0\,0\,0\,1\,1\,1\,1\,0\,0\,0\,1\,0\,0\,0\,1\,0 \\
1\,1\,0\,0\,1\,0\,1\,0\,1\,0\,1\,1\,1\,0\,1\,0\,0\,1\,0\,1\,0\,1\,1\,1\,1\,1\,1\,0 \\
1\,0\,0\,1\,1\,1\,1\,0\,0\,1\,1\,0\,1\,0\,1\,1\,1\,1\,1\,0\,0\,1\,0\,1\,0\,1\,1\,0\,1\,1 \\
0\,0\,1\,0\,1\,1\,0\,1\,0\,0\,1\,1\,0\,1\,0\,0\,1\,1\,0\,1\,0\,0\,1\,1\,0\,1\,0\,0\,0\,1\,1 \\
0\,0\,0\,0\,1\,0\,1\,1\,0\,1\,1\,0\,0\,1\,0\,0\,0\,0\,0\,0\,0\,1\,0\,1\,0\,0\,0\,1\,0 \\
0\,0\,1\,0\,0\,1\,0\,0\,0\,0\,1\,0\,1\,0\,0\,0\,1\,1\,0\,1\,1\,0\,0\,0\,1\,1\,0\,0\,0\,1 \\
0\,1\,1\,0\,0\,1\,0\,0\,0\,0\,0\,1\,0\,0\,0\,1\,0\,1\,0\,0\,0\,1\,0\,0\,0\,1\,0\,0\,0\,0\,1 \\
0\,1\,0\,1\,0\,0\,1\,0\,1\,0\,1\,1\,0\,0\,1\,0\,0\,1\,1\,0\,1\,1\,0\,1\,0\,1\,1\,1\,0\,1\,1 \\
0\,0\,0\,1\,0\,1\,1\,0\,1\,0\,0\,1\,0\,0\,1\,0\,0\,1\,1\,0\,0\,1\,0\,0\,0\,0\,0\,0\,0\,0 \\
1\,1\,1\,0\,0\,1\,0\,0\,0\,1\,0\,0\,1\,1\,0\,1\,0\,0\,1\,0\,0\,1\,1\,0\,0\,0\,0\,1\,0\,1\,0 \\
1\,0\,0\,1\,0\,1\,1\,1\,0\,1\,1\,1\,0\,0\,1\,0\,0\,0\,1\,0\,1\,0\,0\,1\,0\,1\,0\,0\,0\,1 \\
1\,1\,0\,0\,0\,0\,1\,1\,1\,0\,1\,1\,0\,1\,1\,1\,0\,0\,0\,0\,0\,1\,0\,1\,1\,0\,1\,0\,1\,1 \\
0\,0\,1\,0\,0\,1\,0\,0\,0\,1\,0\,0\,0\,0\,0\,0\,1\,0\,0\,1\,1\,1\,0\,1\,0\,0\,1\,0\,0\,1\,0 \\
0\,0\,0\,0\,0\,0\,0\,0\,1\,1\,1\,0\,0\,1\,1\,1\,0\,1\,0\,0\,0\,0\,0\,0\,0\,0\,0\,0\,0 \\
\end{array}
\right]
, \quad (16.4)
$$

which is nonsingular. Therefore, the sampling times α_i's in (16.3) are predictable, so this DSS is synchronizable.

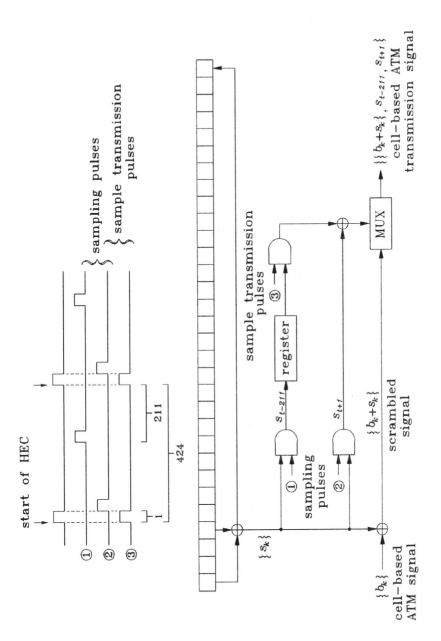

Figure 16.3. A DSS scrambler for scrambling of the cell-based ATM transmission signal.

For the correction of the descrambler SRG, the correction times β_i's are supposed to be uniformly spaced such that

$$\beta_i = 212i + 1, \, i = 0, 1, \cdots, 30. \tag{16.5}$$

If we use the SRG in Fig. 16.1 as a descrambler SRG, by (16.1), (16.3) and (12.19) we obtain the matrix

$$
\tilde{\mathbf{H}} =
\begin{bmatrix}
1 0 1 0 0 1 1 1 0 1 1 0 0 1 1 0 0 0 0 1 1 1 1 1 1 1 0 0 0 1 1 \\
0 1 0 0 1 0 0 0 0 0 0 0 0 0 0 0 0 0 0 0 0 0 0 0 0 0 0 0 0 0 0 \\
0 0 1 0 0 0 0 0 0 0 0 0 0 0 0 0 0 0 0 1 0 0 0 0 0 0 1 0 0 1 \\
0 0 1 1 0 0 1 0 1 1 0 0 1 0 0 1 0 0 1 0 0 1 1 0 1 0 0 1 1 0 1 \\
0 1 0 1 1 0 1 0 0 1 0 0 1 0 0 1 0 1 1 0 1 1 0 0 1 0 0 0 0 1 0 \\
0 1 0 1 1 1 1 0 1 1 1 0 0 0 0 0 1 1 0 0 1 0 1 0 1 0 0 0 1 1 0 \\
0 0 0 0 0 0 0 1 0 0 0 0 0 1 0 0 0 0 0 0 0 1 0 0 0 0 0 1 0 0 \\
0 0 1 0 0 1 0 0 0 0 1 0 0 0 0 0 1 1 0 0 1 0 0 0 0 0 1 0 0 1 0 \\
1 1 0 0 1 0 0 0 0 0 0 0 1 0 1 0 0 1 0 0 0 0 0 0 0 0 0 0 1 1 0 \\
0 0 1 0 1 1 0 0 0 1 0 0 0 0 0 0 1 1 0 1 1 0 1 1 1 1 0 0 1 1 0 \\
0 1 0 0 0 0 0 0 0 0 0 1 0 0 1 0 0 1 1 0 1 0 0 1 0 0 1 0 0 1 1 \\
0 0 0 0 0 0 0 0 0 0 0 1 1 0 0 1 0 0 1 0 0 1 0 1 0 0 0 0 0 0 \\
1 1 0 0 1 0 0 0 1 0 0 0 1 1 1 0 1 0 0 0 0 0 0 0 0 0 0 0 0 0 0 \\
0 0 1 1 0 0 1 0 1 0 0 0 0 0 0 0 0 1 1 0 1 0 0 1 1 0 1 0 0 \\
0 1 0 0 0 0 0 1 1 0 1 0 0 1 0 1 1 0 1 0 1 1 1 0 1 0 0 0 1 1 0 \\
0 1 0 0 1 1 0 0 1 0 0 1 0 0 1 0 0 1 0 0 1 0 0 0 1 0 1 1 0 0 0 \\
1 0 0 0 0 1 1 0 1 0 0 0 0 0 1 1 0 0 0 1 1 1 0 0 0 0 0 0 0 0 1 \\
0 1 1 1 0 0 0 1 1 0 1 0 1 1 1 0 0 1 1 1 1 0 1 1 1 1 0 0 1 1 0 \\
0 0 0 0 1 0 1 1 1 1 1 1 1 0 1 1 0 1 0 0 0 1 1 1 0 1 0 0 0 0 1 \\
1 1 1 0 0 0 0 1 1 0 1 0 0 1 0 1 1 0 1 0 0 1 1 1 1 0 1 1 1 0 0 0 0 \\
0 1 0 0 0 1 0 0 1 0 0 1 0 0 1 0 0 1 0 0 1 0 0 1 0 0 1 0 1 1 1 \\
0 1 0 0 0 0 1 0 0 0 0 0 1 0 0 0 1 0 0 0 0 0 0 1 0 1 1 0 1 1 0 \\
0 0 0 0 0 1 0 0 1 0 0 1 0 1 1 0 1 1 1 0 0 0 0 1 1 1 1 0 1 1 1 \\
0 1 0 0 1 1 0 0 1 0 0 1 0 0 1 1 0 1 1 0 1 1 0 0 1 1 0 0 1 0 1 \\
0 0 1 1 0 1 0 0 1 1 1 0 0 1 0 0 0 0 1 0 0 0 0 0 1 1 0 1 1 0 0 \\
0 0 0 1 0 1 0 0 0 1 0 0 0 0 0 0 0 0 1 0 1 0 0 0 1 0 0 0 0 0 0 \\
1 0 1 1 0 0 0 0 1 1 0 0 0 1 0 0 1 0 0 0 0 0 1 0 1 1 0 1 0 1 1 \\
1 0 1 0 0 0 0 1 1 0 0 0 1 0 0 1 1 0 1 1 0 1 1 0 0 0 1 1 0 1 1 \\
1 0 1 1 0 1 1 1 1 1 0 0 0 0 0 1 1 1 1 0 0 0 0 1 0 1 0 0 0 1 1 0 \\
0 1 1 0 1 0 0 0 1 1 0 0 1 0 0 0 1 0 0 0 1 1 1 0 1 0 0 0 0 0 0 \\
0 0 0 0 0 0 0 0 1 1 1 1 1 0 1 1 1 0 1 0 1 0 0 0 0 0 0 0 0 0 0
\end{bmatrix}
\tag{16.6}
$$

Inserting this along with $L = 31$, $\alpha = 212$ and $\beta = 1$ into (12.28), we get the common correction vector

$$
\begin{aligned}
\mathbf{c}_0 &= \mathbf{c}_1 = \cdots = \mathbf{c}_{30} \\
&= [\,0 1 1 0 1 0 0 1 1 0 1 1 1 0 0 1 1 0 0 1 1 1 0 1 1 0 0 0 1 0 0\,]^t.
\end{aligned}
$$

Using these parameters, we can obtain the descrambler in Fig. 16.4.[4] In this descrambler, three different timing signals are required for gen-

---

[4]This is the descrambler recommended by the ITU-T via Recommendation I.432.

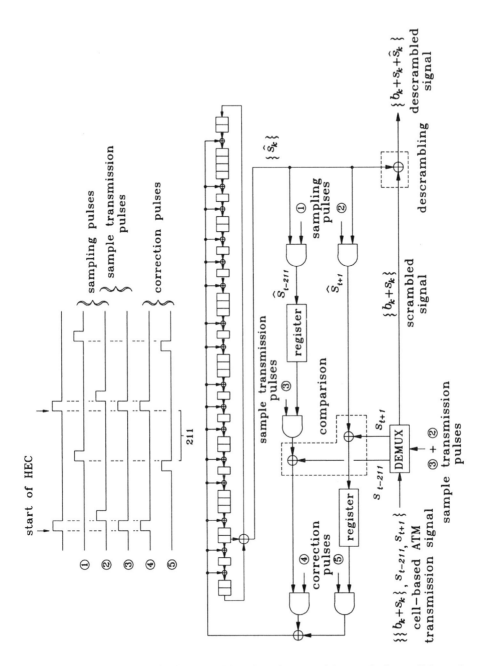

Figure 16.4. A DSS descrambler for descrambling of the cell-based ATM transmission signal(recommended by ITU-T Recommendation I.432).

erating the sample transmission times γ_i's in (16.2), the sampling times α_i's in (16.3), and the correction times β_i's in (16.5), as indicated in the figure.

If we choose another descrambler SRG whose state transition matrix and generating vectors are respectively

$$\mathbf{T} = \mathbf{A}^t_{x^{31}+x^3+1}, \tag{16.7a}$$
$$\mathbf{h} = \mathbf{e}_4 + \mathbf{e}_{25} + \mathbf{e}_{30}, \tag{16.7b}$$

then we obtain, in a similar manner to the case of the SRG in Fig. 16.1, the common correction vector

$$\begin{aligned}
\mathbf{c}_0 &= \mathbf{c}_1 = \cdots = \mathbf{c}_{30} \\
&= [\,0\,0\,0\,0\,0\,0\,0\,0\,0\,1\,0\,1\,0\,0\,0\,1\,0\,0\,1\,0\,0\,0\,0\,0\,0\,0\,0\,1\,0\,0\,0\,]^t.
\end{aligned}$$

for the correction times β_i's in (16.5). The resulting descrambler in this case is as depicted in Fig. 16.5. Note that this descrambler requires less exclusive-OR gates than the descrambler in Fig. 16.4

16.2   Equivalent DSS with Minimized Timing Circuitry

As discussed in Section 12.4, an equivalent DSS can be obtained by employing the concurrent sampling and the immediate correction, and the resulting DSS has a minimized timing circuitry as the timing circuits for generating the sampling times and the correction times are all eliminated.

For the scrambler in Fig. 16.3, we move the sampling times α_i's in (16.3) to the sample transmission times γ_i's in (16.2). Then inserting (16.1), (16.2) and (16.3) to (12.35), we get the sampling vectors

$$\left\{ \begin{aligned}
\mathbf{v}_0 &= \mathbf{v}_2 = \cdots = \mathbf{v}_{30} = (\mathbf{T}^t)^{-211} \cdot \mathbf{h} \\
&= [\,1\,0\,1\,0\,0\,1\,1\,1\,0\,1\,1\,0\,0\,1\,1\,0\,0\,0\,0\,1\,1\,1\,1\,1\,1\,1\,0\,0\,0\,1\,1\,]^t, \\
\mathbf{v}_1 &= \mathbf{v}_3 = \cdots = \mathbf{v}_{29} = \mathbf{h} \\
&= [\,1\,0\,0\,1\,0\,0\,0\,0\,0\,0\,0\,0\,0\,0\,0\,0\,0\,0\,0\,0\,0\,0\,0\,0\,0\,0\,0\,0\,0\,0\,0\,]^t,
\end{aligned} \right. \tag{16.8}$$

for the concurrent sampling. The resulting DSS scrambler is as shown in Fig. 16.6. This scrambler requires more exclusive-OR gates than the

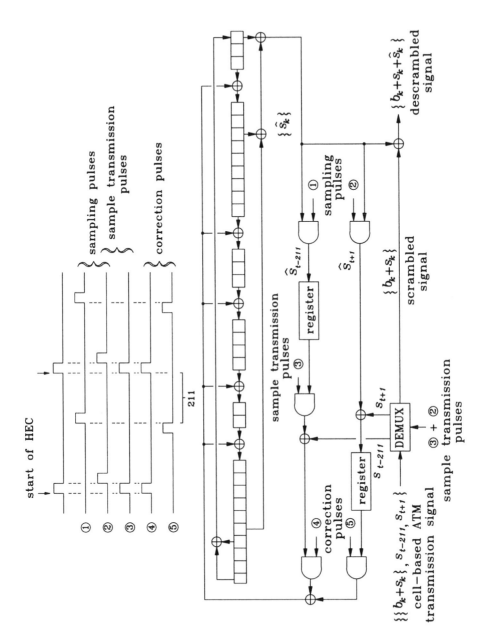

Figure 16.5. Another DSS descrambler for descrambling of the cell-based ATM transmission signal.

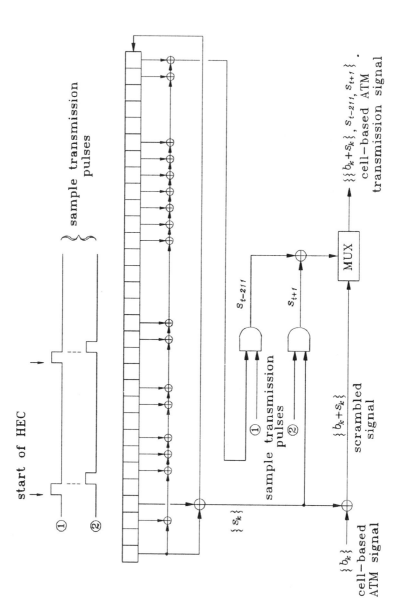

Figure 16.6. An equivalent DSS scrambler employing the concurrent sampling.

scrambler in Fig. 16.3 does, but does not require the timing circuit for generating sampling times.

In order to eliminate the timing circuits of the descrambler in Fig. 16.4, we adopt the immediate correction. For this, we choose the correction times

$$\begin{cases} \beta_{2i} = 424i + 1, i = 0, 1, \cdots, 15, \\ \beta_{2i+1} = 424i + 2, i = 0, 1, \cdots, 14. \end{cases} \tag{16.9}$$

Then, inserting (16.6) and (16.9) to (12.26), we can obtain the correction vectors

$$\begin{cases} \mathbf{c}_0 = \mathbf{c}_2 = \cdots = \mathbf{c}_{30} \\ \quad = [0\,1\,1\,0\,1\,0\,0\,1\,1\,0\,1\,1\,1\,0\,0\,1\,1\,0\,0\,1\,1\,1\,0\,1\,1\,0\,0\,0\,1\,0\,0]^t, \\ \mathbf{c}_1 = \mathbf{c}_3 = \cdots = \mathbf{c}_{29} \\ \quad = [0\,0\,0\,0\,1\,1\,0\,0\,0\,0\,0\,1\,0\,1\,1\,1\,1\,1\,1\,1\,0\,0\,1\,0\,1\,0\,1\,1\,1\,1\,0\,1]^t. \end{cases}$$

The resulting DSS descrambler is as depicted in Fig. 16.7. From the figure, we observe that the timing pulses generating the sampling and correction times in Fig. 16.4 are not necessary any longer. Note that the scrambler in Fig. 16.6 and the descrambler in Fig. 16.7 are respectively equivalent to those in Fig. 16.3 and Fig. 16.4.

16.3 An Optimal DSS Design

The DSS scrambler and descrambler circuits in Figs. 16.6 and 16.7 are minimal in the sense of minimized timing circuitry, but are not truly minimal in the overall sense. It is because a part of the circuit is consumed to realize concurrent sampling for the independently designed sampling scheme. Therefore, in this section we demonstrate how to optimize the overall DSS design by taking the sampling times to be identical to the sample transmission times from the beginning, which we may call *immediate sampling*. The resulting scrambler and the descrambler are not equivalent to those in Figs. 16.3, 16.4, 16.6 and 16.7. Nonetheless, they do envision a guideline for an optimal DSS design with the minimized circuitry.

We choose the sampling times α_i's to be identical to the sample transmission times γ_i's in (16.2). Then, inserting (16.2) and $\Psi(x) =$

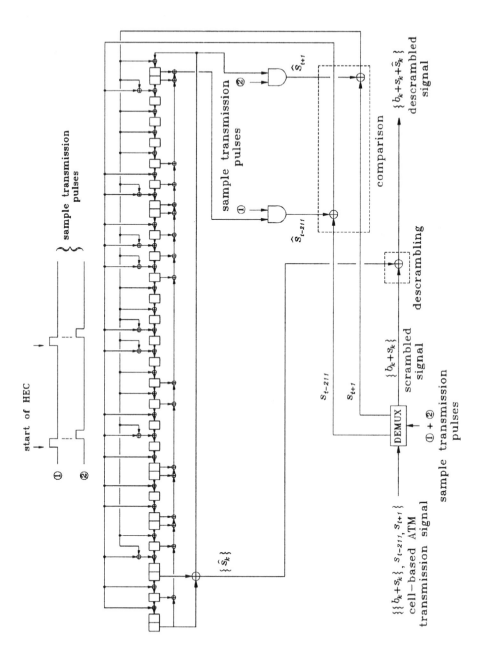

Figure 16.7. An equivalent descrambler employing the concurrent sampling and the immediate correction.

HEC$_8$ + s_t	HEC$_7$ + s_{t+1}	HEC$_6$	HEC$_5$	HEC$_4$	HEC$_3$	HEC$_2$	HEC$_1$

Figure 16.8. Modified data structure for sample conveyance within the ATM cell.

$x^{31} + x^3 + 1$ into (11.16), we obtain the discrimination matrix

$$\Delta_\alpha =
\begin{bmatrix}
1000000000000000000000000000000\\
0100000000000000000000000000000\\
0110100010000010000011001010000\\
0011010001000001000001100101000\\
0010100011011010110011001000100\\
0001010001101101011001100100010\\
0100000001001000100000000100000\\
0010000000100100000100000010000\\
0101001011011010111010101100000\\
0010100101101101011101010110000\\
0000000000110100110100000000000\\
0000000000011010011010000000000\\
0110100000000000011001010000000\\
0011010000000000011001010000000\\
0000100000100000100000010100100\\
0000010000010000010000001010010\\
0111110011000100011110001000100\\
0011111001100010001111000100010\\
0001110011010111100010101101111\\
1001111001101011110001010110011\\
0001011011001000000001010001000\\
0000101011001000000010100010\\
1100100000100010100010001000010\\
0110010000010001010001000100001\\
0010110100100100110010000000000\\
0001011010010010011001000000000\\
0000111011001010001010010100011\\
1001011101100101000101001010001\\
0100100010000001001110100100100\\
0010010001000000100111010010010\\
0111011000010010010010010001101
\end{bmatrix},$$

$$(16.10)$$

which is also nonsingular. So, the new sampling times are also predictable. In this case, the data structure for the sample conveyance is modified as shown in Fig. 16.8. For these sampling times, if we choose the SRG in Fig. 16.1 as the scrambler SRG, we obtain the scrambler shown in Fig. 16.9. This scrambler turns out to be an optimal scrambler as it requires neither the additional timing circuits for generating the timing clocks needed in the scrambler of Fig. 16.3 nor the additional exclusive-OR gates appearing in the scrambler of Fig. 16.6.

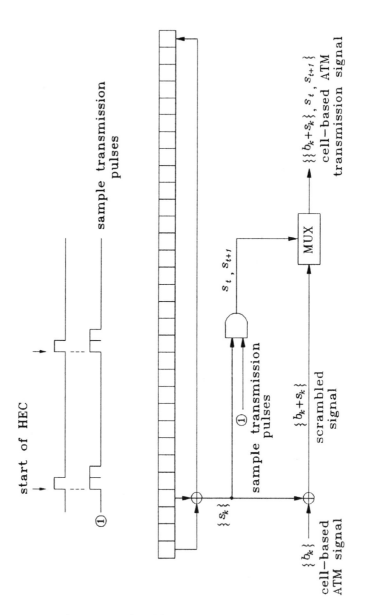

Figure 16.9. An optimal DSS scrambler employing the immediate sampling.

For the relevant optimal descrambler design, we choose the immediate correction times β_i's in (16.9). Then, inserting (16.1) and (16.2) into (12.19), we obtain the matrix

$$
\tilde{H} =
\begin{bmatrix}
1\,0\,0\,1\,0\,0\,0\,0\,0\,0\,0\,0\,0\,0\,0\,0\,0\,0\,0\,0\,0\,0\,0\,0\,0\,0\,0\,0\,0\,0\,0\,0 \\
0\,1\,0\,0\,1\,0\,0\,0\,0\,0\,0\,0\,0\,0\,0\,0\,0\,0\,0\,0\,0\,0\,0\,0\,0\,0\,0\,0\,0\,0\,0\,0 \\
0\,1\,1\,0\,0\,1\,0\,1\,1\,0\,0\,1\,0\,0\,1\,0\,0\,1\,0\,0\,1\,1\,0\,1\,0\,0\,1\,1\,0\,1\,0 \\
0\,0\,1\,1\,0\,0\,1\,0\,1\,1\,0\,0\,1\,0\,0\,1\,0\,0\,1\,0\,0\,1\,1\,0\,1\,0\,0\,1\,1\,0\,1 \\
1\,0\,1\,1\,1\,1\,0\,1\,1\,1\,0\,0\,0\,0\,1\,1\,0\,0\,1\,0\,1\,0\,1\,0\,0\,0\,1\,1\,0\,0 \\
0\,1\,0\,1\,1\,1\,1\,0\,1\,1\,1\,0\,0\,0\,0\,0\,1\,1\,0\,0\,1\,0\,1\,0\,1\,0\,0\,0\,1\,1\,0 \\
0\,1\,0\,0\,1\,0\,0\,0\,0\,1\,0\,0\,0\,0\,0\,1\,1\,0\,0\,1\,0\,0\,0\,0\,0\,1\,0\,0\,1\,0\,0 \\
0\,0\,1\,0\,0\,1\,0\,0\,0\,0\,1\,0\,0\,0\,0\,0\,1\,1\,0\,0\,1\,0\,0\,0\,0\,0\,1\,0\,0\,1\,0 \\
0\,1\,0\,1\,1\,0\,0\,0\,1\,0\,0\,0\,0\,0\,0\,1\,1\,0\,1\,1\,0\,1\,1\,1\,1\,0\,0\,1\,1\,0\,0 \\
0\,0\,1\,0\,1\,1\,0\,0\,0\,1\,0\,0\,0\,0\,0\,0\,1\,1\,0\,1\,1\,0\,1\,1\,1\,1\,0\,0\,1\,1\,0 \\
0\,0\,0\,0\,0\,0\,0\,0\,0\,0\,1\,1\,0\,0\,1\,0\,0\,1\,0\,0\,1\,0\,1\,0\,0\,0\,0\,0\,0\,0 \\
0\,0\,0\,0\,0\,0\,0\,0\,0\,0\,1\,1\,0\,0\,1\,0\,0\,1\,0\,0\,1\,0\,1\,0\,0\,0\,0\,0\,0\,0 \\
0\,1\,1\,0\,0\,1\,0\,1\,0\,0\,0\,0\,0\,0\,0\,0\,1\,1\,0\,1\,0\,0\,1\,1\,0\,1\,0\,0\,0 \\
0\,0\,1\,1\,0\,0\,1\,0\,1\,0\,0\,0\,0\,0\,0\,0\,0\,1\,1\,0\,1\,0\,0\,1\,1\,0\,1\,0\,0 \\
1\,0\,0\,1\,1\,0\,0\,1\,0\,0\,1\,0\,0\,1\,0\,0\,1\,0\,0\,1\,0\,0\,0\,1\,0\,1\,1\,0\,0 \\
0\,1\,0\,0\,1\,1\,0\,0\,1\,0\,0\,1\,0\,0\,1\,0\,0\,1\,0\,0\,1\,0\,0\,0\,1\,0\,1\,1\,0\,0\,0 \\
1\,1\,1\,0\,0\,0\,1\,1\,0\,1\,0\,1\,1\,1\,0\,0\,1\,1\,1\,1\,0\,1\,1\,1\,1\,0\,0\,1\,1\,0\,0 \\
0\,1\,1\,1\,0\,0\,0\,1\,1\,0\,1\,0\,1\,1\,1\,0\,0\,1\,1\,1\,1\,0\,1\,1\,1\,1\,0\,0\,1\,1\,0 \\
1\,1\,1\,0\,0\,0\,1\,1\,0\,1\,0\,0\,1\,1\,0\,1\,0\,0\,1\,1\,1\,1\,0\,1\,1\,1\,0\,0\,0\,0\,1 \\
1\,1\,1\,0\,0\,0\,0\,1\,1\,0\,1\,0\,0\,1\,1\,0\,1\,0\,0\,1\,1\,1\,1\,0\,1\,1\,1\,0\,0\,0\,0 \\
1\,0\,0\,0\,0\,1\,0\,0\,0\,0\,0\,1\,0\,0\,0\,1\,0\,0\,0\,0\,0\,0\,1\,0\,1\,1\,0\,1\,1\,0\,0 \\
0\,1\,0\,0\,0\,0\,1\,0\,0\,0\,0\,0\,1\,0\,0\,0\,1\,0\,0\,0\,0\,0\,0\,1\,0\,1\,1\,0\,1\,1\,0 \\
1\,0\,0\,1\,1\,0\,0\,1\,0\,0\,1\,0\,0\,1\,1\,0\,1\,1\,0\,1\,1\,0\,0\,1\,1\,0\,0\,1\,0\,1\,0 \\
0\,1\,0\,0\,1\,1\,0\,0\,1\,0\,0\,1\,0\,0\,1\,1\,0\,1\,1\,0\,1\,1\,0\,0\,1\,1\,0\,0\,1\,0\,1 \\
0\,0\,1\,0\,1\,0\,0\,0\,1\,0\,0\,0\,0\,0\,0\,1\,0\,1\,0\,0\,0\,1\,0\,0\,0\,0\,1\,0 \\
0\,0\,0\,1\,0\,1\,0\,0\,0\,1\,0\,0\,0\,0\,0\,0\,0\,1\,0\,1\,0\,0\,0\,1\,0\,0\,0\,0\,0\,0 \\
0\,1\,1\,0\,0\,0\,1\,1\,0\,0\,0\,1\,0\,0\,1\,1\,0\,1\,1\,0\,1\,1\,0\,0\,0\,1\,1\,0\,1\,1\,1 \\
1\,0\,1\,0\,0\,0\,0\,1\,1\,0\,0\,0\,1\,0\,0\,1\,1\,0\,1\,1\,0\,1\,1\,0\,0\,0\,1\,1\,0\,1\,1 \\
1\,1\,0\,1\,0\,0\,0\,1\,1\,0\,0\,1\,0\,0\,0\,1\,0\,0\,0\,1\,1\,1\,0\,1\,0\,0\,0\,0\,0\,0\,0 \\
0\,1\,1\,0\,1\,0\,0\,0\,1\,1\,0\,0\,1\,0\,0\,0\,1\,0\,0\,0\,1\,1\,1\,0\,1\,0\,0\,0\,0\,0\,0 \\
1\,1\,0\,0\,1\,1\,0\,0\,1\,1\,0\,1\,0\,0\,0\,0\,0\,0\,0\,0\,0\,0\,0\,0\,1\,1\,1\,0\,0
\end{bmatrix},
$$

and inserting this and (16.9) to (12.26), we can obtain the correction vectors

$$
\begin{cases}
c_0 = c_2 = \cdots = c_{30} \\
\quad = [1\,0\,0\,1\,0\,1\,0\,1\,0\,0\,0\,0\,0\,0\,0\,1\,1\,1\,0\,0\,1\,0\,0\,1\,1\,0\,0\,1\,1\,0\,1\,1\,1]^t, \\[4pt]
c_1 = c_3 = \cdots = c_{29} \\
\quad = [1\,1\,1\,0\,0\,1\,0\,0\,1\,0\,0\,1\,1\,0\,0\,1\,1\,1\,0\,0\,1\,0\,0\,0\,0\,0\,0\,0\,0\,0\,1]^t.
\end{cases}
$$

The resulting descrambler takes the shape shown in Fig. 16.10. Again, we find that it is an optimal descrambler as it does not require the additional timing circuits for generating various timing pulses needed in the descrambler of Fig. 16.4, yet with the required number of exclusive-OR gates reduced far below that of the descrambler in Fig. 16.7. Note that

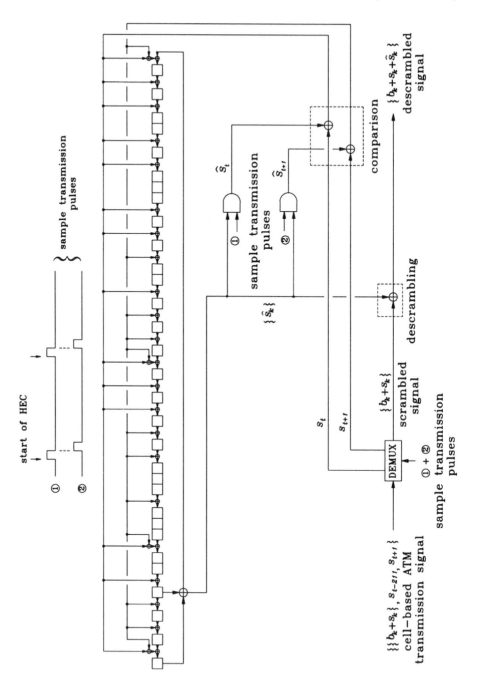

Figure 16.10. An optimal DSS descrambler employing the immediate correction as well as the immediate sampling.

the optimal descrambler in Fig. 16.10 is compatible with the optimal scrambler in Fig. 16.9 but not with those in Figs. 16.3 and 16.6.[5]

16.4   Parallel DSS for Cell-Based ATM Signals

We now consider how to apply the parallel scrambling techniques developed in Chapter 13 to the scrambling of the cell-based ATM signal.

In order to apply the parallel DSS techniques, we regard the cell-based ATM signal $\{ b_k \}$ as the bit-interleaved sequence of the eight parallel sequences $\{ b_k^j \}$, $j = 0, 1, \cdots, 7$.[6] If we employ this parallel scrambling scheme, the serial scrambling rate of 155.520(or 622.080) Mbps drops to an eighth, which is 19.44(or 77.76) Mbps.

To realize a PSRG generating the eight parallel scrambling sequences $\{ s_k^j \}$, $j = 0, 1, \cdots, 7$, we choose the nonsingular matrix

$$\mathbf{R} = [\, \mathbf{e}_0 \quad \mathbf{A}_{x^{31}+x^3+1}^8 \cdot \mathbf{e}_0 \quad \mathbf{A}_{x^{31}+x^3+1}^{8\times 2} \cdot \mathbf{e}_0 \quad \cdots \quad \mathbf{A}_{x^{31}+x^3+1}^{8\times 30} \cdot \mathbf{e}_0 \,]^t,$$

$$(16.11)$$

and insert this to (13.1a) and (13,1c) respectively. Then, we obtain the state transition matrix

$$\mathbf{T} = \mathbf{A}_{x^{31}+x^3+1}^t \qquad (16.12a)$$

and the generating vectors

$$\begin{cases}
\mathbf{h}_0 = [\, 1\,0\,0\,0\,0\,0\,0\,0\,0\,0\,0\,0\,0\,0\,0\,0\,0\,0\,0\,0\,0\,0\,0\,0\,0\,0\,0\,0\,0\,0\,0 \,]^t, \\
\mathbf{h}_1 = [\, 0\,0\,1\,0\,1\,0\,0\,0\,0\,0\,0\,0\,0\,0\,0\,0\,1\,0\,0\,0\,0\,0\,0\,0\,0\,0\,0\,0\,0\,0\,0 \,]^t, \\
\mathbf{h}_2 = [\, 0\,1\,0\,0\,0\,0\,0\,0\,1\,0\,0\,0\,0\,0\,0\,0\,0\,0\,0\,0\,0\,0\,0\,0\,0\,0\,0\,0\,0\,0\,0 \,]^t, \\
\mathbf{h}_3 = [\, 0\,0\,0\,1\,0\,1\,0\,0\,0\,0\,1\,0\,1\,0\,0\,0\,0\,1\,0\,0\,0\,0\,0\,0\,1\,0\,0\,0\,0\,0\,0 \,]^t, \\
\mathbf{h}_4 = [\, 0\,0\,1\,0\,0\,0\,0\,0\,0\,0\,0\,0\,0\,0\,0\,0\,1\,0\,0\,0\,0\,0\,0\,0\,0\,0\,0\,0\,0\,0\,0 \,]^t, \\
\mathbf{h}_5 = [\, 0\,1\,0\,0\,0\,0\,1\,0\,0\,0\,0\,0\,0\,0\,0\,0\,0\,0\,1\,0\,0\,0\,0\,0\,0\,0\,0\,0\,0\,0\,0 \,]^t, \\
\mathbf{h}_6 = [\, 0\,0\,0\,1\,0\,0\,0\,0\,0\,0\,1\,0\,0\,0\,0\,0\,0\,1\,0\,0\,0\,0\,0\,0\,1\,0\,0\,0\,0\,0\,0 \,]^t, \\
\mathbf{h}_7 = [\, 0\,0\,1\,0\,0\,0\,0\,1\,0\,1\,0\,0\,0\,0\,1\,0\,0\,0\,0\,0\,0\,1\,0\,0\,0\,0\,0\,0\,1\,0\,0 \,]^t.
\end{cases}$$

$$(16.12b)$$

The resulting PSRG is as shown in Fig. 16.11.

---

[5]It is because the optimal scrambler employs a different data structure(compare Figs. 16.2 and 16.8). If we had had enough knowledge on the DSS(as discussed in this book) at the time the data structure in Fig. 16.2 was recommended, we must have rather opted for the data structure in Fig. 16.8.

[6]In general, we can regard the cell-based ATM signal $\{ b_k \}$ as the bit-interleaved sequence of the N parallel sequences $\{ b_k^j \}$, $j = 0, 1, \cdots, N - 1$. Among various possible choices of N, we take $N = 8$ as it matches the byte-level parallel processing.

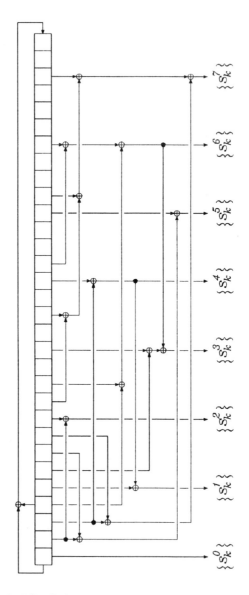

Figure 16.11. A PSRG for parallel scrambling of the cell-based ATM transmission signal.

By Theorem 13.2, the parallel sampling times $\hat{\alpha}_i$'s and the sampled parallel sequence numbers l_i's respectively become

$$\begin{cases} \hat{\alpha}_{2i} = 53i - 27, & i = 0, 1, \cdots, 15, \\ \hat{\alpha}_{2i+1} = 53i, & i = 0, 1, \cdots, 14, \end{cases} \tag{16.13a}$$

$$\begin{cases} l_{2i} = 5, & i = 0, 1, \cdots, 15, \\ l_{2i+1} = 1, & i = 0, 1, \cdots, 14 \ ; \end{cases} \tag{16.13b}$$

and by Theorem 13.3, the parallel sample transmission times $\hat{\gamma}_i$'s and the transmitted signal numbers m_i's respectively become

$$\hat{\gamma}_{2i} = \hat{\gamma}_{2i+1} = 53i, \ i = 0, 1, \cdots, 15, \tag{16.14a}$$

$$\begin{cases} m_{2i} = 0, & i = 0, 1, \cdots, 15, \\ m_{2i+1} = 1, & i = 0, 1, \cdots, 14. \end{cases} \tag{16.14b}$$

Therefore, if we move the parallel sampling times $\hat{\alpha}_i$'s in (16.13a) to the parallel sample transmission times $\hat{\gamma}_i$'s in (16.14a) for the concurrent sampling, then by Theorem 13.4 we obtain the sampling vectors

$$\begin{cases} \mathbf{v}_0 = \mathbf{v}_2 = \cdots = \mathbf{v}_{30} = (\mathbf{T}^t)^{-27} \cdot \mathbf{h}_5 \\ \quad = [\, 1\,1\,1\,0\,0\,1\,0\,0\,1\,0\,1\,1\,0\,1\,1\,0\,1\,1\,0\,1\,1\,0\,1\,1\,1\,1\,1\,1\,1\,1\,1 \,]^t, \\ \mathbf{v}_1 = \mathbf{v}_3 = \cdots = \mathbf{v}_{29} = \mathbf{h}_1 \\ \quad = [\, 0\,0\,1\,0\,1\,0\,0\,0\,0\,0\,0\,0\,0\,0\,0\,1\,0\,0\,0\,0\,0\,0\,0\,0\,0\,0\,0\,0\,0\,0 \,]^t. \end{cases} \tag{16.15}$$

Using these, we can draw the parallel scrambler for the concurrent sampling as shown in Fig. 16.12.[7]

We finally consider the parallel correction. Due to (16.14a), the parallel sample transmission times are identical for the even-indexed sample and the odd-indexed sample. Therefore, the relevant parallel correction becomes a double correction. Inserting (16.12a), (16.14a)

---

[7]Note that the sample transmission pulses are spaced 53 bit-times apart, where the bit-time itself is lengthened to eight times that of the serial scrambling case, thus making the clock rate for overall operation drop to an eighth.

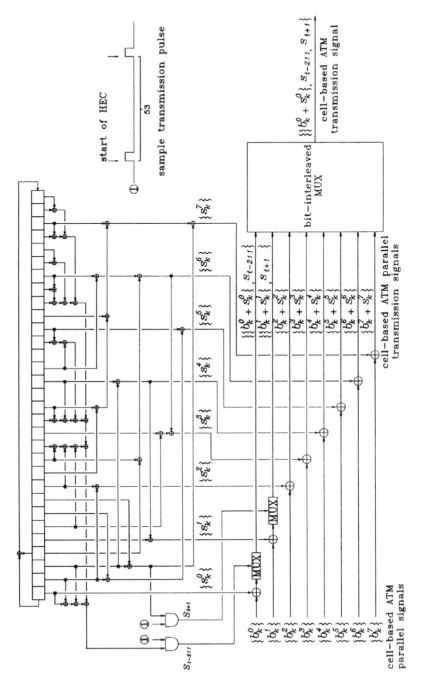

Figure 16.12. A PDSS scrambler for the cell-based ATM transmission signal.

and (16.15) to (13.7), we obtain the matrix

$$
\tilde{\mathbf{V}} =
\begin{bmatrix}
1 1 1 0 0 1 0 0 1 0 1 1 0 1 1 0 1 1 0 1 1 0 1 1 1 1 1 1 1 1 1 1 \\
0 0 1 0 1 0 0 0 0 0 0 0 0 0 0 1 0 0 0 0 0 0 0 0 0 0 0 0 0 0 0 0 \\
0 1 0 0 1 0 0 0 0 0 0 0 0 0 1 0 0 1 0 0 0 0 0 0 0 1 0 0 0 \\
0 0 0 0 0 0 1 0 0 0 0 1 0 0 0 0 0 0 0 0 1 0 1 1 0 1 0 \\
0 0 0 0 0 1 0 0 1 0 0 1 0 0 1 0 0 1 0 0 0 0 1 1 0 0 0 0 1 0 \\
0 1 0 0 0 0 0 0 0 1 0 0 0 0 1 0 1 1 0 1 1 0 1 1 0 1 0 0 1 0 \\
0 0 0 0 0 0 0 0 0 0 1 0 1 0 0 0 0 0 1 0 0 0 0 1 0 1 0 0 \\
0 1 0 0 0 0 1 1 1 1 0 1 0 0 0 0 0 1 0 0 1 0 0 0 0 0 0 0 0 \\
0 0 0 1 0 1 0 0 0 1 0 1 1 0 0 0 0 1 1 1 0 0 0 1 0 1 0 0 0 0 0 \\
0 1 0 1 1 1 1 0 1 1 0 0 1 0 0 1 0 0 1 0 0 0 0 1 0 0 1 0 1 1 1 \\
1 0 1 1 0 0 1 0 0 0 1 0 0 0 0 1 0 0 0 0 1 0 1 0 0 1 0 1 1 0 1 \\
0 0 1 0 0 1 0 0 0 0 0 0 1 0 1 1 0 1 0 1 0 1 0 0 0 0 1 0 0 1 0 \\
1 1 0 1 0 0 1 1 0 0 0 1 1 1 0 0 1 1 1 0 0 1 0 0 0 1 0 0 0 0 0 \\
0 0 0 1 0 1 1 0 1 1 1 1 0 0 1 0 1 1 1 0 1 0 0 1 1 0 1 0 0 0 1 \\
1 1 1 0 0 0 0 1 0 0 1 0 1 1 1 1 1 0 1 0 0 0 0 1 0 0 1 0 0 1 0 \\
0 1 0 1 0 0 1 0 0 1 0 0 0 1 0 0 0 0 0 0 1 0 0 1 0 1 0 0 0 0 0 \\
0 0 0 1 0 1 1 0 0 0 1 1 1 1 1 0 1 0 0 0 0 1 0 1 1 0 1 0 1 0 \\
0 0 0 1 1 0 1 0 0 0 1 1 0 0 1 0 1 1 0 0 1 0 1 1 0 1 1 0 0 0 0 \\
1 1 1 0 1 1 1 1 0 1 1 1 1 1 1 1 0 1 1 1 1 1 0 1 1 1 1 0 0 1 \\
1 1 1 1 0 1 0 0 0 0 0 0 0 0 1 0 0 1 1 0 1 1 0 1 1 0 0 1 0 \\
0 0 1 0 0 1 0 0 1 0 0 0 0 0 0 0 0 1 0 0 0 0 1 0 1 1 0 1 \\
0 0 0 0 0 0 1 0 0 1 1 0 0 1 0 0 0 0 1 0 0 0 0 1 0 0 0 1 0 1 \\
0 0 1 0 0 1 0 0 0 0 1 0 0 0 0 0 0 0 1 1 0 1 1 0 0 1 0 1 0 0 0 \\
0 0 1 0 1 1 0 1 1 0 1 1 0 0 1 1 0 0 1 0 0 0 0 0 1 0 1 0 1 0 \\
0 1 0 0 0 0 0 1 0 1 1 0 1 1 0 1 0 0 0 1 0 0 1 0 0 0 0 0 0 0 1 \\
0 0 0 0 1 1 1 0 1 0 1 0 1 1 0 0 0 0 1 0 1 0 1 0 0 0 0 0 \\
1 0 0 1 0 0 0 1 0 1 0 0 1 1 0 0 1 0 0 1 0 1 0 1 0 0 1 1 0 1 0 \\
1 1 1 0 1 1 0 1 0 0 1 1 0 1 0 1 0 1 0 1 0 0 0 0 0 1 1 1 1 0 \\
1 1 0 1 0 0 1 1 0 1 0 0 1 1 1 0 1 1 0 0 0 0 1 1 1 0 1 0 1 1 0 \\
0 0 1 0 0 0 1 1 0 0 0 0 1 0 1 0 1 1 1 1 0 1 0 0 0 1 0 0 0 1 \\
0 1 0 0 0 1 1 1 1 1 1 1 0 0 0 1 0 1 0 0 0 1 1 0 0 1 0 0 1 0 \\
\end{bmatrix}
,
$$

(16.16)

and inserting this and (16.15) along with $J = 16$ and $\hat{\gamma} = 53$ into (13.40), we get

$$
\begin{cases}
a_0 = 1, \\
a_1 = 0, \\
a_{30} = 0.
\end{cases}
$$

(16.17)

Therefore, by (13.39), we obtain the correction vectors

$$
\begin{cases}
\mathbf{c}_0 = \mathbf{c}_2 = \cdots = \mathbf{c}_{30} \\
\quad = [\, 0 1 1 0 1 1 1 0 0 0 1 1 1 0 0 1 0 1 1 1 0 1 1 0 1 1 1 0 0 1 1 \,]^t, \\
\mathbf{c}_1 = \mathbf{c}_3 = \cdots = \mathbf{c}_{31} \\
\quad = [\, 1 0 1 0 1 0 0 1 0 0 1 0 0 0 1 1 1 0 1 1 1 1 1 1 0 0 0 1 1 0 0 \,]^t,
\end{cases}
$$

(16.18a)

$$\begin{cases} \mathbf{c}_0 = \mathbf{c}_2 = \cdots = \mathbf{c}_{30} \\ \quad = [\,0\,1\,1\,0\,1\,1\,1\,0\,0\,0\,1\,1\,1\,0\,0\,1\,0\,1\,1\,1\,0\,1\,1\,0\,1\,1\,1\,0\,0\,1\,1\,]^t, \\ \mathbf{c}_1 = \mathbf{c}_3 = \cdots = \mathbf{c}_{31} \\ \quad = [\,0\,0\,1\,0\,1\,1\,1\,0\,1\,1\,1\,1\,0\,1\,1\,1\,0\,1\,0\,1\,1\,0\,1\,1\,1\,0\,1\,1\,1\,0\,1\,]^t, \end{cases}$$

$$(16.18\text{b})$$

for immediate correction. If we choose the correction vectors in (16.18b), we obtain the parallel descrambler shown in Fig. 16.13.

16.5 Design of Three-State Synchronization Mechanisms

In order to complete the design of DSSs for use in the cell-based ATM transmission, we need to design the three-state synchronization mechanisms fitting to the ATM application. The three-state synchronization mechanism, as described in Section 15.1, consists of three states $--$ the acquisition state, the verification state, and the steady state $--$ and the transition among the three states depends on the specific state transition scheme taken for the design and the involved parameter setting. In this section, we will consider two state transition schemes $--$ the *windowed-observation scheme* and the *thresholded-counting scheme*. For these two state transition schemes we will consider the parameter settings and the accompanied performances. Also, in support of the performance comparison, we will discuss the criteria for performance evaluations.

16.5.1 Windowed-Observation State Transition Scheme

The windowed-observation state transition scheme refers to the scheme that relies on the windowed observation samples to make decision for state transition in each state. If described in terms of the state transition diagram for the three-state synchronization mechanism shown in Fig. 10.2, this scheme corresponds to the case in which the observation sample lengths L, L_V and L_S are all fixed along with their relevant thresholds N_V and N_S.

In the acquisition state, cell delineation is done based on the last six bits in the HEC field (See Fig. 16.2). If the involved CRC reveals that those six bits are correct, then the other two bits, HEC_8 and HEC_7 are also assumed correct. Under this assumption, the two samples $(s_{t-211}$

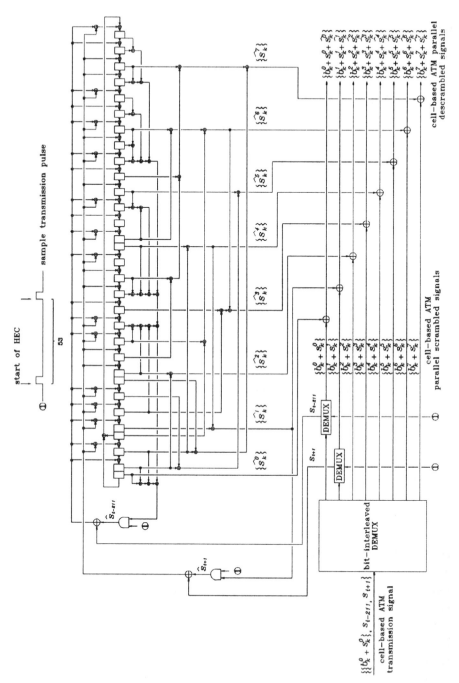

Figure 16.13. A PDSS descrambler for the cell-based ATM transmission signal.

and s_{t+1}) embedded on those two bits can be extracted and thus be used for the SRG synchronization. State transition occurs to the verification state when the number of consecutive cells with the correct HEC field reaches the threshold value X_A. This threshold value is 16 in the ATM application since the number of correct samples required for synchronization is $L = 31$ and two samples are carried in one cell.

Operations in the verification state and the steady state are as described in Section 15.1. If the number of 1's in the window of L_V sample-comparison data is not greater than the predetermined threshold value N_V, the descrambler enters the steady state ; and otherwise it returns to the acquisition state. In the steady state, the descrambler repeats comparison processing in units of L_S sample pairs, and stays in the steady state unless the number of 1's in the window of L_S sample-comparison data exceeds the predetermined threshold value N_S, and returns back to the acquisition state otherwise.

If we redraw the three-state synchronization diagram in Fig. 10.2 by filling in each state with the operations described above, we obtain the state transition diagram in Fig. 16.14. The three large circles in the diagram represent the three states and the small circles indicate the counter values. In the acquisition state the counter value increments by one at each correct HEC and is reset to 0 otherwise. As the X_Ath consecutive occurrence of the correct HEC, for $X_A = \lceil \frac{L}{2} \rceil$, the state transition occurs to the verification state.[8] In the verification and the steady states, the counter increments(or decrements) by one at each sample match(or sample mismatch) in the comparison of the received and the descrambler-generated samples. For the state transition, the solid line indicates the transition to be decided at every sample of cell, and the dashed line indicates the transition to be decided at every window of L_V or L_S samples. State transition occurs from the verification state to the steady state if the counter value lies between $L_V - 2N_V$ and L_V at the end of L_V sample comparisons, as it indicates that the number of sample mismatches does not exceed the threshold N_V ; and the state transition occurs to the acquisition state otherwise. Likewise, state transition occurs from the steady state to itself if the counter value lies between $L_S - 2N_S$ and L_S at the end of L_S sample comparisons; and transition occurs to the acquisition state otherwise. Note that there is no particular counter value fixed for conditioning the relevant state tran-

---

[8]Note that $\lceil x \rceil$ indicates the smallest integer larger than or equal to x.

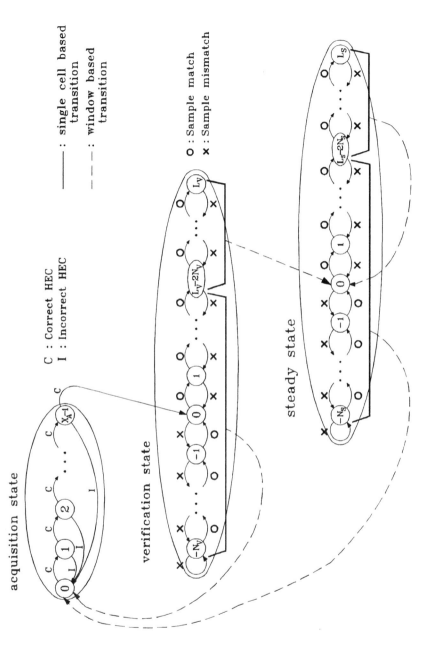

Figure 16.14. State transition diagram of the windowed-observation scheme.

Table 16.1. Values of L_V conforming to Theorem 15.7

$N_V \backslash N_A$	1	2	3	4	5
0	11	11	12	14	16
1	13	14	15	18	20
2	16	16	19	23	25
3	18	20	22	24	32
4	22	23	25	30	36
5	25	28	29	32	37

sition. The state transition rather depends on the observation windows of sizes L_V and L_S.

In the design of the windowed-observation state transition mechanism, there are four parameters to determine -- the observation sample numbers L_V and L_S, and the threshold values N_V and N_S. As mentioned in Chapter 15, a larger observation sample number yields a lower probability of wrong decision, but at the cost of a slowed synchronization speed. This happens because the descrambler has to wait longer period of time to make the synchronization decision with a longer observation window than with a smaller one.

The two parameters L_V and N_V can be determined according to Theorem 15.7.[9] Table 16.1 lists the resulting values of L_V for various N_A and N_V pairs. The table can be interpreted in such a manner that the observation sample number L_V can guarantee correct decision unless the number of sample errors does not exceed the corresponding threshold N_A in the acquisition state and the threshold N_V in the verification state.

For the choice of observation length L_V, the probability to make a wrong decision in the verification state, on the condition that N_A or less errors occur in the acquisition state, is in the order of $p_e^{N_V+1}$ for the corresponding threshold N_V, where p_e denotes the bit error probability of the transmission line.

EXAMPLE 16.1 For a correct decision in the verification state in the presence of one sample error in the acquisition state, i.e., $W[\mathcal{E}_A] \leq 1$ and one sample error in the verification state, i.e., $W[\mathcal{E}_V] \leq 1$, we choose $L_V =$

---

[9]It is also possible to determine N_V and L_V according to Theorem 15.6. In this case, however, we need to investigate { ζ_i }'s for all possible patterns of $\mathcal{E}_A$'s, whose number is 2^{31} for the cell-based ATM transmission. Therefore, for such a large value of L, it is not practical to employ Theorem 15.6.

Table 16.2. Values of L_S conforming to Theorem 15.9

N_S	1	2	3	4	5
L_S	44	48	56	63	69

13, which is the entry in the slot $(N_A, N_V) = (1,1)$. Then the probability that the descrambler makes a wrong decision in the verification state, with one or no error occurring in the acquisition state, is in the order of p_e^2. ♣

Since state transition occurs at every window of L_V sample comparisons, the state transition time from the verification state to either the steady state or the acquisition state is $\lceil \frac{L_V}{2} \rceil$ cell times.[10] However, transition time to the acquisition state can be curtailed by modifying the state transition scheme such that the state jumps to the acquisition state on detecting $N_V + 1$ mismatches of samples(i.e., 1's in the sample-comparison sequence) even before all the L_V comparisons are made. This modification results in different transition times for different patterns of $\mathcal{E}_A$, but the average transition time gets reduced.

For the operation in the steady state, the necessary parameters L_S and N_S can be determined based on Theorem 15.9. Table 16.2 lists the pairs of L_S and N_S thus found for the ATM transmission parameters. For this choice of observation length L_S, the probability that a descrambler makes a false decision in the steady state is in the order of $p_e^{N_S+1}$ for the corresponding threshold N_S.

EXAMPLE 16.2 For a correct decision in steady state in the presence of one sample error the sample observation number $L_S = 44$. Then, the resulting probability to make false decision in the steady state lies in the order of p_e^2. ♣

State transition takes place if there occur more than N_S sample errors in any of the observation windows of size L_S. Therefore, the average time that the descrambler remains in the steady state is in the order of $\lceil \frac{L_S}{2} \rceil / p_e^{N_S+1}$ cell times. Considering that the errored sample number could exceed the threshold N_S due to a temporary transmission error,

---

[10]Cell time, which we use as the unit of time in this section, is the time for the duration of a cell. In the 155.520 Mbps cell-based ATM transmission, one cell time equals 2.73 μs.

not due to loss of synchronization, we can modify the state transition scheme in such a way that the state transition takes place after several more windows of examining. In this situation the above average in-synchronization time could become much longer.

16.5.2   Thresholded-Counting State Transition Scheme

The thresholded-counting state transition scheme refers the scheme that employs fixed thresholds for counting the events for state transition.[11] We consider this scheme based on the state transition diagram in Fig. 16.15, which is a redrawing of Fig. 10.2 in view of the thresholded-counting state transition scheme. The notation or symbols for the diagram is the same as for the windowed-observation scheme in Fig. 16.14. The main difference of the thresholded-counting scheme lies in that every state transition occurs counter-threshold based. In other words, state transition occurs from the verification state to the steady state if the counter value reaches the threshold Y_V and to the acquisition state if the counter value drops below the threshold V_V. Likewise, state transition occurs from the steady state to itself if the counter value grows beyond the threshold Z_S and to the acquisition state if the counter value drops below the threshold W_S.

In designing the three-state synchronization mechanism based on the thresholded-counting state transition scheme, there are four threshold parameters to be determined, which are V_V, Y_V, W_S and Z_S. In the following, we consider how to determine these four parameters.

We first consider the case of the threshold V_V by taking an example. Suppose that we have chosen $V_V = 14$ and that the descrambler is synchronized. Then, state transition will not take place unless three consecutive cells come in with each cell containing at least one corrupted sample. The probability of such an occurrence is in the order of p_e^3, for a low bit error probability (for example, $p_e = 10^{-9}$). Therefore, we can choose $V_V = 14$ to make a correct decision in the verification state with the error vector $\mathcal{E}_A = \mathbf{0}$, in the presence of up to 2 sample errors in the verification state.

Differently from the case of V_V, the threshold Y_V can not be determined through a simple calculation. Since the counter threshold values are determined based on the sample-comparison sequence η_i, there could

---

[11]The thresholded-counting state transition scheme is the scheme introduced by ITU-T via Recommendation I.432.

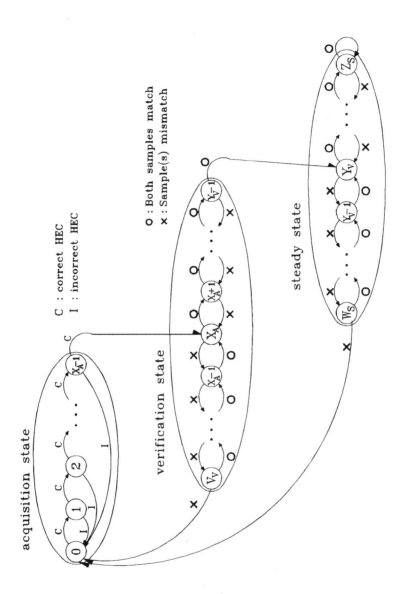

Figure 16.15. State transition diagram of the thresholded-counting scheme.

be many different values depending on the error patterns in the acquisition and the verification states. Therefore, in order to determine the value Y_V that prevents false synchronization with some limitations on $W[\mathcal{E}_A]$ and $W[\mathcal{E}_V]$, we need to determine the maximum counter value for all possible patterns of $\mathcal{E}_A$ and $\mathcal{E}_V$. To make it worse, the dimension of the vector $\mathcal{E}_V$ itself is variable since the length of the verification state is not fixed. For illustration purpose, therefore, we focus on the case when $W[\mathcal{E}_A] = 1$. We denote, in this case, by $\hat{Y}_V$ the maximum value of the threshold Y_V with $\mathcal{E}_V$ set to $\mathbf{0}$. If we choose Y_V such that $Y_V = \hat{Y}_V + 1$, then the resulting (conditional) probability of false synchronization with $\mathcal{E}_A \neq \mathbf{0}$ is in the order of p_e. Therefore, in general, we can choose Y_V such that $Y_V = \hat{Y}_V + y$ to prevent false synchronization when $W[\mathcal{E}_A] \leq 1$ and $W[\mathcal{E}_V] \leq y$. For the descrambler of the cell-based ATM transmission system, we obtain the value $\hat{Y}_V = 21$, and therefore we can prevent false synchronization by taking $\hat{Y}_V = 22$ in case a single error occurs in the acquisition state.

In the steady state, the counter is made not to increase over the threshold Z_S and the state transition occurs when the counter value drops below the lower threshold W_S. Therefore, only the difference of the two thresholds, $Z_S - W_S$, affects the performance in the steady state. This implies that we can put $Z_S = Y_V$, and determine W_S in compliance with the desired level of performance. For a proper choice of the threshold W_S, we consider the case when the initial error vector σ is zero. In this case, the worst-case scenario is that the first w consecutive cells contain the corrupted samples from the beginning of the steady state. In order to remain in the steady state in this situation, W_S should be chosen to be less than Y_V but not less than $Y_V - w$. It is also possible to choose W_S based on some averaged performance measures, rather than by the worst-case principles. In this case, a performance analysis is needed on the related parameters to get the necessary data.[12]

16.5.3   Performance Evaluation

We now compare the performances of the two state transition schemes for the three-state synchronization mechanism in the cell-based ATM transmission environment. For the performance evaluations, we take the following performance parameters.

---

[12]The thresholds suggested by ITU-T are $V_V = 8$, $Y_V = 24$, $W_S = 16$ and $Z_S = 24$.

DEFINITION 16.1 (Performance Parameters in Verification State) *Average synchronization time from the verification state, T_{VS},* refers to the state transition time from the verification state to the steady state when the descrambler is synchronized to the scrambler.

Probability of false reinitialization from the verification state, P_{VA}, refers to the probability that the descrambler falsely goes back to the acquisition state from the verification state when the descrambler is actually synchronized to the scrambler.

Average reinitialization time from the verification state, $T_{\tilde{V}A}$, refers to the average state transition time from the verification state to the acquisition state when the descrambler is not synchronized to the scrambler.

Probability of false synchronization from the verification state, $P_{\tilde{V}\tilde{S}}$, refers to the probability that the descrambler goes to the steady state from the verification state even if the descrambler is not synchronized to the scrambler.

Average synchronization time, T_{AS}, refers to the average state transition time from the beginning of the acquisition state to the true steady state.

Probability of false synchronization, $P_{A\tilde{S}}$, refers to the probability of false synchronization, starting from the initial acquisition state.

DEFINITION 16.2 (Performance Parameters in Steady State) *Average false reinitialization time from the steady state, T_{SA},* refers to the average state transition time from the steady state to the acquisition state while the descrambler is synchronized to the scrambler.

Maximum average reinitialization time from the steady state, $T_{\tilde{S}A}$, refers to the maximum average state transition time from the false steady state to the acquisition state.

Fig. 16.16 illustrates the performance parameters of the three-state synchronization mechanism. In the figure, π denotes the probability of getting into the verification state with the descrambler synchronized, which is $(1 - p_e)^L$. The symbol A, V and S respectively denote the acquisition, the verification and the steady states ; $\tilde{V}$ and $\tilde{S}$ respectively denote the *false* verification and steady states. In other words, V denotes the verification state for the case when the descrambler is truly synchronized, and $\tilde{V}$ the case when it is not. Similarly, S denotes the true steady state, and $\tilde{S}$ denotes the false steady state.

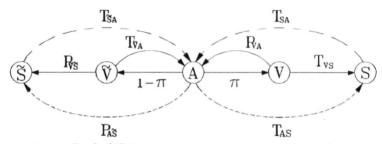

Figure 16.16. Probabilities or average transition times for state transitions in the three-state synchronization mechanism(Tilde($\sim$) indicates the false verification or steady state).

There are two requirements to be imposed on the state transition schemes to yield a desirable performance, which are speed and accuracy. To be more specific, the descrambler should be fast in executing right transitions, and should avoid wrong transitions as much as possible. For instance, if there was no error during the acquisition state, transition time from the verification state to the steady state should be as short as possible. However, in case there came in any errored sample(s) during the acquisition state, the probability that the descrambler declares the steady state should be low. In terms of performance parameters, it is desirable to have small values for T_{VS}, P_{VA}, $T_{\tilde{V}A}$, $P_{\tilde{V}\tilde{S}}$, T_{AS}, $P_{A\tilde{S}}$, $T_{\tilde{S}A}$, and large values for T_{SA}. The most representative parameter among them is T_{AS}, as the descrambler is expected to be in synchronization state after this amount of time.

The four parameters L_V, N_V, L_S, N_S for the windowed-observation state transition scheme are to be determined such that the performance parameters T_{VS}, P_{VA}, $T_{\tilde{V}A}$, $P_{\tilde{V}\tilde{S}}$, T_{AS}, $P_{A\tilde{S}}$, T_{SA}, $T_{\tilde{S}A}$ fall within some satisfactory ranges. Likewise, the four parameters V_V, Y_V, W_S, Z_S for the thresholded-counting state transition scheme are to be determined such that the eight parameters fall within some satisfactory ranges. Detailed discussion on determining the state transition parameters and evaluating the relevant performance parameters is a tedious and simulation-intensive work, and is beyond the scope of the book. Therefore, we conclude the section by introducing the performance parameters in Table 16.3 readily evaluated for the chosen state transition parameters $L_V = 18$, $N_V = 2$, $L_S = 44$, $N_S = 1$ of the windowed-observation scheme and $V_V = 14$, $Y_V = 24$, $W_S = 23$, $Z_S = 24$ of the thresholded-counting scheme. We observe that the performances of the two schemes

Table 16.3. Performance comparison at the bit error probability 10^{-9}

	windowed-observation scheme with $L_V = 18, N_V = 2,$ $L_S = 44, N_S = 1$	thresholded-counting scheme with $V_V = 14, Y_V = 24,$ $W_S = 23, Z_S = 24$
T_{VS}	9.00	8.00
P_{VA}	8.16×10^{-25}	8.00×10^{-27}
$T_{\tilde{V}A}$	4.95	10.23
$P_{\tilde{V}\tilde{S}}$	2.58×10^{-19}	5.16×10^{-19}
T_{AS}	24.50	24.00
$P_{A\tilde{S}}$	8.00×10^{-27}	1.60×10^{-26}
T_{SA}	2.33×10^{16}	2.56×10^{17}
$T_{\tilde{S}A}$	22.00	74.00

are comparable in general but the windowed-observation scheme outperforms the thresholded-counting scheme in the average reinitialization time $T_{\tilde{V}A}$ and the maximum average reinitialization time $T_{\tilde{S}A}$.

16.6 Applications to High-Speed Data Networks

Finally, we demonstrate how to apply the DSS techniques to high-speed data networks.

In the high-speed data networks which provide isochronous services (or real-time services) in addition to non-isochronous data services, it is desirable to use fixed-size packets, or equivalently, small-size frames as the means for information transfer. By doing so, the high-speed data network can possibly deliver the real-time services with a tolerable amount of delays. The resulting signal structure in the high-speed data network therefore takes the form shown in Fig. 16.17. The transmission signal becomes a concatenated stream of the fixed-size packets of length F which consists of the header of length F_H and the user information field of length F_U. In this aspect, the high-speed networks can be regarded as an extension of the ATM network, at least in concept, and therefore the DSS scrambling techniques can be efficiently applied for transmission of the constituent high-speed data signals. Consequently,

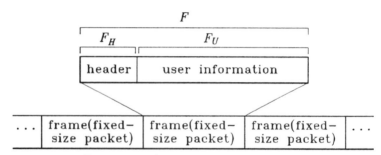

Figure 16.17. Structure of transmission signal for high-speed data networks.

no Manchester or other sophisticated line coding is necessary, and no "surcharge" of transmission rate applies to this case.

In order to apply the DSS technique in the high-speed data network with the transmission signal structure in Fig. 16.17, we have to first choose the scrambling space $V[\Psi(x)]$ and the sampling times α_i's. For the scrambling space $V[\Psi(x)]$, it is desirable to choose its characteristic polynomial $\Psi(x)$ to be a primitive one, since any scrambling sequence $\{ s_k \}$ in the primitive scrambling space becomes a PRBS. The dimension L of the scrambling space $V[\Psi(x)]$ is better to be large for a better scrambling effect, but is better to be small for a short sample transmission time or synchronization time. As far as the synchronization time is concerned, the choice of L also depends on the number of this time slots in the header available for sample conveyance. If only a single time slot is available for this, it is not adequate to choose a large L since the synchronization time can exceed the threshold value. In general, if K time slots are available, L can be chosen much larger since the synchronization time becomes approximately L/K in this case. Once the dimension L of the scrambling space $V[\Psi(x)]$ is fixed, we choose a particular characteristic polynomial $\Psi(x)$ and K sampling slots such that the sample-transmission times become predictable sampling times, since the immediate sampling does not require additional timing circuitry. For the correction, we choose the immediate correction times so that additional timing circuits can be eliminated.

For example, we consider a high-speed data network employing the transmission signal of rate 1 Gbps and the frame composed of 12 bytes of header and 1680 bytes of user information, i.e., $F_H = 96$, $F_I = 13440$ and $F = 13536$. We assume, for convenience, that the first three bits in the header are available for sample conveyance($K = 3$), which means that the sample transmission times are[13]

$$\gamma_{3i+j} = iF + j, \quad i = 0, 1, \cdots, \quad j = 0, 1, 2. \tag{16.19}$$

If we choose $L = 15$ for the dimension of the scrambling space, then we can synchronize the descrambler after the first five frames. Among many possible choices of the 15-dimensional scrambling spaces, we choose the primitive space $V[x^{15} + x + 1]$, for which the concurrent sampling times are predictable. That is, if we choose the sampling times

$$\alpha_{3i+j} = iF + j, \quad i = 0, 1, 2, 3, 4, \quad j = 0, 1, 2, \tag{16.20}$$

and insert this along with $\Psi(x) = x^{15} + x + 1$ into (11.16), we obtain the discrimination matrix

$$\Delta_\alpha = \begin{bmatrix} 1 0 0 0 0 0 0 0 0 0 0 0 0 0 0 \\ 0 1 0 0 0 0 0 0 0 0 0 0 0 0 0 \\ 0 0 1 0 0 0 0 0 0 0 0 0 0 0 0 \\ 1 1 1 0 0 1 1 0 0 0 0 1 0 1 0 \\ 0 1 1 1 0 0 1 1 0 0 0 0 1 0 1 \\ 1 1 1 1 1 0 0 1 1 0 0 0 0 1 0 \\ 1 0 1 0 1 0 0 1 1 0 1 1 0 0 0 \\ 0 1 0 1 0 1 0 0 1 1 0 1 1 0 0 \\ 0 0 1 0 1 0 1 0 0 1 1 0 1 1 0 \\ 0 0 0 0 0 0 1 1 1 1 0 1 1 1 1 \\ 1 1 0 0 0 0 0 1 1 1 1 0 1 1 1 \\ 1 0 1 0 0 0 0 0 1 1 1 1 0 1 1 \\ 1 1 1 0 1 1 1 1 0 0 0 0 0 0 1 \\ 1 0 1 1 0 1 1 1 1 0 0 0 0 0 0 \\ 0 1 0 1 1 0 1 1 1 1 0 0 0 0 0 \end{bmatrix}, \tag{16.21}$$

which is nonsingular. If we choose the scrambler SRG to be the one with the state transition matrix $\mathbf{T} = \mathbf{A}_{x^{15}+x+1}^t$ and the generating vector $\mathbf{h} = \mathbf{e}_0$, we obtain the scrambler depicted in Fig. 16.18.

To realize the counterpart descrambler, we choose the immediate correction times

$$\beta_{3i+j} = iF + j + 1, \quad i = 0, 1, 2, 3, 4, \quad j = 0, 1, 2. \tag{16.22}$$

Then, for the state transition matrix $\mathbf{T} = \mathbf{A}_{x^{15}+x+1}^t$ and the generating

---

[13]This is just for convenience of notation. In reality, it does not matter whether or not the three sample-conveying bits are actually positioned in the front part, as the timing notation can be rearranged to make those three bits come first.

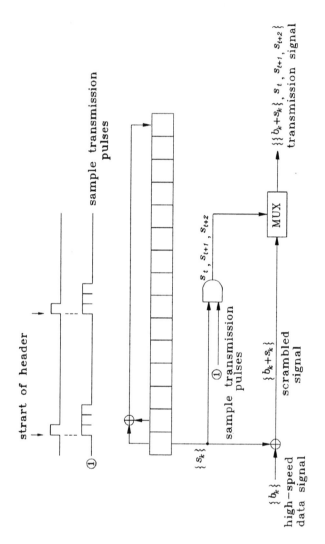

Figure 16.18. An example of scrambler for high-speed data transmission.

vector $\mathbf{h} = \mathbf{e}_0 + \mathbf{e}_1 + \mathbf{e}_6$, we obtain, by (12.19), the matrix

$$\tilde{\mathbf{H}} = \begin{bmatrix}
1 1 0 0 0 0 1 0 0 0 0 0 0 0 0 \\
0 1 1 0 0 0 0 1 0 0 0 0 0 0 0 \\
0 0 1 1 0 0 0 0 1 0 0 0 0 0 0 \\
1 0 1 0 1 0 1 0 1 0 0 0 0 1 1 \\
1 0 0 1 0 1 0 1 0 1 0 0 0 0 1 \\
1 0 0 0 1 0 1 0 1 0 1 0 0 0 0 \\
1 0 1 0 1 1 1 1 1 0 0 1 1 1 \\
1 0 0 1 0 1 1 1 1 1 0 0 1 1 \\
1 0 0 0 1 0 1 1 1 1 1 1 0 0 1 \\
0 0 1 0 0 0 0 0 0 0 1 1 1 1 1 \\
1 1 0 1 0 0 0 0 0 0 0 1 1 1 1 \\
1 0 1 0 1 0 0 0 0 0 0 0 1 1 1 \\
0 1 0 1 1 1 0 1 0 0 1 1 1 1 1 \\
1 1 1 0 1 1 1 0 1 0 0 1 1 1 1 \\
1 0 1 1 0 1 1 1 0 1 0 0 1 1 1
\end{bmatrix},$$

and inserting this along with $\mathbf{T} = \mathbf{A}_{x^{15}+x+1}^{t}$ and (16.22) into (12.26) we can obtain the correction vectors

$$\begin{cases}
\mathbf{c}_0 = \mathbf{c}_3 = \cdots = \mathbf{c}_{12} = [\,1\,0\,1\,0\,1\,0\,0\,1\,1\,0\,0\,0\,0\,0\,1\,]^t, \\
\mathbf{c}_1 = \mathbf{c}_4 = \cdots = \mathbf{c}_{13} = [\,0\,1\,0\,1\,0\,1\,0\,0\,0\,0\,0\,0\,1\,0\,0\,]^t, \\
\mathbf{c}_2 = \mathbf{c}_5 = \cdots = \mathbf{c}_{14} = [\,0\,1\,0\,1\,0\,1\,0\,0\,0\,0\,0\,0\,1\,0\,0\,]^t.
\end{cases}$$

Using these parameters, we can obtain the descrambler shown in Fig. 16.19.

Now we consider the parallel realizations of the scrambler and the descrambler in Figs. 16.18 and 16.19. In order to apply the parallel DSS techniques to this, we regard the transmission signal $\{\,b_k\,\}$ as the bit-interleaved signal of the eight parallel sequences $\{\,b_k^j\,\}$, $j = 0, 1, \cdots,$ 7. If we employ this parallel scrambling scheme, the serial scrambling rate of 1 Gbps drops to $125(=1000/8)$ Mbps.

Inserting $N = 8$, $\Psi_I(x) = x^{15} + x + 1$ and $\mathbf{R} = \mathbf{I}$ into (13.1a) and (13.1b), we obtain the PSRG with the state transition matrix

$$\mathbf{T} = \begin{bmatrix}
0 0 0 0 0 0 0 0 1 0 0 0 0 0 0 \\
0 0 0 0 0 0 0 0 0 1 0 0 0 0 0 \\
0 0 0 0 0 0 0 0 0 0 1 0 0 0 0 \\
0 0 0 0 0 0 0 0 0 0 0 1 0 0 0 \\
0 0 0 0 0 0 0 0 0 0 0 0 1 0 0 \\
0 0 0 0 0 0 0 0 0 0 0 0 0 1 0 \\
0 0 0 0 0 0 0 0 0 0 0 0 0 0 1 \\
1 1 0 0 0 0 0 0 0 0 0 0 0 0 0 \\
0 1 1 0 0 0 0 0 0 0 0 0 0 0 0 \\
0 0 1 1 0 0 0 0 0 0 0 0 0 0 0 \\
0 0 0 1 1 0 0 0 0 0 0 0 0 0 0 \\
0 0 0 0 1 1 0 0 0 0 0 0 0 0 0 \\
0 0 0 0 0 1 1 0 0 0 0 0 0 0 0 \\
0 0 0 0 0 0 1 1 0 0 0 0 0 0 0 \\
0 0 0 0 0 0 0 1 1 0 0 0 0 0 0
\end{bmatrix}, \qquad (16.23a)$$

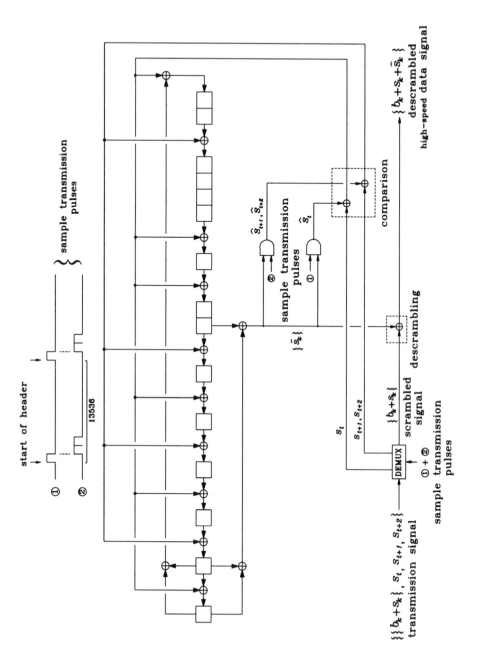

Figure 16.19. An example of descrambler for high-speed data transmission.

and the generating vectors

$$\mathbf{h}_j = \mathbf{e}_j, \quad j = 0, 1, \cdots, 7. \tag{16.23b}$$

By Theorem 13.2, the parallel sampling times $\hat{\alpha}_i$'s and the sampled parallel sequence numbers l_i's respectively become

$$\hat{\alpha}_{3i} = \hat{\alpha}_{3i+1} = \hat{\alpha}_{3i+2} = iF/N = 1692i, \quad i = 0, 1, 2, 3, 4, \tag{16.24a}$$

$$l_{3i+j} = j, \quad i = 0, 1, 2, 3, 4, \quad j = 0, 1, 2, \tag{16.24b}$$

which respectively correspond to the sample transmission times $\tilde{\gamma}_i$'s and the transmitted signal numbers m_i's. Using these parameters, we can draw the parallel scrambler as shown in Fig. 16.20.

We finally consider the parallel correction. Due to (16.24a), the parallel sample transmission times for three consecutive samples, which are identical to the parallel sampling times, are all the same. Therefore, the relevant parallel correction becomes a triple correction, that is, $K = 3$ and hence $J = L/K = 5$. Since $\mathbf{R} = \mathbf{I}$ for the PSRG with $\mathbf{T}$ in (16.23a) and $\mathbf{h}_j$'s in (16.23b), the matrix $\tilde{\mathbf{V}}$ in (13.7) becomes, by (13.8), $\tilde{\mathbf{V}} = \Delta_\alpha$. Therefore, inserting these parameters with $\hat{\gamma} = 1692$ and $\hat{\beta} = 1$ into (13.62), we obtain the correction vectors

$$\begin{cases} \mathbf{c}_0 = [\,0\,1\,1\,0\,1\,0\,0\,1\,0\,0\,1\,0\,1\,1\,1\,]^t, \\ \mathbf{c}_1 = [\,0\,1\,0\,1\,0\,1\,0\,1\,1\,0\,0\,0\,0\,0\,0\,]^t, \\ \mathbf{c}_2 = [\,1\,1\,0\,0\,1\,1\,0\,0\,1\,0\,0\,0\,0\,0\,0\,]^t, \end{cases}$$

for immediate correction. The resulting parallel descrambler is as shown in Fig. 16.21.

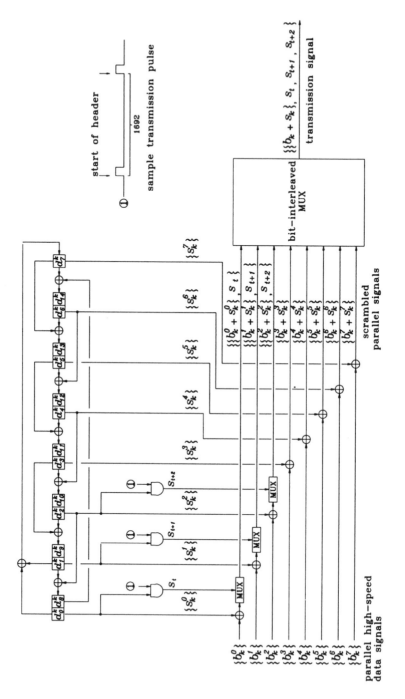

Figure 16.20. An example of parallel scrambler for high-speed data transmission.

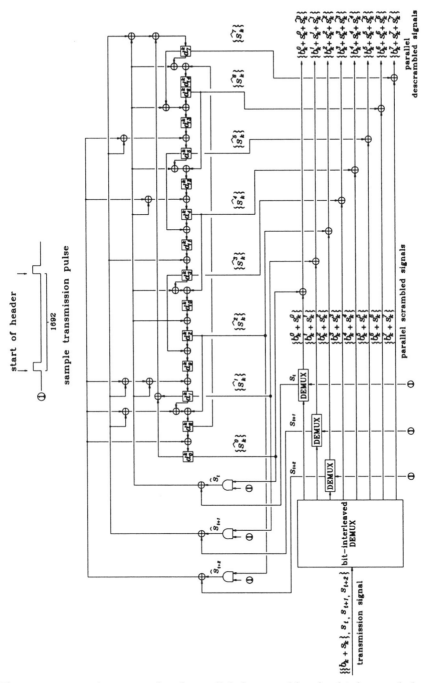

Figure 16.21. An example of parallel descrambler for high-speed data transmission.

Part IV

SELF SYNCHRONOUS SCRAMBLING

Chapter 17

Introduction to Self Synchronous Scrambling

Self synchronous scrambling (SSS) is a scrambling technique which employs a kind of SRGs for use in scrambling as well as in descrambling of data bitstream. The transmission data stream is scrambled by passing through a scrambler composed of shift register generators and exclusive-OR gates, and the scrambled signal is descrambled by passing through a descrambler which is the same as the scrambler but the input and the output are reversed. In the SSS, the states of the scrambler and descrambler shift registers are automatically synchronized without any additional synchronization process. Scrambling effect of the SSS is good in general, but a bit error in the scrambled bitstream can cause multibit errors after descrambling. For the SSS, parallel scrambling techniques are available in the multiplexed environment, as for the FSS and the DSS. However, differently from these two cases, the *parallel SSS* (PSSS) can be applied to the parallel input signals without requiring any over-laid multiplexing frame. Instead, it requires an appropriate means to properly align the order of the descrambled parallel output signals.

In this part we will concentrate our discussion on the SSS and the related signal alignment issues, and in this introductory chapter of the part, we will introduce the operation of the SSS and the concept of signal alignment.

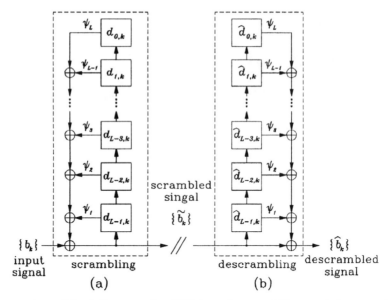

Figure 17.1. Blockdiagram for SSS based scrambling and descrambling functions. (a) Scrambling, (b) descrambling.

17.1 Operation of SSS

The SSS is the oldest and the simplest one among the three scrambling techniques we have discussed so far.[1] In the SSS, scrambling is already done when a data stream *passes through* a sort of SRG, and descrambling is automatically done as the scrambled data stream passes through an input/output-reversed replica of the scrambling SRG. The SSS does not care whatever the data stream to be scrambled is, whether it is framed or not, and it does not require any processing for the synchronization of the descrambler.

A blockdiagram of the SSS is shown in Fig. 17.1. The configurations of the scrambler and the descrambler are uniquely characterized by a *characteristic polynomial* $\Psi(x)$. The characteristic polynomial for the scrambler and descrambler in Fig. 17.1 is $\sum_{i=0}^{L} \psi_i x^i$ with $\psi_0 = 1$.

In the SSS, the input sequence $\{ b_k \}$ itself controls the state of shift registers in the scrambler, and the scrambled sequence $\{ \tilde{b}_k \}$ controls

---

[1] Among the PDH signals recommended in the 1970's via ITU-T Recommendation G.702, the 32.064 Mbps signal composed of five DS-2 signals, and the 97.728 Mbps signal composed of three DS-3 signals are to be scrambled via the SSS before transmission.

the state of shift registers in the descrambler. The input sequence $\{\,b_k\,\}$, in the process of scrambling, takes information on the states of the shift registers, and the scrambled sequence $\{\,\tilde{b}_k\,\}$ delivers it to the descrambler for use in synchronization. Therefore, differently from the FSS or the DSS which requires special synchronization treatments in the scrambler and descrambler, the SSS automatically acquires synchronization when the number of received data reaches the size of the embedded SRG length. But if a bit error occurs in the scrambled signal $\{\,\tilde{b}_k\,\}$ during the transmission, it stays in the shift registers of the descrambler for a while, causing multibit errors in the descrambled signal $\{\,\hat{b}_k\,\}$. We call this *error-multiplication* phenomenon, and it is the only drawback the SSS has.[2]

As for other scrambling techniques, it is possible to convert the serial SSS to the PSSS. That is, there exists a parallel scrambling technique that can generate scrambled multiple data streams which, when multiplexed, become identical to the multiplexed and serially scrambled data stream of the original parallel data streams. However, such parallel scrambling technique is applicable only for the bit-interleaved multiplexing in the SSS case. In other words, there is no multibit-parallel scrambling technique available for multibit-interleaved environment, as it violates the input-output causal relation (refer to Section 19.1).

17.2   Signal Alignment

One of the salient features of the SSS is that is does not look into the internal contents of the data stream to be scrambled. The SSS scrambler simply takes in the input stream and lets it pass through the scrambler, and then scrambling is done. If applied to parallel scrambling, this property enables multiple base-rate bitstreams to be parallel-scrambled without additional framing, but still making the output bitstream correctly retrievable in the parallel descrambler. In the following, we examine this property more closely.

We consider the blockdiagrams of the serial and the parallel SSS in Fig. 17.2. In the case of the serial SSS, the N base-rate input signals $\{\,b_k^j\,\}$'s are bit-interleaved multiplexed first, a frame is overlaid on the multiplexed signal $\{\,b_k\,\}$, and then it is serial-scrambled. In the de-

---

[2]A more detailed description on self-synchronization and error-multiplication in the SSS follows in the next chapter.

scrambler the reverse processings are done for a correct retrieval of the N base-rate output signals { b_k^j }'s. The role of frame-formatting and frame-alignment functions in this case is to secure proper *signal alignments*, that is, to properly distribute the jth input signal { b_k^j }, $j = 0$, $1, \cdots, N - 1$, to the jth output line. In case there are no such functions, it is still possible to do correct scrambling and descrambling for the SSS but it is impossible to achieve proper signal alignments due to lack of reference points for distribution. In fact, the frame overlaid in Fig. 17.2(a) provides such a reference point. Note that the frame-formatting and the frame-alignment functions in the serial SSS are performed at the transmission rate.

In the case of the parallel SSS, it is possible to replace the frame-formatting and the frame-alignment functions with the permuting and depermuting functions as illustrated in Fig. 17.2(b). The permuting and depermuting functions are nothing more than rearranging functions of connections which can be done in hard-wired form. The N base-rate signals { b_k^j }'s are frame-aligned at the base rate in search for the frame-alignment words readily existing in those base-rate signals. In case the frame-alignments are not done as expected, the 1:N bit-interleaved demultiplexer shifts the received data by one bit and lets the frame-aligning processes follow on the rearranged base-rate signals. This procedure repeats until the signal-alignments are done as expected. The permuting and the depermuting functions rearrange the data in such a manner that the expected frame alignments are done only when all base-rate signals are properly distributed to the desired output lines.[3] Note that in the parallel SSS, the transmission rate is exactly N times the base rate because there is no overlaid frame, and all the processings are done at the base rate.

17.3   Organization of the Part

The part is organized as follows : In Chapter 18, we examine the serial SSS techniques, investigating its self-synchronization and error-multiplication properties, and in Chapter 19, we investigate the parallel SSS techniques, discussing how to realize the parallel scrambler and parallel descrambler. In Chapter 20, we demonstrate how to apply these SSS techniques for low-rate realizations of the SDH-based ATM cell scram-

---

[3]A more detailed description on the signal alignment follows in Chapter 21.

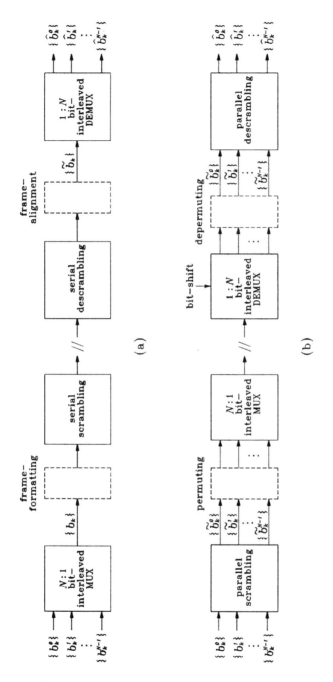

Figure 17.2. Arrangements for signal alignment. (a) Serial SSS, (b) parallel SSS

bling. Finally, in Chapter 21, we consider the signal-alignment issue in conjunction with the parallel SSS.

Chapter 18

Serial Self Synchronous Scrambling

In the *self synchronous scrambling* (SSS), the input data bitstream controls the state of shift registers in the scrambler, and the scrambled signal controls the state of shift registers in the descrambler. We investigate these behaviors in this chapter, and examine the self-synchronization and the error-multiplication properties of the SSS. We first consider how to obtain the scrambled and the descrambled signals respectively from the input and the scrambled signals. Then, we examine how the scrambler and descrambler are automatically synchronized without any additional synchronization processing. Finally, we consider to what extent the transmission error multiplies through the self synchronous descrambling.

18.1   Scrambled Signals

We consider the self synchronous scrambler in Fig. 17.1(a) whose characteristic polynomial is $\Psi(x) = \sum_{i=0}^{L} \psi_i x^i$, The state of the jth shift register at time k, of $d_{j,k}$, $j = 0, 1, \cdots, L-1$, is determined by the previous states $d_{j,k-1}$, $j = 0, 1, \cdots, L-1$, through the relation

$$d_{j,k} = \begin{cases} d_{j+1.k-1}, & j = 0, 1, \cdots, L-2, \\ \sum_{i=0}^{L-1} \psi_{L-i} d_{i,k-1} + b_{k-1}, & j = L-1 \end{cases} \quad (18.1)$$

In vector form, the kth state vectors $\mathbf{d}_k$, $k = 1, 2, \cdots$, is represented by the $(k-1)$th state vector $\mathbf{d}_{k-1}$ and the kth input signal element b_k such that

$$\mathbf{d}_k = \mathbf{A}^t_{\Psi_r(x)} \cdot \mathbf{d}_{k-1} + b_{k-1}\mathbf{e}_{L-1}, \qquad (18.2)$$

where $\mathbf{A}_{\Psi_r(x)}$ is the companion matrix for the reciprocal polynomial $\Psi_r(x) \equiv x^L \Psi(x^{-1})$ of the characteristic polynomial $\Psi(x)$, i.e.,

$$\mathbf{A}_{\Psi_r(x)} = \begin{bmatrix} 0 & 0 & \cdots & 0 & \psi_L \\ 1 & 0 & \cdots & 0 & \psi_{L-1} \\ 0 & 1 & \cdots & 0 & \psi_{L-2} \\ \vdots & \vdots & \ddots & \vdots & \vdots \\ 0 & 0 & \cdots & 1 & \psi_1 \end{bmatrix} ; \qquad (18.3)$$

Noting that the scrambled signal element $\tilde{b}_k$ moves into the last shift register in the scrambler to become its next state $d_{L-1,k+1}$, or equivalently $\tilde{b}_k = \mathbf{e}^t_{L-1} \cdot \mathbf{d}_{k+1}$, we get the following expression for the scrambled signal element $\tilde{b}_k$:

$$\tilde{b}_k = \mathbf{e}^t_{L-1} \cdot \mathbf{A}^t_{\Psi_r(x)} \cdot \mathbf{d}_k + b_k. \qquad (18.4)$$

EXAMPLE 18.1 We consider the scrambler in Fig. 18.1, whose characteristic polynomial is $\Psi(x) = x^7 + x + 1$. The companion matrix $\mathbf{A}_{\Psi_r(x)}$ for the reciprocal polynomial $\Psi_r(x) = x^7 \Psi(x^{-1}) = x^7 + x^6 + 1$ is

$$\mathbf{A}_{\Psi_r(x)} = \begin{bmatrix} 0 & 0 & 0 & 0 & 0 & 0 & 1 \\ 1 & 0 & 0 & 0 & 0 & 0 & 0 \\ 0 & 1 & 0 & 0 & 0 & 0 & 0 \\ 0 & 0 & 1 & 0 & 0 & 0 & 0 \\ 0 & 0 & 0 & 1 & 0 & 0 & 0 \\ 0 & 0 & 0 & 0 & 1 & 0 & 0 \\ 0 & 0 & 0 & 0 & 0 & 1 & 1 \end{bmatrix} . \qquad (18.5)$$

If the initial state vector $\mathbf{d}_0$ is $[\,1\,0\,0\,0\,1\,0\,1\,]^t$, and the input signal $\{\,b_k\,\}$ is

$$\{\,b_k\,\} = \{\,1, 0, 1, 1, 1, 0, 1, 0, 1, 1, 1, 0, 1, 0, 0, 0, 0, 0, 1, 1, \cdots\,\}, \quad (18.6)$$

then by (18.2) the state vector $\mathbf{d}_k$'s are as listed in Table 18.1. Therefore, by (18.4), the scrambled signal $\{\,\tilde{b}_k\,\}$ becomes

$$\{\,\tilde{b}_k\,\} = \{\,1, 1, 0, 1, 1, 1, 1, 0, 0, 1, 1, 0, 0, 1, 1, 1, 0, 1, 0, 1, \cdots\,\}. \; \clubsuit \quad (18.7)$$

Table 18.1. The state vector $\mathbf{d}_k$ and the scrambled signal $\{\,\tilde{b}_k\,\}$ of the scrambler in Fig. 18.1 for the input signal $\{\,b_k\,\}$ in (18.6)

k	$\mathbf{d}_0$	$\{\,b_k\,\}$	$\{\,\tilde{b}_k\,\}$
0	$[\,1\,0\,0\,0\,1\,0\,1\,]^t$	1	1
1	$[\,0\,0\,0\,1\,0\,1\,1\,]^t$	0	1
2	$[\,0\,0\,1\,0\,1\,1\,1\,]^t$	1	0
3	$[\,0\,1\,0\,1\,1\,1\,0\,]^t$	1	1
4	$[\,1\,0\,1\,1\,1\,0\,1\,]^t$	1	1
5	$[\,0\,1\,1\,1\,0\,1\,1\,]^t$	0	1
6	$[\,1\,1\,1\,0\,1\,1\,1\,]^t$	1	1
7	$[\,1\,1\,0\,1\,1\,1\,1\,]^t$	0	0
8	$[\,1\,0\,1\,1\,1\,1\,0\,]^t$	1	0
9	$[\,0\,1\,1\,1\,1\,0\,0\,]^t$	1	1
10	$[\,1\,1\,1\,1\,0\,0\,1\,]^t$	1	1
11	$[\,1\,1\,1\,0\,0\,1\,1\,]^t$	0	0
12	$[\,1\,1\,0\,0\,1\,1\,0\,]^t$	1	0
13	$[\,1\,0\,0\,1\,1\,0\,0\,]^t$	0	1
14	$[\,0\,0\,1\,1\,0\,0\,1\,]^t$	0	1
15	$[\,0\,1\,1\,0\,0\,1\,1\,]^t$	0	1
16	$[\,1\,1\,0\,0\,1\,1\,1\,]^t$	0	0
17	$[\,1\,0\,0\,1\,1\,1\,0\,]^t$	0	1
18	$[\,0\,0\,1\,1\,1\,0\,1\,]^t$	1	0
19	$[\,0\,1\,1\,1\,0\,1\,0\,]^t$	1	1
$\vdots$	$\vdots$	$\vdots$	$\vdots$

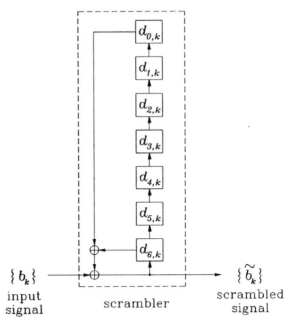

Figure 18.1. The self synchronous scrambler whose characteristic poly-nomial is $\Psi(x) = x^7 + x + 1$.

As there come in enough number of input signal elements b_k's, the state vector $\mathbf{d}_k$ in (18.2) and the scrambled output signal $\tilde{b}_k$ in (18.4) can be expressed in terms of b_k. The resulting expressions are as described in the following two theorems :

THEOREM 18.1 *For a scrambler whose characteristic polynomial is*
$\Psi(x) = \sum_{i=0}^{L} \psi_i x^i$, *the kth state vector $\mathbf{d}_k$, $k = L, L+1, \cdots$, is repre-sented by*

$$\mathbf{d}_k = [\,\tilde{b}_{k-L} \quad \tilde{b}_{k-L+1} \quad \cdots \quad \tilde{b}_{k-1}\,]^t. \qquad (18.8)$$

Proof : It suffices to show that $\mathbf{e}_{L-i}^t \cdot \mathbf{d}_k = \tilde{b}_{k-i}$, $i = 1, 2, \cdots, L$. We show this by induction. By (18.2), we have the relation $\mathbf{e}_{L-1}^t \cdot \mathbf{d}_k = \mathbf{e}_{L-1}^t \cdot \mathbf{A}_{\Psi_r(x)}^t \cdot \mathbf{d}_{k-1} + b_{k-1}$, which becomes $\tilde{b}_{k-1}$ due to (18.4). This proves the case $i = 1$. Now we assume that $\mathbf{e}_{L-i}^t \cdot \mathbf{d}_k = \tilde{b}_{k-i}$ for an $i = 1, 2, \cdots, L - 1$, and prove $\mathbf{e}_{L-(i+1)}^t \cdot \mathbf{d}_k = \tilde{b}_{k-(i+1)}$. By (18.2), we get the relation $\mathbf{e}_{L-(i+1)}^t \cdot \mathbf{d}_k = \mathbf{e}_{L-(i+1)}^t \cdot \mathbf{A}_{\Psi_r(x)}^t \cdot \mathbf{d}_{k-1}$, which become $\mathbf{e}_{L-i}^t \cdot \mathbf{d}_{k-1}$ by (18.3). Therefore, $\mathbf{e}_{L-(i+1)}^t \cdot \mathbf{d}_k = \tilde{b}_{k-(i+1)}$ due to the assumption. This completes the proof. ∎

THEOREM 18.2 *For a scrambler whose characteristic polynomial is* $\Psi(x) = \sum_{i=0}^{L} \psi_i x^i$, *the* k*th scrambled signal element* $\tilde{b}_k$, $k = L$, $L+1$, $\cdots$, *is represented by*

$$\tilde{b}_k = b_k + \sum_{i=1}^{L} \psi_i \tilde{b}_{k-i}. \tag{18.9}$$

Proof : This theorem is directly obtained by inserting (18.8) to (18.4). ∎

Theorem 18.1 states that the kth state vector $\mathbf{d}_k$, $k = L$, $L + 1$, $\cdots$, is determined by the L previously scrambled signal elements $\tilde{b}_{k-i}$, $i = 1, 2, \cdots, L$. In fact, this is an obvious result that can be intuitively obtained from the scrambler in Fig. 17.1(a). The significance of this theorem is in that it visualizes how the scrambled signal $\{ \tilde{b}_k \}$ carries the information on the states of the shift registers in the scrambler. Theorem 18.2 describes how the kth scrambled signal element $\tilde{b}_k$, $k = L$, $L + 1$, $\cdots$, can be determined from the kth input signal element b_k and the L previously scrambled signal elements $\tilde{b}_{k-i}$, $i = 1, 2, \cdots, L$. As a consequence, if we know the first L scrambled signal elements $\tilde{b}_j$, $j = 0, 1, \cdots, L - 1$, then we can obtain the scrambled signal $\{ \tilde{b}_k \}$ from the input signal $\{ b_k \}$ by inserting $k = L$, $L+1$, $\cdots$, into (18.9) repeatedly.

EXAMPLE 18.2 For the scrambler in Fig. 18.1, whose characteristic polynomial is $\Psi(x) = x^7 + x + 1$, we can confirm that the state vector $\mathbf{d}_k$, $k = 7, 8, \cdots$, in Table 18.1 meets (18.8). In addition, by (18,9) the kth scrambled signal element $\tilde{b}_k$, $k = 7, 8, \cdots$, becomes $b_k + \tilde{b}_{k-1} + \tilde{b}_{k-7}$. For the initial state vector $\mathbf{d}_0 = [\, 1\ 0\ 0\ 0\ 1\ 0\ 1\,]^t$ and the input signal $\{ b_k \}$ in (18.6), the first seven scrambled signal elements are $\tilde{b}_0 = 1$, $\tilde{b}_1 = 1$, $\tilde{b}_2 = 0$, $\tilde{b}_3 = 1$, $\tilde{b}_4 = 1$, $\tilde{b}_5 = 1$, $\tilde{b}_6 = 1$. Inserting $k = 7$ into the relation $\tilde{b}_k = b_k + \tilde{b}_{k-1} + \tilde{b}_{k-7}$, we obtain $\tilde{b}_7 = b_7 + \tilde{b}_6 + \tilde{b}_0 = 0$. In a similar manner, applying $k = 8, 9, \cdots$, repeatedly, we can get the scrambled signal $\{ \tilde{b}_k \}$ in (18.7). ♣

18.2 Descrambled Signals

In a similar manner, we consider the self synchronous descrambler in Fig. 17.1(b) whose characteristic polynomial is also $\Psi(x) = \sum_{i=0}^{L} \psi_i x^i$. Since the state of the jth shift register at time k, $\hat{d}_{j,k}$, $j = 0, 1, \cdots$.

$L - 1$, is determined by the previous states $\hat{d}_{j,k-1}$, $j = 0, 1, \cdots, L - 1$ such that

$$\hat{d}_{j,k} = \begin{cases} \hat{d}_{j+1.k-1}, & j = 0, 1, \cdots, L - 2, \\ \tilde{b}_{k-1}, & j = L - 1, \end{cases} \qquad (18.10)$$

we get the vector expression

$$\hat{\mathbf{d}}_k = \tilde{\mathbf{O}}_L \cdot \hat{\mathbf{d}}_{k-1} + \tilde{b}_{k-1} \mathbf{e}_{L-1} \qquad (18.11)$$

for the state vector $\hat{\mathbf{d}}_k$, where $\tilde{\mathbf{O}}$ is the *upper-diagonal matrix* defined by

$$\tilde{\mathbf{O}}_L \equiv \begin{bmatrix} \mathbf{0}_{(L-1)\times 1} & \mathbf{I}_{(L-1)\times(L-1)} \\ 0 & \mathbf{0}_{1\times(L-1)} \end{bmatrix} \qquad (18.12)$$

The kth descrambled signal element $\hat{b}_k$ is obtained in the exactly same manner as the scrambled signal element $\tilde{b}_k$ was obtained from the scrambler, so we get the expression

$$\hat{b}_k = \mathbf{e}_{L-1}^t \cdot \mathbf{A}_{\Psi_r(x)}^t \cdot \hat{\mathbf{d}}_k + \tilde{b}_k. \qquad (18.13)$$

EXAMPLE 18.3 We consider the descrambler in Fig. 18.2, whose characteristic polynomial is $\Psi(x) = x^7 + x + 1$. If its initial state vector $\hat{\mathbf{d}}_0$ is $[\,0\ 1\ 1\ 0\ 0\ 1\ 0\,]^t$, then by (18.11) the state vector $\hat{\mathbf{d}}_k$'s for the scrambled signal $\{\ \tilde{b}_k\ \}$ in (18.7) become as listed in Table 18.2. Therefore, by (18.13), the descrambled signal $\{\ \hat{b}_k\ \}$ becomes

$$\{\hat{b}_k\} = \{1, 1, 0, 1, 0, 1, 0, 0, 1, 1, 1, 0, 1, 0, 0, 0, 0, 0, 1, 1, \cdots\}. \clubsuit \quad (18.14)$$

As for the scrambler's case, the state vector $\hat{d}_k$ of the descrambler as well as the descrambled signal element $\hat{b}_k$ can be expressed in terms of the input scrambled signal element $\tilde{b}_k$ if there come in enough number of $\tilde{b}_k$'s. The resulting expressions are as described in the following two theorems :

THEOREM 18.3 *For a descrambler whose characteristic polynomial is* $\Psi(x) = \sum_{i=0}^{L} \psi_i x^i$, *the kth state vector* $\hat{\mathbf{d}}_k$, $k = L, L+1, \cdots$, *is represented by*

$$\hat{\mathbf{d}}_k = [\,\tilde{b}_{k-L}\ \tilde{b}_{k-L+1}\ \cdots\ \tilde{b}_{k-1}\,]^t \qquad (18.15)$$

regardless of the initial state vector $\hat{\mathbf{d}}_0$.

Table 18.2. The state vector $\hat{\mathbf{d}}_k$ and the descrambled signal $\{\,\hat{b}_k\,\}$ of the descrambler in Fig. 18.2 for the scrambled signal $\{\,\tilde{b}_k\,\}$ in (18.7) (The asterisks(*) indicate the initial 7 states in which $\hat{\mathbf{d}}_k \neq \mathbf{d}_k$ and $\hat{b}_k \neq b_k$ for the corresponding $\mathbf{d}_k$ and b_k in Table 18.1)

k	$\hat{\mathbf{d}}_0$	$\{\,\tilde{b}_k\,\}$	$\{\,\hat{b}_k\,\}$
* 0	$[\,0\,1\,1\,0\,0\,1\,0\,]^t$	1	1
* 1	$[\,1\,1\,0\,0\,1\,0\,1\,]^t$	1	1
* 2	$[\,1\,0\,0\,1\,0\,1\,1\,]^t$	0	0
* 3	$[\,0\,0\,1\,0\,1\,1\,0\,]^t$	1	1
* 4	$[\,0\,1\,0\,1\,1\,0\,1\,]^t$	1	0
* 5	$[\,1\,0\,1\,1\,0\,1\,1\,]^t$	1	1
* 6	$[\,0\,1\,1\,0\,1\,1\,1\,]^t$	1	0
7	$[\,1\,1\,0\,1\,1\,1\,1\,]^t$	0	0
8	$[\,1\,0\,1\,1\,1\,1\,0\,]^t$	0	1
9	$[\,0\,1\,1\,1\,1\,0\,0\,]^t$	1	1
10	$[\,1\,1\,1\,1\,0\,0\,1\,]^t$	1	1
11	$[\,1\,1\,1\,0\,0\,1\,1\,]^t$	0	0
12	$[\,1\,1\,0\,0\,1\,1\,0\,]^t$	0	1
13	$[\,1\,0\,0\,1\,1\,0\,0\,]^t$	1	0
14	$[\,0\,0\,1\,1\,0\,0\,1\,]^t$	1	0
15	$[\,0\,1\,1\,0\,0\,1\,1\,]^t$	1	0
16	$[\,1\,1\,0\,0\,1\,1\,1\,]^t$	0	0
17	$[\,1\,0\,0\,1\,1\,1\,0\,]^t$	1	0
18	$[\,0\,0\,1\,1\,1\,0\,1\,]^t$	0	1
19	$[\,0\,1\,1\,1\,0\,1\,0\,]^t$	1	1
$\vdots$	$\vdots$	$\vdots$	$\vdots$

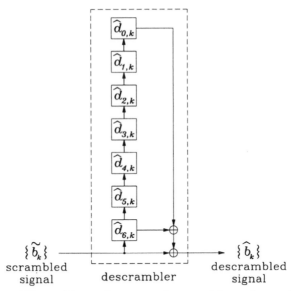

Figure 18.2. The self synchronous descrambler whose characteristic polynomial is $\Psi(x) = x^7 + x + 1$.

Proof : It suffices to show that $\mathbf{e}_{L-i}^t \cdot \hat{\mathbf{d}}_k = \tilde{b}_{k-i}$, $i = 1, 2, \cdots, L$. We show this by induction. By (18.11), we have the relation $\mathbf{e}_{L-1}^t \cdot \hat{\mathbf{d}}_k = \mathbf{e}_{L-1}^t \cdot \tilde{\mathbf{O}}_L \cdot \hat{\mathbf{d}}_{k-1} + \tilde{b}_{k-1}$, which becomes $\tilde{b}_{k-1}$ due to (18.12). This proves the case $i = 1$. Now we assume that $\mathbf{e}_{L-i}^t \cdot \hat{\mathbf{d}}_k = \tilde{b}_{k-i}$ for an $i = 1$, $2, \cdots, L - 1$, and prove $\mathbf{e}_{L-(i+1)}^t \cdot \mathbf{d}_k = \tilde{b}_{k-(i+1)}$. By (18.11), we get the relation $\mathbf{e}_{L-(i+1)}^t \cdot \hat{\mathbf{d}}_k = \mathbf{e}_{L-(i+1)}^t \cdot \tilde{\mathbf{O}}_L \cdot \hat{\mathbf{d}}_{k-1}$, which becomes $\mathbf{e}_{L-i}^t \cdot \hat{\mathbf{d}}_{k-1}$ by (18.12). Therefore, $\mathbf{e}_{L-(i+1)}^t \cdot \hat{\mathbf{d}}_k = \tilde{b}_{k-(i+1)}$ due to the assumption. This completes the proof. ∎

THEOREM 18.4 *For a descrambler whose characteristic polynomial is* $\Psi(x) = \sum_{i=0}^{L} \psi_i x^i$, *the kth descrambled element* $\hat{b}_k$, $k = L, L+1, \cdots$, *is represented by*

$$\hat{b}_k = \sum_{i=0}^{L} \psi_i \tilde{b}_{k-i} \tag{18.16}$$

regardless of the initial state vector $\hat{\mathbf{d}}_0$.

Proof : This theorem is directly obtained by inserting (18.15) to (18.13). ∎

Theorem 18.3 states that the kth state vector $\hat{\mathbf{d}}_k$, $k = L, L+1, \cdots$, is determined by the L previously scrambled signal elements $\tilde{b}_{k-i}$, $i = 1, 2, \cdots, L$, and, again, this is obvious if viewed from the descrambler in Fig. 17.1(b). It is important to note that the scrambled signal $\{ \tilde{b}_k \}$ itself controls the states of the shift registers in the descrambler. Theorem 18.4 describes how the kth descrambled signal element $\hat{b}_k$, $k = L, L+1, \cdots$, can be determined from the $(L+1)$ scrambled signal elements $\tilde{b}_{k-i}$, $i = 0, 1, \cdots, L$. As a consequence, we can obtain the descrambled signal $\{ \hat{b}_k \}$ from the scrambled signal $\{ \tilde{b}_k \}$ by inserting $k = L, L+1, \cdots$, to (18.16) repeatedly.

EXAMPLE 18.4 For the scrambler in Fig. 18.2, whose characteristic polynomial is $\Psi(x) = x^7 + x + 1$, we can confirm that the state vector $\hat{\mathbf{d}}_k$, $k = 7, 8, \cdots$, in Table 18.2 meets (18.15). In addition, by (18.16) the kth descrambled element $\hat{b}_k$, $k = 7, 8, \cdots$, becomes $\tilde{b}_k + \tilde{b}_{k-1} + \tilde{b}_{k-7}$. For $k = 7$, we obtain $\hat{b}_7 = \tilde{b}_7 + \tilde{b}_6 + \tilde{b}_0$, which becomes 0 for the scrambled signal $\{ \tilde{b}_k \}$ in (18.7). In a similar manner, applying $k = 8, 9, \cdots$, repeatedly, we can get the descrambled signal $\{ \hat{b}_k \}$ in (18.14). ♣

18.3 Self-Synchronization

In order to examine how the SSS acquires synchronization, we compare Theorems 18.1 and 18.2 respectively with Theorems 18.3 and 18.4. Then we can deduce the following two theorems on its self-synchronization property :

THEOREM 18.5 *For a scrambler and descrambler pair whose characteristic polynomial is* $\Psi(x) = \sum_{i=0}^{L} \psi_i x^i$, *the kth state vector* $\hat{\mathbf{d}}_k$, $k = L$, $L+1, \cdots$, *of the descrambler is identical to the scrambler kth state vector* $\mathbf{d}_k$ *of the scrambler, regardless of the scrambler initial state vector* $\mathbf{d}_0$ *and the descrambler initial state vector* $\hat{\mathbf{d}}_0$.

Proof : The theorem is obvious since (18.8) and (18.15) are the same. ∎

THEOREM 18.6 *For a scrambler and descrambler pair whose characteristic polynomial is* $\Psi(x) = \sum_{i=0}^{L} \psi_i x^i$, *the kth descrambled signal element* $\hat{b}_k$, $k = L, L+1, \cdots$, *is identical to the kth input signal element* b_k, *regardless of the scrambler initial state vector* $\mathbf{d}_0$ *and the descrambler initial state vector* $\hat{\mathbf{d}}_0$.

Proof : This theorem is directly obtained by inserting (18.9) into (18.16).   ∎

The theorems mean that if the number of received scrambled data reaches the length of the scrambler and descrambler, the descrambler state vectors are automatically synchronized to the scrambler state vectors and, as a consequence, the descrambled output signal becomes identical to the original input signal. That is, synchronization is acquired without any additional processing, or self-synchronization is working, in the SSS. In fact, the mediator for this self-synchronization is the scrambled signal itself. Note that the scrambled signal totally dictates the states of the scrambler as well as the descrambler.

EXAMPLE 18.5 For the scrambler and the descrambler pair in Figs. 18.1 and 18.2, whose characteristic polynomial is $\Psi(x) = x^7 + x + 1$, we can confirm from Tables 18.1 and 18.2 that for the input signal $\{ b_k \}$ in (18.6), the scrambler state vector $\mathbf{d}_k$, $k = 7, 8, \cdots$, is identical to the descrambler state vector $\hat{\mathbf{d}}_k$, and the descrambled signal element $\hat{b}_k$, $k = 7, 8, \cdots$, is identical to the original input signal b_k. The asterisks($*$) in Table 18.2 indicate the initial 7 states($k = 0, 1, \cdots, 6$) in which $\hat{\mathbf{d}}_k \neq \mathbf{d}_k$ and $\hat{b}_k \neq b_k$ for the corresponding $\mathbf{d}_k$ and b_k in Table 18.1. ♣

18.4   Error-Multiplication

We finally consider the error-multiplication property of the SSS. That is, we examine how many errors occur in the descrambled signal $\{ \hat{b}_k \}$ caused by a bit error in the scrambled signal $\{ \tilde{b}_k \}$ that occurred during the transmission. We describe this in terms of the weight of the characteristic polynomial $\Psi(x)$ defined below :

DEFINITION 18.7 (Weight of Characteristic Polynomial) For a characteristic polynomial $\Psi(x) = \sum_{i=0}^{L} \psi_i x^i$, we define the *weight* $W[\Psi(x)]$ to be the number of the coefficients ψ_i's which are 1.

EXAMPLE 18.6 In the case of the characteristic polynomial $\Psi(x) = x^7 + x + 1$, there are three unity coefficients ψ_0, ψ_1 and ψ_7. Therefore, its weight $W[x^7 + x + 1]$ is 3. ♣

THEOREM 18.8 *For a scrambler and descrambler pair whose characteristic polynomial is $\Psi(x) = \sum_{i=0}^{L} \psi_i x^i$, let the weight $W[\Psi(x)]$ be w*

such that $\psi_{ij} = 1$ for $j = 0, 1, \cdots, w\text{-}1$. Then, if the kth scrambled signal element $\tilde{b}_k$, for a $k = L, L+1, \cdots$, is errored, the w descrambled signal elements $\hat{b}_{k+i_j}$, $j = 0, 1, \cdots, w\text{-}1$, are all errored.

Proof : This theorem is obvious in view of (18.16). ∎

The theorem indicates that the number of errored signal elements in the descrambled signal, caused by a single bit error in the scrambled signal, is identical to the weight of the characteristic polynomial.

EXAMPLE 18.7 In the case of the scrambler and descrambler pair in Figs. 18.1 and 18.2, the weight of the characteristic polynomial $\Psi(x) = x^7 + x + 1$ is 3. Therefore, in this SSS, error-multiplication factor is 3. If $\tilde{b}_8 = 0$ in the scrambled signal $\{\tilde{b}_k\}$ in (18.7) gets errored to become 1, then the three descrambled data $\hat{b}_8$, $\hat{b}_9$ and $\hat{b}_{15}$ get errored to become $\hat{b}_8 = \hat{b}_9 = 0$, $\hat{b}_{15} = 1$, We can confirm this using (18.16). ♣

Chapter 19

Parallel Self Synchronous Scrambling

In the *parallel self synchronous scrambling* (PSSS), the parallel input signals are scrambled before multiplexing, and the scrambled signal is descrambled after demultiplexing, as in other parallel scrambling techniques. In this chapter, we discuss the behaviors of parallel-scrambled and parallel-descrambled signals for the SSS, and consider how to realize parallel scramblers and parallel descramblers generating these signals. We first consider the relation between the serial scrambling and the parallel scrambling, and examine the parallel-scrambled signals and the parallel-descrambled signals. Then, we mathematically model parallel scramblers and parallel descramblers, and consider how to realize them for use in the PSSS.

19.1   Parallel-Scrambled   and   -Descrambled Signals

In the serial SSS, N parallel input signals $\{ b_k^j \}$, $j = 0, 1, \cdots, N - 1$, are multiplexed, and then the multiplexed signal $\{ b_k \}$ is scrambled by passing through a serial scrambler. In the receiving part, the scrambled signal $\{ \tilde{b}_k \}$ is descrambled by passing through a serial descrambler, and then the descrambled signal $\{ \hat{b}_k \}$ is demultiplexed for the recovery of the parallel input signals $\{ b_k^i \}$'s. On the other hand, in the parallel SSS, N parallel input signals $\{ b_k^j \}$, $j = 0, 1, \cdots, N - 1$, are scrambled before multiplexing by passing through a parallel scrambler, and the scrambled signal $\{\ \tilde{b}_k\ \}$ is descrambled after demultiplexing by passing through a parallel descrambler. They are embodied in the functional blockdiagrams in Fig. 19.1(a) and (b).[1]

As in the case of the FSS and the DSS, we can consider two categories of the parallel SSS $--$ the (bit-)*parallel SSS* (PSSS) which employs a bit-interleaved multiplexing, and the *multibit-parallel SSS* (MPSSS) which employs a multibit-interleaved multiplexing. However, in the case of the MPSSS, the parallel input data do not come in the order they were originally generated, while the states of shift registers in the scrambler and the descrambler are determined by the input data coming in the order of their generation (refer to (8.1)).[2] This implies that some of the shift register states must be determined by the input data yet to come in the future. Therefore, *the MPSSS system becomes a non-causal system, which can not be realized in the form of simple delayless linear system.* So we concentrate our discussion on the PSSS in this section.

In the PSSS, for the same parallel input signals $\{\ b_k^i\ \}$'s, the scrambled signal $\{\ \tilde{b}_k\ \}$ and the parallel-descrambled signals $\{\ \hat{b}_k^j\ \}$'s in Fig. 19.1(b) should be identical respectively to the serial-scrambled signal $\{\ \tilde{b}_k\ \}$ and the descrambled signals $\{\ \hat{b}_k^j\ \}$'s in Fig. 19.1(a). Therefore, the multiplexed signals of the parallel-scrambled signals $\{\ \tilde{b}_k^j\ \}$'s and the parallel-descrambled signals $\{\ \hat{b}_k^j\ \}$'s should be identical respectively to the serial-scrambled signal $\{\ \tilde{b}_k\ \}$ and the serial-descrambled signal $\{\ \hat{b}_k\ \}$. In other words, the parallel-scrambled signals $\{\ \tilde{b}_k^j\ \}$ and the parallel-descrambled signals $\{\ \hat{b}_k^j\ \}$, $j = 0, 1, \cdots, N - 1$, are the N-decimated signals respectively of the serial-scrambled signal $\{\ \tilde{b}_k\ \}$ and the serial-descrambled signal $\{\ \hat{b}_k\ \}$, that is,

$$\{\ \tilde{b}_k^j\ \} \ = \ \{\ \tilde{b}_{j+kN}\ \}, \tag{19.1a}$$

$$\{\ \hat{b}_k^j\ \} \ = \ \{\ \hat{b}_{j+kN}\ \}. \tag{19.1b}$$

The following two theorems respectively describe the relation between the parallel input signals $\{\ b_k^j\ \}$'s and the parallel-scrambled sig-

---

[1] We assume in this chapter that the multiplexing and the demultiplexing functions in both Figs. 19.1(a) and (b) are furnished with a proper means, such as frame-formatting and frame alignment, to properly distribute the N descrambled signals to the destined output channels.

[2] It is the same for the cases of the MPFSS and the MPDSS as far as the input data ordering is concerned. However it does not cause the non-causality problem for them because the states of their shift registers are determined independently of the input data.

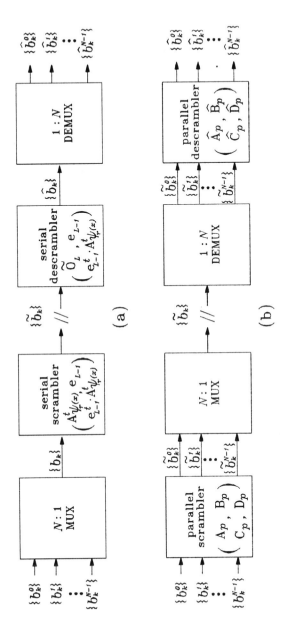

Figure 19.1. Functional blockdiagrams for serial and parallel self synchronous scrambling. (a) Serial scrambling, (b) parallel scrambling.

nals { $\tilde{b}_k^j$ }'s of parallel scramblers, and the relation between the parallel-scrambled signals { $\tilde{b}_k^j$ }'s and the parallel-descrambled signals { $\hat{b}_k^j$ }'s of parallel descramblers.[3]

THEOREM 19.1 *Let* { b_k^j } *and* { $\tilde{b}_k^j$ }, $j = 0, 1, \cdots, N\text{-}1$, *be respectively the parallel input signals and the parallel-scrambled signals of a parallel scrambler for the serial scrambler whose characteristic polynomial is* $\Psi(x) = \sum_{i=0}^{L} \psi_i x^i$. *Then, the parallel-scrambled data* $\tilde{b}_k^j$'s *are represented as follows : If $N > L$,*

$$
\tilde{b}_k^j =
\begin{cases}
b_k^j + \sum_{i=1}^{j} \psi_i \tilde{b}_k^{j-i} + \sum_{i=j+1}^{L} \psi_i \tilde{b}_{k-1}^{N+j-i}, \\
\qquad j = 0,\ 1,\ \cdots,\ L\text{-}1,\ k = 1,\ 2,\ \cdots, \\[2mm]
b_k^j + \sum_{i=1}^{L} \psi_i \tilde{b}_k^{j-i}, \\
\qquad j = L,\ L\text{+}1,\ \cdots,\ N\text{-}1,\ k = 0,\ 1,\ \cdots;
\end{cases}
\tag{19.2a}
$$

and if $N \leq L$,

$$
\tilde{b}_k^j =
\begin{cases}
b_k^j + \sum_{i=1}^{j} \psi_i \tilde{b}_k^{j-i} + \sum_{l=1}^{q} \sum_{i=j+1}^{N+j} \psi_{i+(l-1)N} \tilde{b}_{k-l}^{N+j-i} \\
\quad + \sum_{i=j+1}^{r} \psi_{i+qN} \tilde{b}_{k-q-1}^{N+j-i}, \\
\qquad j = 0,\ 1,\ \cdots,\ r\text{-}1,\ k = q\text{+}1,\ q\text{+}2,\ \cdots, \\[2mm]
b_k^j + \sum_{i=1}^{j} \psi_i \tilde{b}_k^{j-i} + \sum_{l=1}^{q-1} \sum_{i=j+1}^{N+j} \psi_{i+(l-1)N} \tilde{b}_{k-l}^{N+j-i} \\
\quad + \sum_{i=j+1}^{N+r} \psi_{i+(q-1)N} \tilde{b}_{k-q}^{N+j-i}, \\
\qquad j = r,\ r\text{+}1,\ \cdots,\ N\text{-}1,\ k = q,\ q\text{+}1,\ \cdots,
\end{cases}
\tag{19.2b}
$$

where q and r are respectively the quotient and the remainder of L divided by N.

Proof : We first prove the theorem for the case $N > L$. By (19.1a) and (18.9), we have the relation $\tilde{b}_k^j = \tilde{b}_{j+kN} = b_{j+kN} + \sum_{i=1}^{L} \psi_i \tilde{b}_{j-i+kN}$ for $j + kN \geq L$. If we rewrite it separately for $j < L$ and $j \geq L$, we get $\tilde{b}_k^j = b_{j+kN} + \sum_{i=1}^{j} \psi_i \tilde{b}_{j-i+kN} + \sum_{i=j+1}^{L} \psi_i \tilde{b}_{N+j-i+(k-1)N}$ for $j = 0, 1, \cdots, L - 1$; $\tilde{b}_k^j = b_{j+kN} + \sum_{i=1}^{L} \psi_i \tilde{b}_{j-i+kN}$ for $j = L, L + 1, \cdots, N - 1$. Transforming these expressions in view of (19.1a), we get (19.2a).

Equation (19.2b) for the case $N \leq L$ can be proved in a similar manner. ∎

---

[3]Note that the first L scrambled or descrambled data are not considered in the theorems. It is because they are the data generated before the SSS acquires synchronization and are therefore meaningless (refer to Sections 18.1 and 18.2).

THEOREM 19.2 *Let* $\{\ \tilde{b}_k^j\ \}$ *and* $\{\ \hat{b}_k^j\ \}$, $j = 0,\ 1,\ \cdots,\ N\text{-}1$, *be respectively the parallel-scrambled signals and the parallel-descrambled signals of a parallel descrambler for the serial descrambler whose characteristic polynomial is* $\Psi(x) = \sum_{i=0}^{L} \psi_i x^i$. *Then, the parallel-descrambled data* $\hat{b}_k^j$'s *are represented as follows : If* $N > L$,

$$
\hat{b}_k^j = \begin{cases}
\sum_{i=0}^{j} \psi_i \tilde{b}_k^{j-i} + \sum_{i=j+1}^{L} \psi_i \tilde{b}_{k-1}^{N+j-i}, \\
\qquad j = 0,\ 1,\ \cdots,\ L\text{-}1,\ k = 1,\ 2,\ \cdots, \\
\\
\sum_{i=0}^{L} \psi_i \tilde{b}_k^{j-i}, \\
\qquad j = L,\ L+1,\ \cdots,\ N\text{-}1,\ k = 0,\ 1,\ \cdots ;
\end{cases}
\tag{19.3a}
$$

and if $N \leq L$,

$$
\hat{b}_k^j = \begin{cases}
\sum_{i=0}^{j} \psi_i \tilde{b}_k^{j-i} + \sum_{l=1}^{q} \sum_{i=j+1}^{N+j} \psi_{i+(l-1)N} \tilde{b}_{k-l}^{N+j-i} \\
\quad + \sum_{i=j+1}^{r} \psi_{i+qN} \tilde{b}_{k-q-1}^{N+j-i}, \\
\qquad j = 0,\ 1,\ \cdots,\ r\text{-}1,\ k = q+1,\ q+2,\ \cdots, \\
\\
\sum_{i=0}^{j} \psi_i \tilde{b}_k^{j-i} + \sum_{l=1}^{q-1} \sum_{i=j+1}^{N+j} \psi_{i+(l-1)N} \tilde{b}_{k-l}^{N+j-i} \\
\quad + \sum_{i=j+1}^{N+r} \psi_{i+(q-1)N} \tilde{b}_{k-q}^{N+j-i}, \\
\qquad j = r,\ r+1,\ \cdots,\ N\text{-}1,\ k = q,\ q+1,\ \cdots,
\end{cases}
\tag{19.3b}
$$

where q *and* r *are respectively the quotient and the remainder of L divided by N.*

Proof : This theorem can be proved, based on (19.1b) and (18.16), in a way similar to that of Theorem 19.1. ■

EXAMPLE 19.1 We consider the parallel scrambling with the parallel input signal number $N = 6$ for the serial scrambling whose characteristic polynomial is $\Psi(x) = x^7 + x + 1$. Then, the quotient q and the remainder r of $L = 7$ divided by $N = 6$ are both 1. Therefore, for $j = 0$, (19.2b) renders the relation $\tilde{b}_k^0 = b_k^0 + \sum_{i=1}^{6} \psi_i \tilde{b}_{k-1}^{6-i} + \psi_7 \tilde{b}_{k-2}^5$ for $k = 2, 3, \cdots$, which turns into $\tilde{b}_k^0 = b_k^0 + \tilde{b}_{k-1}^5 + \tilde{b}_{k-2}^5$ for the coefficients $\psi_0 = \psi_1 = \psi_7 = 1$ and $\psi_2 = \psi_3 = \psi_4 = \psi_5 = \psi_6 = 0$. If we repeat this procedure for $j = 1, 2, 3, 4, 5$, we obtain the following parallel-scrambled data $\tilde{b}_k^j$'s

$$
\begin{cases}
\tilde{b}_k^0 = b_k^0 + \tilde{b}_{k-1}^5 + \tilde{b}_{k-2}^5, & k = 2, 3, \cdots, \\
\tilde{b}_k^1 = b_k^1 + \tilde{b}_k^0 + \tilde{b}_{k-1}^0, & k = 1, 2, \cdots, \\
\tilde{b}_k^2 = b_k^2 + \tilde{b}_k^1 + \tilde{b}_{k-1}^1, & k = 1, 2, \cdots, \\
\tilde{b}_k^3 = b_k^3 + \tilde{b}_k^2 + \tilde{b}_{k-1}^2, & k = 1, 2, \cdots, \\
\tilde{b}_k^4 = b_k^4 + \tilde{b}_k^3 + \tilde{b}_{k-1}^3, & k = 1, 2, \cdots, \\
\tilde{b}_k^5 = b_k^5 + \tilde{b}_k^4 + \tilde{b}_{k-1}^4, & k = 1, 2, \cdots.
\end{cases}
\tag{19.4}
$$

For the parallel-descrambled data $\hat{b}_k^j$'s, we can repeat a similar procedure based on (19.3b) to obtain the following expressions :

$$
\begin{cases}
\hat{b}_k^0 = \tilde{b}_k^0 + \tilde{b}_{k-1}^5 + \tilde{b}_{k-2}^5, & k = 2, 3, \cdots, \\
\hat{b}_k^1 = \tilde{b}_k^1 + \tilde{b}_k^0 + \tilde{b}_{k-1}^0, & k = 1, 2, \cdots, \\
\hat{b}_k^2 = \tilde{b}_k^2 + \tilde{b}_k^1 + \tilde{b}_{k-1}^1, & k = 1, 2, \cdots, \\
\hat{b}_k^3 = \tilde{b}_k^3 + \tilde{b}_k^2 + \tilde{b}_{k-1}^2, & k = 1, 2, \cdots, \\
\hat{b}_k^4 = \tilde{b}_k^4 + \tilde{b}_k^3 + \tilde{b}_{k-1}^3, & k = 1, 2, \cdots, \\
\hat{b}_k^5 = \tilde{b}_k^5 + \tilde{b}_k^4 + \tilde{b}_{k-1}^4, & k = 1, 2, \cdots. \quad \clubsuit
\end{cases}
\tag{19.5}
$$

19.2   Parallel Self Synchronous Scrambler

We consider how to realize parallel self synchronous scramblers generating the parallel-scrambled signals $\{\ \tilde{b}_k^j\ \}$'s meeting the relations in (19.2).

For this, we first mathematically model the parallel self synchronous scrambler. In the parallel SSS, the parallel input signals $\{\ b_k^j\ \}$'s are scrambled while passing through the parallel scrambler to generate the parallel-scrambled signals $\{\ \tilde{b}_k^j\ \}$'s. This means that a parallel scrambler is a *multi-input-multi-output* (MIMO) system whose inputs are the parallel input signals $\{\ b_k^j\ \}$'s and the outputs are the parallel-scrambled signals $\{\ \tilde{b}_k^j\ \}$'s. To describe such a parallel scrambler system, we introduce the concept of signal vector.

DEFINITION 19.3 (Signal Vector) For N parallel signals $\{\ b_k^j\ \}$, $j = 0, 1, \cdots, N - 1$, we define the kth *signal vector* $\mathbf{b}_k$ to be the N-vector

$$
\mathbf{b}_k \equiv [\ b_k^0\ \ b_k^1\ \ \cdots\ \ b_k^{N-1}\ ]^t.
\tag{19.6}
$$

In accordance to this definition, we denote by $\tilde{\mathbf{b}}_k$ the kth signal vector for signals $\{ \tilde{b}_k^j \}$, $j = 0, 1, \cdots, N - 1$, that is,

$$\tilde{\mathbf{b}}_k = [\, \tilde{b}_k^0 \;\; \tilde{b}_k^1 \;\; \cdots \;\; \tilde{b}_k^{N-1} \,]^t. \tag{19.7}$$

Then, the state vector $\mathbf{d}_k$ of a parallel scrambler and the parallel-scrambled signal vector $\tilde{\mathbf{b}}_k$ can be respectively represented by

$$\mathbf{d}_k = \mathbf{A}_p \cdot \mathbf{d}_{k-1} + \mathbf{B}_p \cdot \mathbf{b}_{k-1}, \tag{19.8a}$$

$$\tilde{\mathbf{b}}_k = \mathbf{C}_p \cdot \mathbf{d}_k + \mathbf{D}_p \cdot \mathbf{b}_k, \tag{19.8b}$$

for the four state matrices $\mathbf{A}_p$, $\mathbf{B}_p$, $\mathbf{C}_p$ and $\mathbf{D}_p$ of sizes $L \times L$, $L \times N$, $N \times L$ and $N \times N$ respectively.[4] Therefore, a parallel scrambler is uniquely determined by the four state matrices.

As a preliminary step to determine these four state matrices for a parallel scrambler, we first consider the relation between the parallel input signal vector $\mathbf{b}_k$ and the parallel-scrambled signal vector $\tilde{\mathbf{b}}_k$.

THEOREM 19.4 *For a serial-scrambled signal $\{ \tilde{b}_k \}$ of a serial scrambler whose characteristic polynomial is $\Psi(x) = \sum_{i=0}^{L} \psi_i x^i$, let $\{ \tilde{b}_k^j \}$, $j = 0, 1, \cdots, N\text{-}1$, be the parallel-scrambled signals meeting the relations in (19.2). Then, the following relations hold for the parallel input signal vector $\mathbf{b}_k$ and the parallel-scrambled signal vector $\tilde{\mathbf{b}}_k$: If $N > L$,*

$$\mathbf{b}_k = \begin{cases} \mathbf{M}_{\Psi(x)} \cdot [\, u_{0,0} \;\; u_{0,1} \;\; \cdots \;\; u_{0,L-1} \;\; \tilde{\mathbf{b}}_0^t \,]^t, & k = 0, \\[2mm] \mathbf{M}_{\Psi(x)} \cdot [\, \tilde{b}_{k-1}^{N-L} \;\; \tilde{b}_{k-1}^{N-L+1} \;\; \cdots \;\; \tilde{b}_{k-1}^{N-1} \;\; \tilde{\mathbf{b}}_k^t \,]^t, & k = 1, 2, \cdots, \end{cases}$$

$$\tag{19.9a}$$

for arbitrary binary numbers $u_{0,i}$, $i = 0, 1, \cdots, L\text{-}1$; and if $N \le L$,

$$\mathbf{b}_k = \begin{cases} \mathbf{M}_{\Psi(x)} \cdot [\, v_{0,0} \;\; v_{0,1} \;\; \cdots \;\; v_{0,r-1} \;\; \tilde{\mathbf{b}}_0^t \;\; \tilde{\mathbf{b}}_1^t \;\; \cdots \;\; \tilde{\mathbf{b}}_q^t \,]^t, \\[1mm] \qquad k = q, \\[2mm] \mathbf{M}_{\Psi(x)} \cdot [\, \tilde{b}_{k-q-1}^{N-r} \;\; \tilde{b}_{k-q-1}^{N-r+1} \;\; \cdots \;\; \tilde{b}_{k-q-1}^{N-1} \;\; \tilde{\mathbf{b}}_{k-q}^t \;\; \tilde{\mathbf{b}}_{k-q+1}^t \;\; \cdots \;\; \tilde{\mathbf{b}}_k^t \,]^t, \\[1mm] \qquad k = q+1, q+2, \cdots, \end{cases}$$

$$\tag{19.9b}$$

---

[4]Note that these equations are the state space expressions for a multi-input multi-output linear system whose input, output and state variables are respectively $\mathbf{b}_k$, $\tilde{\mathbf{b}}_k$ and $\mathbf{d}_k$. In fact, (18.2) and (18.4) are a special example of these equations applied to a single-input and single-output system having multiple internal states. The subscript p in the state matrices $\mathbf{A}_p$, $\mathbf{B}_p$, $\mathbf{C}_p$ and $\mathbf{D}_p$ indicates that they are for the *parallel* SSS system.

for arbitrary binary numbers $v_{0,i}$, $i = 0, 1, \cdots, r\text{-}1$, *where* q *and* r *are respectively the quotient and the remainder of* L *divided by* N *and* $\mathbf{M}_{\Psi(x)}$ *is the* $N \times (L + N)$ *linear convolutional matrix defined by*

$$
\mathbf{M}_{\Psi(x)} \equiv
\begin{bmatrix}
\psi_L & \psi_{L-1} & \cdots & \psi_0 & 0 & 0 & \cdots & 0 \\
0 & \psi_L & \psi_{L-1} & \cdots & \psi_0 & 0 & \cdots & 0 \\
\vdots & \ddots & \ddots & \ddots & \cdots & \ddots & \ddots & \vdots \\
0 & \cdots & 0 & \ddots & \ddots & \cdots & \ddots & 0 \\
0 & \cdots & 0 & 0 & \psi_L & \psi_{L-1} & \cdots & \psi_0
\end{bmatrix}. \quad (19.10)
$$

Proof : This theorem is obtained by rewriting (19.2) in vector expressions and representing the residual terms for the meaningless first L input data by the arbitrary binary numbers $u_{0,i}$'s and $v_{0,i}$'s. ∎

EXAMPLE 19.2 In the case of the parallel scrambling with the parallel signal number $N = 6$ for the serial scrambling whose characteristic polynomial is $\Psi(x) = x^7 + x + 1$, the quotient and the remainder of $L = 7$ divided by N are $q = 1$ and $r = 1$, and the matrix $\mathbf{M}_{x^7+x+1}$ in (19.10) becomes

$$
\mathbf{M}_{x^7+x+1} =
\begin{bmatrix}
1 & 0 & 0 & 0 & 0 & 0 & 1 & 1 & 0 & 0 & 0 & 0 & 0 \\
0 & 1 & 0 & 0 & 0 & 0 & 0 & 1 & 1 & 0 & 0 & 0 & 0 \\
0 & 0 & 1 & 0 & 0 & 0 & 0 & 0 & 1 & 1 & 0 & 0 & 0 \\
0 & 0 & 0 & 1 & 0 & 0 & 0 & 0 & 0 & 1 & 1 & 0 & 0 \\
0 & 0 & 0 & 0 & 1 & 0 & 0 & 0 & 0 & 0 & 1 & 1 & 0 \\
0 & 0 & 0 & 0 & 0 & 1 & 0 & 0 & 0 & 0 & 0 & 1 & 1
\end{bmatrix}. \quad (19.11)
$$

Therefore, by (19.9b) we have the relation

$$
\mathbf{b}_k =
\begin{cases}
\mathbf{M}_{x^7+x+1} \cdot [\, v_{0,0} \;\; \tilde{\mathbf{b}}_0^t \;\; \tilde{\mathbf{b}}_1^t \,]^t, & k = 1, \\
\mathbf{M}_{x^7+x+1} \cdot [\, \tilde{\mathbf{b}}_{k-2}^5 \;\; \tilde{\mathbf{b}}_{k-1}^t \;\; \tilde{\mathbf{b}}_k^t \,]^t, & k = 2, 3, \cdots,
\end{cases}
\quad (19.12)
$$

where $v_{0,0}$ is either 0 or 1. We can confirm that this relation is identical to that in (19.4). ♣

We now consider how to determine the four state matrices $\mathbf{A}_p$, $\mathbf{B}_p$, $\mathbf{C}_p$ and $\mathbf{D}_p$ for the parallel scrambler.

THEOREM 19.5 *For a serial-scrambled signal* $\{\, \tilde{b}_k \,\}$ *of a serial scrambler whose characteristic polynomial is* $\Psi(x) = \sum_{i=0}^{L} \psi_i x^i$, *let* $\{\, \tilde{b}_k^j \,\}$,

$j = 0, 1, \cdots, N\text{-}1$, be the parallel-scrambled signals meeting the relations in (19.2), and let $\mathbf{M}^{+}_{\Psi(x)}$ and $\mathbf{M}^{-}_{\Psi(x)}$ be the $N \times L$ and the $N \times N$ submatrices of $\mathbf{M}_{\Psi(x)}$ such that

$$[\,\mathbf{M}^{+}_{\Psi(x)}\ \ \mathbf{M}^{-}_{\Psi(x)}\,] \equiv \mathbf{M}_{\Psi(x)}. \tag{19.13}$$

Then, the parallel scrambler whose state matrices are

$$\mathbf{A}_p = \mathbf{U} \cdot (\mathbf{A}^{t}_{\Psi_r(x)})^{N} \cdot \mathbf{U}^{-1}, \tag{19.14a}$$

$$\mathbf{B}_p = \mathbf{U} \cdot [\,(\mathbf{A}^{t}_{\Psi_r(x)})^{N-1} \cdot \mathbf{e}_{L-1}\ \ (\mathbf{A}^{t}_{\Psi_r(x)})^{N-2} \cdot \mathbf{e}_{L-1}\ \ \cdots\ \ \mathbf{e}_{L-1}\,], \tag{19.14b}$$

$$\mathbf{C}_p = (\mathbf{M}^{-}_{\Psi(x)})^{-1} \cdot \mathbf{M}^{+}_{\Psi(x)} \cdot \mathbf{U}^{-1}, \tag{19.14c}$$

$$\mathbf{D}_p = (\mathbf{M}^{-}_{\Psi(x)})^{-1}, \tag{19.14d}$$

for an $L \times L$ nonsingular matrix $\mathbf{U}$ generates the parallel-scrambled signals $\{\,\tilde{b}^{i}_{k}\,\}$'s.

Proof : Inserting (19.14a) and (19.14b) into (19.8a), we get

$$\mathbf{d}_k = \mathbf{U} \cdot (\mathbf{A}^{t}_{\Psi_r(x)})^{N} \cdot \mathbf{U}^{-1} \cdot \mathbf{d}_{k-1}$$

$$+\mathbf{U} \cdot [\,(\mathbf{A}^{t}_{\Psi_r(x)})^{N-1} \cdot \mathbf{e}_{L-1}\ \ (\mathbf{A}^{t}_{\Psi_r(x)})^{N-2} \cdot \mathbf{e}_{L-1}\ \ \cdots\ \ \mathbf{e}_{L-1}\,] \cdot \mathbf{b}_{k-1},$$

which can be rewritten in the form

$$\mathbf{U}^{-1} \cdot \mathbf{d}_k = (\mathbf{A}^{t}_{\Psi_r(x)})^{N} \cdot (\mathbf{U}^{-1} \cdot \mathbf{d}_{k-1})$$

$$+ [\,(\mathbf{A}^{t}_{\Psi_r(x)})^{N-1} \cdot \mathbf{e}_{L-1}\ \ (\mathbf{A}^{t}_{\Psi_r(x)})^{N-2} \cdot \mathbf{e}_{L-1}\ \ \cdots\ \ \mathbf{e}_{L-1}\,] \cdot \mathbf{b}_{k-1}\,; \tag{19.15}$$

and inserting (19.14c) and (19.14d) into (19.8b), we get

$$\tilde{\mathbf{b}}_k = (\mathbf{M}^{-}_{\Psi(x)})^{-1} \cdot \mathbf{M}^{+}_{\Psi(x)} \cdot \mathbf{U}^{-1} \cdot \mathbf{d}_k + (\mathbf{M}^{-}_{\Psi(x)})^{-1} \cdot \mathbf{b}_k, \tag{19.16}$$

which can be rewritten in the form

$$\mathbf{b}_k = \mathbf{M}^{+}_{\Psi(x)} \cdot \mathbf{U}^{-1} \cdot \mathbf{d}_k + \mathbf{M}^{-}_{\Psi(x)} \cdot \tilde{\mathbf{b}}_k$$

$$= \mathbf{M}_{\Psi(x)} \cdot \begin{bmatrix} \mathbf{U}^{-1} \cdot \mathbf{d}_k \\ \tilde{\mathbf{b}}_k \end{bmatrix} \tag{19.17}$$

due to (19.13).

We first prove the theorem for the case $N > L$. Inserting $k = 0$ into (19.17), we obtain the relation in (19.9a) for $k = 0$. If $N > L$, the matrices composed of the last L rows of $(\mathbf{M}^-_{\Psi(x)})^{-1} \cdot \mathbf{M}^+_{\Psi(x)}$ and $(\mathbf{M}^-_{\Psi(x)})^{-1}$ are respectively identical to the matrices $(\mathbf{A}^t_{\Psi_r(x)})^N$ and $[(\mathbf{A}^t_{\Psi_r(x)})^{N-1} \cdot \mathbf{e}_{L-1} \ (\mathbf{A}^t_{\Psi_r(x)})^{N-2} \cdot \mathbf{e}_{L-1} \ \cdots \ \mathbf{e}_{L-1}]$.[5] Hence, by (19.15) and (19.16), we obtain $\mathbf{U}^{-1} \cdot \mathbf{d}_k = [\tilde{b}^{N-L}_{k-1} \ \tilde{b}^{N-L+1}_{k-1} \ \cdots \ \tilde{b}^{N-1}_{k-1}]^t$, $k = 1, 2, \cdots$. Therefore, inserting this to (19.17), we get the relation in (19.9a) for $k = 1, 2, \cdots$.

Now we prove the case $N \leq L$. Let q and r be respectively the quotient and the remainder of L divided by N, and let

$$\mathbf{U}^{-1} \cdot \mathbf{d}_k = [\tilde{d}_{0,k} \ \tilde{d}_{1,k} \ \cdots \ \tilde{d}_{L-1,k}]^t. \tag{19.18}$$

In case $N \leq L$, the matrices composed of the last N rows of $(\mathbf{A}^t_{\Psi_r(x)})^N$ and $[(\mathbf{A}^t_{\Psi_r(x)})^{N-1} \cdot \mathbf{e}_{L-1} \ (\mathbf{A}^t_{\Psi_r(x)})^{N-2} \cdot \mathbf{e}_{L-1} \ \cdots \ \mathbf{e}_{L-1}]$ are respectively identical to the matrices $(\mathbf{M}^-_{\Psi(x)})^{-1} \cdot \mathbf{M}^+_{\Psi(x)}$ and $(\mathbf{M}^-_{\Psi(x)})^{-1}$.[6] Hence by (19.15) and (19.16), we obtain the relation

$$\tilde{d}_{L-i,k} = \tilde{b}^{N-i}_{k-1}, \quad i = 1, 2, \cdots, N. \tag{19.19a}$$

In addition, by (19.15) we have the relation

$$\tilde{d}_{i,k} = \tilde{d}_{i+N,k-1}, \quad i = 0, 1, \cdots, L - N - 1. \tag{19.19b}$$

Inserting (19.9a) and (19.9b) to (19.18), we get

$$\mathbf{U}^{-1} \cdot \mathbf{d}_k = \begin{cases} [\tilde{d}_{L-r,0} \ \tilde{d}_{L-r+1,0} \ \cdots \ \tilde{d}_{L-1,0} \ \tilde{b}^t_0 \ \tilde{b}^t_1 \ \cdots \ \tilde{b}^t_{q-1}]^t, \\ \quad k = q, \\ [\tilde{b}^{N-r}_{k-q-1} \ \tilde{b}^{N-r+1}_{k-q-1} \ \cdots \ \tilde{b}^{N-1}_{k-q-1} \ \tilde{b}^t_{k-q} \ \tilde{b}^t_{k-q+1} \ \cdots \ \tilde{b}^t_{k-1}]^t, \\ \quad k = q+1, \ q+2, \ \cdots, \end{cases}$$

Finally, returning this to (19.17), we obtain the relation in (19.9b).[7] ∎

According to the theorem, there are a number of ways to build a parallel scrambler depending on the choice of the nonsingular matrix $\mathbf{U}$. Once the matrix is chosen, it is straightforward to determine the four state matrices for any desired characteristic polynomial.

---

[5] Refer to Property B.16 in Appendix B.

[6] Refer to Property B.16 in Appendix B.

[7] Note that $u_{0,0}, u_{0,1}, \cdots, u_{0,L-1}$ in (19.9a) and $v_{0,0}, v_{0,1}, \cdots, v_{0,r-1}$ in (19.9b) are all arbitrary binary numbers.

EXAMPLE 19.3 We realize parallel scramblers with $N = 6$ for the serial scrambler in Fig. 18.1 whose characteristic polynomial is $\Psi(x) = x^7 + x + 1$. In this case, by (19.11) and (19.13) we have

$$
\mathbf{M}^+_{x^7+x+1} = \begin{bmatrix}
1 & 0 & 0 & 0 & 0 & 0 & 1 \\
0 & 1 & 0 & 0 & 0 & 0 & 0 \\
0 & 0 & 1 & 0 & 0 & 0 & 0 \\
0 & 0 & 0 & 1 & 0 & 0 & 0 \\
0 & 0 & 0 & 0 & 1 & 0 & 0 \\
0 & 0 & 0 & 0 & 0 & 1 & 0
\end{bmatrix}, \tag{19.20a}
$$

$$
\mathbf{M}^-_{x^7+x+1} = \begin{bmatrix}
1 & 0 & 0 & 0 & 0 & 0 \\
1 & 1 & 0 & 0 & 0 & 0 \\
0 & 1 & 1 & 0 & 0 & 0 \\
0 & 0 & 1 & 1 & 0 & 0 \\
0 & 0 & 0 & 1 & 1 & 0 \\
0 & 0 & 0 & 0 & 1 & 1
\end{bmatrix}. \tag{19.20b}
$$

If we insert these along with $\mathbf{U} = \mathbf{I}_{7\times7}$ and the companion matrix (18.3) to (19.14), we get

$$
\mathbf{A}_p = \begin{bmatrix}
0 & 0 & 0 & 0 & 0 & 0 & 1 \\
1 & 0 & 0 & 0 & 0 & 0 & 1 \\
1 & 1 & 0 & 0 & 0 & 0 & 1 \\
1 & 1 & 1 & 0 & 0 & 0 & 1 \\
1 & 1 & 1 & 1 & 0 & 0 & 1 \\
1 & 1 & 1 & 1 & 1 & 0 & 1 \\
1 & 1 & 1 & 1 & 1 & 1 & 1
\end{bmatrix}, \tag{19.21a}
$$

$$
\mathbf{B}_p = \begin{bmatrix}
0 & 0 & 0 & 0 & 0 & 0 \\
1 & 0 & 0 & 0 & 0 & 0 \\
1 & 1 & 0 & 0 & 0 & 0 \\
1 & 1 & 1 & 0 & 0 & 0 \\
1 & 1 & 1 & 1 & 0 & 0 \\
1 & 1 & 1 & 1 & 1 & 0 \\
1 & 1 & 1 & 1 & 1 & 1
\end{bmatrix}, \tag{19.21b}
$$

$$
\mathbf{C}_p = \begin{bmatrix}
1 & 0 & 0 & 0 & 0 & 0 & 1 \\
1 & 1 & 0 & 0 & 0 & 0 & 1 \\
1 & 1 & 1 & 0 & 0 & 0 & 1 \\
1 & 1 & 1 & 1 & 0 & 0 & 1 \\
1 & 1 & 1 & 1 & 1 & 0 & 1 \\
1 & 1 & 1 & 1 & 1 & 1 & 1
\end{bmatrix}, \tag{19.21c}
$$

$$\mathbf{D}_p = \begin{bmatrix} 1 & 0 & 0 & 0 & 0 & 0 \\ 1 & 1 & 0 & 0 & 0 & 0 \\ 1 & 1 & 1 & 0 & 0 & 0 \\ 1 & 1 & 1 & 1 & 0 & 0 \\ 1 & 1 & 1 & 1 & 1 & 0 \\ 1 & 1 & 1 & 1 & 1 & 1 \end{bmatrix}. \tag{19.21d}$$

Using these state matrices, we can build the parallel scrambler shown in Fig. 19.2. ♣

19.3   Parallel Self Synchronous Descrambler

We now consider how to realize parallel self synchronous descramblers generating the parallel descrambled signals { $\hat{b}_k^j$ }'s meeting the relations in (19.3).

In a similar manner to the case of the parallel scrambler, we mathematically model the parallel descrambler as a multi-input-multi-output system. For this, we denote by $\hat{\mathbf{b}}_k$ the parallel-descrambled signal vector, i.e.,

$$\hat{\mathbf{b}}_k = [\, \hat{b}_k^0 \;\; \hat{b}_k^1 \;\; \cdots \;\; \hat{b}_k^{N-1} \,]^t. \tag{19.22}$$

Then, the state vector $\hat{\mathbf{d}}_k$ of a parallel descrambler and the parallel-descrambled signal vector $\hat{\mathbf{b}}_k$ can be respectively represented by

$$\hat{\mathbf{d}}_k = \hat{\mathbf{A}}_p \cdot \hat{\mathbf{d}}_{k-1} + \hat{\mathbf{B}}_p \cdot \tilde{\mathbf{b}}_{k-1}, \tag{19.23a}$$

$$\hat{\mathbf{b}}_k = \hat{\mathbf{C}}_p \cdot \hat{\mathbf{d}}_k + \hat{\mathbf{D}}_p \cdot \tilde{\mathbf{b}}_k, \tag{19.23b}$$

for the state matrices $\hat{\mathbf{A}}_p, \hat{\mathbf{B}}_p, \hat{\mathbf{C}}_p$ and $\hat{\mathbf{D}}_p$ of sizes $L \times L, L \times N, N \times L$ and $N \times N$ respectively.[8] Therefore, a parallel descrambler is also uniquely determined by the four state matrices.

In order to determine the four state matrices for a parallel descrambler, we first consider the relation between the parallel-scrambled signal vector $\tilde{\mathbf{b}}_k$ and the parallel-descrambled signal vector $\hat{\mathbf{b}}_k$.

---

[8] As for the case of the parallel scrambler, these equations form the state space expressions for the parallel descrambler whose input, output, and state variables are respectively $\tilde{\mathbf{b}}_k$, $\hat{\mathbf{b}}_k$ and $\hat{\mathbf{d}}_k$. The two equations (18.11) and (18.13) are a special example of these equations applied to a single-input single-output serial descrambler having multiple internal states.

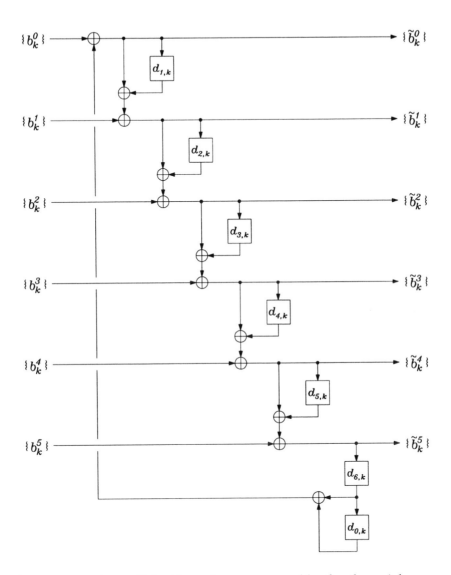

Figure 19.2. A parallel self synchronous scrambler for the serial scrambler in Fig. 18.1 whose characteristic polynomial is $\Psi(x) = x^7 + x + 1$.

THEOREM 19.6 *For a serial-descrambled signal* $\{\hat{b}_k\}$ *of a serial descrambler whose characteristic polynomial is* $\Psi(x) = \sum_{i=0}^{L} \psi_i x^i$, *let* $\{\hat{b}_k^j\}$, $j = 0, 1, \cdots, N$-1, *be the parallel-descrambled signals meeting the relations in (19.3). Then, the following relations hold for the parallel-scrambled signal vector* $\tilde{\mathbf{b}}_k$ *and the parallel-descrambled signal vector* $\hat{\mathbf{b}}_k$: *If* $N > L$,

$$\hat{\mathbf{b}}_k = \begin{cases} \mathbf{M}_{\Psi(x)} \cdot [\, u_{0,0} \ \ u_{0,1} \ \ \cdots \ \ u_{0,L-1} \ \ \tilde{\mathbf{b}}_0^t \,]^t, & k = 0, \\ \mathbf{M}_{\Psi(x)} \cdot [\, \tilde{b}_{k-1}^{N-L} \ \ \tilde{b}_{k-1}^{N-L+1} \ \ \cdots \ \ \tilde{b}_{k-1}^{N-1} \ \ \tilde{\mathbf{b}}_k^t \,]^t, & k = 1, \ 2, \ \cdots, \end{cases}$$

$$(19.24a)$$

for arbitrary binary numbers $u_{0,i}$, $i = 0, 1, \cdots, L$-1 ; *and if* $N \leq L$,

$$\hat{\mathbf{b}}_k = \begin{cases} \mathbf{M}_{\Psi(x)} \cdot [\, v_{0,0} \ \ v_{0,1} \ \ \cdots \ \ v_{0,r-1} \ \ \tilde{\mathbf{b}}_0^t \ \ \tilde{\mathbf{b}}_1^t \ \ \cdots \ \ \tilde{\mathbf{b}}_q^t \,]^t, \\ \qquad k = q, \\ \mathbf{M}_{\Psi(x)} \cdot [\, \tilde{b}_{k-q-1}^{N-r} \ \ \tilde{b}_{k-q-1}^{N-r+1} \ \ \cdots \ \ \tilde{b}_{k-q-1}^{N-1} \ \ \tilde{\mathbf{b}}_{k-q}^t \ \ \tilde{\mathbf{b}}_{k-q+1}^t \ \ \cdots \ \ \tilde{\mathbf{b}}_k^t \,]^t, \\ \qquad k = q+1, \ q+2, \ \cdots, \end{cases}$$

$$(19.24b)$$

for arbitrary binary numbers $v_{0,i}$, $i = 0, 1, \cdots, r$-1, *where* q *and* r *are respectively the quotient and the remainder of* L *divided by* N, *and* $\mathbf{M}_{\Psi(x)}$ *is the* $N \times (L + N)$ *linear convolutional matrix in (19.10).*

Proof : This theorem is obtained by rewriting (19.3) in vector expressions and representing the residual terms for the meaningless first L input data by arbitrary binary numbers. ∎

EXAMPLE 19.4 In the case of the parallel descrambling with the parallel signal number $N = 6$ for the serial descrambling whose characteristic polynomial is $\Psi(x) = x^7 + x + 1$, the quotient q and the remainder r of L divided by N are both 1. Therefore, by (19.24b) and (19.11) we obtain the relation

$$\hat{\mathbf{b}}_k = \begin{cases} \mathbf{M}_{x^7+x+1} \cdot [\, v_{0,0} \ \ \tilde{\mathbf{b}}_0^t \ \ \tilde{\mathbf{b}}_1^t \,]^t, & k = 1, \\ \mathbf{M}_{x^7+x+1} \cdot [\, \tilde{b}_{k-2}^5 \ \ \tilde{\mathbf{b}}_{k-1}^t \ \ \tilde{\mathbf{b}}_k^t \,]^t, & k = 2, \ 3, \ \cdots, \end{cases} \qquad (19.25)$$

where $v_{0,0}$ is either 0 or 1. We can confirm that this relation is identical to the relation in (19.5). ♣

We finally consider how to determine the four state matrices $\hat{\mathbf{A}}_p$, $\hat{\mathbf{B}}_p$, $\hat{\mathbf{C}}_p$ and $\hat{\mathbf{D}}_p$ for the parallel descrambler.

THEOREM 19.7 *For a serial-descrambled signal $\{\hat{b}_k\}$ of a serial descrambler whose characteristic polynomial is $\Psi(x) = \sum_{i=0}^{L} \psi_i x^i$, let $\{\hat{b}_k^j\}$, $j = 0, 1, \cdots, N$-1, be the parallel-descrambled signals meeting the relations in (19.3), and let $\mathbf{Q}^+$ and $\mathbf{Q}^-$ be respectively the $L \times L$ and the $L \times N$ submatrices of the $L \times (L + N)$ matrix $\mathbf{Q} \equiv [\mathbf{0}_{L \times N} \quad \mathbf{I}_{L \times L}]$ such that*

$$[\mathbf{Q}^+ \quad \mathbf{Q}^-] = \mathbf{Q}, \tag{19.26}$$

where $\mathbf{0}_{L \times N}$ denotes an $L \times N$ null matrix. Then, the parallel descrambler whose state matrices are

$$\hat{\mathbf{A}}_p = \mathbf{U} \cdot \mathbf{Q}^+ \cdot \mathbf{U}^{-1}. \tag{19.27a}$$

$$\hat{\mathbf{B}}_p = \mathbf{U} \cdot \mathbf{Q}^-, \tag{19.27b}$$

$$\hat{\mathbf{C}}_p = \mathbf{M}_{\Psi(x)}^+ \cdot \mathbf{U}^{-1}, \tag{19.27c}$$

$$\hat{\mathbf{D}}_p = \mathbf{M}_{\Psi(x)}^-, \tag{19.27d}$$

for an $L \times L$ nonsingular matrix $\mathbf{U}$ generates the parallel-descrambled signals $\{\hat{b}_k^j\}$'s.[9]

Proof : Inserting (19.27a) and (19.27b) into (19.23a), we get

$$\hat{\mathbf{d}}_k = \mathbf{U} \cdot \mathbf{Q}^+ \cdot \mathbf{U}^{-1} \cdot \hat{\mathbf{d}}_{k-1} + \mathbf{U} \cdot \mathbf{Q}^- \cdot \tilde{\mathbf{b}}_{k-1},$$

which can be rewritten in the form

$$\mathbf{U}^{-1} \cdot \hat{\mathbf{d}}_k = \mathbf{Q}^+ \cdot (\mathbf{U}^{-1} \cdot \hat{\mathbf{d}}_{k-1}) + \mathbf{Q}^- \cdot \tilde{\mathbf{b}}_{k-1} ; \tag{19.28}$$

and inserting (19.27c) and (19.27d) into (19.23b), we get

$$\hat{\mathbf{b}}_k = \mathbf{M}_{\Psi(x)}^+ \cdot \mathbf{U}^{-1} \cdot \hat{\mathbf{d}}_k + \mathbf{M}_{\Psi(x)}^- \cdot \tilde{\mathbf{b}}_k$$

$$= \mathbf{M}_{\Psi(x)} \cdot \begin{bmatrix} \mathbf{U}^{-1} \cdot \hat{\mathbf{d}}_k \\ \tilde{\mathbf{b}}_k \end{bmatrix} \tag{19.29}$$

due to (19.13).

We first prove the theorem for the case $N > L$. In this case, by (19.26) we have

$$\mathbf{Q}^+ = \mathbf{0}_{L \times L}, \tag{19.30a}$$

$$\mathbf{Q}^- = [\mathbf{0}_{L \times (N-L)} \quad \mathbf{I}_{L \times L}]. \tag{19.30b}$$

---

[9] Note that $\mathbf{U}$ is another arbitrary nonsingular matrix which is irrelevant to the matrix $\mathbf{U}$ in Theorem 19.5.

Inserting $k = 0$ into (19.29), we obtain the relation in (19.24a) for $k = 0$. In the case of $k = 1, 2, \cdots$, by (19.28) and (19.30), we have $\mathbf{U}^{-1} \cdot \hat{\mathbf{d}}_k = [\, \tilde{b}_{k-1}^{N-L} \ \tilde{b}_{k-1}^{N-L+1} \ \cdots \ \tilde{b}_{k-1}^{N-1} \,]^t$. Therefore, inserting this into (19.29), we obtain the relation in (19.24a) for $k = 1, 2, \cdots$.

Now we prove the theorem for the case $N \leq L$. Let q and r be respectively the quotient and the remainder of L divided by N, and let

$$\mathbf{U}^{-1} \cdot \hat{\mathbf{d}}_k = [\, \tilde{d}_{0,k} \ \tilde{d}_{1,k} \ \cdots \ \tilde{d}_{L-1,k} \,]^t . \tag{19.31}$$

In case $N \leq L$, by (19.26) we have

$$\mathbf{Q}^+ = \begin{bmatrix} \mathbf{0}_{(L-N) \times N} & \mathbf{I}_{(L-N) \times (L-N)} \\ \mathbf{0}_{N \times N} & \mathbf{0}_{N \times (L-N)} \end{bmatrix}, \tag{19.32a}$$

$$\mathbf{Q}^- = \begin{bmatrix} \mathbf{0}_{(L-N) \times L} \\ \mathbf{I}_{N \times N} \end{bmatrix}, \tag{19.32b}$$

and applying these to (19.28), we get the relation

$$\tilde{d}_{i,k} = \begin{cases} \tilde{d}_{i+N,k-1}, & i = 0, 1, \cdots, L - N - 1, \\ \tilde{b}_{k-1}^{N-L+i}, & i = L - N, L - N + 2, \cdots, L - 1. \end{cases} \tag{19.33}$$

Now, inserting (19.33) to (19.31), we get

$$\mathbf{U}^{-1} \cdot \hat{\mathbf{d}}_k = \begin{cases} [\, \tilde{d}_{L-r,0} \ \tilde{d}_{L-r+1,0} \ \cdots \ \tilde{d}_{L-1,0} \ \tilde{\mathbf{b}}_0^t \ \tilde{\mathbf{b}}_1^t \ \cdots \ \tilde{\mathbf{b}}_{q-1}^t \,]^t, \\ \qquad k = q, \\ [\, \tilde{b}_{k-q-1}^{N-r} \ \tilde{b}_{k-q-1}^{N-r+1} \ \cdots \ \tilde{b}_{k-q-1}^{N-1} \ \tilde{\mathbf{b}}_{k-q}^t \ \tilde{\mathbf{b}}_{k-q+1}^t \ \cdots \ \tilde{\mathbf{b}}_{k-1}^t \,]^t, \\ \qquad k = q+1, q+2, \cdots, \end{cases}$$

and returning this to (19.29), we finally obtain the relation in (19.24b). ∎

The theorem describes how to build parallel descramblers to match the parallel scrambler designed through Theorem 19.5. As in the case of the parallel scrambler design, there are a number of ways to build a parallel descrambler depending on the choice of $\mathbf{U}$. However, once a matrix is chosen for $\mathbf{U}$, the remaining process is straightforward.

EXAMPLE 19.5 We realize parallel descramblers with $N = 6$ for the serial descrambler in Fig. 18.2 whose characteristic polynomial is $\Psi(x) =$

$x^7 + x + 1$. In this case, by (19.26) we have

$$\mathbf{Q}^+ = \begin{bmatrix} 0 & 0 & 0 & 0 & 0 & 0 & 1 \\ 0 & 0 & 0 & 0 & 0 & 0 & 0 \\ 0 & 0 & 0 & 0 & 0 & 0 & 0 \\ 0 & 0 & 0 & 0 & 0 & 0 & 0 \\ 0 & 0 & 0 & 0 & 0 & 0 & 0 \\ 0 & 0 & 0 & 0 & 0 & 0 & 0 \\ 0 & 0 & 0 & 0 & 0 & 0 & 0 \end{bmatrix}, \qquad (19.34a)$$

$$\mathbf{Q}^- = \begin{bmatrix} 0 & 0 & 0 & 0 & 0 & 0 \\ 1 & 0 & 0 & 0 & 0 & 0 \\ 0 & 1 & 0 & 0 & 0 & 0 \\ 0 & 0 & 1 & 0 & 0 & 0 \\ 0 & 0 & 0 & 1 & 0 & 0 \\ 0 & 0 & 0 & 0 & 1 & 0 \\ 0 & 0 & 0 & 0 & 0 & 1 \end{bmatrix}. \qquad (19.34b)$$

If we insert these along with $\mathbf{U} = \mathbf{I}_{7 \times 7}$ and the linear convolutional matrix in (19.20) to (19.27), we obtain

$$\hat{\mathbf{A}}_p = \begin{bmatrix} 0 & 0 & 0 & 0 & 0 & 0 & 1 \\ 0 & 0 & 0 & 0 & 0 & 0 & 0 \\ 0 & 0 & 0 & 0 & 0 & 0 & 0 \\ 0 & 0 & 0 & 0 & 0 & 0 & 0 \\ 0 & 0 & 0 & 0 & 0 & 0 & 0 \\ 0 & 0 & 0 & 0 & 0 & 0 & 0 \\ 0 & 0 & 0 & 0 & 0 & 0 & 0 \end{bmatrix}, \qquad (19.35a)$$

$$\hat{\mathbf{B}}_p = \begin{bmatrix} 0 & 0 & 0 & 0 & 0 & 0 \\ 1 & 0 & 0 & 0 & 0 & 0 \\ 0 & 1 & 0 & 0 & 0 & 0 \\ 0 & 0 & 1 & 0 & 0 & 0 \\ 0 & 0 & 0 & 1 & 0 & 0 \\ 0 & 0 & 0 & 0 & 1 & 0 \\ 0 & 0 & 0 & 0 & 0 & 1 \end{bmatrix}, \qquad (19.35b)$$

$$\hat{\mathbf{C}}_p = \begin{bmatrix} 1 & 0 & 0 & 0 & 0 & 0 & 1 \\ 0 & 1 & 0 & 0 & 0 & 0 & 0 \\ 0 & 0 & 1 & 0 & 0 & 0 & 0 \\ 0 & 0 & 0 & 1 & 0 & 0 & 0 \\ 0 & 0 & 0 & 0 & 1 & 0 & 0 \\ 0 & 0 & 0 & 0 & 0 & 1 & 0 \end{bmatrix}, \qquad (19.35c)$$

$$
\hat{\mathbf{D}}_p \;=\; \begin{bmatrix}
1 & 0 & 0 & 0 & 0 & 0 \\
1 & 1 & 0 & 0 & 0 & 0 \\
0 & 1 & 1 & 0 & 0 & 0 \\
0 & 0 & 1 & 1 & 0 & 0 \\
0 & 0 & 0 & 1 & 1 & 0 \\
0 & 0 & 0 & 0 & 1 & 1
\end{bmatrix} . \tag{19.35d}
$$

Using these state matrices, we can build the parallel descrambler shown in Fig. 19.3. Note that this descrambler can be used in conjunction with the parallel scrambler in Fig. 19.2. ♣

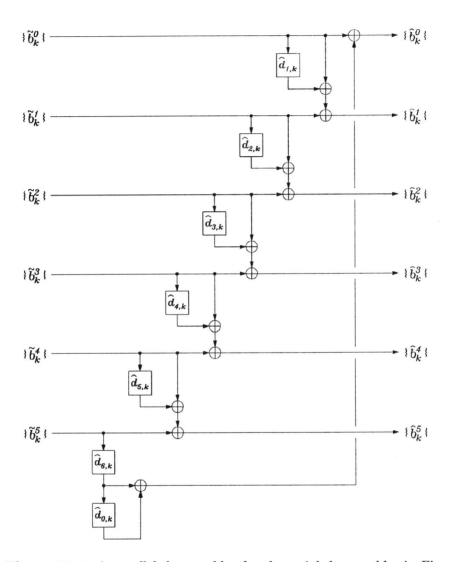

Figure 19.3. A parallel descrambler for the serial descrambler in Fig. 18.2 whose characteristic polynomial is $\Psi(x) = x^7 + x + 1$.

Chapter 20

Applications to Scrambling in SDH-Based ATM Transmission

In this chapter, we consider how to apply the serial and the parallel SSS techniques developed in the previous chapters to the ATM-cell scrambling in the SDH-based ATM transmission, which is the most typical lightwave transmission employing the SSS. We first describe the scrambling operations used for cell scrambling in the SDH-based ATM transmission. Then, we consider how to apply the parallel SSS techniques to the ATM-cell scrambling.

20.1 Scrambling of SDH-Based ATM Signals

As described in Chapter 16, in the BISDN, the information data are carried in the form of ATM cells shown in Fig. 16.2, which can be transmitted via two ways of transmission −− the *cell-based ATM transmission* and the *SDH-based ATM transmission*. In the cell-based ATM transmission, the ATM-cell stream is scrambled through the DSS, and then directly transmitted at the rate of 155.520(or 622.080) Mbps. However, in the SDH-based ATM transmission, two kinds of scramblings are applied, as illustrated in Fig. 20.1. The ATM-cell stream is first scrambled through the SSS, and then mapped to the payload space of STM-1(or STM-4), which is again scrambled by the FSS. In this case, the employed FSS is identical to that for the SDH/SONET scrambling described in Chapter 9.

In the SDH-based ATM transmission, the SSS employs the scrambler and the descrambler whose characteristic polynomial is $\Psi(x) = x^{43} + 1$,

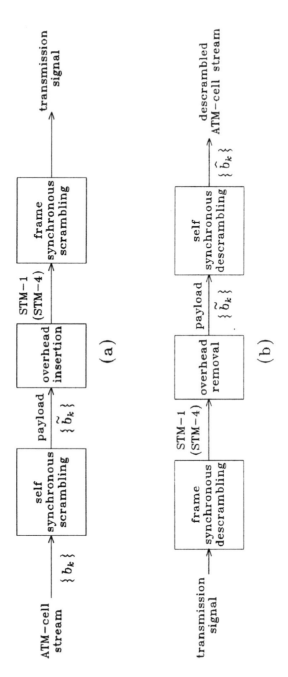

Figure 20.1. Scrambling-related blockdiagram of the SDH-based ATM transmission system. (a) Transmitter, (b) receiver.

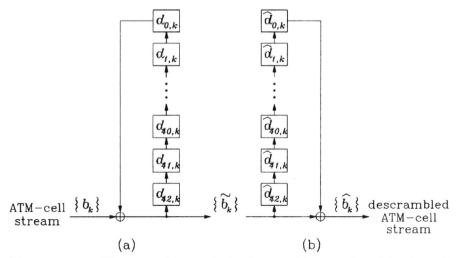

Figure 20.2. The scrambler and the descrambler employed in the self synchronous scrambling of the ATM cells in the SDH-based ATM-cell stream. (a) Scrambler, (b) descrambler.

as shown in Fig. 20.2. So the descrambler in this case gets automatically synchronized to the scrambler after receiving the first 43 scrambled data. This SSS has been selected to minimize the error-multiplication in the ATM-cell stream, which is two by Theorem 18.8. Further, in order to avoid error-multiplication in the cell header, the header is excluded from scrambling and descrambling. That is, during the clock time for the five byte header, the scrambler and descrambler operations are suspended and their states are retained. From the ATM cell payload's point of view, it appears as if the SSS is applied to the connected stream of the cell payloads without intermission. In this sense, we need to interpret Fig. 20.1 such that the sequence $\{ b_k \}$ indicates the cell payload stream, and the overhead in the overhead insertion and overhead removal functional blocks indicates the two-fold overheads −− the cell header and the SDH overhead.

20.2 Parallel SSS for Cell Scrambling in SDH-Based ATM Transmission

We consider how to apply the parallel scrambling techniques developed in Chapter 19 to the cell scrambling in the SDH-based ATM transmission.

In a similar manner to the case of the cell-based ATM signal, we regard the SDH-based ATM signal $\{\, b_k \,\}$ as the bit-interleaved signal of the eight parallel signals $\{\, b_k^j \,\}$, $j = 0, 1, \cdots, 7$, that is, $N = 8$, and apply the parallel scrambling techniques to this. Then, the SDH-based transmission system in Fig. 20.1 can be redrawn, by employing the PSSS and the PFSS, as shown in Fig. 20.3. If we employ these parallel scrambling schemes, their corresponding serial scrambling rate of 155.520(or 622.080) Mbps drops to an eight, which is 19.44(or 77.76) Mbps.

In the case of the PFSS, since the serial FSS operation is identical to that for scrambling of the SDH/SONET signal, the PFSS described in Chapter 9 can be equally applied. That is, the PSRGs in Figs. 9.6 and 9.7 can be used for scrambling of the SDH-based ATM transmission signal STM-1(or STM-4).

We consider the PSSS for the cell payloads $\{\, b_k^j \,\}$, $j = 0, 1, \cdots, 7$, in the SDH-based ATM parallel signals. Inserting $\Psi(x) = x^{43} + 1$ to (19.10), we obtain the 8×51 matrix $\mathbf{M}_{x^{43}+1}$

$$\mathbf{M}_{x^{43}+1} = [\,\mathbf{I}_{8\times8}\ \ \mathbf{0}_{8\times35}\ \ \mathbf{I}_{8\times8}\,], \tag{20.1}$$

which is decomposed into the 8×43 matrix $\mathbf{M}_{x^{43}+1}^{+}$ and the 8×8 matrix $\mathbf{M}_{x^{43}+1}^{-}$ such that

$$\mathbf{M}_{x^{43}+1}^{+} = [\,\mathbf{I}_{8\times8}\ \ \mathbf{0}_{8\times35}\,], \tag{20.2a}$$

$$\mathbf{M}_{x^{43}+1}^{-} = \mathbf{I}_{8\times8}. \tag{20.2b}$$

If we choose the nonsingular matrix $\mathbf{U} = \mathbf{I}_{8\times8}$, and insert it along with (20.2) and $\Psi_r(x) = x^{43}\Psi(x^{-1}) = x^{43} + 1$ into (19.14), then we obtain

$$\mathbf{A}_p = [\,\mathbf{e}_{35}\ \mathbf{e}_{36}\ \cdots\ \mathbf{e}_{42}\ \mathbf{e}_0\ \mathbf{e}_1\ \cdots\ \mathbf{e}_{34}\,], \tag{20.3a}$$

$$\mathbf{B}_p = [\,\mathbf{e}_{35}\ \mathbf{e}_{36}\ \cdots\ \mathbf{e}_{42}\,], \tag{20.3b}$$

$$\mathbf{C}_p = [\,\mathbf{I}_{8\times8}\ \ \mathbf{0}_{8\times35}\,], \tag{20.3c}$$

$$\mathbf{D}_p = \mathbf{I}_{8\times8}. \tag{20.3d}$$

Employing these parameters, we can design the parallel scrambler shown in Fig. 20.4.

The 43×43 matrix $\mathbf{Q}^{+}$ and the 43×8 matrix $\mathbf{Q}^{-}$ in (19.26) become respectively

$$\mathbf{Q}^{+} = [\,\mathbf{0}_{43\times8}\ \mathbf{e}_0\ \mathbf{e}_1\ \cdots\ \mathbf{e}_{34}\,], \tag{20.4a}$$

$$\mathbf{Q}^{-} = [\,\mathbf{e}_{35}\ \mathbf{e}_{36}\ \cdots\ \mathbf{e}_{42}\,]. \tag{20.4b}$$

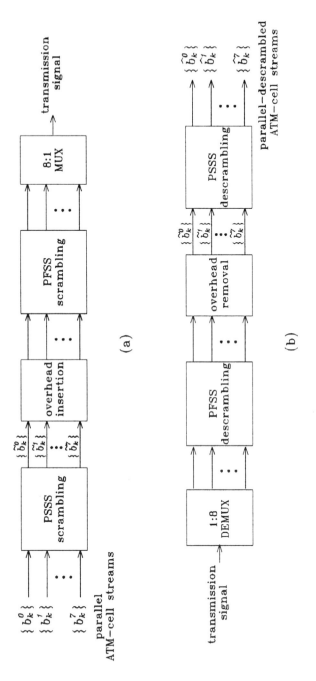

Figure 20.3. Blockdiagram of the SDH-based ATM transmission system employing the parallel scramblings. (a) Transmitter, (b) receiver.

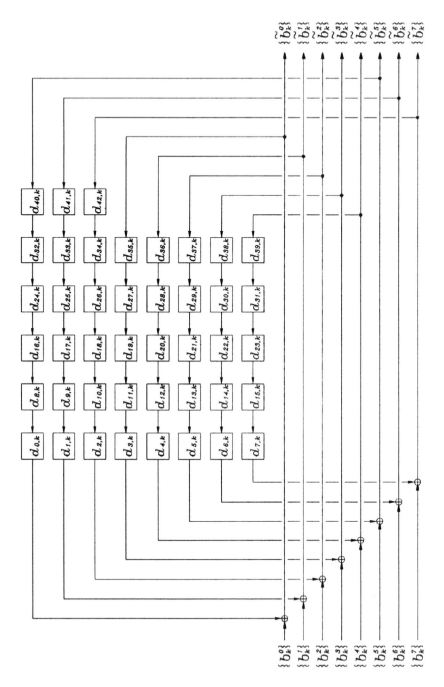

Figure 20.4. A PSSS scrambler for the SDH-based ATM-cell stream.

If we choose the nonsingular matrix $\mathbf{U} = \mathbf{I}_{8 \times 8}$, by (20.2) and (20.4) we obtain

$$\hat{\mathbf{A}}_p = [\, \mathbf{0}_{43 \times 8} \quad \mathbf{e}_0 \quad \mathbf{e}_1 \quad \cdots \quad \mathbf{e}_{34} \,], \qquad (20.5a)$$

$$\hat{\mathbf{B}}_p = [\, \mathbf{e}_{35} \quad \mathbf{e}_{36} \quad \cdots \quad \mathbf{e}_{42} \,], \qquad (20.5b)$$

$$\hat{\mathbf{C}}_p = [\, \mathbf{I}_{8 \times 8} \quad \mathbf{0}_{8 \times 35} \,], \qquad (20.5c)$$

$$\hat{\mathbf{D}}_p = \mathbf{I}_{8 \times 8}. \qquad (20.5d)$$

Using these parameters, we can obtain the parallel descrambler shown in Fig. 20.5.

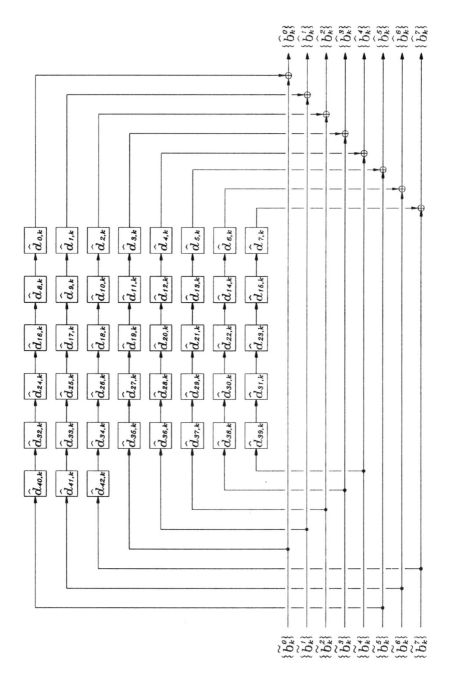

Figure 20.5. A PSSS descrambler for the SDH-based ATM-cell stream.

Chapter 21

Signal-Alignment with Parallel Self Synchronous Scrambling

The SSS has the unique property that it can scramble and descramble data sequences without looking into the internal contents of the data. Inherited from this property, the PSSS has the capability to scramble and descramble multiple base-rate signals without overlaying any frame structure. However, there is one problem appearing in this case, which is the proper distribution of the base-rate signals to their destined output lines, or the proper signal alignment. In this chapter, we consider such an signal-alignment issue in the transmission systems employing the PSSS, and discuss how to resolve the issue by employing the concept of permuting. We first represent three basic building blocks of the systems -- multiplexer, scrambler and permuter -- in terms of mathematical operators, which are directly translated into number-operators. Based on these operators, we consider how to identify the received signals in the form of a table called the signal-detection table, and examine how to characterize the signal-detection table by employing the signal-detection table characteristic expression. For various system configurations, we determine their signal-detection tables, and investigate their properties. Finally, we demonstrate how to apply the signal-detection table to the signal-alignments of the PSSS-scrambled transmission signals.

21.1 Operators for Functional Processors

As depicted in Fig. 17.2(b), a transmission system employing the parallel self synchronous scrambling without overlaid frames is composed of three basic functional processors —— multiplexer, scrambler and permuter. In this section, we consider how to represent these functional processors in terms of mathematical operators, and investigate their properties.

DEFINITION 21.1 (Operators) We denote by D, S and P the *operators that respectively represent the operations of unit delay, scrambling and permuting.*[1] For these operators, we define the operation symbols $+$ and $*$ such that they respectively imply

$$[O_1 + O_2](\mathbf{b}_k) = O_1(\mathbf{b}_k) + O_2(\mathbf{b}_k), \qquad (21.1a)$$

$$[O_1 * O_2](\mathbf{b}_k) = O_1[O_2(\mathbf{b}_k)], \qquad (21.1b)$$

for a signal vector $\mathbf{b}_k \equiv [\, b_k^0 \ b_k^1 \ \cdots \ b_k^{N-1} \,]^t$ representing the N parallel signals $\{\, b_k^i \,\}$'s, where O_1 and O_2 represent the operators D, S and P.

Each of the three operators D, S and P is time-invariant and is defined with respect to signal vectors of parallel signals. Repeated applications of an operator O can be expressed in terms of exponents, that is,

$$O^j = \overbrace{O * O * \cdots * O}^{j}. \qquad (21.2)$$

For the *delay operator* indicating the unit-time delay in a serial signal, we have the following three properties :

THEOREM 21.2 *Let* $\mathbf{b}_k$ *and* $\hat{\mathbf{b}}_k$ *be respectively the input and the output signal vectors of the delay element* D. *Then,*

$$\hat{\mathbf{b}}_k = D(\mathbf{b}_k)$$

$$= [\, b_{k-1}^{N-1} \ b_k^0 \ \cdots \ b_k^{N-3} \ b_k^{N-2} \,]^t. \qquad (21.3)$$

Proof : This theorem is directly obtained by the definition of delay operator D. ∎

THEOREM 21.3 *For a signal vector* $\mathbf{b}_k$,

$$D^N(\mathbf{b}_k) = \mathbf{b}_{k-1}. \qquad (21.4)$$

Proof : This theorem is directly obtained by (21.2) and (21.3). ∎

---

[1] Note that the term scrambling in this chapter refers to the self synchronous scrambling.

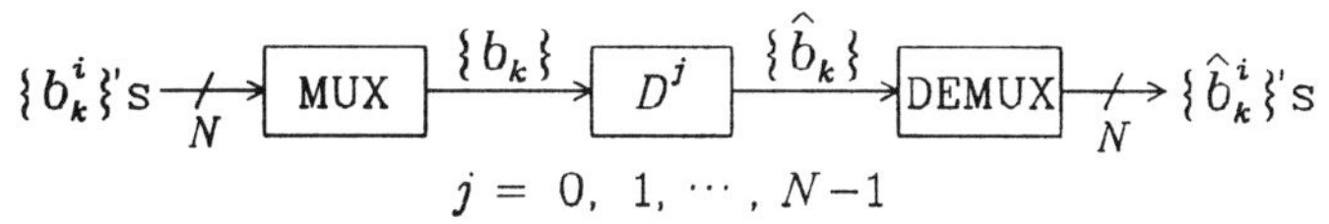

$$j = 0, 1, \cdots, N-1$$

Figure 21.1. A transmission medium model.

THEOREM 21.4 *For a time-invariant operator O,*

$$D^N * O = O * D^N. \tag{21.5}$$

Proof : Let $\hat{\mathbf{b}}_k = O(\mathbf{b}_k)$. Then, by (21.1b), $[D^N * O](\mathbf{b}_k) = D^N[O(\mathbf{b}_k)] = D^N(\hat{\mathbf{b}}_k)$, which becomes $\hat{\mathbf{b}}_{k-1}$ due to (21.4). On the other hand, by (21.1b) and (21.4), $[O * D^N](\mathbf{b}_k) = O[D^N(\mathbf{b}_k)] = O(\mathbf{b}_{k-1})$, which becomes $\hat{\mathbf{b}}_{k-1}$ due to the time-invariance of O. Therefore, $[D^N * O](\mathbf{b}_k) = [O * D^N](\mathbf{b}_k)$, which yields (21.5). ∎

Theorem 21.2 describes how to obtain the output signal vector of the unit-time delay element ; Theorem 21.3 means that the time-length of one delay for the parallel signals corresponds to the time-length of N delays for the serial signal ; and Theorem 21.4 means that an arbitrary operator is commutative to the N-time delay operator D^N.

If we assume that there is no transmission error, transmission medium can be represented by D^i for an unknown $i = 0, 1, 2, \cdots$, since the received signal is a delayed replica of the transmitted signal but with unknown number of delays. But since $D^{Nm+j}(\mathbf{b}_k) = D^j(\mathbf{b}_{k-m})$ by (21.4), the case of $Nm + j$ delays is equivalent to the case of j delays for any integer m as far as the signal-alignment is concerned. That is, the demultiplexed parallel signals are identical in both cases. Therefore, the *transmission medium* can be modeled into D^j, $j = 0, 1, \cdots, N - 1$, as depicted in Fig. 21.1. Also shown in the figure are a multiplexer and a demultiplexer respectively attached in front of and behind the transmission medium. Since those functional processors do not change the essential situation, we can express the combination of multiplexer-transmission medium-demultiplexer in the form $\hat{\mathbf{b}}_k = D^j(\mathbf{b}_k)$, $j = 0, 1, \cdots, N - 1$.

We denote by S and S^{-1} respectively the *scrambling operator* and the *descrambling operator*, since descrambling is the inverse process of scrambling. Then, for the input and output signal vectors $\mathbf{b}_k$ and $\hat{\mathbf{b}}_k$,

$\hat{\mathbf{b}}_k = S(\mathbf{b}_k)$, if $\mathbf{b}_k = S^{-1}(\hat{\mathbf{b}}_k)$, and vice versa.[2]

THEOREM 21.5 *Let S^{-1} be the descrambling operator for the descrambler or parallel descrambler whose characteristic polynomial is $\Psi(x)$.*[3] *Then,*

$$S^{-1} = \Psi(D). \tag{21.6}$$

Proof : This theorem is obvious in view of the descrambler shown in Fig. 17.1(b). ∎

The theorem means that the descrambling operator S^{-1} is directly obtained by replacing x in the characteristic polynomial $\Psi(x)$ with the delay operator D. If we denote by the symbol Θ the set of integers taken from the exponents of the characteristic polynomial $\Psi(x)$, then the equation (21.6) can be written in the form of

$$S^{-1} = \sum_{j \in \Theta} D^j. \tag{21.7}$$

Note that the set Θ always contains the element 0, which implies that $\Psi(x)$ always includes the term 1 (or x^0).

EXAMPLE 21.1 In the case of the descrambler in Fig. 18.2 or the parallel descrambler in Fig. 19.3, its characteristic polynomial is $\Psi(x) = x^7 + x + 1$. Therefore, by (21.6), the descrambling operator becomes $S^{-1} = D^7 + D^1 + D^0$, where D^0 indicates no-delay or identity operator. In this case, the set Θ is $\{ 0, 1, 7 \}$, since the exponents of $\Psi(x)$ are 0, 1 and 7. ♣

Since the descrambling operator is the sum of delay operators, we have the following commutative property.

THEOREM 21.6 *For a descrambling operator S^{-1} and an integer j,*

$$S^{-1} * D^j = D^j * S^{-1}. \tag{21.8}$$

---

[2]In this section we put hat(ˆ) to denote the output of the corresponding signal vector(or its elements) when it is the output from a functional processor.

[3]Note that a scrambler(or descrambler) and its corresponding parallel scrambler(or descrambler) are identical in operations.

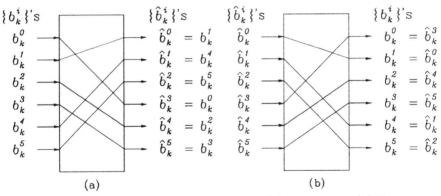

Figure 21.2. An example of a permuter and depermuter. (a) Permuter, (b) depermuter.

We denote by P and P^{-1} respectively the *permuting operator* and the *depermuting operator*, since depermuting is the inverse process of permuting. Then, for the input and output signal vectors $\mathbf{b}_k$ and $\hat{\mathbf{b}}_k$, $\hat{\mathbf{b}}_k = P(\mathbf{b}_k)$, if $\mathbf{b}_k = P^{-1}(\hat{\mathbf{b}}_k)$, and vice versa.

EXAMPLE 21.2 For the permuter and depermuter pair in Fig. 21.2, the permuting and depermuting operator expressions are

$$\begin{aligned}
\hat{\mathbf{b}}_k &= P([\, b_k^0 \ b_k^1 \ b_k^2 \ b_k^3 \ b_k^4 \ b_k^5 \,]^t) \\
&= [\, b_k^1 \ b_k^4 \ b_k^5 \ b_k^0 \ b_k^2 \ b_k^3 \,]^t, \\
\mathbf{b}_k &= P^{-1}([\, \hat{b}_k^0 \ \hat{b}_k^1 \ \hat{b}_k^2 \ \hat{b}_k^3 \ \hat{b}_k^4 \ \hat{b}_k^5 \,]^t) \\
&= [\, \hat{b}_k^3 \ \hat{b}_k^0 \ \hat{b}_k^4 \ \hat{b}_k^5 \ \hat{b}_k^1 \ \hat{b}_k^2 \,]^t. \quad \clubsuit
\end{aligned}$$

21.2 Number-Operators for Operators

A signal element b_{k-m}^i in the parallel signals $\{\, b_k^i \,\}$, $i = 0, 1, \cdots, N-1$, can be represented by i and m only, since the signal name b and the reference time k are common to all signal elements in $\{\, b_k^i \,\}$'s. So we represent b_{k-m}^i by $d^m i$, calling it the *number-based* notation. In the number-based notation, i is the signal number and d indicates the unit-time delay with respect to the reference time k, thus making d^m imply m delays. Note that d corresponds to D^N, since the time-length of one delay for the parallel signals $\{\, b_k^i \,\}$'s corresponds to the time of N delays for the serial signal $\{\, b_k \,\}$.

The input-output relation for an operator O can be translated into a number-based expression by employing the number-based notation. We use the number-operator $\hat{O}$ to indicate this. More specifically,

DEFINITION 21.7 (Number-Operator) For an operator O with the input and the output signal vectors $\mathbf{b}_k$ and $\hat{\mathbf{b}}_k$ which are related with

$$\sum_{(i,m) \in \Omega_{out}} \hat{b}^i_{k-m} = \sum_{(j,l) \in \Omega_{in}} b^j_{k-l},$$

we define the *number-operator* $\hat{O}$ to be an operator for number-based notations of signal elements such that

$$\hat{O}\left(\sum_{(i,m) \in \Omega_{out}} d^m i\right) = \sum_{(j,l) \in \Omega_{in}} d^l j,$$

where Ω_{in} and Ω_{out} are two-dimensional sets describing how the signal elements combine to perform the operation for O, and i, $j = 0, 1, \cdots$, $N-1$, and l, m are integers.[4]

EXAMPLE 21.3 For an operator O, if $\hat{b}^0_k + \hat{b}^1_{k-1} = b^0_k + b^5_{k-1} + b^5_{k-2}$, then its corresponding number-operator expression is $\hat{O}(d^0 0 + d^1 1) = d^0 0 + d^1 5 + d^2 5$, or equivalently, $\hat{O}(0 + d1) = 0 + d5 + d^2 5$. ♣

For number-operators, we have the following three properties.

THEOREM 21.8 *If* $\hat{\mathbf{b}}_k = O(\mathbf{b}_k)$ *and* $\hat{O}(i) = d^m j$ *for an operator* O, *then* $\{\,\hat{b}^i_k\,\} = \{\,b^j_k\,\}$.

    Proof : By definition of number-operator, $\hat{O}(i) = d^m j$ means that $\hat{b}^i_k = b^j_{k-m}$. Therefore, $\{\,\hat{b}^i_k\,\} = \{\,b^j_k\,\}$. ∎

THEOREM 21.9 *For an operator* O, *the relation*

$$\hat{O}\left(\sum_{(i,l) \in \Omega} d^l i\right) = \sum_{(i,l) \in \Omega} d^l \hat{O}(i).$$

holds for a two-dimensional set Ω *with* $i = 0, 1, \cdots, N-1$, *and integer* l.

---

[4] Note that the number-based expression takes the form "$\hat{O}$ (combination of output signals) = combination of input signals". This expression is convenient in defining the terms related with signal-alignment, as it appears in the next section.

Proof : Let $\hat{\mathbf{b}}_k = O(\mathbf{b}_k)$ with

$$\hat{b}^i_k = \sum_{(j,m) \in \Omega_i} b^j_{k-m} \tag{21.9}$$

for $i = 0, 1, \cdots, N - 1$. Then, we have, due to the time-invariance of operator,

$$\sum_{(i,l) \in \Omega} \hat{b}^i_{k-l} = \sum_{(i,l) \in \Omega} \sum_{(j,m) \in \Omega_i} b^j_{k-l-m},$$

and the corresponding number-operator expression is

$$\hat{O}(\sum_{(i,l) \in \Omega} d^l i) = \sum_{(i,l) \in \Omega} \sum_{(j,m) \in \Omega_i} d^{l+m} j$$

$$= \sum_{(i,l) \in \Omega} d^l \sum_{(j,m) \in \Omega_i} d^m j.$$

But since the number-operator expression for (21.9) is $\hat{O}(i) = \sum_{(j,m) \in \Omega_i} d^m j$, we have the theorem. ∎

THEOREM 21.10 *Let $O = O_2 * O_1$ for operators O_1 and O_2. If we denote the corresponding number-operator expression by $\hat{O} = \hat{O}_2 * \hat{O}_1$, then, for $i = 0, 1, \cdots, N - 1$, $\hat{O}(i) = [\hat{O}_2 * \hat{O}_1](i) = \hat{O}_1(\hat{O}_2(i))$.*

Proof : Let $\hat{\mathbf{b}}_k = O_1(\mathbf{b}_k)$ and $\tilde{\mathbf{b}}_k = O_2(\hat{\mathbf{b}}_k)$ with

$$\hat{b}^j_k = \sum_{(n,m) \in \Omega_{1_j}} b^n_{k-m} \tag{21.10}$$

and

$$\tilde{b}^i_k = \sum_{(j,l) \in \Omega_{2_i}} \hat{b}^j_{k-l}. \tag{21.11}$$

Then, their number-operator based expressions are, respectively,

$$\hat{O}_1(j) = \sum_{(n,m) \in \Omega_{1_j}} d^m n \tag{21.12}$$

and

$$\hat{O}_2(i) = \sum_{(i,l) \in \Omega_{2_i}} d^l j. \tag{21.13}$$

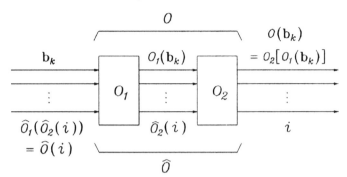

Figure 21.3. Relation between operators and number-operators.

Inserting (21.10) into (21.11), we obtain, by time-invariance of operators,

$$\tilde{b}_k^i = \sum_{(j,l) \in \Omega_{2_i}} \sum_{(n,m) \in \Omega_{1_j}} b_{k-l-m}^n,$$

whose number-operator based expression is

$$\hat{O}(i) = \sum_{(j,l) \in \Omega_{2_i}} \sum_{(n,m) \in \Omega_{1_j}} d^{l+m} n$$

$$= \sum_{(j,l) \in \Omega_{2_i}} d^l \sum_{(n,m) \in \Omega_{1_j}} d^m n.$$

Applying (21.12) to this, we obtain

$$\hat{O}(i) = \sum_{(j,l) \in \Omega_{2_i}} d^l \hat{O}_1(j),$$

and applying Theorem 21.9 and (21.13) to this, we finally obtain $\hat{O}(i) = \hat{O}_1(\sum_{(j,l) \in \Omega_{2_i}} d^l j) = \hat{O}_1(\hat{O}_2(i))$. ∎

   Theorem 21.8 means that if $\hat{O}(i) = d^m j$, the ith output signal { $\hat{b}_k^i$ } is identical to the jth input signal { b_k^j } ; and Theorem 21.9 implies that number-operators behave like linear operators with respect to +. Theorem 21.10 is illustrated by Fig. 21.3. We observe from the figure that the operator O provides input-oriented expressions, while the number-operator $\hat{O}$ provides output number-based expressions.

EXAMPLE 21.4 For an operator O, we have, by Theorem 21.9, the relation $\hat{O}(0 + d5 + d^2 5) = \hat{O}(0) + d\hat{O}(5) + d^2\hat{O}(5)$. ♣

The *number-operator* $\hat{D}$ corresponding to the delay operator D is, by (21.3), written in the form

$$\hat{D}(i) = \begin{cases} d(N-1), & i = 0, \\ i-1, & i = 1, 2, \cdots, N-1. \end{cases}$$

In general, number-operator $\hat{D}^j$ corresponding to D^j, $j = 0, 1, \cdots, N-1$, the jth repetition of delay operator D, is

$$\hat{D}^j(i) = \begin{cases} d(N+i-j), & i = 0, 1, \cdots, j-1 \\ i-j, & i = j, j+1, \cdots, N-1. \end{cases} \tag{21.14}$$

The *number-operator* $\hat{S}^{-1}$ corresponding to the descrambling operator S^{-1} is, by (21.7), written in the form

$$\hat{S}^{-1}(i) = \sum_{j \in \Theta} \hat{D}^j(i). \tag{21.15}$$

For a permuter with the permuting operator P, if $\hat{\mathbf{b}}_k = P(\mathbf{b}_k)$ with $\hat{b}_k^i = b_k^{j_i}$ for i, $j_i = 0, 1, \cdots, N-1$, then the corresponding *number-operator* $\hat{P}$ renders the expression $\hat{P}(i) = j_i$. Note that there is no delay involved with the permuting operations. If we write this into a combined expression, we obtain

$$\hat{P}(0, 1, \cdots, N-1) = (j_0, j_1, \cdots, j_{N-1}),$$

or, in short,

$$\hat{P} = (j_0, j_1, \cdots, j_{N-1}).$$

Here, $\hat{P}$ is called the *permuter characteristic expression*. As a scrambler is characterized by its characteristic polynomial, a permuter is characterized by its characteristic expression. In a formal form,

DEFINITION 21.11 (Permuter Characteristic Expression) For a permuter with the operator P, the *permuter characteristic expression* $\hat{P}$ is defined to be

$$\hat{P} = (\hat{P}(0), \hat{P}(1), \cdots, \hat{P}(N-1)). \tag{21.16}$$

EXAMPLE 21.5 In the case of the permuter and depermuter in Fig. 21.2, the permuting and depermuting number-operators $\hat{P}$ and $\hat{P}^{-1}$ are

$$\begin{cases} \hat{P}(0) = 1, & \hat{P}(1) = 4, & \hat{P}(2) = 5, \\ \hat{P}(3) = 0, & \hat{P}(4) = 2, & \hat{P}(5) = 3, \end{cases}$$

$$\begin{cases} \hat{P}^{-1}(0) = 3, & \hat{P}^{-1}(1) = 0, & \hat{P}^{-1}(2) = 4, \\ \hat{P}^{-1}(3) = 5, & \hat{P}^{-1}(4) = 1, & \hat{P}^{-1}(5) = 2. \end{cases}$$

Therefore, the permuter and depermuter characteristic expressions are

$$\hat{P} = (1, 4, 5, 0, 2, 3), \qquad (21.17a)$$

$$\hat{P}^{-1} = (3, 0, 4, 5, 1, 2). \; \clubsuit \qquad (21.17b)$$

21.3   Signal-Detection Tables

A transmission system is composed of the transmitting part, the transmission medium, and the receiving part, and, from the signal-alignment point of view, it can be depicted as shown in Fig. 21.4. The transmitting part consists of a series of functional processors, which is represented by the operator R, and a multiplexer, whereas the receiving part consists of a demultiplexer and the inverse functional processors for R^{-1}. The functional processors for R are scramblers, permuters, or combinations of them. As discussed in Section 21.1, the cascade of multiplexer, transmission medium and demultiplexer can be represented by D^j, $j = 0$, 1, $\cdots$, $N - 1$, or equivalently, $\hat{\mathbf{c}}_k = D^j(\mathbf{c}_k)$.[5] Therefore, the overall transmission system in Fig. 21.4 can be described by $\hat{\mathbf{b}}_k = R_j(\mathbf{b}_k)$ with

$$R_j = R^{-1} * D^j * R \qquad (21.18)$$

for $j = 0$, 1, $\cdots$, $N - 1$. Based on this, we define the terms signal-detection, signal-detection value $T(i, j)$, signal-detection table T, and signal-alignment state as follows :

DEFINITION 21.12 (Signal-Detection) If the output signal $\{ \hat{b}^i_k \}$, $i = 0, 1, \cdots, N - 1$, is equal to an input signal $\{ b^l_k \}$, $l = 0, 1, \cdots, N - 1$, we say *signal-detection* is made, or more specifically, input signal $\{ b^l_k \}$ is *detected* at output signal $\{ \hat{b}^i_k \}$.

Based on Theorem 21.8, we can paraphrase the definition as follows : For a given set of i and j, if there exists some set of l and m such that

$$\hat{R}_j(i) = d^m l, \qquad (21.19)$$

---

[5]Note that the delay block D^j is introduced in Fig. 21.4 to represent the transmission medium since the SSS system under consideration has no overlaid frames and consequently the starting point of the descrambler could be off from that of the scrambler by j serial bit-times.

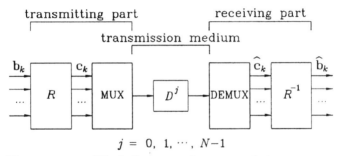

$$j = 0, 1, \cdots, N-1$$

Figure 21.4. Blockdiagram of a transmission system.

then the input signal number l is detected at the output signal number i. Here, $i, j, l = 0, 1, \cdots, N - 1$, and m is a non-negative integer.

DEFINITION 21.13 (Signal-Detection Value) We name the value in (21.19) *signal-detection value* and denote it by $T(i,j)$. In case signal-detection does not occur, we designate the signal detection value by $\times$ to indicate that the signal-detection value is an *invalid* one.

In terms of equations, the definition can be organized as

$$T(i,j) = \begin{cases} d^m l, & \text{if signal is detected and } \hat{R}_j(i) = d^m l, \\ \times, & \text{otherwise.} \end{cases} \quad (21.20)$$

DEFINITION 21.14 (Signal-Detection Table) The $N \times N$ table which lists N^2 possible signal-detection values $T(i,j)$, $i, j = 0, 1, \cdots, N - 1$, is called the *signal-detection table* and is designated by T.[6]

DEFINITION 21.15 (Signal-Alignment) If, for $j = j_0$, signal is detected at every output signal $\{ \hat{b}_k^i \}$, $i = 0, 1, \cdots, N - 1$, and if each detected signal number is equal to the output signal number, then we say *signal-alignment* is made, or signals are *properly aligned*, and we call the state j_0 a *signal-alignment state*.

In terms of signal-detection values, we can describe the signal-alignment state j_0 as

$$T(i, j_0) = d^m i, \quad i = 0, 1, \cdots, N - 1.$$

---

[6]Note that T denotes the signal-detection table in this chapter, while it indicates the periodicity of the sequence space elsewhere.

Table 21.1. An example of the signal-detection table

$i\backslash j$	0	1	2	3	4	5
0	0	$d5$	$d4$	$d3$	$d2$	$d1$
1	1	0	$d5$	$d4$	$d3$	$d2$
2	2	1	0	$d5$	$d4$	$d3$
3	3	2	1	0	$d5$	$d4$
4	4	3	2	1	0	$d5$
5	5	4	3	2	1	0

This implies that if signals are properly aligned for $j = j_0$, then the j_0th column of the signal-detection table contains the signal numbers 0 through $N - 1$ in the increasing order.

EXAMPLE 21.6 We consider the transmission system with $N = 6$ whose transmitting part consists of the parallel scrambler in Fig. 19.2 and a 6:1 multiplexer, and whose receiving part consists of a 1:6 demultiplexer and the parallel descrambler in Fig. 19.3. Then for the system the operator R in Fig. 21.4 is S with $S^{-1} = D^7 + D^1 + D^0$ (refer to Example 21.1), and hence the operator R_j in (21.18) becomes $R_j = S^{-1} * D^j * S$. Applying (21.8) to this, we have $R_j = D^j * S^{-1} * S = D^j$, and thus $\hat{R}_j(i) = \hat{D}^j(i)$. In case $i = 0$, we obtain, by (21.14), $\hat{R}_0(0) = 0$, $\hat{R}_1(0) = d5$, $\hat{R}_2(0) = d4$, $\hat{R}_3(0) = d3$, $\hat{R}_4(0) = d2$, $\hat{R}_5(0) = d1$; and we obtain, by (21.20), $T(0,0) = 0$, $T(0,1) = d5$, $T(0,2) = d4$, $T(0,3) = d3$, $T(0,4) = d2$, $T(0,5) = d1$. Repeating the same procedure for $i = 1, 2, \cdots, 5$, we obtain the signal-detection table in Table 21.1. We observe from the table that the state $j = 0$ is the signal-alignment state. ♣

THEOREM 21.16 *For an operator R, if we define the characteristic operator C_R as*

$$C_R = R^{-1} * D * R, \tag{21.21}$$

then,

$$R_j = C_R^j, \tag{21.22}$$
$$C_R^N = D^N, \tag{21.23}$$

where R_j is the operator corresponding to the overall transmission system in (21.18).

Proof : From (21.18) and (21.21), we obtain $R_j = R^{-1} * D^j * R$
$= (R^{-1} * D * R)^j = C_R^j$. Also, by (21.5), $C_R^N = (R^{-1} * D * R)^N =$
$R^{-1} * D^N * R = R^{-1} * R * D = D^N$. ∎

Based on this theorem, the expression for signal-detection values in
(21.20) can be rewritten as

$$T(i,j) = \begin{cases} d^m l, & \text{if signal is detected and } \hat{C}_R^j(i) = d^m l, \\ \times, & \text{otherwise.} \end{cases} \qquad (21.24)$$

Note that the subscript R in C_R refers the operator R for which the
characteristic operator is defined.

For signal-detection tables, we have the following three properties.

THEOREM 21.17 *For $i = 0, 1, \cdots, N$-1, $T(i,j) = i$ if and only if $j = 0$.*

Proof : If $T(i,j) = i$ for a j, then $\hat{C}_R^j(i) = i$. Thus, we have $\hat{C}_R^{jN}(i)$
$= \hat{C}_R^{j(N-1)}[\hat{C}_R^j(i)] = \hat{C}_R^{j(N-1)}(i) = \hat{C}_R^{j(N-2)}[\hat{C}_R^j(i)] = \hat{C}_R^{j(N-2)}(i) = \cdots =$
i. But from (21.4) and (21.23), $d^j i = \hat{D}^{jN}(i) = \hat{C}_R^{jN}(i)$. Therefore $d^j i$
$= i$, and hence we have $j = 0$.

Conversely, if $j = 0$, then C_R^0 becomes an identity operator. Thus
we have, from (21.24), $T(i,0) = \hat{C}_R^0(i) = i$. ∎

THEOREM 21.18 *If a signal-detection value $T(i,j) = d^m l$, then $m \in \{$
$0, 1 \}$ for causal systems.*

Proof : First, if $j = 0$, then we have $T(i,0) = i$ by Theorem 21.17,
which implies $m = 0$.

Now let $j \neq 0$. If $T(i,j) = d^m l$, then we have, from (21.24), $\hat{C}_R^j(i)$
$= d^m l$. Applying the operator $\hat{C}_R^{-j}$ to this, we obtain, from Theorem
21.9, $i = \hat{C}_R^{-j}(d^m l) = d^m \hat{C}_R^{-j}(l)$. And applying (21.4) and (21.23) to
this, we get $i = d^{m-1} \hat{D}^N(\hat{C}_R^{-j}(l)) = d^{m-1} \hat{C}_R^N(\hat{C}_R^{-j}(l)) = d^{m-1} \hat{C}_R^{N-j}(l)$.
Thus we have $\hat{C}_R^{N-j}(l) = d^{1-m} i$, and, by (21.24), $T(l, N-j) = d^{1-m} i$.
For causal systems, the delay in $d^m l$ as well as the delay in $d^{1-m} i$ must
be non-negative. Therefore, we have $m \in \{ 0, 1, 2, \cdots \}$ and $1-m \in \{$
$0, 1, 2, \cdots \}$, and hence $m \in \{ 0, 1 \}$. ∎

THEOREM 21.19 *If $T(i,j_1) = d^{m_1} l_1$ and $T(i,j_2) = d^{m_2} l_2$ for $j_1 < j_2$,
then $(m_1, m_2) \in \{ (0,0), (0,1), (1,1) \}$.*

Proof : Since $(m_1, m_2) \in \{ (0,0), (0,1), (1,0), (1,1) \}$ by Theorem 21.18, it suffices to show that (m_1, m_2) can not be (1,0). To prove it by contradiction, we suppose $(m_1, m_2) = (1,0)$. Then by (21.24) we have $\hat{C}_R^{j_1}(i) = dl_1$ and $\hat{C}_R^{j_2}(i) = l_2$. This and Theorem 21.9 yield $\hat{C}_R^{j_2-j_1}(l_1)$ $= d^{-1}\hat{C}_R^{j_2-j_1}(dl_1) = d^{-1}\hat{C}_R^{j_2}[\hat{C}_R^{-j_1}(dl_1)] = d^{-1}\hat{C}_R^{j_2}(i) = d^{-1}l_2$, and hence $T(j_2 - j_1, l_1) = d^{-1}l_2$, contradicting Theorem 21.18. ∎

Theorem 21.17 implies that signals are properly aligned when $j = 0$ and only when $j = 0$; and Theorem 21.18 means that the time delay part of any signal-detection value is either d^0 or d^1. That is, the detected signals have either no delay or unit time delay. In fact this theorem applies to all transmission systems within our interest, since all the constituent functional processors as well as the transmission medium in them are causal. Theorem 21.19 means that every valid signal-detection value on the right hand side of a delayed detection value is delayed.

EXAMPLE 21.7 From the signal-detection table T in Table 21.1, we observe that $j = 0$ is a signal-alignment state ; the time delay part of signal-detection values are either d^0 or d^1 ; and every valid signal-detection values on the right hand side of a delayed detection value is delayed. ♣

The following theorem describes how the signal-detection values are correlated.

THEOREM 21.20 *If $T(i, j_0) = l$, then for $j = 0, 1, \cdots, N\text{-}1$,*

$$T(l,j) = \begin{cases} T(i, j + j_0), & \text{if } j + j_0 < N, \\ dT(i, j + j_0 - N), & \text{if } j + j_0 \geq N \; ; \end{cases} \quad (21.25a)$$

and if $T(i, j_0) = dl$, then for $j = 0, 1, \cdots, N\text{-}1$,

$$T(l,j) = \begin{cases} d^{-1}T(i, j + j_0), & \text{if } j + j_0 < N, \\ T(i, j + j_0 - N), & \text{if } j + j_0 \geq N. \end{cases} \quad (21.25b)$$

Proof : We first prove (21.25a). If $T(i, j_0) = l$, we have, from (21.24), $\hat{C}_R^{j_0}(i) = l$. Applying $\hat{C}_R^j$ to this we obtain $\hat{C}_R^j(l) = \hat{C}_R^{j+j_0}(i)$. Therefore, in case $j + j_0 < N$, we have, from (21.24), the relation $T(l, j) = T(i, j+j_0)$. In case $j + j_0 \geq N$, $\hat{C}_R^j(l) = \hat{C}_R^{j+j_0}(i) = \hat{C}_R^{j+j_0-N}[\hat{C}_R^N(i)] = \hat{C}_R^{j+j_0-N}[\hat{D}^N(i)]$ by (21.23). But, by Theorems 21.3 and 21.9, it becomes $\hat{C}_R^j(l) = \hat{C}_R^{j+j_0-N}(di) = d\hat{C}_R^{j+j_0-N}(i)$. Hence we have $T(l, j) = dT(i, j + j_0 - N)$.

Equation (21.25b) can be proved in a similar manner. ∎

According to the theorem, if signal number l or dl is detected at the ith row of a signal-detection table, then the lth row of the table is determined by the ith row.

EXAMPLE 21.8 In the case of the signal-detection table T in Table 21.1, $T(1,1) = 0$ and $T(1,2) = d5$. Therefore, by (21.25a) the signal-detection values of the 0th row become $T(0,0) = T(1,1) = 0$, $T(0,1) = T(1,2) = d5$, $T(0,2) = T(1,3) = d4$, $T(0,3) = T(1,4) = d3$, $T(0,4) = T(1,5) = d2$, $T(0,5) = dT(1,0) = d1$; and by (21.25b) the signal-detection values of the 5th row become $T(5,0) = d^{-1}T(1,2) = 5$, $T(5,1) = d^{-1}T(1,3) = 4$, $T(5,2) = d^{-1}T(1,4) = 3$, $T(5,3) = d^{-1}T(1,5) = 2$, $T(5,4) = T(1,0) = 1$, $T(5,5) = T(1,1) = 0$. We can confirm these from Table 21.1. ♣

21.4 Signal-Detection Table Characteristic Expressions

In this section, we consider how to characterize signal-detection tables.

DEFINITION 21.21 (Signal-Detection Table Characteristic Expression) For a signal-detection table T, we define the *signal-detection table characteristic expression* $\hat{T}$ to be a collection of rows in the signal-detection table T which contain more than one valid signal-detection values but contain no delay terms. More specifically, if such rows are denoted by $\hat{T}^{(l)}$, $l = 0, 1, \cdots$, then

$$\hat{T} \equiv \{\,\hat{T}^{(1)},\ \hat{T}^{(2)},\ \cdots\,\},$$

so $\hat{T} = \phi$ in case no such row exists.

EXAMPLE 21.9 In the case of the signal-detection table T in Table 21.1, the characteristic expression is $\hat{T} = \{\,(5,\ 4,\ 3,\ 2,\ 1,\ 0)\,\}$. ♣

Then, for signal-detection table characteristic expressions we have the following theorem.

THEOREM 21.22 *Let $\hat{T}$ be the characteristic expression for a signal-detection table T. Then, the table T is uniquely determined from $\hat{T}$, but not from any subset of it.*

Proof : To prove the theorem, we need the following two lemmas.

LEMMA 21.23 *A valid signal number in* $\hat{T}$ *is not identical to any other signal number in* $\hat{T}$.

Proof : We denote by $\hat{T}^{(l)}(j)$ the jth element in $\hat{T}^{(l)}$ and prove the lemma by contradiction.

First, we suppose $\hat{T}^{(l)}(j_1) = \hat{T}^{(l)}(j_2) = v$ and $j_1 < j_2$ for an l. Then we have $T(i,j_1) = T(i,j_2) = v$, where $i = \hat{T}^{(l)}(0)$ by Theorem 21.17. Applying Theorem 21.20 to this, we obtain $T(v, j_2 - j_1) = T(i, (j_2 - j_1) + j_1) = T(i,j_2) = v$. But it is a contradiction to Theorem 21.17, because $j_2 - j_1 \neq 0$.

Next, we suppose $\hat{T}^{(l_1)}(j_1) = \hat{T}^{(l_2)}(j_2)$ and $j_1 \leq j_2$ for l_1 and $l_2 \neq l_1$. Then we have $T(i_1,j_1) = T(i_2,j_2) = v$, where $i_1 = \hat{T}^{(l_1)}(0)$ and $i_2 = \hat{T}^{(l_2)}(0)$ by Theorem 21.17. Applying (21.24) to this, we obtain $\hat{C}_R^{j_1}(i_1) = \hat{C}_R^{j_2}(i_2) = v$, and thus $\hat{C}_R^{j_2 - j_1}(i_2) = i_1$. Since $i_1 \neq i_2$, we have $j_1 \neq j_2$ by Theorem 21.17. But in case $j_1 < j_2$, this yields $\hat{C}_R^{N-(j_2-j_1)}(i_1) = \hat{C}_R^N[\hat{C}_R^{-(j_2-j_1)}(i_1)] = \hat{C}_R^N(i_2)$. Applying (21.23) and (21.4) to this, we obtain $\hat{C}_R^{N-(j_2-j_1)}(i_1) = \hat{D}^N(i_2) = di_2$, and thus $T(i_1, N - (j_2 - j_1)) = di_2$. This implies that the i_1th row in T, which corresponds to $\hat{T}^{(l_1)}$ in $\hat{T}$, contains a delayed element. But it is a contradiction to the definition of $\hat{T}$. ∎

LEMMA 21.24 *Let i be a signal number that is not contained in* $\hat{T}$. *Then*

$$T(i,j) = \begin{cases} i, & \text{if } j = 0, \\ \times, & \text{if } j = 1,\ 2,\ \cdots,\ N\text{-}1. \end{cases}$$

Proof : We prove this also by contradiction. We suppose that the ith row of T contains two or more valid signal-detection values. Then, there exists at least one valid signal-detection value with unit-time delay, since the ith row is not contained in $\hat{T}$. Let $T(i,j) = dl$ be the first delayed signal-detection value in the ith row. Then, by Theorem 21.19, all the subsequent valid signal-detection values are delayed values. Therefore, by (21.25b), all the signal-detection values in the lth row are delayed-free. Thus the lth row must be an element of $\hat{T}$, and $T(l, N - j) = T(i, 0) = i$ by (21.25b). This means that i is contained in $\hat{T}$, which is a contradiction. ∎

By Lemma 21.23, each valid signal number in $\hat{T}$ appears only once, so the rows in $\hat{T}$ corresponding to those signal numbers are uniquely determined by Theorem 21.20. For the signal numbers that do not appear in $\hat{T}$, Lemma 21.24 proves that the corresponding rows are also uniquely determined.

To prove the theorem, therefore, it suffices to show that the signal-detection table T can not be correctly determined from any subset of $\hat{T}$. We consider the expression $\hat{T} - \hat{T}^{(l)}$, which is a proper subset of $\hat{T}$. Each signal number in $\hat{T}^{(l)}$ does not appear in $\hat{T} - \hat{T}^{(l)}$ by Lemma 21.23, and the corresponding row in T is reconstructed to contain single valid signal-detection value as Lemma 21.24 describes. Therefore the resulting T differs from the T obtained from $\hat{T}$. This proves the theorem. ∎

The theorem means that a signal-detection table is uniquely characterized by its characteristic expression. Combining this theorem with Theorems 21.17 and 21.20 along with Lemma 21.24, we can establish the signal-detection table from its characteristic expression, as the following example illustrates.

EXAMPLE 21.10 The signal-detection table T in Table 21.1 can be reconstructed from its characteristic expression $\hat{T} = \{ (5, 4, 3, 2, 1, 0) \}$ as follows : The 5th row is $(5, 4, 3, 2, 1, 0)$ by Theorem 21.17. The 0th row is $(0, d5, d4, d3, d2, d1)$ by Theorem 21.20. Rows 1 through 4 are reconstructed in a similar manner. ♣

21.5 Signal-Detection Tables for Various System Configurations

Now we examine signal-detection tables for various configurations of transmission systems such as scrambler, permuter, permuter-scrambler, scrambler-permuter, scrambler-permuter-scrambler, and permuter-scrambler-permuter based systems. The operators representing those systems, which correspond to R in Fig. 21.4, are respectively S, P, $P * S$, $S * P$, $S_1 * P * S_2$, and $P_1 * S * P_2$, and their characteristic operators are respectively

$$
\begin{align}
C_S &= S^{-1} * D * S, & \text{(21.26a)} \\
C_P &= P^{-1} * D * P, & \text{(21.26b)} \\
C_{PS} &= P^{-1} * S^{-1} * D * S * P, & \text{(21.26c)} \\
C_{SP} &= S^{-1} * P^{-1} * D * P * S, & \text{(21.26d)}
\end{align}
$$

Table 21.2. The signal-detection table for scrambler based systems

$i \backslash j$	0	1	$\cdots$	$N - 2$	$N - 1$
0	0	$d(N-1)$	$\cdots$	$d2$	$d1$
1	1	0	$\cdots$	$d3$	$d2$
2	2	1	$\cdots$	$d4$	$d3$
$\vdots$	$\vdots$	$\vdots$	$\vdots$	$\vdots$	$\vdots$
$N-1$	$N-1$	$N-2$	$\cdots$	1	0

$$C_{S_1 P S_2} = S_1^{-1} * P^{-1} * S_2^{-1} * D * S_2 * P * S_1, \qquad (21.26\mathrm{e})$$

$$C_{P_1 S P_2} = P_1^{-1} * S^{-1} * P_2^{-1} * D * P_2 * S * P_1. \qquad (21.26\mathrm{f})$$

We denote the corresponding signal-detection tables by T_S, T_P, T_{PS}, T_{SP}, $T_{S_1 P S_2}$, and $T_{P_1 S P_2}$ respectively.

21.5.1 Scrambler Based Systems

For scrambler based systems, the characteristic operator becomes, by (21.26a) and (21.8), $C_S = D * S^{-1} * S = D$. Hence we obtain, by (21.24), $T_S(i, j) = \hat{C}_S^j(i) = \hat{D}^j(i)$, and thus the signal-detection table T_S in Table 21.2. We observe from the table that its characteristic expression $\hat{T}_S$ is

$$\hat{T}_S = \{ (N - 1, N - 2, \cdots 1, 0) \}. \qquad (21.27)$$

EXAMPLE 21.11 Employing the parallel scrambler in Fig. 19.2 to the scrambler based system, we obtain, by (21.27), the characteristic expression $\hat{T}_S = \{ (5, 4, 3, 2, 1, 0) \}$. Applying Theorem 21.20, we can obtain the table T_S in Table 21.1. ♣

21.5.2 Permuter Based Systems

In the case of permuter based systems, we have, by (21.24) and (21.26b), the signal-detection value $T_P(i, j) = \hat{C}_P^j(i) = [\hat{P}^{-1} * \hat{D}^j * \hat{P}](i)$. Applying Theorem 21.10 to this, we obtain $T_P(i, j) = [\hat{D}^j * \hat{P}](\hat{P}^{-1}(i)) = \hat{P}\{\hat{D}^j(\hat{P}^{-1}(i))\}$. Since $\hat{P}$ is a delay-free operator, the characteristic expression $\hat{T}_P$ is determined by $\hat{D}^j(\hat{P}^{-1}(i))$. From (21.14) we find that the $\hat{P}(N - 1)$th row is the only element qualified for $\hat{T}_P$, and therefore

$$\hat{T}_P = \{ (\hat{P}(N - 1), \hat{P}(N - 2), \cdots, \hat{P}(0)) \}. \qquad (21.28)$$

Table 21.3. A signal-detection table for a permuter based system

$i\backslash j$	0	1	2	3	4	5
0	0	5	4	3	$d3$	$d2$
1	1	$d3$	$d2$	$d0$	$d5$	$d4$
2	2	0	5	4	1	$d3$
3	3	2	0	5	4	1
4	4	1	$d3$	$d2$	$d0$	$d5$
5	5	4	1	$d3$	$d2$	$d0$

EXAMPLE 21.12 Employing the permuter in Fig. 21.2 to the permuter based system, we obtain, by (21.28) and (21.17a), $\hat{T}_P = \{$ (3, 2, 0, 5, 4, 1) $\}$. Applying Theorem 21.20, we can complete the table T_P as in Table 21.3. ♣

21.5.3   Permuter-Scrambler Based Systems

In the case of permuter-scrambler based systems, the characteristic operator becomes, by (21.26c) and (21.8), $C_{PS} = P^{-1} * D * S^{-1} * S * P = P^{-1} * D * P$. Therefore $C_{PS} = C_P$ and $T_{PS} = T_P$, and consequently

$$\hat{T}_{PS} = \hat{T}_P = \{ (\hat{P}(N-1), \hat{P}(N-2), \cdots, \hat{P}(0)) \}. \qquad (21.29)$$

EXAMPLE 21.13 We employ the permuter in Fig. 21.2 and the parallel scrambler in Fig. 19.2 to form a permuter-scrambler based system. Then, by (21.29) and Example 21.12, we obtain $\hat{T}_{PS} = \hat{T}_P = \{$ (3, 2, 0, 5, 4, 1) $\}$, and hence the signal-detection table T_{PS} becomes identical to Table 21.3. ♣

21.5.4   Scrambler-Permuter Based Systems

In the case of scrambler-permuter based systems, the formulations of the characteristic expression $\hat{T}_{SP}$ is rather involved. The signal-detection table T_{SP} must be determined first in this case, and then $\hat{T}_{SP}$ can be extracted from T_{SP} by following the definition.

Let the scramblers in the scrambler-permuter based systems have the number-operator expression in (21.15). We define set $\Lambda_{i,j}$, i, $j = 0$,

$1, \cdots, N - 1$, by

$$
\begin{aligned}
\Lambda_{i,j} &= \{ \hat{C}_P^j[\hat{D}^l(i)] : l \in \Theta \} \\
&= \{ T_P(\hat{D}^l(i), j) : l \in \Theta \},
\end{aligned}
\tag{21.30a}
$$

and we denote by $v_{i,j}$ the "latest" signal number of $\Lambda_{i,j}$. The term "late" here means late from the serial signal's point of view. For example, 3 is later than 1, and 1 is later than $d6$, since b_{k-m}^i is interpreted as $d^m i$. We define set $\Lambda'_{i,j}$ based on $v_{i,j}$ such that

$$
\Lambda'_{i,j} = \{ \hat{D}^l(v_{i,j}) : l \in \Theta \}.
\tag{21.30b}
$$

Then, the signal-detection value $T_{SP}(i,j)$ is determined by the following theorem.

THEOREM 21.25 *The signal-detection value $T_{SP}(i,j)$ for scrambler-permuter based system is*

$$
T_{SP}(i,j) = \left\{
\begin{array}{ll}
v_{i,j}, & \text{if } \Lambda_{i,j} = \Lambda'_{i,j}, \\
\times, & \text{otherwise.}
\end{array}
\right.
\tag{21.31}
$$

Proof : We prove the theorem in the following manner : We first show that if signal is detected, then the signal detection value $T_{SP}(i,j)$ is $v_{i,j}$. Then we show that $T_{SP}(i,j) = v_{i,j}$ if and only if $\Lambda_{i,j} = \Lambda'_{i,j}$.

From (21.26b) and (21.26d), we have the relation $C_{SP} = S^{-1} * \hat{C}_P * S$, and thus $\hat{C}_{SP}^j(i) = [\hat{S}^{-1} * \hat{C}_P^j * S](i)$. Applying Theorem 21.10 to this, we obtain $\hat{C}_{SP}^j(i) = \hat{S}\{\hat{C}_P^j[\hat{S}^{-1}(i)]\}$, and hence $\hat{S}^{-1}[\hat{C}_{SP}^j(i)] = \hat{C}_P^j[\hat{S}^{-1}(i)]$. Applying (21.15) and Theorem 21.9 to this, we obtain $\sum_{l \in \Theta} \hat{D}^l[\hat{C}_{SP}^j(i)] = \hat{C}_P^j[\sum_{l \in \Theta} \hat{D}^l(i)] = \sum_{l \in \Theta} \hat{C}_P^j[\hat{D}^l(i)]$, and the last term equals $\sum_{l \in \Lambda_{i,j}} l$ by the definition of $\Lambda_{i,j}$. Therefore,

$$
\sum_{l \in \Theta} \hat{D}^l[\hat{C}_{SP}^j(i)] = \sum_{l \in \Lambda_{i,j}} l.
\tag{21.32}
$$

If signal is detected and signal-detection value $T_{SP}(i,j)$ is $u_{i,j}$, then we have, from (21.24) and (21.32), $\sum_{l \in \Theta} \hat{D}^l(u_{i,j}) = \sum_{l \in \Lambda_{i,j}} l$. Since the set Θ is composed of 0 and positive integers, the latest signal number in the left-hand side of this equation is $u_{i,j}$, and the latest signal number in the right-hand side of the equation is, by definition, $v_{i,j}$. Therefore $T_{SP}(i,j) = u_{i,j} = v_{i,j}$, which establishes the first part of the proof.

Table 21.4. A signal-detection table for a scrambler-permuter based system

$i\backslash j$	0	1	2	3	4	5
0	0	×	×	×	×	×
1	1	×	×	×	×	×
2	2	×	×	×	×	×
3	3	×	×	5	×	×
4	4	×	×	×	×	×
5	5	×	×	$d3$	×	×

Now, we prove the second part. If $T_{SP}(i,j) = v_{i,j}$, then, by (21.24) and (21.32), $\sum_{l \in \Theta} \hat{D}^l(v_{i,j}) = \sum_{l \in \Lambda_{i,j}} l$. But the left-hand side equals $\sum_{l \in \Lambda'_{i,j}} l$, by the definition of $\Lambda'_{i,j}$. Therefore, $\Lambda_{i,j} = \Lambda'_{i,j}$. Conversely, if $\Lambda_{i,j} = \Lambda'_{i,j}$, then we obtain, from (21.32), $\sum_{l \in \Theta} \hat{D}^l[\hat{C}^j_{SP}(i)] = \sum_{l \in \Lambda'_{i,j}} l$. Applying (21.15) to this, we obtain $\hat{S}^{-1}[\hat{C}^j_{SP}(i)] = \hat{S}^{-1}(v_{i,j})$, and thus $T_{SP}(i,j) = \hat{C}^j_{SP}(i) = v_{i,j}$. ∎

EXAMPLE 21.14 We employ the parallel scrambler in Fig. 19.2 and the permuter in Fig. 21.2 to form a scrambler-permuter based system. Then, $\Theta = \{0, 1, 7\}$ (refer to Example 21.1), and T_P is as in Table 21.3 (refer to Example 21.12). So, we obtain, for $i = 1$, $j = 0, 1, 2$, $\Lambda_{0,0} = \{0, d5, d^2 5\}$, $\Lambda_{0,1} = \{5, d4, d^2 4\}$, $\Lambda_{0,2} = \{4, d1, d^2 2\}$, and thus $v_{0,0} = 0$, $v_{0,1} = 5$, $v_{0,2} = 4$, and $\Lambda'_{0,0} = \{0, d5, d^2 5\}$, $\Lambda'_{0,1} = \{5, 4, d4\}$, $\Lambda'_{0,2} = \{4, 3, d3\}$. Therefore, $\Lambda_{0,0} = \Lambda'_{0,0}$, $\Lambda_{0,1} \neq \Lambda'_{0,1}$, and $\Lambda_{0,2} \neq \Lambda'_{0,2}$, and, by Theorem 21.25, $T_{SP}(0,0) = v_{0,0} = 0$, $T_{SP}(0,1) = T_{SP}(0,2) = \times$. In a similar way, we can evaluate $T_{SP}(i,j)$ for all $i, j = 0, 1, \cdots, N-1$, completing the table T_{SP} as in Table 21.4. From the table, we find the characteristic expression

$$\hat{T}_{SP} = \{(3, \times, \times, 5, \times, \times)\}. \quad \clubsuit \qquad (21.33)$$

21.5.5 Scrambler-Permuter-Scrambler Based Systems

In the case of scrambler-permuter-scrambler based systems, the characteristic operator is, by (21.26e) and (21.8), $C_{S_1 P S_2} = S_1^{-1} * P^{-1} * D *$

$S_2^{-1} * S_2 * S_1 = S_1^{-1} * P^{-1} * D * P * S_1$. Therefore $C_{S_1 P S_2} = C_{S_1 P}$, $T_{S_1 P S_2}$ $= T_{S_1 P}$ and consequently

$$\hat{T}_{S_1 P S_2} = \hat{T}_{S_1 P}. \qquad (21.34)$$

EXAMPLE *21.15* Employing the permuter in Fig. 21.2 and the parallel scramblers whose characteristic polynomials are $\Psi_1(x) = x^7 + x + 1$ and $\Psi_2(x) = x^5 + x^3 + 1$, we obtain, by (21.34) and (21.33), $\hat{T}_{S_1 P S_2} = \hat{T}_{S_1 P}$ $= \{ (3, \times, \times, 5, \times, \times) \}$, and hence the signal-detection table $T_{S_1 P S_2}$ becomes identical to Table 21.4. ♣

21.5.6   Permuter-Scrambler-Permuter Based Systems

In the case of permuter-scrambler-permuter based systems, the characteristic operator is, by (21.26d) and (21.26f), $C_{P_1 S P_2} = P_1^{-1} * C_{S P_2} * P_1$, or equivalently, $\hat{C}^j_{P_1 S P_2}(i) = [\hat{P}_1^{-1} * \hat{C}_{S P_2} * \hat{P}_1](i)$. Applying Theorem 21.10 to this, we obtain $\hat{C}_{P_1 S P_2}(i) = \hat{P}_1[\{\hat{C}_{S P_2}(\hat{P}_1^{-1}(i))\}]$. For convenience, we define

$$\hat{P}_1(\times) = \times. \qquad (21.35)$$

Then, from (21.24) we have the relation $T_{P_1 S P_2}(i, j) = \hat{P}_1\{T_{S P_2}(\hat{P}_1^{-1}(i), j)\}$ since the number-operator $\hat{P}_1$ now converts only signal numbers. Inserting $i = \hat{P}_1(l)$ to this, we obtain $T_{P_1 S P_2}(\hat{P}_1(l), j) = \hat{P}_1\{T_{S P_2}(l, j)\}$. Therefore, if the lth row in $T_{S P_2}$ is an element of $\hat{T}_{S P_2}$, then the row i $= \hat{P}_1(l)$ in $T_{P_1 S P_2}$ is an element of $\hat{T}_{P_1 S P_2}$, and thus

$$\hat{T}_{P_1 S P_2} = \hat{P}_1(\hat{T}_{S P_2}). \qquad (21.36)$$

EXAMPLE *21.16* Employing the parallel scrambler in Fig. 19.2 and the permuters whose characteristic expressions are $\hat{P}_1 = (2, 3, 4, 1, 5, 0)$ and $\hat{P}_2 = (1, 4, 5, 0, 2, 3)$, we obtain, from (21.33), $\hat{T}_{S P_2} = (3, \times, \times, 5, \times, \times)$. Therefore, by (21.36) and (21.35), we have $\hat{T}_{P_1 S P_2} = (\hat{P}_1(3), \hat{P}_1(\times), \hat{P}_1(\times), \hat{P}_1(5), \hat{P}_1(\times), \hat{P}_1(\times)) = \{ (1, \times, \times, 0, \times, \times) \}$. By this and Theorem 21.20, we can complete the signal-detection table $T_{P_1 S P_2}$ as in Table 21.5. ♣

21.5.7   Summary of Signal-Detection Tables

We summarize the signal-detection table characteristic expressions in (21.27), (21.28), (21.29), (21.34) and (21.36), into the following theorem.

Table 21.5. A signal-detection table for a permuter-scrambler-permuter based system

$i\backslash j$	0	1	2	3	4	5
0	0	×	×	$d1$	×	×
1	1	×	×	0	×	×
2	2	×	×	×	×	×
3	3	×	×	×	×	×
4	4	×	×	×	×	×
5	5	×	×	×	×	×

THEOREM 21.26 *The characteristic expressions* $\hat{T}_S$, $\hat{T}_P$, $\hat{T}_{PS}$, $\hat{T}_{S_1PS_2}$, $\hat{T}_{P_1SP_2}$ *meet the following relations :*

$$\hat{T}_S = \{(N-1, N-2, \cdots, 0)\}, \tag{21.37a}$$
$$\hat{T}_P = \{(\hat{P}(N-1), \hat{P}(N-2), \cdots, \hat{P}(0))\}, \tag{21.37b}$$
$$\hat{T}_{PS} = \hat{T}_P, \tag{21.37c}$$
$$\hat{T}_{S_1PS_2} = \hat{T}_{S_1P}, \tag{21.37d}$$
$$\hat{T}_{P_1SP_2} = \hat{P}_1(\hat{T}_{SP_2}). \tag{21.37e}$$

According to the theorem, the signal-detection table characteristic expressions of scrambler, permuter, and permuter-scrambler based systems are directly obtained from the constituent permuter characteristic expressions. The signal-detection values for the three systems are always valid, so no × mark appears in the corresponding signal-detection tables. It should be noted that the constituent scrambler does not, while permuter does, influence on the characteristic expressions in those cases. The signal-detection table characteristic expressions of scrambler-permuter-scrambler(S-P-S) and permuter-scrambler-permuter(P-S-P) based systems are directly obtained from the signal-detection table characteristic expression of the embedded scrambler-permuter(S-P) configurations. The signal-detection table of an S-P configuration can be obtained using Theorem 21.25, and its characteristic expression is extracted from the table. Since the signal-detection values of the S-P configuration can include invalid signal numbers, the signal-detection tables of the S-P-S and P-S-P configurations can contain × marks in them.

21.6   Applications of Signal-Detection Tables to Signal-Alignment

We finally demonstrate how to apply the signal-detection tables to signal alignments in the transmission systems employing the self synchronous scrambling.

If the transmission system in Fig. 17.1(b) employs only a parallel scrambler without permuters as the functional processor, or equivalently $R = S$, then the signal-detection table takes the form of Table 21.2. Unfortunately, the signal-detection table itself does not provide any clue to identifying the signal-alignment state $(j = 0)$, since the receiving part of the base-rate processor which does frame-formatting and frame-aligning functions on the individual base-rate signal is not capable of recognizing the signal numbers. It simply checks whether or not frames are aligned, issuing *out-of-frame* (OOF) indication in case the frame cannot be aligned. So, every detected-signal number looks the same to the base-rate processor. Therefore, we conclude that once the signal-detection table is filled with all valid signal numbers, it is impossible to identify the properly aligned state.

In order to make the signal-alignment state distinguishable from others, therefore, the transmission system must be arranged in such a way that the signal-detection table include some invalid signal-detection values marked with ×. One simple means to achieve this is the signal-inversion technique adopted by some conventional systems. The essence of it is in inverting the output of one particular base-rate processor at the transmitting part, inverting it back at the receiving part. This generates regularly located × marks at every state other than the properly aligned-state. However, it could malfunction in case base-rate processors are partially equipped with no auxiliary idle channel drivers installed for the unequipped slots. One specific example of this is the case when one or more base-rate processors are unequipped and the inverted base-rate signal is damaged by line error or by the corresponding base-rate processor failure.

A fundamental and complete solution can be found in employing permuters in addition to scramblers. Examining the signal-detection tables T_S, T_P, T_{PS}, T_{SP}, $T_{S_1 P S_2}$, $T_{P_1 S P_2}$ in Tables 21.2 through 21.5, we observe that the latter three tables can contain × marked entries. Since the table T_{SP} is the basic one from which the tables $T_{S_1 P S_2}$ and $T_{P_1 S P_2}$ are derived, we concentrate our discussions on the scrambler-permuter

Table 21.6. An example of "perfect" signal-detection table for $N = 6$

$i\backslash j$	0	1	2	3	4	5
0	0	×	×	×	×	×
1	1	×	×	×	×	×
2	2	×	×	×	×	×
3	3	×	×	×	×	×
4	4	×	×	×	×	×
5	5	×	×	×	×	×

configuration.

If, for example, we choose a parallel scrambler with the characteristic polynomial $\Psi(x) = x^7 + x + 1$, and a permuter with the characteristic expression $\hat{P} = (1, 0, 3, 2, 5, 4)$, we obtain, using Theorem 21.25, the signal-detection table in Table 21.6 whose characteristic expression is $\hat{T} = \phi$. The signal-alignment state is clearly distinguished from others in this case, since the signals are detected only when $j = 0$. Therefore, this scrambler-permuter configuration, when used along with properly chosen rolling or bit-shift logics, ensures a perfect signal-alignment even at the partially equipped situations. In this sense, the signal-detection table with its characteristic expression of the form $\hat{T} = \phi$ can be referred as a "perfect" table.

In general, the *maximum signal-alignment time* (MSAT) is N times the *maximum average reframe time*(MART) of the base-rate processors.[7] This means that the MSAT could increase beyond the allowed limit in case N is large. In such a situation, therefore, it is not desirable to have a perfect signal-detection table. As a solution to this MSAT problem, we employ "slightly imperfect" signal-detection tables. That is, we choose scrambler-permuter pairs such that the resulting signal-detection table contains a few regularly spaced columns which include valid signal-detection values with the corresponding characteristic expression $\hat{T} \neq \phi$.

To illustrate the above scheme, we choose the scrambler having the characteristic equation $\Psi(x) = x^{15} + x^{14} + 1$ and the permuter having the

---

[7]MSAT refers to the worst-case signal-alignment time to have all N base-rate signals properly distributed to their destined output lines, and MART refers to the worst-case average time for a base-rate signal to get frame-aligned.

characteristic expression $\hat{P} = (2, 4, 5, 0, 3, 1, 8, 10, 11, 6, 9, 7)$. Then, we obtain the signal-detection table in Table 21.7, whose characteristic expression is

$$\hat{T} = \{ (6, \times, \times, \times, \times, \times, 0, \times, \times, \times, \times, \times),$$
$$(7, \times, \times, \times, \times, \times, 1, \times, \times, \times, \times, \times),$$
$$(8, \times, \times, \times, \times, \times, 2, \times, \times, \times, \times, \times),$$
$$(9, \times, \times, \times, \times, \times, 3, \times, \times, \times, \times, \times),$$
$$(10, \times, \times, \times, \times, \times, 4, \times, \times, \times, \times, \times),$$
$$(11, \times, \times, \times, \times, \times, 5, \times, \times, \times, \times, \times)\}.$$

With the given signal-detection table, the receiving part of base-rate processors can not distinguish the two states $j = 0$ and $j = 6$, since signal-detection is made for the whole 12 signals on both states. But if we apply, in addition, the aforementioned signal-inversion technique to some base-rate processors, then the two states become distinguishable. For example, if the signal-inversion is applied to the 0th base-rate processor, then the two entries $T(0,6)$ and $T(6,6)$ change to $\times$; and if signal-inversion process is applied to the 0th and the 1-st base-rate processors, then the entries $T(0,6)$, $T(1,6)$, $T(6,6)$ and $T(7,6)$ change to $\times$. The state $j = 6$ then becomes distinct from the state $j = 0$ as well as from the other ten states. So we can add an additional rolling logic to detect the state $j = 6$, and to make the rolling process jump six states upon detecting the state. The MSAT then reduces to seven times the MART from the original 12 times the MART.

The above scheme can be further improved if we employ the signal-detection table in Table 21.8, whose characteristic expression is

$$\hat{T} = \{ (8, \times, \times, \times, 4, \times, \times, \times, 0, \times, \times, \times),$$
$$(9, \times, \times, \times, 5, \times, \times, \times, 1, \times, \times, \times),$$
$$(10, \times, \times, \times, 6, \times, \times, \times, 2, \times, \times, \times),$$
$$(11, \times, \times, \times, 7, \times, \times, \times, 3, \times, \times, \times)\}.$$

The table can be obtained from the scrambler having the characteristic polynomial $\Psi(x) = x^{15} + x^{14} + 1$, and the permuter having the characteristic expression $\hat{P} = (0, 2, 1, 3, 4, 6, 5, 7, 8, 10, 9, 11)$. By applying signal-inversion and adding an additional rolling logic to detect the two states $j = 4$ and $j = 8$ and to jump four states upon detection, the MSAT now reduces to six times the MART.

Table 21.7. An example of "slightly imperfect" signal-detection table for $N = 12$

$i\backslash j$	0	1	2	3	4	5	6	7	8	9	10	11
0	0	×	×	×	×	×	$d6$	×	×	×	×	×
1	1	×	×	×	×	×	$d7$	×	×	×	×	×
2	2	×	×	×	×	×	$d8$	×	×	×	×	×
3	3	×	×	×	×	×	$d9$	×	×	×	×	×
4	4	×	×	×	×	×	$d10$	×	×	×	×	×
5	5	×	×	×	×	×	$d11$	×	×	×	×	×
6	6	×	×	×	×	×	0	×	×	×	×	×
7	7	×	×	×	×	×	1	×	×	×	×	×
8	8	×	×	×	×	×	2	×	×	×	×	×
9	9	×	×	×	×	×	3	×	×	×	×	×
10	10	×	×	×	×	×	4	×	×	×	×	×
11	11	×	×	×	×	×	5	×	×	×	×	×

Table 21.8. Another example of "slightly imperfect" signal-detection table for $N = 12$

$i\backslash j$	0	1	2	3	4	5	6	7	8	9	10	11
0	0	×	×	×	$d8$	×	×	×	$d4$	×	×	×
1	1	×	×	×	$d9$	×	×	×	$d5$	×	×	×
2	2	×	×	×	$d10$	×	×	×	$d6$	×	×	×
3	3	×	×	×	$d11$	×	×	×	$d7$	×	×	×
4	4	×	×	×	0	×	×	×	$d8$	×	×	×
5	5	×	×	×	1	×	×	×	$d9$	×	×	×
6	6	×	×	×	2	×	×	×	$d10$	×	×	×
7	7	×	×	×	3	×	×	×	$d11$	×	×	×
8	8	×	×	×	4	×	×	×	0	×	×	×
9	9	×	×	×	5	×	×	×	1	×	×	×
10	10	×	×	×	6	×	×	×	2	×	×	×
11	11	×	×	×	7	×	×	×	3	×	×	×

For the generalization of discussion, we assume that a scrambler-permuter configuration is designed such that valid signal-detection values appear at every Mth state, or equivalently, every Mth column of the signal-detection table contains valid signal numbers. We also assume that a rolling logic is additionally installed to detect those $M - 1$ states, excluding the properly-aligned state $j = 0$, and to jump M states upon detecting them. Then, the MSAT reduces to $(N/M + M - 1)$ times the MART. The optimal choice of M is thus $\sqrt{N}$, and the resulting MSAT turns out to be about $2M$ times the MART. This amounts to a $M/2$-to-1 reduction.

Therefore, the proper signal-alignment problem and the signal-aligning time problem in the SSS based transmission systems can be completely resolved by employing the concept of scrambling-permuting. More sophisticated resolutions might be also available if variations of the scrambler-permuter combinations such as S-P-S, P-S-P or others are employed.

APPENDICES

Appendix A
Facts from Abstract Algebra

A.1 Field

A *field* $\mathcal{F}$ is a non-empty set composed of *scalars*, with addition and multiplication, satisfying the following conditions :

F1. To every pair of scalars α, $\beta \in \mathcal{F}$, there is associated a *sum* $\alpha_1 + \alpha_2$ in $\mathcal{F}$.

F2. Addition is associative, i.e., $(\alpha_1 + \alpha_2) + \alpha_3 = \alpha_1 + (\alpha_2 + \alpha_3)$.

F3. There exists a scalar, which we denote by 0, such that $\alpha + 0 = \alpha$ for all $\alpha \in \mathcal{F}$.

F4. For each $\alpha \in \mathcal{F}$ there exists a scalar, which we denote by $-\alpha$, such that $\alpha + (-\alpha) = 0$.

F5. Addition is commutative, i.e., $\alpha_1 + \alpha_2 = \alpha_2 + \alpha_1$.

F6. To every pair of scalars α_1, $\alpha_2 \in \mathcal{F}$, there is associated a *product* $\alpha_1 \alpha_2$ in $\mathcal{F}$.

F7. Multiplication is associative, i.e., $(\alpha_1 \alpha_2)\alpha_3 = \alpha_1(\alpha_2 \alpha_3)$.

F8. There exists a scalar different from 0, which we denote by 1, such that $\alpha 1 = \alpha$ for all $\alpha \in \mathcal{F}$.

F9. For each $\alpha \in \mathcal{F}$, $\alpha \neq 0$, there exists a scalar, which we denote by α^{-1}, such that $\alpha \alpha^{-1} = 1$.

F10. Multiplication is commutative, i.e., $\alpha_1 \alpha_2 = \alpha_2 \alpha_1$.

F11. Multiplication is distributive with respect to addition, i.e., $(\alpha_1 + \alpha_2)\alpha_3 = \alpha_1 \alpha_3 + \alpha_2 \alpha_3$.

A field having finite elements is called a *Galois field*, which we denote by $GF(q)$, where q is the number of elements in $GF(q)$. Then, we have the following property in the Galois field $GF(q)$.

PROPERTY A.1 *For a non-zero element α in $GF(q)$, $\alpha^q = \alpha$.*

Proof : Since $GF(q)$ has finite elements, for a non-zero element α in $GF(q)$ there exists a positive integer m such that $\alpha^m = 1$. We rearrange the elements in $GF(q)$ based on α in the following manner : We first put $1, \alpha, \alpha^2, \cdots, \alpha^{m-1}$ in the first row. Then, we take a non-zero element β_1 in $GF(q)$ but not in the first row, multiply each element in the first row by β_1, and put this in the second row. In a similar manner, for $i = 1, 2,$ $\cdots, n - 1$, we make the ith row by choosing a non-zero element β_{i-1} in $GF(q)$ not contained in the previous $(i - 1)$ rows, and multiplying each element in the first row by β_{i-1}. If there does not exist such a non-zero element β_i with $i = n$, then we stop this procedure. This rearrangement of the elements can be depicted as

$$
\begin{array}{ccccc}
1, & \alpha, & \alpha^2, & \cdots, & \alpha^{m-1}, \\
\beta_1, & \alpha\beta_1 & \alpha^2\beta_1, & \cdots, & \alpha^{m-1}\beta_1, \\
\beta_2, & \alpha\beta_2 & \alpha^2\beta_2, & \cdots, & \alpha^{m-1}\beta_2, \\
\cdots, & \cdots, & \cdots, & \cdots, & \cdots, \\
\beta_{n-1}, & \alpha\beta_{n-1} & \alpha^2\beta_{n-1}, & \cdots, & \alpha^{m-1}\beta_{n-1}.
\end{array}
$$

Then, it suffices to prove that a non-zero element in $GF(q)$ appear exactly once in the above rearrangement, since $mn = q - 1$ implies that $\alpha^q = \alpha^{mn+1} = \alpha$. We show this by contradiction. Suppose $\alpha^{i_1}\beta_{j_1} = \alpha^{i_2}\beta_{j_2}$ for integers i_1, i_2, j_1, j_2 with $j_1 < j_2$. Then, $\alpha^{i_1-i_2}\beta_{j_1} = \beta_{j_2}$, and so $i_1 < i_2$ due to the rearrangement procedure. However, if $i_1 < i_2$, we have, by $\alpha^m = 1$, the relation $\alpha^{m+i_1-i_2}\beta_{j_1} = \beta_{j_2}$, which contradicts the rearrangement procedure. This completes the proof. ∎

A.2   Extension of Fields

For every Galois field $GF(q)$, $q = p^L$ for a prime number p and a positive integer L. A Galois field $GF(q)$ with a prime power q is a *subfield* of $GF(q^L)$ for any positive integer L, and $GF(q^L)$ is an *extension field* of $GF(q)$. The following property describes how to extend $GF(q)$ to $GF(q^L)$.

PROPERTY A.2 *For an irreducible polynomial $\Psi_I(x)$ of degree L over $GF(q)$, let α be a root of $\Psi_I(x)$. Then,*

$$
GF(q^L) = \left\{ \sum_{i=0}^{L-1} a_i\alpha^i \; : \; a_i \in GF(q), \; i = 0, 1, \cdots, L - 1 \right\}. \quad \text{(A.1)}
$$

Proof : We can easily confirm that $GF(q^L)$ in (A.1) meets the conditions for field except for the existence of inverse. So we prove only the existence of inverse. Since 1 is its own inverse, we consider the existence of the inverse of an element $\sum_{i=1}^{L-1} a_i \alpha^i$, which is neither 0 nor 1. Let $\Psi(x) = \sum_{i=0}^{L-1} a_i x^i$. Then, since $\Psi_I(x)$ is an irreducible polynomial, and since the degree of $\Psi(x)$ is less than that of $\Psi_I(x)$, there exist polynomials $P_1(x)$ and $P_2(x)$ such that $P_1(x)\Psi(x) - P_2(x)\Psi_I(x) = 1$. Inserting $x = \alpha$ into this, we obtain $P_1(\alpha)\Psi(\alpha) = 1$ since $\Psi_I(\alpha) = 0$. Therefore, $P_1(\alpha)$ is the inverse of $\Psi(\alpha) = \sum_{i=0}^{L-1} a_i \alpha^i$. This completes the proof. ∎

For elements in the extension field $GF(p^L)$ with a prime number p, we have the following property.

PROPERTY A.3 *For an extension field $GF(p^L)$ for a prime number p and a positive integer L,*

$$(\alpha_1 + \alpha_2)^{p^m} = \alpha_1^{p^m} + \alpha_2^{p^m} \tag{A.2}$$

for any positive integer m and any elements α_1 and α_2 in $GF(p^L)$.

Proof : We first prove (A.2) for $m = 1$. Expanding $(\alpha_1 + \alpha_2)^p$, we obtain $(\alpha_1 + \alpha_2)^p = \sum_{i=0}^{p} \{p!/[i!(p-i)!]\}\alpha_1^i \alpha_2^{p-i}$. However, since $p = 0$ in $GF(p^L)$, which is the same in the subfield $GF(p)$, we have the relation $p!/[i!(p-i)!] = 0$ for $i = 1, 2, \cdots, p-1$. Therefore, $(\alpha_1 + \alpha_2)^p = \alpha_1^p + \alpha_2^p$.

Applying this relation to $(\alpha_1 + \alpha_2)^{p^m}$ repeatedly, we can obtain (A.2) for $m = 2, 3, \cdots$. ∎

A.3 Irreducible Polynomials

A polynomial $\Psi_I(x)$ of degree L whose coefficients are elements of a Galois field $GF(q)$ is called an *irreducible polynomial* over $GF(q)$ if it is not divisible by any polynomial of degree less than L but is greater than 0. For irreducible polynomials, we have the following properties.

PROPERTY A.4 *An irreducible polynomial $\Psi_I(x)$ of degree L over $GF(q)$ divides $x^{q^L} - x$.*[1]

---

[1] In the case of $GF(2)$, $q = 2$ and the inverse symbol "−" for addition is equivalent to the addition "+". Therefore, the properties in this appendix can be applied to the sequence space and SRG theory in this book with q and "−" replaced by 2 and "+".

Proof : If $\Psi_I(x) = x$, the property is obvious, so we prove the case $\Psi_I(x) \neq x$. Let α be a root of $\Psi_I(x)$. Then, α is an element of the extension field $GF(q^L)$ in (A.1), and by Property A.1, $\alpha^{q^L} = \alpha$. This means that α is also a root of $x^{q^L} - x$. Let $x^{q^L} - x = Q(x)\Psi_I(x) + R(x)$ for the quotient polynomial $Q(x)$ and the remainder polynomial $R(x)$. Inserting $x = \alpha$ into this, we obtain the relation $R(\alpha) = 0$. However, $R(\alpha)$ in $GF(q^L)$ becomes 0 only if $R(x) = 0$. Therefore, $R(x) = 0$, and hence $\Psi_I(x)$ divides $x^{q^L} - x$.   ∎

PROPERTY A.5 *Every irreducible factor $\Psi_I(x)$ of the polynomial $x^{q^L} - x$ which is irreducible over $GF(q)$ has degree L or less.*

Proof : Let $\Psi_I(x)$ of degree n be an irreducible factor of $x^{q^L} - x$ over $GF(q)$, and consider the extension field $GF(q^n)$ in (A.1) for a root α of $\Psi_I(x)$ (Note that in (A.1) L is replaced by n in this case). Then, for an arbitrary element $\beta = \sum_{i=0}^{n-1} a_i\alpha^i$ in $GF(q^n)$, we have, by Properties A.3 and A.1, the relation $\beta^{q^L} = \sum_{i=0}^{n-1} a_i^{q^L} \alpha^{iq^L} = \sum_{i=0}^{n-1} a_i\alpha^i = \beta$, which means that all the q^n elements in $GF(q^n)$ are roots of $x^{q^L} - x$. However, $x^{q^L} - x$ has no more than q^L roots. Therefore, $q^L \geq q^n$, and hence $L \geq n$.   ∎

A.4   Primitive Polynomials

An element α in $GF(q)$ is called a *primitive element* if every non-zero element in $GF(q)$ can be represented by a power of α, and an irreducible polynomial $\Psi_p(x)$ of degree L over $GF(q)$ is called a *primitive polynomial* if it has a primitive element in $GF(q^L)$ as a root. For primitive polynomials, we have the following property.

PROPERTY A.6 *A polynomial $\Psi_p(x)$ of degree L over $GF(q)$ is primitive, if the smallest integer T for which $\Psi_p(x)$ divides $x^T - 1$ is $q^L - 1$.*

Proof : Let α be a root of $\Psi_p(x)$ in the extension field $GF(q^L)$ in (A.1). Then, since $\alpha^i \neq 1$ for $i = 1, 2, \cdots, q^L - 2$, the $(q^L - 1)$ elements α^i, $i = 0, 1, \cdots, q^L - 2$, respectively correspond to the $(q^L - 1)$ non-zero elements of $GF(q^L)$ in (A.1). Therefore, α is a primitive element of $GF(q^L)$, and so $\Psi_p(x)$ is a primitive polynomial.   ∎

A.5 Others

For polynomials over $GF(q)$, we have the following properties.

PROPERTY A.7 *For a polynomial $\Psi(x)$ over $GF(p^L)$ with a prime number p, the relation $\Psi(x^{p^m}) = [\Psi(x)]^{p^m}$ holds for a positive integer m.*

Proof : The property can be proved in a way similar to that of Property A.3. ∎

PROPERTY A.8 *The polynomial $x^m - 1$ divides $x^n - 1$, if and only if m divides an integer n.*

Proof : Let $n = Qm$ for an integer Q. Then, since $y = 1$ is a root of $y^Q - 1$, $y - 1$ divides $y^Q - 1$. Therefore, inserting $y = x^m$ into this, $x^m - 1$ divides $x^n - 1$, which prove the "if" part.

To prove the "only if", let $n = Qm + R$, where Q and R are respectively the quotient and the remainder of n divided by m. Then, x^n can be written in the form of $x^n - 1 = x^R(x^{Qm} - 1) + x^R - 1$. If $x^m - 1$ divides $x^n - 1$, the remainder polynomial $x^R - 1$ should be 0, so $R = 0$. Therefore, m divides n. ∎

PROPERTY A.9 *For two polynomials $\Psi_1(x)$ and $\Psi_2(x)$, $GCD[\Psi_1(x), \Psi_2(x)] = GCD[\Psi_1(x), GCD[\Psi_1(x), \Psi_2(x)]]$.*

Proof : Let $GCD[\Psi_1(x), \Psi_2(x)] = g(x)$, and let $\Psi_1(x) = a_1(x)g(x)$, $\Psi_2(x) = a_2(x)g(x)$, where $a_1(x)$ and $a_2(x)$ are relatively prime. Then, $GCD[\Psi_1(x), GCD[\Psi_1(x), \Psi_2(x)]] = GCD[a_1(x)g(x), g(x)] = g(x)$. Therefore, we have the property. ∎

Appendix B
Facts from Linear Algebra

B.1 Vector Space

A *vector space* V over a field $\mathcal{F}$ is a non-empty set composed of *vectors*, with *vector addition* and *scalar multiplication*, satisfying the following conditions :

A1. To every pair of vectors v_1, $v_2 \in V$, there is associated a *sum* $v_1 + v_2$ in V.

A2. Addition is associative, i.e., $(v_1 + v_2) + v_3 = v_1 + (v_2 + v_3)$.

A3. There exists a vector, which we denote by 0, such that $v + 0 = v$ for all $v \in V$.

A4. For each $v \in V$, there exists a vector, which we denote by $-v$, such that $v + (-v) = 0$.

A5. Addition is commutative, i.e., $v_1 + v_2 = v_2 + v_1$.

B1. To every scalar $\alpha \in \mathcal{F}$ and vector $v \in V$, there is associated a *product* αv in V.

B2. Scalar multiplication is associative, i.e., $\alpha_1(\alpha_2 v) = (\alpha_1 \alpha_2)v$.

B3. Scalar multiplication is distributive with respect to vector addition, i.e., $\alpha(v_1 + v_2) = \alpha v_1 + \alpha v_2$.

B4. Scalar multiplication is distributive with respect to scalar addition, i.e., $(\alpha_1 + \alpha_2)v = \alpha_1 v + \alpha_2 v$.

B5. For a scalar $1 \in \mathcal{F}$, $1v = v$.

For a vector space V, a linearly independent set of vectors spanning the vector space V is called a *basis* of V, and the number of elements in a basis is called the *dimension*, which we denote by $dim(V)$. Then, the dimension of a sum space obtained by linear sums of two finite-dimensional vector spaces is determined by the following property.

PROPERTY B.1 *For two finite-dimensional vector spaces V_1 and V_2, let the sum space $V_1 + V_2$ be the vector space obtained by linear sums of V_1 and V_2, i.e.,*

$$V_1 + V_2 = \{\, \mathbf{v}_1 + \mathbf{v}_2 \; : \; \mathbf{v}_1 \in V_1 \text{ and } \mathbf{v}_2 \in V_2 \,\}.$$

Then, the dimension of the sum space $V_1 + V_2$ becomes

$$dim(V_1 + V_2) = dim(V_1) + dim(V_2) - dim(V_1 \cap V_2). \qquad \text{(B.1)}$$

Proof : Let $dim(V_1 \cap V_2) = l$, $dim(V_1) = l + m$, $dim(V_2) = l + n$, and let $\{\, \mathbf{v}_1, \cdots, \mathbf{v}_l \,\}$ be a basis of $V_1 \cap V_2$. Then, this basis can be extended to a basis $\{\, \mathbf{v}_1, \cdots, \mathbf{v}_l, \hat{\mathbf{v}}_1, \cdots, \hat{\mathbf{v}}_m \,\}$ of V_1 and also to a basis $\{\, \mathbf{v}_1, \cdots, \mathbf{v}_l, \tilde{\mathbf{v}}_1, \cdots, \tilde{\mathbf{v}}_n \,\}$ of V_2. Therefore, it suffices to show that the set $\{\, \mathbf{v}_1, \cdots, \mathbf{v}_l, \hat{\mathbf{v}}_1, \cdots, \hat{\mathbf{v}}_m, \tilde{\mathbf{v}}_1, \cdots, \tilde{\mathbf{v}}_n \,\}$ is a basis of $V_1 + V_2$. It is clear that this set spans $V_1 + V_2$. To prove the linear independence, let $\sum_{i=1}^{l} \alpha_i \mathbf{v}_i + \sum_{j=1}^{m} \beta_j \hat{\mathbf{v}}_j + \sum_{k=1}^{n} \gamma_k \tilde{\mathbf{v}}_k = \mathbf{0}$ for scalars α_i, β_j, $\gamma_k \in \mathcal{F}$. Then, $\sum_{i=1}^{l} \alpha_i \mathbf{v}_i + \sum_{j=1}^{m} \beta_j \hat{\mathbf{v}}_j = -\sum_{k=1}^{n} \gamma_k \tilde{\mathbf{v}}_k$. The left-hand side is in V_1 and the right-hand side is in V_2, and hence both are in $V_1 \cap V_2$. Therefore, noting that $\{\, \mathbf{v}_1, \cdots, \mathbf{v}_l \,\}$ is a basis of $V_1 \cap V_2$, β_j's and γ_k's are all 0, and hence α_i's are also 0. This completes the proof. ∎

B.2   Minimal Polynomials of Matrices

For a matrix $\mathbf{M}$, a non-zero lowest-degree monic polynomial $\Psi_{\mathbf{M}}(x)$ such that $\Psi_{\mathbf{M}}(\mathbf{M}) = \mathbf{0}$ is said the *minimal polynomial* of $\mathbf{M}$. For minimal polynomials, we have the following properties.

PROPERTY B.2 *The minimal polynomial $\Psi_{\mathbf{M}}(x)$ of a matrix $\mathbf{M}$ is unique.*

Proof : Let two polynomials $\Psi_1(x)$ and $\Psi_2(x)$ be minimal polynomials of a matrix $\mathbf{M}$. Then, since $\Psi_1(x)$ and $\Psi_2(x)$ have the same degree, $\Psi_1(x)$ can be represented by $\Psi_1(x) = \Psi_2(x) + R(x)$ for a polynomial $R(x)$ of degree less than $\Psi_1(x)$. Inserting $x = \mathbf{M}$ into this, we have the relation $R(\mathbf{M}) = \mathbf{0}$ since $\Psi_1(\mathbf{M}) = \Psi_2(\mathbf{M}) = \mathbf{0}$. However, since the degree of $R(x)$ is less than that of the minimal polynomial $\Psi_1(x)$, $R(x)$ should be a zero polynomial. Therefore, $\Psi_1(x) = \Psi_2(x)$. This proves the property. ∎

PROPERTY B.3 *Let* $\Psi_M(x)$ *be the minimal polynomial of a matrix* M. *Then, the relation* $\Psi(M) = 0$ *holds for a polynomial* $\Psi(x)$, *if and only if* $\Psi_M(x)$ *divides* $\Psi(x)$.

Proof : We first show the "if" part of the property. If $\Psi(x)$ is a multiple of $\Psi_M(x)$, $\Psi(x)$ can be represented by $\Psi(x) = Q(x)\Psi_M(x)$ for a polynomial $Q(x)$. Therefore, inserting $x = M$ into this, we obtain $\Psi(M) = Q(M) \cdot \Psi_M(M) = 0$ since $\Psi_M(M) = 0$.

Now we prove the "only if" part. Let $\Psi(x) = Q(x)\Psi_M(x) + R(x)$, where $Q(x)$ and $R(x)$ are respectively the quotient and remainder polynomials of $\Psi(x)$ divided by $\Psi_M(x)$. Inserting $x = M$ into this, we have the relation $\Psi(M) = Q(M) \cdot \Psi_M(M) + R(M)$. If $\Psi(M) = 0$, $R(M) = 0$, so $R(x)$ should be 0 since the degree of $R(x)$ is less than that of the minimal polynomial $\Psi_M(x)$. Therefore, $\Psi_M(x)$ divides $\Psi(x)$. ∎

PROPERTY B.4 *The degree of the minimal polynomial* $\Psi_M(x)$ *of an* $L \times L$ *matrix* M *can not be larger than* L.

Proof : Let $\Psi(x)$ be the characteristic polynomial of an $L \times L$ matrix M. Then, $\Psi(M) = 0$, and hence by Property B.3, the minimal polynomial $\Psi_M(x)$ divides $\Psi(x)$. However, since the degree of $\Psi(x)$ is L, the degree of $\Psi_M(x)$ can not be larger than L. ∎

PROPERTY B.5 *For an* $L \times L$ *matrix* M *and an* L-*vector* v, *let* $\Psi_{M,v}(x)$ *be the non-zero lowest-degree polynomial meeting* $\Psi_{M,v}(M) \cdot v = 0$. *Then,* $\Psi(M) \cdot v = 0$ *for a polynomial* $\Psi(x)$, *if and only if* $\Psi_{M,v}(x)$ *divides* $\Psi(x)$.

Proof : The proof of the "if" part is trivial. To prove the "only if" part, let $\Psi(x) = Q(x)\Psi_{M,v}(x) + R(x)$, where $Q(x)$ and $R(x)$ are the quotient and the remainder polynomials of $\Psi(x)$ divided by $\Psi_{M,v}(x)$. Inserting $x = M$, and then multiplying the result by v, we have the relation $\Psi(M) \cdot v = Q(M) \cdot \Psi_{M,v}(M) \cdot v + R(M) \cdot v$, and applying $\Psi(M) \cdot v = 0$ and $\Psi_{M,v}(M) \cdot v = 0$ to this, we obtain $R(M) \cdot v = 0$. But, since the degree of $R(x)$ is less than that of $\Psi_{M,v}(x)$, $R(x)$ should be a zero polynomial, and therefore $\Psi_{M,v}(x)$ divides $\Psi(x)$. ∎

PROPERTY B.6 *For an* $L \times L$ *matrix* M, *let* $\Psi_M(x)$ *be the minimal polynomial of* M. *Then, there exists an* L-*vector* v *such that the non-zero lowest-degree polynomial* $\Psi_{M,v}(x)$ *meeting* $\Psi_{M,v}(M) \cdot v = 0$ *is* $\Psi_M(x)$.

Proof : Since $\Psi_M(M) = 0$, by Property B.5 $\Psi_{M,e_i}(x)$ divides $\Psi_M(x)$ for each basis vector e_i, $i = 0, 1, \cdots, L-1$. Thus, the least common multiple $\Psi(x)$ of the polynomials $\Psi_{M,e_i}(x)$, $i = 0, 1, \cdots, L-1$, divides $\Psi_M(x)$. However, $\Psi(M) = 0$, so by Property B.3 $\Psi_M(x)$ divides $\Psi(x)$. Therefore, $\Psi(x) = \Psi_M(x)$.

We first prove the property in the case when $\Psi_M(x) = [\Psi_I(x)]^w$ for an irreducible polynomial $\Psi_I(x)$ and an integer w. Then, since $\Psi_{M,e_i}(x)$, $i = 0, 1, \cdots, L-1$, divides $\Psi_M(x)$, $\Psi_{M,e_i}(x)$ is represented by $\Psi_{M,e_i}(x) = [\Psi_I(x)]^{w_i}$, where $w_i \leq w$. Since the least common multiple of $[\Psi_I(x)]^{w_i}$, $i = 0, 1, \cdots, L-1$, is identical to $\Psi_M(x) = [\Psi_I(x)]^w$, there exists an w_l which is w. This means that $\Psi_{M,e_l}(x) = [\Psi_I(x)]^w = \Psi_M(x)$.

It is simple to generalize the above special case to the case when $\Psi_M(x) = \prod_j [\Psi_I^j(x)]^{w_j}$ for irreducible polynomials $\Psi_I^j(x)$'s and integers w_j's.   ∎

B.3   Properties of Companion Matrix

For the companion matrix $A_{\Psi(x)}$ in (4.7), we have the following properties.

PROPERTY B.7 *For the companion matrix $A_{\Psi(x)}$ with a degree-L characteristic polynomial $\Psi(x)$,*

$$A_{\Psi(x)} \cdot e_i = \begin{cases} e_{i+1}, & i = 0, 1, \cdots, L-2, \\ \sum_{j=0}^{L-1} \psi_j e_j, & i = L-1, \end{cases} \tag{B.2a}$$

$$e_i^t \cdot A_{\Psi(x)} = \begin{cases} e_{L-1}^t, & i = 0, \\ e_{i-1}^t + \psi_i e_{L-1}^t, & i = 1, 2, \cdots, L-1, \end{cases} \tag{B.2b}$$

$$A_{\Psi(x)}^t \cdot e_i = \begin{cases} e_{L-1}, & i = 0, \\ e_{i-1} + \psi_i e_{L-1}, & i = 1, 2, \cdots, L-1, \end{cases} \tag{B.2c}$$

$$e_i^t \cdot A_{\Psi(x)}^t = \begin{cases} e_{i+1}^t, & i = 0, 1, \cdots, L-2, \\ \sum_{j=0}^{L-1} \psi_j e_j^t, & i = L-1. \end{cases} \tag{B.2d}$$

Proof : This property is directly obtained by (4.7).   ∎

PROPERTY B.8 *For the companion matrix $A_{\Psi(x)}$ with a degree-L characteristic polynomial $\Psi(x)$,*

$$A_{\Psi(x)}^i \cdot e_0 = e_i, \quad i = 0, 1, \cdots, L-1, \tag{B.3a}$$

$$e_0^t \cdot (A_{\Psi(x)}^t)^i = e_i^t, \quad i = 0, 1, \cdots, L-1. \tag{B.3b}$$

Proof : This property is a direct outcome of (B.2a) and (B.2d). ■

PROPERTY B.9 *The minimal polynomial of the companion matrix* $\mathbf{A}_{\Psi(x)}$ *is* $\Psi(x)$.

Proof : We first show that $\Psi(\mathbf{A}_{\Psi(x)}) = \mathbf{0}$. Let $\Psi(x) = \sum_{i=0}^{L} \psi_i x^i$. Then, $\Psi(\mathbf{A}_{\Psi(x)}) = \sum_{i=0}^{L} \psi_i \mathbf{A}_{\Psi(x)}^i$, and hence by (B.3a), we have the relations $\Psi(\mathbf{A}_{\Psi(x)}) \cdot \mathbf{e}_j = \mathbf{A}_{\Psi(x)}^j \cdot \sum_{i=0}^{L} \psi_i \mathbf{A}_{\Psi(x)}^i \cdot \mathbf{e}_0$, $j = 0, 1, \cdots, L - 1$. Applying (B.3a) to this again, we obtain $\Psi(\mathbf{A}_{\Psi(x)}) \cdot \mathbf{e}_j = \mathbf{A}_{\Psi(x)}^j \cdot (\sum_{i=0}^{L-1} \psi_i \mathbf{e}_i + \mathbf{A}_{\Psi(x)} \cdot \mathbf{e}_{L-1})$, which is $\mathbf{0}$ due to (B.2a). Therefore, $\Psi(\mathbf{A}_{\Psi(x)}) = \mathbf{0}$. To complete the proof, it suffices to show that there does not exists a non-zero polynomial $P(x)$ of degree less than $\Psi(x)$ such that $P(\mathbf{A}_{\Psi(x)}) = \mathbf{0}$. We show this by contradiction. Suppose that for a polynomial $P(x) = \sum_{i=0}^{\hat{L}} p_i x^i$, $\hat{L} < L$, $P(\mathbf{A}_{\Psi(x)}) = \mathbf{0}$. Then, $P(\mathbf{A}_{\Psi(x)}) \cdot \mathbf{e}_0 = \mathbf{0}$, and hence $\sum_{i=0}^{\hat{L}} p_i \mathbf{A}_{\Psi(x)}^i \cdot \mathbf{e}_0 = \mathbf{0}$. Applying (B.3a) to this, we obtain the relation $\sum_{i=0}^{\hat{L}} p_i \mathbf{e}_i = \mathbf{0}$. Therefore, $p_i = 0$ for $i = 0, 1, \cdots, \hat{L}$, that is, $P(x) = 0$. However, this is a contradiction to the assumption that $P(x)$ is a non-zero polynomial. This completes the proof. ■

PROPERTY B.10 *For the companion matrix* $\mathbf{A}_{\hat{\Psi}(x)}$ *with a characteristic polynomial* $\hat{\Psi}(x)$, *the relation* $\Psi(\mathbf{A}_{\hat{\Psi}(x)}) = \mathbf{0}$ *holds for a polynomial* $\Psi(x)$, *if and only if* $\hat{\Psi}(x)$ *divides* $\Psi(x)$.

Proof : By Property B.9, the minimal polynomial of the companion matrix $\mathbf{A}_{\hat{\Psi}(x)}$ is $\hat{\Psi}(x)$. Therefore, by Property B.3, we have the property. ■

PROPERTY B.11 *For the transform matrices* $\mathbf{B}_{\Psi(x)}$ *and* $\mathbf{C}_{\Psi(x)}$ *in (4.12),*

$$\mathbf{C}_{\Psi(x)} = \mathbf{B}_{\Psi(x)}^{-1}. \tag{B.4}$$

Proof : It suffices to show that $\mathbf{C}_{\Psi(x)} \cdot \mathbf{B}_{\Psi(x)} = \mathbf{I}$. Let $\mathbf{r}_i^t$, $i = 0, 1, \cdots, L - 1$, be the ith row of $\mathbf{C}_{\Psi(x)} \cdot \mathbf{B}_{\Psi(x)}$. Then, by (4.12),

$$\mathbf{r}_i^t = \sum_{j=i+1}^{L} \psi_j \mathbf{e}_0^t \cdot \mathbf{A}_{\Psi(x)}^{j-i}, \tag{B.5}$$

so it suffices to show that $\mathbf{r}_i^t = \mathbf{e}_i^t$, $i = 0, 1, \cdots, L - 1$. We show this by induction. For $i = L - 1$, we have, by (B.5), $\mathbf{r}_{L-1}^t = \psi_L \mathbf{e}_0^t \cdot \mathbf{A}_{\Psi(x)}$, and so by (B.2b), $\mathbf{r}_{L-1}^t = \mathbf{e}_{L-1}^t$. Now we assume that $\mathbf{r}_i^t = \mathbf{e}_i^t$ for an i $= L - 1, L - 2, \cdots, 1$, and prove that $\mathbf{r}_{i-1}^t = \mathbf{e}_{i-1}^t$. By (B.5), we have the relation $\mathbf{r}_{i-1}^t = \psi_i \mathbf{e}_0^t \cdot \mathbf{A}_{\Psi(x)} + \sum_{j=i+1}^{L} \psi_j \mathbf{e}_0^t \cdot \mathbf{A}_{\Psi(x)}^{j-i+1} = \psi_i \mathbf{e}_0^t \cdot \mathbf{A}_{\Psi(x)}$ $+ \mathbf{r}_i^t \cdot \mathbf{A}_{\Psi(x)}$. Therefore, by the assumption $\mathbf{r}_i^t = \mathbf{e}_i^t$, $\mathbf{r}_{i-1}^t = \psi_i \mathbf{e}_0^t \cdot \mathbf{A}_{\Psi(x)}$ $+ \mathbf{e}_i^t \cdot \mathbf{A}_{\Psi(x)}$, and hence by (B.2b) $\mathbf{r}_{i-1}^t = \psi_i \mathbf{e}_{L-1}^t + (\mathbf{e}_{i-1}^t + \psi_i \mathbf{e}_{L-1}^t) =$ $\mathbf{e}_{i-1}^t$. This completes the proof. ∎

PROPERTY B.12 *For the transform matrices* $\mathbf{B}_{\Psi(x)}$ *and* $\mathbf{C}_{\Psi(x)}$,

$$\mathbf{B}_{\Psi(x)} = \mathbf{B}_{\Psi(x)}^t, \tag{B.6a}$$

$$\mathbf{C}_{\Psi(x)} = \mathbf{C}_{\Psi(x)}^t. \tag{B.6b}$$

Proof : In view of (4.12b), we have (B.6b). To prove (B.6a), we consider the relation $\mathbf{B}_{\Psi(x)} \cdot \mathbf{C}_{\Psi(x)} = \mathbf{I}$, which is directly obtained by (B.4). Then, this can be rewritten in the form of $\mathbf{C}_{\Psi(x)}^t \cdot \mathbf{B}_{\Psi(x)}^t = \mathbf{I}$, and hence by (B.6b), $\mathbf{C}_{\Psi(x)} \cdot \mathbf{B}_{\Psi(x)}^t = \mathbf{I}$. Therefore, by (B.4), $\mathbf{B}_{\Psi(x)}^t = \mathbf{B}_{\Psi(x)}$. ∎

PROPERTY B.13 *For the companion matrix* $\mathbf{A}_{\Psi(x)}$ *and the transform matrix* $\mathbf{C}_{\Psi(x)}$,

$$\mathbf{A}_{\Psi(x)} \cdot \mathbf{C}_{\Psi(x)} = (\mathbf{A}_{\Psi(x)} \cdot \mathbf{C}_{\Psi(x)})^t. \tag{B.7}$$

Proof : Multiplying $\mathbf{A}_{\Psi(x)}$ in (4.7) and $\mathbf{C}_{\Psi(x)}$ in (4.12b), we obtain

$$\mathbf{A}_{\Psi(x)} \cdot \mathbf{C}_{\Psi(x)} = \begin{bmatrix} \psi_L & 0 & 0 & \cdots & 0 & 0 \\ 0 & \psi_2 & \psi_3 & \cdots & \psi_{L-1} & \psi_L \\ 0 & \psi_3 & \psi_4 & \cdots & \psi_L & 0 \\ \vdots & \vdots & \vdots & \vdots & \vdots & \vdots \\ 0 & \psi_{L-1} & \psi_L & \cdots & 0 & 0 \\ 0 & \psi_L & 0 & \cdots & 0 & 0 \end{bmatrix},$$

which is symmetric. Therefore, (B.7) holds. ∎

PROPERTY B.14 *For the companion matrix* $\mathbf{A}_{\Psi(x)}$ *and the transform matrices* $\mathbf{B}_{\Psi(x)}$ *and* $\mathbf{C}_{\Psi(x)}$,

$$\mathbf{A}_{\Psi(x)}^i = \mathbf{C}_{\Psi(x)} \cdot (\mathbf{A}_{\Psi(x)}^t)^i \cdot \mathbf{B}_{\Psi(x)}, \quad i = 0, 1, \cdots, \tag{B.8a}$$

$$(\mathbf{A}_{\Psi(x)}^t)^i = \mathbf{B}_{\Psi(x)} \cdot \mathbf{A}_{\Psi(x)}^i \cdot \mathbf{C}_{\Psi(x)}, \quad i = 0, 1, \cdots. \tag{B.8b}$$

Proof : By (B.7) and (B.6b), $\mathbf{A}_{\Psi(x)} \cdot \mathbf{C}_{\Psi(x)} = \mathbf{C}_{\Psi(x)} \cdot \mathbf{A}_{\Psi(x)}^t$, and applying (B.4) to this, we obtain $\mathbf{A}_{\Psi(x)} = \mathbf{C}_{\Psi(x)} \cdot \mathbf{A}_{\Psi(x)}^t \cdot \mathbf{B}_{\Psi(x)}$. Therefore, $\mathbf{A}_{\Psi(x)}^i = (\mathbf{C}_{\Psi(x)} \cdot \mathbf{A}_{\Psi(x)}^t \cdot \mathbf{B}_{\Psi(x)})^i = \mathbf{C}_{\Psi(x)} \cdot (\mathbf{A}_{\Psi(x)}^t)^i \cdot \mathbf{B}_{\Psi(x)}$ due to (B.4). This proves (B.8a).

Equation (B.8b) is directly obtained by (B.8a) and (B.4) ∎

PROPERTY B.15 *For the companion matrix* $\mathbf{A}_{\Psi(x)}$ *with a degree-L characteristic polynomial* $\Psi(x)$, *and an L-vector* $\mathbf{v}$,

$$[\,\mathbf{v} \quad \mathbf{A}_{\Psi(x)}^t \cdot \mathbf{v} \quad \cdots \quad (\mathbf{A}_{\Psi(x)}^t)^{L-1} \cdot \mathbf{v}\,]$$
$$= [\,\mathbf{v} \quad \mathbf{A}_{\Psi(x)}^t \cdot \mathbf{v} \quad \cdots \quad (\mathbf{A}_{\Psi(x)}^t)^{L-1} \cdot \mathbf{v}\,]^t, \qquad (B.9a)$$

$$[\,\mathbf{v} \quad \mathbf{A}_{\Psi(x)} \cdot \mathbf{v} \quad \cdots \quad \mathbf{A}_{\Psi(x)}^{L-1} \cdot \mathbf{v}\,] \cdot \mathbf{C}_{\Psi(x)}$$
$$= \mathbf{C}_{\Psi(x)} \cdot [\,\mathbf{v} \quad \mathbf{A}_{\Psi(x)} \cdot \mathbf{v} \quad \cdots \quad \mathbf{A}_{\Psi(x)}^{L-1} \cdot \mathbf{v}\,]^t. \qquad (B.9b)$$

Proof : To prove (B.9a), let $[\,\mathbf{v} \quad \mathbf{A}_{\Psi(x)}^t \cdot \mathbf{v} \quad \cdots \quad (\mathbf{A}_{\Psi(x)}^t)^{L-1} \cdot \mathbf{v}\,] = [\,a_{i,j}\,]_{L\times L}$. Then, $a_{i,j} = \mathbf{e}_i^t \cdot (\mathbf{A}_{\Psi(x)}^t)^j \cdot \mathbf{v}$ and $a_{j,i} = \mathbf{e}_j^t \cdot (\mathbf{A}_{\Psi(x)}^t)^i \cdot \mathbf{v}$, $i, j = 0, 1, \cdots, L-1$. Therefore, by (B.3b) we have the relation $a_{i,j} = \mathbf{e}_0^t \cdot (\mathbf{A}_{\Psi(x)}^t)^{i+j} \cdot \mathbf{v} = a_{j,i}$. This proves (B.9a).

Now we prove (B.9b). By (B.4) and (B.8a), we have the relation $[\,\mathbf{v} \quad \mathbf{A}_{\Psi(x)} \cdot \mathbf{v} \quad \cdots \quad \mathbf{A}_{\Psi(x)}^{L-1} \cdot \mathbf{v}\,] \cdot \mathbf{C}_{\Psi(x)} = \mathbf{C}_{\Psi(x)} \cdot [\,\mathbf{B}_{\Psi(x)} \cdot \mathbf{v} \quad \mathbf{A}_{\Psi(x)}^t \cdot \mathbf{B}_{\Psi(x)} \cdot \mathbf{v} \quad \cdots \quad (\mathbf{A}_{\Psi(x)}^t)^{L-1} \cdot \mathbf{B}_{\Psi(x)} \cdot \mathbf{v}\,] \cdot \mathbf{C}_{\Psi(x)}$, and applying (B.9a) to this, we have the relation $[\,\mathbf{v} \quad \mathbf{A}_{\Psi(x)} \cdot \mathbf{v} \quad \cdots \quad \mathbf{A}_{\Psi(x)}^{L-1} \cdot \mathbf{v}\,] \cdot \mathbf{C}_{\Psi(x)} = \mathbf{C}_{\Psi(x)} \cdot [\,\mathbf{B}_{\Psi(x)} \cdot \mathbf{v} \quad \mathbf{A}_{\Psi(x)}^t \cdot \mathbf{B}_{\Psi(x)} \cdot \mathbf{v} \quad \cdots \quad (\mathbf{A}_{\Psi(x)}^t)^{L-1} \cdot \mathbf{B}_{\Psi(x)} \cdot \mathbf{v}\,]^t \cdot \mathbf{C}_{\Psi(x)}$. Therefore, by (B.6b), $[\,\mathbf{v} \quad \mathbf{A}_{\Psi(x)} \cdot \mathbf{v} \cdots \mathbf{A}_{\Psi(x)}^{L-1} \cdot \mathbf{v}\,] \cdot \mathbf{C}_{\Psi(x)} = \mathbf{C}_{\Psi(x)} \cdot [\,\mathbf{C}_{\Psi(x)} \cdot \mathbf{B}_{\Psi(x)} \cdot \mathbf{v} \quad \mathbf{C}_{\Psi(x)} \cdot \mathbf{A}_{\Psi(x)}^t \cdot \mathbf{B}_{\Psi(x)} \cdot \mathbf{v} \quad \cdots \quad \mathbf{C}_{\Psi(x)} \cdot (\mathbf{A}_{\Psi(x)}^t)^{L-1} \cdot \mathbf{B}_{\Psi(x)} \cdot \mathbf{v}\,]^t$, and hence by (B.4) and (B.8a), we have (B.9b). ∎

PROPERTY B.16 *For the two submatrices* $\mathbf{M}_{\Psi(x)}^+$ *and* $\mathbf{M}_{\Psi(x)}^-$ *in (19.13),*

$$(\mathbf{M}_{\Psi(x)}^-)^{-1} \cdot \mathbf{M}_{\Psi(x)}^+ = \begin{bmatrix} \mathbf{e}_{L-1}^t \cdot \mathbf{A}_{\Psi_r(x)}^t \\ \mathbf{e}_{L-1}^t \cdot (\mathbf{A}_{\Psi_r(x)}^t)^2 \\ \vdots \\ \mathbf{e}_{L-1}^t \cdot (\mathbf{A}_{\Psi_r(x)}^t)^N \end{bmatrix}, \qquad (B.10a)$$

$$(\mathbf{M}_{\Psi(x)}^-)^{-1} = [\,c_{i,j}\,]_{N \times N},$$

$$c_{i,j} = \begin{cases} \mathbf{e}_{L-1}^t \cdot (\mathbf{A}_{\Psi_r(x)}^t)^{i-j} \cdot \mathbf{e}_{L-1}, & i \geq j, \\ 0, & \text{otherwise,} \end{cases} \quad \text{(B.10b)}$$

where $\Psi_r(x)$ is the reciprocal polynomial of $\Psi(x)$, that is, $\Psi_r(x) = x^L \Psi(x^{-1})$.

Proof : We first prove (B.10) for the case $N \leq L$. In this case, by (19.10) the two submatrices $\mathbf{M}_{\Psi(x)}^+$ and $\mathbf{M}_{\Psi(x)}^-$ in (19.13) become respectively

$$\mathbf{M}_{\Psi(x)}^+ = \begin{bmatrix} \psi_L & \psi_{L-1} & \cdots & \cdots & \cdots & \psi_1 \\ 0 & \psi_L & \cdots & \cdots & \cdots & \psi_2 \\ \vdots & \ddots & \ddots & \cdots & \cdots & \vdots \\ 0 & \cdots & 0 & \psi_L & \cdots & \psi_N \end{bmatrix}, \quad \text{(B.11a)}$$

$$\mathbf{M}_{\Psi(x)}^- = \begin{bmatrix} \psi_0 & 0 & \cdots & 0 & 0 \\ \psi_1 & \psi_0 & \cdots & 0 & 0 \\ \vdots & \vdots & \ddots & \vdots & \vdots \\ \psi_{N-2} & \psi_{N-3} & \cdots & \psi_0 & 0 \\ \psi_{N-1} & \psi_{N-2} & \cdots & \psi_1 & \psi_0 \end{bmatrix}. \quad \text{(B.11b)}$$

For (B.10a), it suffices to show that $\mathbf{M}_{\Psi(x)}^+ = \mathbf{M}_{\Psi(x)}^- \cdot [\,(\mathbf{e}_{L-1}^t \cdot \mathbf{A}_{\Psi_r(x)}^t)^t \ (\mathbf{e}_{L-1}^t \cdot (\mathbf{A}_{\Psi_r(x)}^t)^2)^t \ \cdots \ (\mathbf{e}_{L-1}^t \cdot (\mathbf{A}_{\Psi_r(x)}^t)^N)^t\,]^t$, or equivalently, $\sum_{j=i+1}^L \psi_j \mathbf{e}_{L+i-j}^t = \mathbf{e}_i^t \cdot \mathbf{M}_{\Psi(x)}^- \cdot [\,(\mathbf{e}_{L-1}^t \cdot \mathbf{A}_{\Psi_r(x)}^t)^t \ (\mathbf{e}_{L-1}^t \cdot (\mathbf{A}_{\Psi_r(x)}^t)^2)^t \ \cdots \ (\mathbf{e}_{L-1}^t \cdot (\mathbf{A}_{\Psi_r(x)}^t)^N)^t\,]^t$, $i = 0, 1, \cdots, N-1$. By (B.2d), we have the relation $\sum_{j=i+1}^L \psi_j \mathbf{e}_{L+i-j}^t = \sum_{j=i+1}^L \psi_j \mathbf{e}_{L-1}^t \cdot (\mathbf{A}_{\Psi_r(x)}^t)^{i+1-j}$. However, for the companion matrix $\mathbf{A}_{\Psi_r(x)}$ of the reciprocal polynomial $\Psi_r(x)$, $\sum_{j=0}^L \psi_j \mathbf{A}_{\Psi_r(x)}^{-j} = \mathbf{0}$, and thus $\sum_{j=i+1}^L \psi_j (\mathbf{A}_{\Psi_r(x)}^t)^{-j} = \sum_{j=0}^i \psi_j (\mathbf{A}_{\Psi_r(x)}^t)^{-j}$. Therefore, $\sum_{j=i+1}^L \psi_j \mathbf{e}_{L+i-j}^t = \sum_{j=0}^i \psi_j \mathbf{e}_{L-1}^t \cdot (\mathbf{A}_{\Psi_r(x)}^t)^{i+1-j}$, and so by (B.11b), $\sum_{j=i+1}^L \psi_j \mathbf{e}_{L+i-j}^t = \mathbf{e}_i^t \cdot \mathbf{M}_{\Psi(x)}^- \cdot [\,(\mathbf{e}_{L-1}^t \cdot \mathbf{A}_{\Psi_r(x)}^t)^t \ (\mathbf{e}_{L-1}^t \cdot (\mathbf{A}_{\Psi_r(x)}^t)^2)^t \ \cdots \ (\mathbf{e}_{L-1}^t \cdot (\mathbf{A}_{\Psi_r(x)}^t)^N)^t\,]^t$.

For (B.10b), it suffices to show that for the matrix $[\,c_{i,j}\,]_{N \times N}$ in (B.10b), $\mathbf{M}_{\Psi(x)}^- \cdot [\,c_{i,j}\,] = \mathbf{I}$. Let $\mathbf{M}_{\Psi(x)}^- \cdot [\,c_{i,j}\,] = [\,a_{i,j}\,]_{N \times N}$. Then, in view of (B.10b) and (B.11b), we obtain $a_{i,j} = 0$ for $i < j$, and $a_{i,i} = 1$ for $i = 0, 1, \cdots, N-1$. So it suffices to prove that $a_{i,j} = 0$ for $i > j$. By (B.10b) and (B.11b), in case $i > j$, $a_{i,j} = \sum_{l=0}^{i-j} \psi_l \mathbf{e}_{L-1}^t \cdot (\mathbf{A}_{\Psi_r(x)}^t)^{i-j-l} \cdot \mathbf{e}_{L-1}$, which can be written in the form of $\sum_{l=i-j+1}^{L} \psi_l \mathbf{e}_{L-1}^t \cdot (\mathbf{A}_{\Psi_r(x)}^t)^{i-j-l} \cdot \mathbf{e}_{L-1}$ due to $\sum_{l=0}^{L} \psi_l (\mathbf{A}_{\Psi_r(x)}^t)^{-l} = \mathbf{0}$. Applying (B.2d) to this, we obtain $a_{i,j} = \sum_{l=i-j+1}^{L} \psi_l \mathbf{e}_{L-1+i-j-l}^t \cdot \mathbf{e}_{L-1} = 0$.

For the case $N > L$, equation (B.10) can be proved in a similar manner. ∎

B.4   Similarity of Matrices

For two matrices $\mathbf{M}_1$ and $\mathbf{M}_2$, if there exists a nonsingular matrix $\mathbf{Q}$ such that $\mathbf{M}_1 = \mathbf{Q}^{-1} \cdot \mathbf{M}_2 \cdot \mathbf{Q}$, the two matrices $\mathbf{M}_1$ and $\mathbf{M}_2$ are said to be *similar*.

PROPERTY B.17 *If two matrices $\mathbf{M}_1$ and $\mathbf{M}_2$ are similar, then their minimal polynomials $\Psi_{\mathbf{M}_1}(x)$ and $\Psi_{\mathbf{M}_2}(x)$ are identical.*

Proof : Let $\mathbf{M}_1 = \mathbf{Q}^{-1} \cdot \mathbf{M}_2 \cdot \mathbf{Q}$ for a nonsingular matrix $\mathbf{Q}$. Then, $\Psi_{\mathbf{M}_2}(\mathbf{M}_1) = \mathbf{Q}^{-1} \cdot \Psi_{\mathbf{M}_2}(\mathbf{M}_2) \cdot \mathbf{Q}$, which is $\mathbf{0}$ due to $\Psi_{\mathbf{M}_2}(\mathbf{M}_2) = \mathbf{0}$. Therefore, by Property B.3, the minimal polynomial $\Psi_{\mathbf{M}_1}(x)$ of $\mathbf{M}_1$ divides $\Psi_{\mathbf{M}_2}(x)$. In a similar manner, we can prove that $\Psi_{\mathbf{M}_2}(x)$ divides $\Psi_{\mathbf{M}_1}(x)$. Therefore, $\Psi_{\mathbf{M}_1}(x) = \Psi_{\mathbf{M}_2}(x)$. ∎

PROPERTY B.18 *For an $L \times L$ matrix $\mathbf{M}$, if the minimal polynomial of $\mathbf{M}$ is a degree-L polynomial $\Psi(x)$, then there exists an $L \times L$ nonsingular matrix*

$$\mathbf{Q} = [\, \mathbf{v} \quad \mathbf{M} \cdot \mathbf{v} \quad \cdots \quad \mathbf{M}^{L-1} \cdot \mathbf{v} \,] \tag{B.12}$$

formed by the matrix $\mathbf{M}$ and an L-vector $\mathbf{v}$. Further, the relation $\mathbf{A}_{\Psi(x)} = \mathbf{Q}^{-1} \cdot \mathbf{M} \cdot \mathbf{Q}$ holds for the nonsingular matrix $\mathbf{Q}$ in (B.12).

Proof : We first show the existence of a nonsingular matrix $\mathbf{Q}$ in (B.12) for an L-vector $\mathbf{v}$. We prove this by contradiction. Suppose that the matrix $\mathbf{Q}$ in (B.12) is singular for an arbitrary L-vector $\mathbf{v}$. Then, the lowest-degree polynomial $\Psi(x)$ such that $\Psi(\mathbf{M}) \cdot \mathbf{v} = \mathbf{0}$ for any L-vector $\mathbf{v}$ has degree less than L. However, this is a contradiction to Property B.6.

Now we prove that $\mathbf{A}_{\Psi(x)} = \mathbf{Q}^{-1} \cdot \mathbf{M} \cdot \mathbf{Q}$ for the nonsingular matrix in (B.12). By (4.7), we obtain the relation $\mathbf{Q} \cdot \mathbf{A}_{\Psi(x)} = [\, \mathbf{M} \cdot \mathbf{v} \quad \mathbf{M}^2 \cdot \mathbf{v} \quad \cdots \quad \mathbf{M}^{L-1} \cdot \mathbf{v} \quad \sum_{i=0}^{L-1} \psi_i \mathbf{M}^i \cdot \mathbf{v} \,]$. However, $\sum_{i=0}^{L-1} \psi_i \mathbf{M}^i = \mathbf{M}^L$ by $\Psi(\mathbf{M}) = \mathbf{0}$. Therefore, $\mathbf{Q} \cdot \mathbf{A}_{\Psi(x)} = [\, \mathbf{M} \cdot \mathbf{v} \quad \mathbf{M}^2 \cdot \mathbf{v} \quad \cdots \quad \mathbf{M}^L \cdot \mathbf{v} \,] = \mathbf{M} \cdot \mathbf{Q}$, so $\mathbf{A}_{\Psi(x)} = \mathbf{Q}^{-1} \cdot \mathbf{M} \cdot \mathbf{Q}$. ∎

References

[1] ANSI T1.105-1991, "Digital hierarchy – optical interface rates and formats specifications(SONET)", 1991.

[2] Bellamy, J. C., *Digital Telephony*, John Wiley & Sons Inc., 2nd ed., New York 1991.

[3] Bellcore, TR-NWT-000253, "Synchronous optical network(SONET) transport systems : Common generic", Issue 2, 1991.

[4] Bellcore, TR-TSP-000496, "SONET Add/Drop multiplex equipment generic criteria", Issue 3, 1992.

[5] Bylanski, P., and D. G. W. Ingram, *Digital Transmission Systems*, Peter Peregrinus Ltd., 1980.

[6] Choi, D. W., "Parallel scrambling techniques for digital multiplexer", *AT&T Tech. J.*, pp.123-136, Sept./Oct. 1986.

[7] Dixon, R. C., *Spread Spectrum Systems*, 2nd. ed., John Wiley and Sons, New York, 1984.

[8] Fredricksen, S., "Pseudo-randomness properties of binary shift register sequences", *IEEE Trans. Inform. Theory.*, vol.21, pp.115-120, 1975.

[9] Gantmacher, F. R., *The Theory of Matrices*, Chelsea Publishing Company, New York, 1960.

[10] Gold, R., "Optimal binary sequences for spread spectrum multiplexing", *IEEE Trans. Inform. Theory*, vol.IT-13, pp.619-621, 1967.

[11] Golomb, S. W., Ed., *Digital Communications with Space Applications*, Prentice-Hall, Englewood Cliffs, N.J., 1964.

[12] Golomb, S. W., *Shift Register Sequences*, 2nd, ed., Aegean Park Press, California, 1982.

[13] Henriksson, U., "On a scrambling property of feedback shift registers", *IEEE Trans. Commun.*, vol.20, pp.998-1001, 1972.

[14] ITU-T Recommendation G.702, "Digital hierarchy bit rates", 1992.

[15] ITU-T Recommendation G.703, "Physical/electrical characteristics of hierarchical digital interfaces", 1992.

[16] ITU-T Recommendation G.707, "Synchronous digital hierarchy bit rates", 1992.

[17] ITU-T Recommendation G.708, "Network node interface for the synchronous digital hierarchy", 1992.

[18] ITU-T Recommendation G.709, "Synchronous multiplexing structure", 1992.

[19] ITU-T Recommendation G.781, "Multiplexing equipment for the SDH", 1992.

[20] ITU-T Recommendation G.782, "Types and general characteristics of synchronous digital hierarchy(SDH) multiplexing equipment", 1992.

[21] ITU-T Recommendation G.783, "Characteristics of synchronous digital hierarchy(SDH) multiplexing equipment functional blocks", 1992.

[22] ITU-T Recommendation G.784, "Synchronous digital hierarchy(SDH) management", 1992.

[23] ITU-T Recommendation I.321, "BISDN protocol reference model and its application", 1990.

[24] ITU-T Recommendation I.361, "BISDN ATM layer specification", 1992.

[25] ITU-T Recommendation I.413, "BISDN user-network interface", 1992.

[26] ITU-T Recommendation I.432, "B-ISDN user-network interface - Physical layer specification", 1992.

[27] Kasai, H., S. Senmoto, and M. Matsushita, "PCM jitter suppression by scrambling", *IEEE Trans. Commun.*, vol.COM-22, pp.1114-1122, Aug. 1974.

[28] Krutz, W. H., Ed., *Linear Sequential Switching Circuits : Selected Papers*, Holden-Day, San Francisco, 1965.

[29] Kim, S. C., and B. G. Lee, "A signal-alignment theory in rolling-based lightwave transmission systems", *IEEE Trans. Commun.*, vol.38, no.12, pp.2119-2130, Dec. 1990.

[30] Kim, S. C., and B. G. Lee, "Sampling and correction time conditions in distributed sample scrambling", in *Proc. IC³N (International Conference on Computer Communication and Networks)*, June 1992, pp.226-229. (San Diego)

[31] Kim, S. C., and B. G. Lee, "Parallel scrambling techniques for multibit-interleaved multiplexing environments", in *Proc. ICC(International Conference on Communications)*, May 1993, pp.1526-1530. (Geneva)

[32] Kim, S. C., and B. G. Lee, "Sampling and correction conditions in general distributed sample scramblers", in *Proc. APCC(Asia-Pacific Conference on Communications)*, Aug. 1993, pp.387-391. (Taejon)

[33] Kim, S. C., and B. G. Lee, "Synchronization of shift register generators in distributed sample scramblers", *IEEE Trans. Commun.*, vol.42, no.3, pp.1400-1408, March 1994.

[34] Kim, S. C., and B. G. Lee, "Parallel shift register generators : Theory and applications to parallel scrambling in multibit-interleaved multiplexing environments", *IEEE Trans. Commun.*, vol.43, no.4, April 1995.

[35] Lee, B. G., and S. C. Kim, "Low-rate parallel scrambling techniques for today's lightwave transmission", *IEEE Commun. Mag.*, vol.33, no.4, pp.84-95, April 1995.

[36] Kim, S. C., and B. G. Lee, "Synchronization of shift register generators in general distributed sample scramblers", to appear in *IEEE Trans. Commun.*

[37] Kim, S. C., and B. G. Lee, "A theory on sequence spaces and shift register generators", to appear in *IEEE Trans. Commun.*

[38] Kim, S. C., and B. G. Lee, "Multibit-parallel scrambling techniques for distributed sample scrambling," to appear in *IEICE Trans. Commun.*

[39] Lee, B. G., "Signal conditioning for lightwave transmission", in *Proc. IEEE TENCON 87*, Aug. 1987, pp.734-738.

[40] Lee, B. G., M. Kang, and J. Lee, *Broadband Telecommunications Technology*, Artech House Inc., Norwood, 1993.

[41] Lee, B. G., and S. C. Kim, "Parallel scrambling system", U. S. Patent 5,241,602, issued Aug. 31, 1993.

[42] Lee, B. G., and S. C. Kim, "Distributed sample scrambling system", U. S. Patent 5,245,661, issued Sep. 14, 1993.

[43] Lee, B. G., and S. C. Kim, "Parallel distributed sample scrambling system", U. S. Patent 5,355,415, issued Oct. 11, 1994..

[44] Oppenheim, A. V., and Schafer R. W., *Discrete-Time Signal Processing*, Prentice-Hall, Englewood Cliffs, N. J., 1989.

[45] Owens, F. F. E., *PCM and Digital Transmission Systems*, McGraw-Hill, New York, 1982.

[46] Peterson, W. W., and E. J. Weldon, *Error Correcting Codes*, 2nd. ed., The MIT Press, Massachusetts, 1972.

[47] Savage, J. E., "Some simple self-synchronizing digital data scramblers", *Bell Syst. Tech. J.*, vol.46, pp.449-487, Feb. 1967.

[48] Schwartz, M., *Telecommunication Networks : Protocols, Modeling and Analysis*, Addison-Wesley Publishing Co., Reading, 1987.

[49] Selmer, E. S., "Linear recurrence relations over finite fields", Dept. of Math., Univ. of Bergen, Norway, 1966.

[50] Sexton, M., and A. Reid, *Transmission Networking : SONET and the Synchronous Digital Hierarchy*, Artech House, Boston, 1992.

[51] Smirnov, N. I., "Applications of m-sequences in asynchronous radio systems", *Telecommun. Radio Eng.*, vol.24, no.10, pp.26-35, (translated from the Russian journal *Elecktrosvayz*) 1970.

[52] Stallings, W., *Data and Computer Communications*, 3rd. ed., MacMillan Publishing Company, New York, 1991.

[53] Tanenbaum, A. S., *Computer Networks*, 2nd. ed., Prentice-Hall, Englewood Cliffs, N.J., 1988.

[54] Technical Personnel, *Telecommunications Transmission Engineering*, vol.1-3, Bellcore, 1990.

[55] Technical Staff, *Transmission Systems for Communications*, AT&T Bell Laboratories, 1982.

[56] Titsworth, R. C., "Optimal ranging codes", *IEEE Trans. Space Electron. Telem.*, vol.SET-10, pp.19-30, 1964.

[57] Zierler, N., "Linear recurring sequences", *J. Soc. Ind. Appl. Math.*, vol.7, pp.31-48, 1959.

[58] Zierler, N., "Linear recurring sequences and error-correcting codes", *Error Correcting Codes*, H. B. Mann, Ed., New York : Wiley, 1969, pp.47-59.

Index